Lecture Notes in Artificial Intelligence   9876

Subseries of Lecture Notes in Computer Science

LNAI Series Editors

Randy Goebel
 University of Alberta, Edmonton, Canada
Yuzuru Tanaka
 Hokkaido University, Sapporo, Japan
Wolfgang Wahlster
 DFKI and Saarland University, Saarbrücken, Germany

LNAI Founding Series Editor

Joerg Siekmann
 DFKI and Saarland University, Saarbrücken, Germany

More information about this series at http://www.springer.com/series/1244

Ngoc-Thanh Nguyen · Yannis Manolopoulos
Lazaros Iliadis · Bogdan Trawiński (Eds.)

Computational Collective Intelligence

8th International Conference, ICCCI 2016
Halkidiki, Greece, September 28–30, 2016
Proceedings, Part II

 Springer

Editors
Ngoc-Thanh Nguyen
Wrocław University of Technology
Wrocław
Poland

Yannis Manolopoulos
Aristotle University of Thessaloniki
Thessaloniki
Greece

Lazaros Iliadis
Department of Forestry and Management
Democritus University of Thrace
Orestiada, Thrace
Greece

Bogdan Trawiński
Wrocław University of Technology
Wrocław
Poland

ISSN 0302-9743   ISSN 1611-3349 (electronic)
Lecture Notes in Artificial Intelligence
ISBN 978-3-319-45245-6   ISBN 978-3-319-45246-3 (eBook)
DOI 10.1007/978-3-319-45246-3

Library of Congress Control Number: 2016949588

LNCS Sublibrary: SL7 – Artificial Intelligence

Printed on acid-free paper

This Springer imprint is published by Springer Nature
The registered company is Springer International Publishing AG Switzerland

Preface

This volume contains the proceedings of the 9th International Conference on Computational Collective Intelligence (ICCCI 2016), held in Halkidiki, Greece, September 28–30, 2016. The conference was co-organized by the Aristotle University of Thessaloniki, Greece, the Democritus University of Thrace, Greece, and the Wrocław University of Science and Technology, Poland. The conference was run under the patronage of the IEEE SMC Technical Committee on Computational Collective Intelligence.

Following the successes of the First ICCCI (2009) held in Wrocław, Poland, the Second ICCCI (2010) in Kaohsiung, Taiwan, the Third ICCCI (2011) in Gdynia, Poland, the 4th ICCCI (2012) in Ho Chi Minh City, Vietnam, the 5th ICCCI (2013) in Craiova, Romania, the 6th ICCCI (2014) in Seoul, South Korea, and the 7th ICCCI (2015) in Madrid, Spain, this conference continues to provide an internationally respected forum for scientific research in the computer-based methods of collective intelligence and their applications.

Computational collective intelligence (CCI) is most often understood as a sub-field of artificial intelligence (AI) dealing with soft computing methods that enable making group decisions or processing knowledge among autonomous units acting in distributed environments. Methodological, theoretical, and practical aspects of CCI are considered as the form of intelligence that emerges from the collaboration and competition of many individuals (artificial and/or natural). The application of multiple computational intelligence technologies such as fuzzy systems, evolutionary computation, neural systems, consensus theory, etc., can support human and other collective intelligence, and create new forms of CCI in natural and/or artificial systems. Three subfields of the application of computational intelligence technologies to support various forms of collective intelligence are of special interest but are not exclusive: Semantic Web (as an advanced tool for increasing collective intelligence), social network analysis (as the field targeted to the emergence of new forms of CCI), and multi-agent systems (as a computational and modeling paradigm especially tailored to capture the nature of CCI emergence in populations of autonomous individuals).

The ICCCI 2016 conference featured a number of keynote talks and oral presentations, closely aligned to the theme of the conference. The conference attracted a substantial number of researchers and practitioners from all over the world, who submitted their papers for the main track and 12 special sessions.

The main track, covering the methodology and applications of CCI, included: multi-agent systems, knowledge engineering and Semantic Web, natural language and text processing, data-mining methods and applications, decision support and control systems, and innovations in intelligent systems. The special sessions, covering some specific topics of particular interest, included cooperative strategies for decision making and optimization, meta-heuristics techniques and applications, Web systems and human–computer interaction, applications of software agents, social media and the Web of linked data, computational swarm intelligence, ambient networks, information technology in biomedicine, impact of smart and intelligent technology on education, big data mining and searching, machine learning in medicine and biometrics, and low-resource language processing.

We received 277 submissions. Each paper was reviewed by two to four members of the international Program Committee of either the main track or one of the special sessions. We selected the 108 best papers for oral presentation and publication in two volumes of the *Lecture Notes in Artificial Intelligence* series.

We would like to express our thanks to the keynote speakers, Plamen Angelov, Heinz Koeppl, Manuel Núñez, and Leszek Rutkowski, for their world-class plenary speeches. Many people contributed toward the success of the conference. First, we would like to recognize the work of the PC co-chairs and special sessions organizers for taking good care of the organization of the reviewing process, an essential stage in ensuring the high quality of the accepted papers. The workshops and special sessions chairs deserve a special mention for the evaluation of the proposals and the organization and coordination of the work of the 12 special sessions. In addition, we would like to thank the PC members, of the main track and of the special sessions, for performing their reviewing work with diligence. We thank the Organizing Committee chairs, liaison chairs, publicity chair, special issues chair, financial chair, Web chair, and technical support chair for their fantastic work before and during the conference. Finally, we cordially thank all the authors, presenters, and delegates for their valuable contribution to this successful event. The conference would not have been possible without their support.

It is our pleasure to announce that the conferences of the ICCCI series continue their close cooperation with the Springer journal *Transactions on Computational Collective Intelligence*, and the IEEE SMC Technical Committee on Transactions on Computational Collective Intelligence.

Finally, we hope that ICCCI 2016 significantly contributes to the academic excellence of the field and leads to the even greater success of ICCCI events in the future.

September 2016

Ngoc-Thanh Nguyen
Yannis Manolopoulos
Lazaros Iliadis
Bogdan Trawiński

Conference Organization

Honorary Chairs

Pierre Levy — University of Ottawa, Canada
Tadeusz Więckowski — Wrocław University of Science and Technology, Poland

General Chairs

Yannis Manolopoulos — Aristotle University of Thessaloniki, Greece
Ngoc Thanh Nguyen — Wrocław University of Science and Technology, Poland

Program Chairs

Lazaros Iliadis — Democritus University of Thrace, Greece
Costin Badica — University of Craiova, Romania
Kazumi Nakamatsu — University of Hyogo, Japan
Piotr Jędrzejowicz — Gdynia Maritime University, Poland

Special Session Chairs

Bogdan Trawiński — Wrocław University of Science and Technology, Poland
Elias Pimenidis — University of the West of England, UK

Organizing Chairs

Apostolos Papadopoulos — Aristotle University of Thessaloniki, Greece

Keynote Speakers

Plamen Angelov — Lancaster University, UK
Heinz Koeppl — Technische Universität Darmstadt, Germany
Manuel Núñez — Universidad Complutense de Madrid, Spain
Leszek Rutkowski — Czestochowa University of Technology, Poland

Special Sessions Organizers

1. *WASA 2016: 6th Workshop on Applications of Software Agents*

Mirjana Ivanovic	University of Novi Sad, Serbia
Costin Badica	University of Craiova, Romania

2. *CSDMO 2016: Special Session on Cooperative Strategies for Decision Making and Optimization*

Piotr Jędrzejowicz	Gdynia Maritime University, Poland
Dariusz Barbucha	Gdynia Maritime University, Poland

3. *RUMOUR 2016: Workshop on Social Media and the Web of Linked Data*

Diana Trandabat	University "Al. I. Cuza" of Iasi, Romania
Daniela Gifu	University "Al. I. Cuza" of Iasi, Romania

4. *WebSys 2016: Special Session on Web Systems and Human–Computer Interaction*

Kazimierz Choroś	Wrocław University of Science and Technology, Poland
Maria Trocan	Institut Supérieur d'Électronique de Paris, France

5. *MHTA 2016: Special Session on Meta-Heuristics Techniques and Applications*

Pandian Vasant	Universiti Teknologi Petronas, Malaysia
Bharat Singh	Big Data Analyst, Hildesheim, Germany
Neel Mani	Dublin City University, Dublin, Ireland
Rajalingam Sokkalingam	Universiti Teknologi Petronas, Malaysia
Junzo Watada	Waseda University, Japan

6. *CSI 2016: Special Session on Computational Swarm Intelligence*

Urszula Boryczka	University of Silesia, Poland
Mariusz Boryczka	University of Silesia, Poland
Jan Kozak	University of Silesia, Poland

7. *AMNET 2016: Special Session on Ambient Networks*

Vladimir Sobeslav	University of Hradec Kralove, Czech Republik
Ondrej Krejcar	University of Hradec Kralove, Czech Republic
Peter Brida	University of Žilina, Czech Republic
Peter Mikulecky	University of Hradec Kralove, Czech Republic

8. *ITiB 2016: Special Session on IT in Biomedicine*

Ondrej Krejcar	University of Hradec Kralove, Czech Republic
Kamic Kuca	University of Hradec Kralove, Czech Republic
Dawit Assefa Halle	Addis Ababa University, Ethiopia
Tanos C.C. Franca	Military Institute of Engineering, Brazil

9. *ISITE 2016: Special Session on Impact of Smart and Intelligent Technology on Education*

Petra Poulova	University of Hradec Kralove, Czech Republic
Ivana Simonova	University of Hradec Kralove, Czech Republic
Katerina Kostolanyova	University of Ostrava, Czech Republic
Tiia Ruutmann	Tallinn University of Technology, Estonia

10. *BigDMS 2016: Special Session on Big Data Mining and Searching*

Rim Faiz	University of Carthage, Tunisia
Nadia Essoussi	University of Carthage, Tunisia

11. *MLMB 2016: Special Session on Machine Learning in Medicine and Biometrics*

Piotr Porwik	University of Silesia, Poland
Agnieszka Nowak-Brzezińska	University of Silesia, Poland
Robert Koprowski	University of Silesia, Poland
Janusz Jeżewski	Institute of Medical Technology and Equipment, Poland

12. *LRLP 2016: Special Session on Low-Resource Language Processing*

Ualsher Tukeyev	al-Farabi Kazakh National University, Kazakhstan
Zhandos Zhumanov	al-Farabi Kazakh National University, Kazakhstan

International Program Committee

Sharat Akhoury	University of Cape Town, South Africa
Ana Almeida	GECAD-ISEP-IPP, Portugal
Orcan Alpar	University of Hradec Kralove, Czech Republic
Bashar Al-Shboul	University of Jordan, Jordan
Thierry Badard	Laval University, Canada
Amelia Badica	University of Craiova, Romania
Costin Badica	University of Craiova, Romania
Hassan Badir	ENSAT, Morocco
Dariusz Barbucha	Gdynia Maritime University, Poland
Nick Bassiliades	Aristotle University of Thessaloniki, Greece
Artur Bąk	Polish-Japanese Academy of Information Technology, Poland
Narjes Bellamine	ISI & Laboratoire RIADI/ENSI, Tunisia
Maria Bielikova	Slovak University of Technology in Bratislava, Slovakia
Pavel Blazek	University of Defence, Czech Republic
Mariusz Boryczka	University of Silesia, Poland
Peter Brida	University of Žilina, Slovakia

Robert Burduk	Wrocław University of Science and Technology, Poland
Krisztian Buza	Budapest University of Technology and Economics, Hungary
Aleksander Byrski	AGH University Science and Technology, Poland
Jose Luis Calvo-Rolle	University of A Coruña, Spain
David Camacho	Universidad Autonoma de Madrid, Spain
Alberto Cano	Virginia Commonwealth University, USA
Frantisek Capkovic	Slovak Academy of Sciences, Slovakia
Dariusz Ceglarek	Poznan High School of Banking, Poland
Amine Chohra	Paris-East University (UPEC), France
Kazimierz Choroś	Wrocław University of Science and Technology, Poland
Mihaela Colhon	University of Craiova, Romania
Jose Alfredo Ferreira Costa	Universidade Federal do Rio Grande do Norte, Brazil
Ireneusz Czarnowski	Gdynia Maritime University, Poland
Paul Davidsson	Malmö University, Sweden
Gayo Diallo	University of Bordeaux, France
Tien V. Do	Budapest University of Technology and Economics, Hungary
Ivan Dolnak	University of Žilina, Slovakia
Olfa Belkahla Driss	Université de Tunis, Tunisia
Atilla Elci	Aksaray University, Turkey
Vadim Ermolayev	Zaporozhye National University, Ukraine
Nadia Essoussi	University of Carthage, Tunisia
Rim Faiz	University of Carthage, Tunisia
Faiez Gargouri	University of Sfax, Tunisia
Mauro Gaspari	University of Bologna, Italy
Antonio Gonzalez-Pardo	Universidad Autonoma de Madrid, Spain
Huu Hanh Hoang	Hue University, Vietnam
Tzung-Pei Hong	National University of Kaohsiung, Taiwan
Josef Horalek	University of Hradec Králové, Czech Republic
Frédéric Hubert	Laval University, Canada
Maciej Huk	Wrocław University of Science and Technology, Poland
Dosam Hwang	Yeungnam University, Korea
Lazaros Iliadis	Democritus University of Thrace, Greece
Agnieszka Indyka-Piasecka	Wrocław University of Science and Technology, Poland
Dan Istrate	Université de Technologie de Compiègne, France
Mirjana Ivanovic	University of Novi Sad, Serbia
Jaroslaw Jankowski	West Pomeranian University of Technology, Poland
Joanna Jędrzejowicz	University of Gdansk, Poland
Piotr Jędrzejowicz	Gdynia Maritime University, Poland
Jason Jung	Chung-Ang University, Korea
Przemysław Juszczuk	University of Silesia, Poland

Ioannis Karydis	Ionian University, Greece
Petros Kefalas	University of Sheffield International Faculty, CITY College, Greece
Rafał Kern	Wroclaw University of Science and Technology, Poland
Marek Kisiel-Dorohinicki	AGH University Science and Technology, Poland
Attila Kiss	Eötvös Loránd University, Hungary
Jitka Komarkova	University of Pardubice, Czech Republic
Marek Kopel	Wrocław University of Science and Technology, Poland
Ivan Koychev	University of Sofia St. Kliment Ohridski, Bulgaria
Jan Kozak	University of Silesia, Poland
Adrianna Kozierkiewicz-Hetmańska	Wrocław University of Science and Technology, Poland
Ondrej Krejcar	University of Hradec Kralove, Czech Republic
Dariusz Król	Wrocław University of Science and Technology, Poland
Elżbieta Kukla	Wrocław University of Science and Technology, Poland
Julita Kulbacka	Wrocław Medical University, Poland
Marek Kulbacki	Polish-Japanese Academy of Information Technology, Poland
Piotr Kulczycki	Systems Research Institute of the Polish Academy of Science, Poland
Kazuhiro Kuwabara	Ritsumeikan University, Japan
Florin Leon	Technical University Gheorghe Asachi of Iasi, Romania
Edwin Lughofer	Johannes Kepler University Linz, Austria
José María Luna	University of Cordoba, Spain
Juraj Machaj	University of Žilina, Slovakia
Bernadetta Maleszka	Wrocław University of Science and Technology, Poland
Marcin Maleszka	Wrocław University of Science and Technology, Poland
Yannis Manolopoulos	Aristotle University of Thessaloniki, Greece
Antonio David Masegosa Arredondo	University of Granada, Spain
Adam Meissner	Poznań University of Technology, Poland
Ernestina Menasalvas	Universidad Politecnica de Madrid, Spain
Héctor Menéndez	Universidad Autonoma de Madrid, Spain
Jacek Mercik	Wrocław School of Banking, Poland
Peter Mikulecky	University of Hradec Kralove, Czech Republic
Alin Moldoveanu	University Politechnica of Bucharest, Romania
Javier Montero	Universidad Complutense de Madrid, Spain
Ahmed Moussa	Abdelmalek Essaadi University, Morocco
Grzegorz J. Nalepa	AGH University of Science and Technology, Poland

Filippo Neri	University of Naples Federico II, Italy
Linh Anh Nguyen	University of Warsaw, Poland
Ngoc-Thanh Nguyen	Wrocław University of Science and Technology, Poland
Adam Niewiadomski	Lodz University of Technology, Poland
Alberto Núñez	Universidad Complutense de Madrid, Spain
Manuel Núñez	Universidad Complutense de Madrid, Spain
Tarkko Oksala	Aalto University, Finland
Tomasz Orczyk	University of Silesia, Poland
Rafael Parpinelli	Santa Catarina State University, Brazil
Marek Penhaker	VSB – Technical University of Ostrava, Czech Republic
Dariusz Pierzchała	Military University of Technology, Poland
Marcin Pietranik	Wrocław University of Science and Technology, Poland
Elias Pimenidis	University of the West of England, UK
Bartłomiej Płaczek	Silesian University of Technology, Poland
Piotr Porwik	University of Silesia, Poland
Radu-Emil Precup	Politehnica University of Timisoara, Romania
Ales Prochazka	Institute of Chemical Technology, Czech Republic
Paulo Quaresma	Universidade de Evora, Portugal
Ewa Ratajczak-Ropel	Gdynia Maritime University, Poland
José Antonio Sáez	University of Granada, Spain
Virgilijus Sakalauskas	Vilnius University, Lithuania
Jose L. Salmeron	University Pablo de Olavide, Spain
Jakub Segen	Polish-Japanese Academy of Information Technology, Poland
Ali Selamat	Universiti Teknologi Malaysia, Malaysia
Natalya Shakhovska	Lviv Polytechnic National University, Ukraine
Andrzej Siemiński	Wrocław University of Science and Technology, Poland
Vladimir Sobeslav	University of Hradec Kralove, Czech Republic
Stanimir Stoyanov	University of Plovdiv Paisii Hilendarski, Bulgaria
Yasufumi Takama	Tokyo Metropolitan University, Japan
Bogdan Trawiński	Wrocław University of Science and Technology, Poland
Maria Trocan	Institut Superieur d'Electronique de Paris, France
Krzysztof Trojanowski	Polish Academy of Sciences, Poland
Ventzeslav Valev	Bulgarian Academy of Sciences, Bulgaria
Izabela Wierzbowska	Gdynia Maritime University, Poland
Michal Woźniak	Wrocław University of Science and Technology, Poland
Krzysztof Wróbel	Uniwersity of Silesia, Poland
Drago Žagar	University of Osijek, Croatia
Danuta Zakrzewska	Lodz University of Technology, Poland

Constantin-Bala Zamfirescu   Lucian Blaga University of Sibiu, Romania
Aleksander Zgrzywa   Wrocław University of Science and Technology,
                               Poland

Program Committees of Special Sessions

WASA 2016: 6th Workshop on Applications of Software Agents

Amelia Badica	University of Craiova, Romania
Olivier Boissier	ENS Mines Saint-Etienne, France
Paolo Bresciani	FBK, Italy
Marius Brezovan	University of Craiova, Romania
Zoran Budimac	University of Novi Sad, Serbia
Mihaela Colhon	University of Craiova, Romania
Weihui Dai	Fudan University, China
Adina Magdá Florea	University Politehnica of Bucharest, Romania
Giancarlo Fortino	University of Calabria, Italy
Daniela Gifu	Alexandru Ioan Cuza University of Iasi, Romania
Adrian Groza	Technical University of Cluj-Napoca, Romania
Sorin Ilie	University of Craiova, Romania
Galina Ilieva	University of Plovdiv Paisii Hilendarsky, Bulgaria
Nicolae Jascanu	Dunarea de Jos University of Galati, Romania
Gordan Jezic	University of Zagreb, Croatia
Systä Kari	Tampere University of Technology, Finland
Petros Kefalas	University of Sheffield International Faculty, Thessaloniki, Greece
Setsuya Kurahashi	University of Tsukuba, Japan
Mario Kusek	University of Zagreb, Croatia
Florin Leon	Technical University Gheorghe Asachi of Iasi, Romania
Marin Lujak	University Rey Juan Carlos, Spain
Viorel Negru	West University of Timisoara, Romania
Andrea Omicini	University of Bologna, Italy
Mihaela Oprea	University Petroleum-Gas of Ploiesti, Romania
Agostino Poggi	University of Parma, Italy
Ilias Sakellariou	University of Macedonia, Greece
Stanimir Stoyanov	University of Plovdiv Paisii Hilendarsky, Bulgaria
Denis Trcek	University of Ljubljana, Slovenia
George Vouros	University of Piraeus, Greece
Constantin-Bala Zamfirescu	University of Sibiu, Romania

CSDMO 2016: Special Session on Cooperative Strategies for Decision-Making and Optimization

Dariusz Barbucha	Gdynia Maritime University, Poland
Ireneusz Czarnowski	Gdynia Maritime University, Poland
Joanna Jędrzejowicz	University of Gdansk, Poland

Piotr Jędrzejowicz   Gdynia Maritime University, Poland
Edyta Kucharska   AGH University of Science and Technology, Poland
Antonio D. Masegosa   University of Deusto, Spain
Javier Montero   Universidad Complutense de Madrid, Spain
Ewa Ratajczak-Ropel   Gdynia Maritime University, Poland
Iza Wierzbowska   Gdynia Maritime University, Poland
Mahdi Zargayouna   IFSTTAR, France

RUMOUR 2016: Workshop on Social Media and the Web of Linked Data

Nuria Bel   Universitat Pompeu Fabra, Spain
Costin Badica   University of Craiova, Romania
Georgeta Bordea   National University of Ireland, Ireland
Steve Cassidy   Macquarie University, Australia
Dragoş Ciobanu   University of Leeds, UK
Mihaela Colhon   University of Craiova, Romania
Dan Cristea   Alexandru Ioan Cuza University of Iaşi, Romania
Thierry Declerck   Universitat des Saarlandes, Saarbrücken, Germany
Daniela Gîfu   Alexandru Ioan Cuza University of Iaşi, Romania
Jorge Gracia   Universidad Politecnica de Madrid, Spain
Radu Ion   Microsoft Ireland, Ireland
John McCray   National University of Ireland, Ireland
Rada Mihalcea   University of Michigan, USA
Andrei Olariu   University of Bucharest, Romania
Octavian Popescu   IBM Research, USA
Dan Ştefănescu   Vantage Labs, USA
Diana Trandabăţ   Alexandru Ioan Cuza University of Iaşi, Romania
Dan Tufiş   Romanian Academy Research Institute for Artificial
                         Intelligence Mihai Drăgănescu, Romania
Piek Vossen   Vrije Universiteit, Amsterdam, The Netherlands
Gabriela Vulcu   National University of Ireland, Ireland
Michael Zock   Aix-Marseille University, France

WebSys 2016: Special Session on Web Systems and Human–Computer Interaction

František Čapkovič   Academy of Sciences, Slovakia
Kazimierz Choroś   Wrocław University of Science and Technology,
                         Poland
Jarosław Jankowski   West Pomeranian University of Technology, Poland
Ondřej Krejcar   University of Hradec Kralove, Czech Republic
Matthieu Manceny   Institut Supérieur d'Électronique de Paris, France
Aleš Procházka   Institute of Chemical Technology, Czech Republic
Andrzej Siemiński   Wrocław University of Science and Technology,
                         Poland
Maria Trocan   Institut Supérieur d'Électronique de Paris, France
Aleksander Zgrzywa   Wrocław University of Science and Technology,
                         Poland

MHTA 2016: Special Session on Meta-Heuristics Techniques and Applications

Gerhard-Wilhelm Weber	Middle East Technical University, Turkey
Kwon-Hee Lee	Dong-A University, Korea
Igor Litvinchev	Nuevo Leon State University, Mexico
Mohammad Abdullah-Al-Wadud	King Saud University, Saudi Arabia
Vo Ngoc Dieu	HCMC University of Technology, Vietnam
Gerardo Maximiliano Mendez	Instituto Tecnologico de Nuevo Leon, Mexico
Leopoldo Eduardo Cárdenas Barrón	Tecnológico de Monterry, Mexico
Denis Sidorov	Irkutsk State University, Russia
Weerakorn Ongsakul	Asian Institute of Technology, Thailand
Goran Klepac	Raiffeisen Bank Austria, Croatia
Herman Mawengkang	The University of Sumatera Utara, Indonesia
Igor Tyukhov	Moscow State University of Mechanical Engineering, Russia
Hayato Ohwada	Tokyo University of Science, Japan
Ugo Fiore	Federico II University, Italy
Leo Mrsic	University College Effectus – College for Law and Finance, Croatia
Nguyen Trung Thang	Ton Duc Thang University, Vietnam
Nikolai Voropai	Energy Systems Institute, Russia
Shiferaw Jufar	Universiti Teknologi Petronas, Malaysia
Xueguan Song	Dalian University of Technology, China
Ruhul A. Sarker	UNSW, Australia
Vipul Sharma	Lovely Professional University, India

CSI 2016: Special Session on Computational Swarm Intelligence

Urszula Boryczka	University of Silesia, Poland
Mariusz Boryczka	University of Silesia, Poland
Miłosław Chodacki	University of Silesia, Poland
Diana Domańska	University of Oslo, Norway
Wojciech Froelich	University of Silesia, Poland
Przemysław Juszczuk	University of Silesia, Poland
Jan Kozak	University of Silesia, Poland
Dariusz Pierzchała	Military University of Technology, Poland
Rafał Skinderowicz	University of Silesia, Poland
Tomasz Staś	University of Economics in Katowice, Poland
Beata Zielosko	University of Silesia, Poland

AMNET 2016: Special Session on Ambient Networks

Ana Almeida	Porto Superior Institute of Engineering, Portugal
Peter Brida	University of Žilina, Slovakia
Ivan Dolnak	University of Žilina, Slovakia

Elsa Gomes	Porto Superior Institute of Engineering, Portugal
Josef Horalek	University of Hradec Kralove, Czech Republic
Josef Janitor	Technical University of Kosice, Slovakia
Ondrej Krejcar	University of Hradec Kralove, Czech Republic
Goreti Marreiros	Porto Superior Institute of Engineering, Portugal
Peter Mikulecký	University of Hradec Kralove, Czech Republic
Juraj Machaj	University of Žilina, Slovakia
Marek Penhaker	Technical University of Ostrava, Czech Republic
José Salmeron	Universidad Pablo de Olavide of Seville, Spain
Ali Selamat	Universiti Teknologi Malaysia, Malaysia
Vladimir Sobeslav	University of Hradec Kralove, Czech Republic
Stylianakis Vassilis	University of Patras, Greece

ITiB 2016: Special Session on IT in Biomedicine

Dawit Assafa Haile	Addis Ababa University, Ethiopia
Peter Brida	University of Žilina, Slovakia
Richard Cimler	University of Hradec Kralove, Czech Republic
Rafael Dolezal	University of Hradec Kralove, Czech Republic
Ricardo J. Ferrari	Federal University of Sao Carlos, Brazil
Tanos C.C. Franca	Military Institute of Engineering, Brazil
Ondrej Krejcar	University of Hradec Kralove, Czech Republic
Kamil Kuca	University of Hradec Kralove, Czech Republic
Juraj Machaj	University of Žilina, Slovakia
Petra Maresova	University of Hradec Kralove, Czech Republic
Marek Penhaker	Technical University of Ostrava, Czech Republic
Jan Plavka	Technical University of Kosice, Slovakia
Teodorico C. Ramalho	Federal University of Lavras, Brazil
Saber Salehi	Universiti Teknologi Malaysia, Malaysia
Ali Selamat	Universiti Teknologi Malaysia, Malaysia

ISITE 2016: Special Session on Impact of Smart and Intelligent Technology on Education

Pavel Doulik	Jan Evangelista Purkyne University, Czech Republic
Blanka Klimova	University of Hradec Kralove, Czech Republic
Katerina Kostolanyova	University of Ostrava, Czech Republic
Silvia Pokrivcakova	Constantine the Philosopher University, Slovakia
Tatiana Polyakova	Moscow State Technical University for Automobile and Road, Russia
Petra Poulova	University of Hradec Kralove, Czech Republic
Maria Teresa Restivo	Universidade do Porto, Portugal
Tiia Ruutmann	Tallinn University of Technology, Estonia
Jiri Skoda	Jan Evangelista Purkyne University, Czech Republic
Marcela Sokolova	University of Hradec Kralove, Czech Republic
Ivana Simonova	University of Hradec Kralove, Czech Republic

Darina Tóthova — European University Information Systems Slovakia
Milan Turcani — Constantine the Philosopher University, Slovakia

BigDMS 2016: Special Session on Big Data Mining and Searching

Ajith Abraham — Machine Intelligence Research Labs, USA
Thierry Badard — Laval University, Canada
Hassan Badir — ENSAT Tangier, Morocco
Chiheb Ben N'cir — Université de la Manouba, Tunisia
Ismaïl Biskri — Université du Québec à Trois-Rivières, Canada
Guillaume Cleuziou — Université d'Orléans, France
Ernesto Damiani — University of Milan, Italy
Gayo Diallo — University of Bordeaux, France
Aymen Elkhlifi — University of Paris Sorbonne, France
Nadia Essoussi — FSEG Nabeul, University of Carthage, Tunisia
Rim Faiz — IHEC, University of Carthage, Tunisia
Sami Faiz — ISAMM, University of Manouba, Tunisia
Riadh Farah — ISAMM, University of Manouba, Tunisia
Faiez Gargouri — ISIMS, University of Sfax, Tunisia
Lamia Hadrich Belguith — FSEGS, University of Sfax, Tunisia
Frédéric Hubert — Laval University, Canada
Ahmed Moussa — ENSA, Abdelmalek Essaadi University, Morocco
Maria Malek — EISTI, France
Gabriella Pasi — University of Milan Bicocca, Italy

MLMB 2016: Special Session on Machine Learning in Medicine and Biometrics

Nabendu Chaki — University of Calcutta, India
Robert Czabański — University of Silesia, Poland
Adam Gacek — Institute of Medical Technology and Equipment, Poland
Marina Gavrilova — University of Calgary, Canada
Manuel Graña — University of the Basque Country, Spain
Alicja Wakulicz-Deja — University of Silesia, Poland
Robert Koprowski — University of Silesia, Poland
Agnieszka Nowak-Brzezińska — University of Silesia, Poland
Nobuyuki Nishiuchi — Tokyo Metropolitan University, Japan
Małgorzata Przybyła-Kasperek — University of Silesia, Poland
Marek Kurzyński — Wrocław University of Science and Technology, Poland
Roman Simiński — University of Silesia, Poland
Janusz Jeżewski — Institute of Medical Technology and Equipment, Poland
Dragan Simic — University of Novi Sad, Serbia
Ewaryst Tkacz — Silesian University of Technology, Poland

Dariusz Mrozek	Silesian University of Technology, Poland
Bożena Małysiak-Mrozek	Silesian University of Technology, Poland
Michał Dramiński	Polish Academy of Sciences, Warsaw, Poland
Michał Kozielski	Silesian University of Technology, Poland
Rafał Deja	Academy of Business in Dabrowa Gornicza, Poland

LRLP 2016: Special Session on Low-Resource Language Processing

Ualsher Tukeyev	al-Farabi Kazakh National University, Kazakhstan
Zhandos Zhumanov	al-Farabi Kazakh National University, Kazakhstan
Francis Tyers	The Arctic University of Norway, Norway
Madina Mansurova	al-Farabi Kazakh National University, Kazakhstan
Altynbek Sharipbay	L.N. Gumilyov Eurasian National University, Kazakhstan
Orken Mamyrbayev	Institute of Information and Computational Technologies, Kazakhstan
Rustam Musabayev	Institute of Information and Computational Technologies, Kazakhstan
Zhenisbek Assylbekov	Nazarbayev University, Kazakhstan
Aibek Makazhanov	Nazarbayev University, Kazakhstan
Jonathan Washington	Indiana University, USA
Djavdet Suleimanov	Institute of Applied Semiotics, Tatarstan, Russia
Sergazy Narynov	Alem Research Company, Kazakhstan
Miquel Esplà-Gomis	University of Alicante, Spain
Altangerel Chagnaa	National University of Mongolia, Mongolia

Additional Reviewers

Barbieri, Francesco	Khusainov, Aydar	Mocanu, Andrei
Bosque-Gil, Julia	Kravari, Kalliopi	Montero, Javier
Cherichi, Soumaya	Kumar, Krishan	Ouamani, Fadoua
Eklund, Ulrik	Labba, Chahrazed	Rodríguez
Hooper, Paul Charles	Mani, Neel	Fernández, Víctor
Jain, Nikita	Martín, Alejandro	Singh, Meeta
Jemal, Dhouha	Missaoui, Sondess	Takáč, Peter

Contents – Part II

Impact of Smart and Intelligent Technology on Education

Big Data Mining and Searching

Machine Learning in Medicine and Biometrics

Low Resource Language Processing

Contents – Part I

Natural Language and Text Processing

Data Mining Methods and Applications

Decision Support and Control Systems

Innovations in Intelligent Systems

Cooperative Strategies for Decision Making and Optimization

Meta-Heuristics Techniques and Applications

Web Systems and Human-Computer Interaction

Applications of Software Agents

Agent-Based Approach for Ship Damage Control

Eugénio Oliveira[✉] and Paulo Martins

LIACC, DEI / Faculdade de Engenharia, Universidade do Porto,
R. Dr. Roberto Frias, 4200-465 Porto, Portugal
eco@fe.up.pt

Abstract. We here introduce a Multi-Agent System (MAS) approach for solving the crew resources assignment problem whenever a ship, under attack, suffers several damages and, thus, priorities must be assigned in order for it to survive. In the designed system, the ship is the MAS environment and the attacker, equipment, crew and officers (these last ones seen as decision makers) are represented through agents. Decisions on resources assignment are taken after a negotiation process and using an utility-based selection process. Agent-based system design was accomplished by following a systematic Agent Oriented Software Engineering approach, called PORTO, leading to the specification and the implementation of the system.

1 Introduction

A ship is a complex system, where the crew interacts with equipment and auxiliary systems, within a closed space which is subjected to internal and external factors, such as environmental conditions, damage originated by equipment malfunction (e.g. fire or flooding on board), or by external agents (e.g. an aircraft launching a missile towards the ship; or another ship that collides with it). Either way, crew on board must act to prevent the total loss of the ship. This is a subject that must not be neglected during the ship design process, while defining its complement (crew numbers) and the allocation of compartment and equipment. Crew effectiveness when dealing with damage is dependent upon crew element numbers, crew technological knowledge and upon ships arrangement (compartments and equipment allocation within the physical boundaries of the ship). This work deals with the first problem, how to decide what to do in case of damage, how to prioritize when several damages occur at the same time, and how to allocate different crew elements taking account their ability to perform the required tasks. The other problems have already been addressed [5]. We are here dealing with a multi-criteria decision problem on how to distribute limited resources without previously knowing all possible alternatives. Therefore, since alternatives are not known in advance, classical multi-attribute methods cannot be used (e.g. trade-off analysis), and multi-objective generation methods (e.g. multi-objective genetic algorithms) would probably require a considerable amount of time to find viable solutions and would provide a set of different pareto efficient solutions instead of a single one. Moreover, none

© Springer International Publishing Switzerland 2016
N.T. Nguyen et al. (Eds.): ICCCI 2016, Part II, LNAI 9876, pp. 3–13, 2016.
DOI: 10.1007/978-3-319-45246-3_1

of these techniques would be able to represent the natural interaction between the different actors in the decision process. Multi-Agents Systems (MAS), on the other hand, provides a natural way of representing the different actors by autonomous computing entities (agents) that perform different tasks (actions), within an environment (ship). Through interacting and, possibly cooperating with each other, the agents will be able to change the state of the environment with an overall objective: ship survivability. Other authors have already used the MAS paradigm for dealing with kind of disastrous situations [6]. Our MAS architecture includes different kinds of agents ranging from simple ones, reactive agents perceiving the environment (the ship) to those who represent decision makers like XO_Officer and the Commander who will decide upon several alternative actions taking the best utility into account. All along the system design, we followed an Agent-Oriented software engineering methodology, PORTO [3,4] going through all its steps: (i) requirements analysis; (ii) analysis; (iii) architectural design; (iv) detailed design; (v) Implementation; and (vi) testing and validation.

Next section goes through the state of the art on this particular subject, Sect. 3 formulates the problem followed by the requirement analysis and the system analysis (Sects. 4 and 5). Sections 6, 7 and 8 briefly specify the system organization, architecture and design. Last sections introduce implementation, testing and conclusions.

2   General Characteristics of the Problem's Class

Many sophisticated problems we usually address can be classified as belonging to the so called 3D class of problems. They reflect a reality that simultaneously is of a Distributed, Decentralized as well as Dynamic nature. This means that, besides input data and output actions being disperse (Distributed) at different nodes, also, and most important, decision-making can be, at least partially, taken at different nodes of the (Decentralized) system. Moreover, the system trying to solve the overall problem at stake, has to deal with a changing, evolving reality (Dynamic). Although Agent Oriented Programming has been pointed out as a natural paradigm to cope with such situations, and because of the intrinsic autonomous property of Agents, it is up to those agents to find out in run-time, according to the current situation, how to interrelate in order to reach their own intended goals. Finally, and most important, for the same purpose, how to coordinate joint work or reach mutual agreements together with the other agents?

3   Related Work

The problem we intend to tackle has two main aspects. First, crew motion simulation and how it should act upon the equipment (e.g. propulsion, radars, weapons) and ship auxiliary systems (e.g. pipes, electrical distribution system); second, the multi-criteria decision process of crew limited resources distribution. The first aspect is not the object of our study. We will focus on the decision making problem on a ship under damage.

As far as we know, the problem was first addressed in reference [7] introducing SINGRAR, a decision support system that has been in use with success in the Portuguese Navy. Its main component is a fuzzy expert system that assists the command (decision making) during battle, for keeping equipment and auxiliary systems operational while the ship is under attacks/damages. The system integrates the information gathered at different locations along the ship and it proposes repair priorities and resources assignment under the scope of logistics activities. The system is not now in use since it is considered too dependent upon human interaction and its decision process is based upon utility functions depending on predetermined weights that may lead to non-efficient decisions.

The use of multi agent systems in similar problems may be found in several references in the literature, such as references [3,4]. This last reference, in particular, deals with how to manage disruptions in airline operations, such as the ones caused by bad weather, malfunctions and crew absenteeism. In order to solve this problem the reference presents a new negotiation protocol entitled Generic Q-Negotiation (GQN), which includes the Q-learning algorithm.

As far as the methodological approaches for the development of software based on agents we will follow the PORTO methodology [3,4], which has its groundings in GAIA methodology [8], which will also be further mentioned in the text.

4 Specific Problem Formulation

The command activities of most war ships include two officers, the commanding officer and the executive-officer, which are advised by other officers with different technical knowledge and expertise, such as the weapons-officer, the engineering-officer, and the tactical-officer. Further, the crew is made off different petty-officers and unlisted men, all of them with their own expertise from the cook to the radar controller, and from the nurse to the electrician. In case of damage, or combat, the ultimate decision maker is the commanding officer; the responsible for the crew is the executive-officer; the other officers are the ones who decide on how to act in order to achieve the goals established by the commanding officer. We aim to develop a system that may be used in design of new ships to assess the platform independently of how good the crew is, so it should always select the best decision possible taking into account the ship and the scenario (environment). The system must be able to reflect the decision making process of the commanding officer, as far as prioritizing the internal battle space actions and crew resources assignment. For this paper we selected a simple scenario where an aircraft attacks the ship and it hits six times the same compartments/equipment. As a consequence, there are three damaged equipment (radar, propulsion and weapon) and a fire in the engines room.

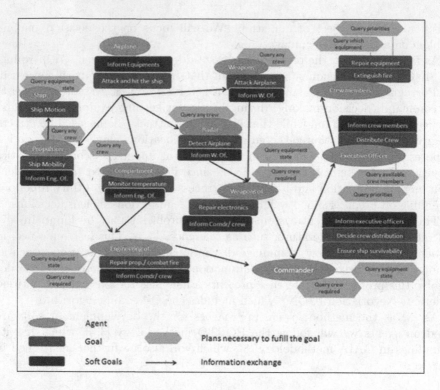

Fig. 1. Actors and Goals diagram

5 Requirement Analysis

As mentioned in introduction, we are going to follow the PORTO methodology that is described in reference [4]. This methodology proposes to start the procedure by goal-oriented early requirements analysis as in the methodology TROPOS [2]. After having selected the different actors and goals, we have built up the actors and goals diagram (Fig. 1), where interactions are presented, and several potential queries have been identified, namely: (1) "Query any crew" meaning that an actor requires intervention from crew members and it asks if there are any crew members available; (2) "Query equipment state" meaning that an actor requires knowing state of equipment; (3) "Query crew required" meaning that an actor requires knowing from equipment or another actor, who is required either to repair equipment or extinguish a fire in a compartment; (4) "Query available crew members" meaning an actor requires knowing the available crew members for the needed repair tasks; (5) "Query priorities" meaning that an actor requires knowing priorities other actor has established; (6) "Query which equipment" meaning that an actor requires knowing to which repair task he was assigned to.

6   Analysis

This stage follows GAIA [5], an Agent-Oriented Software Engineering method, and it includes five sub-phases that will be presented separately.

6.1   Subdividing the System into Sub-organizations

It primarily consists on looking to the problem trying to find sub-goals and sub-organizations dedicated only to achieve those goals. There are three distinct organizations, namely:

(1) Internal Battle state identification sub-organization;
(2) Decision making sub-organization;
(3) Crew distribution sub-organization.

6.2   Environment Model

We here distinguish between resources and active components. The first ones are seen as variables or tuples made available to the agents. The second ones are components and services capable of performing tasks with which agents must interact. Here, resources are: Aircraft information, Ship information, Crew information, Equipment information and Task requirements; Active components are: Damage Manager and Crew manager.

6.3   Preliminary Role Model

Preliminary roles relate with functionality and competences required to achieve the intended goals, independently of the organizational structure that will be further selected. Accordingly, by analyzing the Actors and goals diagram (Fig. 1) we identified the following roles: AttackAction, DamageMonitor, NeedsMonitor, NeedsAuction (associated with Officers demands on needed personnel to the Commander), CrewMonitor (monitoring which tasks the crew is assigned to), AssignCrew (assigning roles to the crew), DecisionMonitor (associated with Commanders decisions on assigning priorities and crew members to tasks), ActionTasks.

6.4   Preliminary Interaction Model

It became then necessary to specify the needed interaction between roles, their dependencies and relationships. To establish the communication protocols we have used FIPA ACL (Agent Comunication Language) protocols. PORTO methodology also indicates that we should build an Enviroment and preliminary roles diagram presented in Fig. 2.

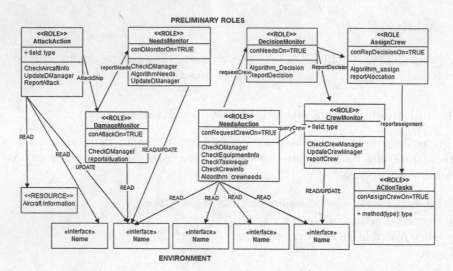

Fig. 2. Environments and preliminary roles diagram

6.5   Organizational Rules

The last task of this stage is to analyze the relationships between roles, as well as between protocols and, also, between roles and protocols. According to the followed methodology, the following constraints and relations must be taken into account: (1) Liveness organizational rules explain how the dynamics of the organization should evolve (relations); (2) Safety organizational rules state rules that are independent of the evolution and always true (constraints). Table 1 shows those above mentioned kind of rules.

7   Architectural Design

The architectural design consists upon translating the previous work into a MAS architecture. This will imply definite decisions about how the next phases will be conducted. Some changes may occur due to implementation difficulties, as it is the case of, due to using a specific ACL protocol, a single agent is replaced by others simpler and more reliable agents. This phase includes: (i) defining the organizational structure; (ii) completing the role and interaction model; (iii) graphical representation using UML 2.0. In our case we will include the final organizational structure and a combined representation of the model reached, after the first two tasks.

7.1   Organizational Structure

The resulting organizational structure may be described by a set of rules mainly derived from the previous analysis leading to three different sub-organizations:

Table 1. Liveness and safety rules

Liveness Organizational Rules	Description
reportattack (AttackAction(attack(x))) $\Longrightarrow$ *reportsituation (DamageMonitor(attack(x)))*	The situation may only be reported after the attack
1- reportsituation (DamageMonitor(attack(x))) $\Longrightarrow$ reportcrew (CrewMonitor(attack(x)))	The crew report may only be done after evaluating the new current situation
2- querycrew (CrewMonitor(attack(x))) $\Longrightarrow$ requestcrew (NeedsAuction(attack(x)))	The auction for crew ressources may only be done after knowing the new needs for crew resources
3- requestcrew (NeedsAuction(attack(x))) $\Longrightarrow$ reportdecision (DecisionMonitor(attack(x)))	The decision may only be done after the requests by the different officers
4- reportdecision(DecisionMonitor(attack(x))) $\Longrightarrow$ reportallocation (AssignCrew(attack(x)))	The allocation of crew resources may only be done after the decision on how to do it has been reported
4- Safety Organizational Rules	Description
4- AttackAction1...n	There will be n attacks by the aircraft
4- DecisionMonitor(Commander)	The decision can only be done by the Commander

"Internal Battle State" including roles "AttackAction", "NeedsMonitor" and "DamageMonitor"; "Decision Making" including the roles: "DecisionMonitor", "CrewMonitor" and "NeedsAuction"; "Crew Distribution" including roles: "AssignCrew" and "AssignTasks".

Roles included in each one of the sub-organizations although tightly interdependent, can also be related with roles of other different sub-organization.

7.2 Graphical Representation

Finally, all previous work can be described in a single diagram where protocols abstractions and role abstractions are added to the previous graphical representation. To understand Fig. 3, it is then necessary to take into account both

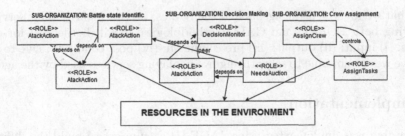

Fig. 3. Final role, interaction and enviroment diagram

protocol abstractions and role abstractions: Protocol abstractions represent roles' interactions as it is the case represented in Fig. 2. Role abstractions include the attributes of the class, and the organizational rules seen as dependencies after recognizing the above mentioned sub-structures. As an example, since Role DamageMonitor relies on information from Role AttackAction (they belong to the same sub-organization, "Internal Battle Station", this can also be seen in Fig. 3 (conAttackOn = true).

8   Detailed Design

Detailed design, according to PORTO methodology, is the stage where both the agent model and service model are produced. The first model is concerned with which agents are going to be implemented and the second is concerned with the services required and who will implement them. We have made the correspondence between agents and roles to the previously identified actors leading to the following Agents Model (Fig. 4):

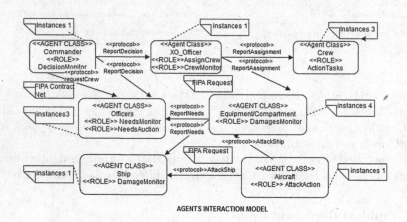

Fig. 4. Agent Model

8.1   Service Model

Now that we identified the agents and their roles, we can identify their services. Following both PORTO and GAIA methodologies, it will be defined for each service: (i) input; (ii) output; (iii) pre-condition; (iv) post condition. Due to lack of space we are not able to show the figure displaying services used by the agents.

9   Implementation

This system was implemented using JADE [1] as support to build the different agents. We here only describe one of the most significant agent: the Commander

Agent. The commander is responsible for the decision process, which is incorporated in the role DecisionMonitor. To implement this we made use of a utility function (U) where the proposal values (p) were multiplied by a set of weights (w) set by the user, in such way that:

$$final\_proposal_i = w_i p_i \text{ and } U = \Sigma_{(i=1...4)} w_i p_i \tag{1}$$

So the best final_proposal will be accepted by the Commander, instead of the best proposal. We intend to improve the decision process by making this decision process adaptable by means of Q-learning algorithm that learns the best possible weights for the intended outcome. At present, the Commander Agent is using the JADE ContractNetInitiator behavior, which is a FIPA-compliant behavior included in its library. Similarly to the ContractNetResponder, this behavior is also defined by handles, namely the handleAllResponses which we had to change in order to incorporate the weighted utility function.

10   Testing

We have tested the system through the analysis of the messages sequence exchanged between agents and decide whether or not it was according to the expected. In fact, after the event Aircraft attacks the Equipment/Compartment and the respective message, all the operational process starts until all the commander decisions on crew and resources assignment to tasks have been made. Since the agents messages exchange graph is very complex and would not be visible clearly enough, we list most of the relevant messages below:

1 Aircraft attacks Equipment/Compartment by sending a message reporting it;
2 Equipment/Compartment sends message with new state to the corresponding Officer and updates the Environment state (equipment/compartment state);
3 Officers send a message to the Evaluation_Agents (one per needed equipment) containing the value to use as a proposal in the contract net (how many crew members are required);
4 Commander initiates a FIPA Contract net protocol in which the Evaluation_Agents participate. Commander selects the best proposal multiplying each one he receives by a "importance factor" (weights from 1 to 10 and $\sum W = 10$ given by the user);
5 The Evaluation_Agent whose proposal was accepted sends a message to XO_Officer;
6 XO_Officer sends message to JADE's DF asking for specific services previously registered by the Crew (fire, mechanical repair and electronics repair);
7 JADE's DF assigns the service to one of three different agents: Firefighter_Agent; Engineer_Agent; or Electronic_Agent, previously registered;
8 After reply from DF, XO_Officer sends a message to the corresponding Equipment/ Compartment indicating crew members already assigned to it. The agent updates its state by changing the number of people acting upon the damage;

9 Next time Aircraft attacks the Equipment/Compartment upgrade their state
and the cycle re-initiates;

10 The Equipment/Compartment that has no more needs (specified in its state)
will send a message to the Officer who will send a message to the Evaluation Agent saying its proposal is 0, i.e. it does not require any other personnel.
After that, when participating in the contract net protocol, this last agent
will refuse to send any proposal;

11 After four runs, there should be no more demand for personnel and all participants in the contract net should be refusing the participation.

11   Future Work and Conclusions

We feel the need to a more adaptive capability to the changing environment, by
sensing exceptional situations in which the system has to immediately react even
before the chain of command does it. We also need to introduce some uncertainty
about the probability with which the ship, the compartments and the equipment
are really hit and damaged under each specific attack. In conclusion we want to
emphasize that this work approaches a complex problem which is adequate to
be solved by using multi-agent systems. A great effort was applied in analyzing
and formulating the problem, applying the agent-oriented software engineering
methodology PORTO, which is based on GAIA and TROPOS. As far as the
implementation is concerned, we followed FIPA guidelines, using ACL protocols,
and JADE. Although not all the intended goals were accomplished, we believe
that the following objectives were achieved:

First and very important is the fact that the system reflects a simplified
version of the existing organization on board and reflects the way limited crew
resources assignment should be dealt with;

Also relevant is that the system takes into account the different kinds of
knowledge of the crew and its respective limitations;

Finally, and decisive, the system is able to be, in an automatic way, close to
the decision making process of the commanding officer;

References

1. Bellifemine, F., Caire, G., Greenwood, D.: Developing Multi-Agent Systems with
   JADE. John Wiley, Chichester (2007)
2. Bresciani, P., Perini, A., Georgini, P., Giunchiglia, F., Myloupoulos, J.: Tropos: An
   agent-oriented software developemnt methodology. J. Auton. Agent. Multi Agent.
   Syst. **8**(4), 203–236 (2004)
3. Castro, A., Oliveira, E.: The rationale behind the development of an airline operations control centre using gaia based methodology. J. Am. Soc. Inf. Sci. Technol.
   2(3), 350–377 (2008)
4. António, J., Castro, M., Rocha, A.P., Oliveira, E.: A New Approach for Disruption
   Management in Airline Operations Control, vol. 562. Springer Verlag, Heidelberg
   (2014). studies in computational intelligence

5. Rossetti R., Brito Carvalho Martins, T.: Ship damage control action simulation using hla. In: Proceedings of The International Conference on Harbour, Maritime and Multimodal Logistics Modelling and Simulation, HMS, pp. 487–499. ACM Press (2013)
6. Scafes, M., Badica, C.: Preliminary design of an agent-based system for human collaboration in chemical incidents response. In: Proceedings of the 7th International Workshop on Modelling, Simulation, Verification and Validation of Enterprise Information Systems, pp. 53–62. Scitepress (2009)
7. Marques, M.S., Pires, J.: Singrar â a fuzzy distributed expert system toassist command and control activities in naval environment. Europ. J. Oper. Res. **145**, 343–362 (2003)
8. Zambonelli, F., Jennings, N., Wooldridge, M.: Developing multiagent systems: the gaia methodology. ACM Trans. Softw. Eng. Method. **12**(3), 317–370 (2003)

Collective Profitability and Welfare in Selling-Buying Intermediation Processes

Amelia Bădică[1], Costin Bădică[1(✉)], Mirjana Ivanović[2], and Ionuţ Buligiu[1]

[1] University of Craiova, A. I. Cuza 13, 200530 Craiova, Romania
ameliabd@yahoo.com, cbadica@software.ucv.ro
[2] University of Novi Sad, Faculty of Sciences, Novi Sad, Serbia
mira@dmi.uns.ac.rs

Abstract. We consider tree-like intermediation business processes that guide the selling-buying activities through a set of transaction chains. A seller is reaching the market of potential buyers that are interested in its products through a set of intermediaries, rather than acting directly on the market. This process generates a tree-structured complex e-commerce transaction. In this paper we propose a formal model of such transactions based on rooted trees and welfare economics. This model enabled us to obtain theoretical results regarding the definition of collectively profitable intermediation transactions and optimal pricing strategies of the transaction participants.

Keywords: Welfare economics · Formal model · Linear algebra · Multi-agent system

1 Introduction

Most often a business does not make its products directly available to the potential customers. Rather, the business is using a complex business process that is responsible for the management of its distribution activities. The distribution sector is in charge with providing the methods, processes and strategies for bringing the products of the business to the market of potential customers that need those products and are interested to buy them.

Typically a manufacturing or wholesale business that is interested in selling its products will use one or more distribution channels. They represent groups of individuals or organizations that are responsible for directing the flow of products from the producers to the market of potential customers that are interested in purchasing them. A distribution channel contains a set of one or more marketing intermediaries. A marketing intermediary is an agent that links a seller to a customer or to another intermediary with the overall goal of linking the initial or root seller to its ultimate buyers. Often a seller can use multiple and different distribution channels simultaneously.

There is a long discussion about the motivation, the functions and the types of distribution channels and marketing intermediaries from an economics perspective [6]. However, in this paper the focus in on defining a simple, yet formal model of intermediation, from a computer science perspective.

N.T. Nguyen et al. (Eds.): ICCCI 2016, Part II, LNAI 9876, pp. 14–24, 2016.
DOI: 10.1007/978-3-319-45246-3_2

In particular, our approach is based on the agent metaphor, here understood in a computational context, as a new model of a "computer system situated in some environment that is capable of flexible autonomous action in order to meet its design objectives" [5].

Our main contribution is the definition of a formal model of intermediation as a multi-agent system (MAS hereafter) containing the producer (or seller), the intermediaries and the customers (or buyers). Here a MAS is a computational system containing a collection of loosely-coupled agents representing the participants of the intermediation process, that interact to solve the given intermediation problem. Then, using the techniques of linear algebra and welfare economics we are able to formulate collective profitability conditions and optimal pricing strategies of the participants to the intermediation transaction.

2   A Formal Model of Intermediation

Let us consider a seller agent denoted with S that is interested to bring its set of products $1, 2, \ldots, k$, $k \geq 1$ to the market of potential customers. The seller can sell the products directly to the customers or it can use a set of intermediaries. Let us denote the customers interested to buy the products with $B_1, B_2, \ldots, B_r$ such that customer B_i is interested to buy a subset $P_i \subseteq \{1, 2, \ldots, k\}$ of the products, for all $1 \leq i \leq r$. We assume that $P_i \cap P_j = \emptyset$ for all $1 \leq i \neq j \leq r$ and that $\cup_{i=1}^{r} P_i = \{1, 2, \ldots, k\}$, so sets $\{P_i\}_{i=1}^{r}$ define a partition of $\{1, 2, \ldots, k\}$.

Furthermore, let us denote with $I_1, I_2, \ldots, I_x$ the intermediaries. We assume that there are $x \geq 0$ intermediaries. An intermediary has a dual role of buyer, as well as of seller. It buys one or more products from a generic seller that can be either S or another intermediary. Then it sells a (subset of) those products to other generic buyers that can be either customers B_j or other intermediaries.

The business process that describes the complex intermediation activity that enables seller S to sell its products to buyers B_i for all $1 \leq i \leq r$ can be modeled as a tree rooted as S. In what follows we call this structure an *intermediation tree*.

Intuitively, an intermediation tree is defined as a rooted tree [3] starting with its root S. Then, we define the children of the root either as intermediaries selling subsets of the set of products or customers that buy directly from the seller S. Now, each intermediary has a role which is similar to the seller S and the process can be continued recursively for each intermediary. Note that the growth of the intermediation tree is potentially infinite as an intermediary I_i selling a subset P of products can decide to sell the whole set P to another intermediary I_j with $i \neq j$. Nevertheless, we require that the intermediation tree is finite, in order to describe a realistic intermediation transaction consisting of a finite number of selling-buying activities.

This intuition is captured by the following definition of an intermediation tree.

Definition 1 (Intermediation Tree). *Let us consider one seller S that is interested to sell $k \geq 1$ products to r customers $B_1, B_2, \ldots, B_r$ such that customer B_i is interested to buy a subset P_i of products for all $1 \leq i \leq r$ and sets $\{P_i\}_{i=1}^r$ define a partition of $\{1, 2, \ldots, k\}$. The* intermediation tree *is a rooted tree defined as follows:*

(i) *The tree contains* internal nodes *representing the seller (the root of the tree) or the other intermediaries, as well as* external nodes *or* leaves, *representing the customers. Therefore sometimes the root of the tree is represented with S (the seller), while a leaf node is represented using the label of the associated customer B_i.*

(ii) *Each tree node N has associated a nonempty subset of the set $\{1, 2, \ldots, k\}$. This set is denoted with $set(N)$.*

(iii) *The root is labeled with set $\{1, 2, \ldots, k\}$, i.e. $set(S) = \{1, 2, \ldots, k\}$.*

(iv) *Let X be an internal node and let us denote with $\mathcal{Y}$ its nonempty set of children. Then the set $\{set(Y) | Y \in \mathcal{Y}\}$ is a partition of $set(X)$.*

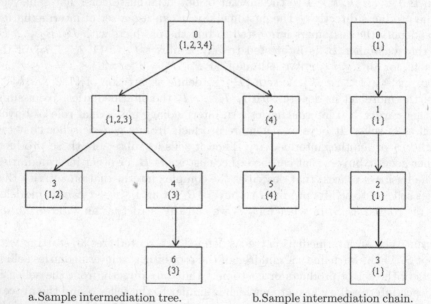

a.Sample intermediation tree.   b.Sample intermediation chain.

Fig. 1. Intermediation trees.

If S denotes the root of an intermediation tree then $set(S)$ denotes the set of products being sold by seller S.

If N is an internal node different from S then N represents an intermediary. $set(N)$ represents the set of products bought by N from the business partner represented by the parent of N and further sold to the partners represented by the children of N.

If N is an external node (or leaf) then N represents an end customer interested in purchasing products $set(N)$.

Note that in an intermediation tree, products are transferred top down from the root to its leaves, while money are transferred bottom-up from the leaves to its root.

Figure 1a illustrates a sample intermediation tree with 7 nodes, such that the root is represented by node 0, and the leafs are denoted by 3, 6 and 5. Here the seller is selling 4 products, i.e. the set associated to the root is $set(0) = \{1, 2, 3, 4\}$. The end customers are represented by node 3 interested to buy products 1 and 2, node 6 interested to buy product 3, and node 5 interested to buy product 4.

Note that in this case there are 3 true intermediaries (excluding the root) represented by internal nodes 1, 2, and 4. For example the intermediary represented by node 1 buys products 1, 2, and 3 from the seller, and sells products 1 and 2 directly to the customer represented by node 3, and sells product 3 to the intermediary represented by node 4. Moreover, the intermediary associated to node 2 simply resells product 4 to the customer associated to leaf node 5.

Each arc of an intermediation tree from node N to a child node M defines a simple transaction where the intermediary or root seller corresponding to N sells the set of products $set(M) \subseteq set(N)$ to the intermediary or end customer corresponding to M.

Proposition 1 (Number of Transactions). *Let us assume that in an intermediation tree there are r nodes representing customers and x internal nodes representing true intermediaries. Then the tree defines a number of $t = r + x$ transactions.*

This result follows almost trivially from the fact that in a free tree with n nodes there are always $n - 1$ edges [3]. For example, referring to the intermediation tree from Fig. 1a, we have $t = 6$, $r = 3$ and $x = 3$, so the equality stated by Proposition 1 holds.

A specific type of intermediation tree is the *intermediation chain*. Here each node of the tree has at most one child, so overall, the tree has a linear shape. It obviously holds that each node of an intermediation chain has associated the same set of products. An example is shown in Fig. 1b.

A simple analysis reveals that any intermediation tree can be decomposed into a set of intermediation chains, following each path from the root to one of its leaves. So, an intermediation tree with r leaves can be decomposed into r intermediation chains. For example, the tree shown in Fig. 1a can be decomposed into 3 chains as follows: (i) $(0, \{1, 2\}) \rightarrow (1, \{1, 2\}) \rightarrow (3, \{1, 2\})$; (ii) $(0, \{3\}) \rightarrow (1, \{3\}) \rightarrow (4, \{3\}) \rightarrow (6, \{3\})$; and (iii) $(0, \{4\}) \rightarrow (2, \{4\}) \rightarrow (5, \{4\})$.

Note that this decomposition is consistent with the observation that we made in the introduction, i.e. that a seller can use multiple and different distribution channels simultaneously.

3 Profitability

An intermediation tree defines rigorously the hierarchical structure of a complex intermediation transaction. The analysis of how such structures get created is beyond the scope of this paper. Nevertheless, we can speculate that seller, intermediation and buyer agents can use the techniques provided by middle-agents and interaction protocols to incrementally define such an intermediation tree [1,2].

In this section we define and assign economic information to an intermediation tree. Then we analyse the property of profitability of an intermediation tree and we define optimal and stable pricing strategies of the participants of an intermediation tree.

Let $Q \subseteq \{1, 2, \ldots, k\}$ be a nonempty subset of products.

We denote with s_Q the limit price of seller S for selling the whole set Q of products. This means that S will agree to sell the whole set Q of products only for a price p such that $p \geq s_Q$.

Similarly, we denote with b_Q the limit price of an end customer B for agreeing to pay and buy the whole set Q of products. This means that B will agree to buy the whole set Q of products only for a price p such that $p \leq b_Q$.

We can assign economic information about limit prices to an intermediation tree as follows:

(i) The root node S is annotated with the seller limit price denoted with s. Actually s should be written as $s_{1,2,\ldots,k}$, but we omit the indices because in this case the meaning is obvious, as s means the limit price of seller S for selling the whole set $\{1, 2, \ldots, k\}$.

(ii) Each leaf node B_i representing an end customer is annotated with the limit price $b_{set(B_i)}$, for each $1 \leq i \leq r$.

For example, referring to the intermediation tree presented in Fig. 1a, the root node 0 is annotated with limit price s, node 3 is annotated with limit price b_{12}, node 6 is annotated with limit price b_3, and node 5 is annotated with limit price b_4.

In what follows we assume that an intermediation tree will also include information about limit prices.

An intermediation tree can be annotated with information about transaction prices as follows. Each arc linking node i to its child j is annotated with the price $p_j > 0$ of the transaction between seller i and buyer j for selling products $set(j)$, for all $1 \leq j \leq t$, where t is the number of tree nodes (excluding the root, with index 0) with indices $1, 2, \ldots, t$.

For example, referring to Fig. 1a, the arc linking node 1 to node 4 is annotated with transaction price p_4.

Let us first consider a potential selling-buying transaction between a generic seller S with limit price s and a generic buyer B with limit price b. Let us also assume that the agreed transaction price is p. The utility gained by seller S is $p - s$ and the utility gained by buyer B is $b - p$. This transaction is profitable if and only if both participants gain, i.e. $p - s \geq 0$ and $b - p \geq 0$. It follows trivially that the transaction is collectively profitable for B and S if and only if $b \geq s$. In this case the transaction price can be fixed to an arbitrary value $p \in [s, b]$.

Using this observation we are interested to derive a necessary and sufficient condition such that an intermediation tree can be collectively profitable for all its participants.

Definition 2 (Collective Profitability). *Let us consider an intermediation tree with $n + 1$ nodes such that the root node is labelled with 0 and the other nodes are labelled with $1, 2, \ldots, n$. The tree is called* collectively profitable *if and only if the tree can be annotated with transaction prices such that each transaction participant is profitable, i.e. it gains by performing the transaction.*

If u_i is the utility of participant represented by node i then i is profitable if and only if $u_i \geq 0$. u_i can be computed as follows:

(i) If $i = 0$, i.e. i is the root node then $u_0 = -s + \sum_{j \in C} p_j$ where C represents the set of children of the root node.
(ii) If i is a true intermediary node, i.e. an internal node different from the root then $u_i = -p_i + \sum_{j \in C} p_j$ where C represents the set of children of node i.
(iii) If i is a leaf node representing an end customer then $u_i = -p_i + b_{set(i)}$.

Let us denote with $children(i)$ the set of children of node i of an intermediation tree. Also let us denote with $\mathcal{L}$ the set of leaves and with $\mathcal{I}$ the set of true intermediary nodes of an intermediation tree.

Let us consider the following system of $t + 1$ inequations with t variables $p_1, p_2, \ldots, p_t$:

$$-s + \sum_{j \in children(0)} p_j \geq 0$$

$$-p_i + \sum_{j \in children(i)} p_j \geq 0 \quad i \in \mathcal{I} \tag{1}$$

$$-p_i + b_{set(i)} \geq 0 \quad i \in \mathcal{L}$$

The following lemma states a necessary and sufficient condition for the collective profitability of an intermediation tree.

Lemma 1. *An intermediation tree is collectively profitable if and only if there exists an annotation with transaction prices that satisfies the system (1) of inequalities.*

Using Lemma 1 we can formulate the following necessary and sufficient condition that states when an intermediation tree is collectively profitable.

Proposition 2 (Necessary and Sufficient Condition for Collective Profitability). *An intermediation tree is collectively profitable if and only if:*

$$\sum_{i \in \mathcal{L}} b_{set(i)} \geq s \tag{2}$$

We are going to prove Proposition 2 for the sample tree presented in Fig. 1a. The proof for the general case is not difficult, it follows the same idea, but the details are more technical, so it is omitted here. Moreover, using an example will be easier to understand for the reader.

Firstly, inequations (1) can be rewritten as equations, for the sample tree presented in Fig. 1a, as follows:

$$
\begin{aligned}
-s + p_1 + p_2 &= \alpha_0 \geq 0 \\
-p_1 + p_3 + p_4 &= \alpha_1 \geq 0 \\
-p_2 + p_5 &= \alpha_2 \geq 0 \\
-p_3 + b_{12} &= \alpha_3 \geq 0 \\
-p_4 + p_6 &= \alpha_4 \geq 0 \\
-p_5 + b_4 &= \alpha_5 \geq 0 \\
-p_6 + b_3 &= \alpha_6 \geq 0
\end{aligned}
\tag{3}
$$

The condition stated by inequality (2) for the sample tree presented in Fig. 1a is defined as follows:

$$
b_{12} + b_3 + b_4 \geq s
\tag{4}
$$

Now, if the tree is collectively profitable, according to Lemma 1, Eq. (3) have a solution. Summing up all the equations we get:

$$
b_{12} + b_3 + b_4 - s = \sum_{i=0}^{6} \alpha_i \geq 0
\tag{5}
$$

so condition 4 follows trivially.

Conversely, we assume that condition (4) is true and we build an annotation of the tree with limit prices $p_i \geq 0$, for all $i = 1, 2, \ldots, 6$ such that inequalities (1) hold. This is reduced to finding $\alpha_i \geq 0$ such that equations (3) are true. To simplify things, we assume that $\alpha_1 = \alpha_2 = \cdots = \alpha_6 = \alpha$ and we look for a suitable value of α.

Solving the last 6 equations of system (3), starting with the last equation, we get:

$$
\begin{aligned}
p_6 &= b_3 - \alpha \geq 0 \\
p_5 &= b_4 - \alpha \geq 0 \\
p_4 &= b_3 - 2\alpha \geq 0 \\
p_3 &= b_{12} - \alpha \geq 0 \\
p_2 &= b_4 - 2\alpha \geq 0 \\
p_1 &= b_{12} + b_3 - 4\alpha \geq 0
\end{aligned}
\tag{6}
$$

α_0 can be computed from the first equation of (3) as follows:

$$
\alpha_0 = b_{12} + b_3 + b_4 - s - 6\alpha \geq 0
\tag{7}
$$

Choosing $0 \leq \alpha \leq \min\{b_3/2, b_4/2, b_{12}, (b_{12}+b_3)/4, (b_{12}+b_3+b_4-s)/6\}$ (this is possible as limit prices are positive and inequality (5) holds), conditions (6) and (7) are satisfied, so the proof is concluded.

4   Welfare Pricing Strategy

In this section we apply some concepts from welfare economics with the goal of defining optimal pricing strategies of the transaction participants. However, we were able to obtain theoretical results only for special cases, that will be outlined here. For the other situations we concluded that either more theoretical investigation is required, or specific computational methods should be employed to determine the optimal pricing strategies of the participants.

Social welfare can be determined using a *collective utility function* [4]. If $\mathcal{A} = \{a, b, \dots\}$ is the set of participant agents, and if each agent $a \in \mathcal{A}$ has an individual utility function $u_a \geq 0$ then a collective utility function U is a positive function defined as follows:

$$U(x) = \ U(u_a(x), u_b(x), \dots) \tag{8}$$

for all $x \in \mathcal{X}$, where $\mathcal{X}$ is the space of possible offers.

Several collective utility functions are proposed in the literature.

Utilitarian social welfare uses the following collective utility function:

$$U_{usw}(x) = \ \sum_{a \in \mathcal{A}} u_a(x) \tag{9}$$

Egalitarian social welfare uses the following collective utility function:

$$U_{esw}(x) = \ \min_{a \in \mathcal{A}} u_a(x) \tag{10}$$

Nash social welfare uses the following collective utility function:

$$U_{nsw}(x) = \ \prod_{a \in \mathcal{A}} u_a(x) \tag{11}$$

In what follows we apply these collective utility functions to determine the optimal pricing strategy for the participants of an intermediation process. The results basically follow from the following lemma.

Lemma 2.   *Let $\mathcal{A}$ be a set of agents and let $\mathcal{X}$ be their space of offers. Let us assume that the utilitarian social welfare function $U_{usw}(x)$ is constant for all offers $x \in \mathcal{X}$ and that there exists an offer $x^* \in \mathcal{X}$ and a constant K such that $u_a(x^*) = K$ for all $a \in \mathcal{A}$. Then the maximum of all the collective utility functions is obtained for $x = x^*$.*

The result stated by Lemma 2 follows from few simple arguments.

Let $U = U_{usw}(x)$. Trivially it follows that $K = U/|\mathcal{A}|$. Using the inequality of arithmetic and geometric means, we have $U_{nsw}(x) \leq (U/|\mathcal{A}|)^{|\mathcal{A}|}$. We get equality if the individual utilities of all the agents are equal, i.e. when $x = x^*$, so $U_{nsw}(x)$ is maximum when $x = x^*$.

Let x be an offer for which at least two agents have distinct utilities. It follows that the agent for which the utility is minimum has a utility that is strictly less then the arithmetic mean $U/|\mathcal{A}| = K$. So $U_{esw}(x) < K$. Then the maximum of $U_{esw}(x)$ is K and it is obtained when $x = x^*$.

Let us now consider an intermediation tree with $t+1$ nodes numbered from 0 (the root node) up to node t. The set of leaf nodes is denoted with $\mathcal{L}$. Let us also assume that the condition stated by inequality (2) holds, so the intermediation tree is collectively profitable. Summing up the utilities of all the participants we obtain:

$$\sum_{i=0}^{t} u_i = \sum_{i \in \mathcal{L}} b_{set(i)} - s \tag{12}$$

According to Eq. (12), the utilitarian social welfare of the participants to the intermediation transaction is constant. So we are under the assumptions of Lemma 2. It follows that the maximum of all the collective utility functions is obtained when $u_i = (\sum_{i \in \mathcal{L}} b_{set(i)} - s)/(t + 1)$ for all $0 \leq i \leq t$.

Let

$$h = (\sum_{i \in \mathcal{L}} b_{set(i)} - s)/(t + 1) \tag{13}$$

We can solve system (14) of equations to find transaction prices $p_1, p_2, \ldots, p_t$. They define the optimal pricing strategy of the participants to maximize their social welfare.

$$-s + \sum_{j \in children(0)} p_j = h$$

$$-p_i + \sum_{j \in children(i)} p_j = h \quad i \in \mathcal{I} \tag{14}$$

$$-p_i + b_{set(i)} = h \quad i \in \mathcal{L}$$

Note that system (14) has always a unique solution. Let us number tree nodes according to the breadth-first traversal [3] (see for example the tree nodes from Fig. 1a). Firstly observe that the first equation is redundant, as it follows by summing up the other t equations, so it can be omitted. Now it follows that (14) is a linear system of t equations. Moreover, the system matrix is upper triangular (i.e. the lower triangle consists only of 0s), while the elements of the diagonal are equal to -1. So the matrix is non-singular with the determinant equal to $(-1)^t$. This proves that (14) has a unique solution.

Proposition 3 (Sufficient Conditions for Maximum Social Welfare). *Let us consider an intermediation transaction such that limit prices satisfy condition (2). If system (14) has a positive solution $p_i \geq 0$ for all $1 \leq i \leq t$ then the maximum equalitarian social welfare is h and the maximum Nash social welfare is h^{t+1}. The optimal pricing strategies are defined by the solution of system (14).*

In what follows we check the application of Proposition 3 to the tree from Fig. 1a. We simplify things by assuming that $b_3 = b_4 = b$ and $b_{12} = 2b$. We obtain system (15) (the first equation of (14) was omitted, as it is redundant).

$$-p_1 + p_3 + p_4 = h$$
$$-p_2 + p_5 = h$$
$$-p_3 + 2b = h$$
$$-p_4 + p_6 = h \tag{15}$$
$$-p_5 + b = h$$
$$-p_6 + b = h$$

Solving system (15) we obtain:

$$
\begin{aligned}
p_6 &= b - h \\
p_5 &= b - h \\
p_4 &= b - 2h \\
p_3 &= 2b - h \\
p_2 &= b - 2h \\
p_1 &= 3b - 4h
\end{aligned}
\tag{16}
$$

Moreover:

$$h = (4b - s)/7 \tag{17}$$

The transaction is collectively profitable, so $h \geq 0$, i.e. $b \geq s/4$. Now, in order to satisfy the assumptions of Proposition 3, using the equation defining p_4 (or p_2) from (16), we obtain:

$$h \leq b/2 \tag{18}$$

Combining (17) and (18) we obtain the following condition:

$$b \leq 2s \tag{19}$$

The conclusion is that if inequality (19) holds then the optimal pricing strategy can be determined using equations (16). Otherwise, if $b > 2s$ we cannot apply Proposition 3. In this case other methods, either theoretical developments or computational approaches must be used, in order to determine the optimal pricing strategy of the participants. These further developments are left as future works.

5   Conclusion

In this paper we proposed a formal model of intermediation business processes that a company can use to distribute its products to the end customers that are interested in purchasing them. The model captures a hierarchically structured intermediation transaction as a rooted tree and it can serve a company with multiple distribution channels working simultaneously. We formulated necessary and sufficient conditions for the collective profitability of such an intermediation transaction. Then we applied the concepts of welfare economics to analyze optimal pricing strategies of the transaction participants. We obtained a theoretical result stating sufficient conditions when the optimal pricing strategy of the participants can be determined by solving a simple system of linear algebraic equations. As future work we plan to strengthen this result, either by formulating tighter optimality conditions or by proposing computational methods to determine optimal pricing strategies in a more general setting. As future work we are also interested to study the stability of pricing strategies of the transaction participants using the concepts of game theory.

References

1. Bădică, C., Budimac, Z., Burkhard, H.-D., Ivanović, M.: Software agents: Languages, tools, platforms. Comput. Sci. Inf. Syst. 8(2), 255–298 (2011). doi:10.2298/CSIS110214013B
2. Bădică, A., Bădică, C.: FSP and FLTL framework for specification and verification of middle-agents. Appl. Math. Comput. Sci. 21(1), 9–25 (2011). doi:10.2478/v10006-011-0001-6
3. Cormen, T.H., Leiserson, C.E., Rivest, R.L., Stein, C.: Introduction to Algorithms, 3rd edn. MIT Press, Cambridge (2009)
4. Jehle, G.A., Reny, P.J.: Advanced Microeconomic Theory, 3rd edn. Pearson Education Limited, UK (2011)
5. Jennings, N.R., Wooldridge, M.: Applications of intelligent agents. In: Jennings, N.R., Wooldridge, M.J. (eds.) Agent Technology, pp. 3–28. Springer, Heidelberg (1998). http://dl.acm.org/citation.cfm?id=277789.277799
6. Kerin, R., Hartley, S., Rudelius, W.: Marketing, 12th edn. McGraw-Hill Education, New York (2014)

Fault-Tolerance in XJAF Agent Middleware

Mirjana Ivanović[1]([⊠]), Jovana Ivković[1], Milan Vidaković[2], Nikola Luburić[2],
and Costin Bădică[3]

[1] University of Novi Sad, Faculty of Sciences,
Department of Mathematics and Informatics, Novi Sad, Serbia
{mira,jovana.ivkovic}@dmi.uns.ac.rs
[2] University of Novi Sad, Faculty of Technical Sciences, Novi Sad, Serbia
{minja,nikola.luburic}@uns.ac.rs
[3] University of Craiova, Craiova, Romania
cbadica@software.ucv.ro

Abstract. This paper presents one solution for enabling the fault-tolera-nce in a particular agent middleware the XJAF agent middleware. The XJAF agent middleware has been actively developing for past seven years at the University of Novi Sad, and is used for both scientific research and educational purposes. One of the most significant features of this middleware is the support for the load-balancing and fault-tolerance. In this paper we present the XJAF architecture and its features and functionalities. The main intention of the paper is to compare results of execution of the same example in two agent frameworks that support clustering: our in-house developed system XJAF and widely known JADE. We shall demonstrate that distributed agent application deployed on the XJAF middleware is capable of surviving the failure of its nodes, while the same application deployed on the JADE is not capable to support it.

Keywords: Software agents · Multi-agent systems · Fault-tolerance

1 Introduction

Distributed artificial intelligence is a field that is constantly developing. One useful and successful approach to implement different distributed artificial intelligence systems is to study, construct and use multi-agent systems [1].

Research in the field of multi-agent systems, and distributed problem solving as well, is considered a part of distributed artificial intelligence. Distributed problem solving involves agents that solve a problem together, while multi-agent systems have a broader definition. They only require that the agents communicate in the environment.

What distinguishes software agents from other concepts and techniques of artificial intelligence is the social aspect of the agents [2,3]. Namely, an agent rarely solves the problems on its own. Instead, agents function in groups, and they usually try to reach their goals through coordination, cooperation and joint actions.

© Springer International Publishing Switzerland 2016
N.T. Nguyen et al. (Eds.): ICCCI 2016, Part II, LNAI 9876, pp. 25–34, 2016.
DOI: 10.1007/978-3-319-45246-3_3

XJAF (*EXtensible Java EE-based Agent Framework*) is a multi-agent system that supports its agents during their whole lifetime [4]. That support includes: maintaining the life cycle of the agent, infrastructure for exchanging messages, etc. Unlike many other multi-agent platforms, XJAF operates on top of computer clusters. That influence development of agents that supports two advanced features: load-balancing and fault-tolerance.

The main goal of this paper is to show that XJAF is able to recover from the loss of a node in a cluster and continue its execution uninterrupted. We will test the fault-tolerance of XJAF against that of JADE using an implementation of the Contract Net protocol as a test case.

The rest of the paper is organized as follows. Section 2 provides a comparison of existing multi-agent systems with the XJAF system. Section 3 provides information about the architecture of XJAF. Section 4 gives details of the Contract Net protocol. Section 5 presents the results of the conducted test. Section 6 outlines the conclusion and future work on the presented topic.

2   Related Works

With the rise of popularity of multi-agent systems there have been a large number of different software tools, systems, platforms and environments that allow development of multi-agent applications.

Agent Developing Framework [5] enables the user to build interoperable, flexible, and scalable applications. Agent Developing Framework uses Java EE technologies such as JNDI, JMS and JMX. Communication is done synchronously or asynchronously through JMS (*Java Message Service*). Although this framework uses the same technology as the XJAF, ADF is no longer maintained. Current downloadable version is from the year 2005.

Voyager [6,7] is middleware software designed for distributed application development, and it does provide the option of developing applications using multi-agent programming, although that is not its main purpose. This framework supports scalability and fault-tolerance, but it is a commercial product.

Cognitive Agent Architecture (Cougaar) [8] is a MAS written in Java and used for developing distributive applications based on software agents. Cougaar allows developers to implement an application with minimum knowledge of the MAS itself. It provides robustness, security and scalability. Agents communicate point-to-point way.

Cougaar consists of a large number of nodes that act like hosts for agents. Every node is executed within its own *Java virtual machine*, and is resilient to failure due to state persistence.

Cougaar provides many advanced features (such as Adaptive Robustness, Adaptive Security, etc.), especially when large applications are in question, but it uses proprietary solutions to implement those features. XJAF, on the other hand, uses industrial-level solutions to achieve the same or similar goals.

Magentix [9] is a MAS that focuses on execution performance. Its implemented in C and installed on the Linux operating system. Agents are represented with a unique Linux process that has 3 layers:

- **main**, which starts when the host is added to the platform
- **service**, which supports services such as Agent Management System and Directory Facilitator
- third layer, which is **dedicated to agents**.

Tests have proven that execution of the programs is fast [9], which is explained by the fact that the operating system is used instead of a mediator between the operating system and the MAS. Communication between agents is point-to-point and based on Transmission Control Protocol (TCP) sockets.

Although slower than Magentix, XJAF provides a larger number of advanced techniques, especially considering its support for computer clusters.

Radiogost [10] is a multi-agent framework that combines agent and web technologies. Radiogost agents and many functioanlities are implemented in JavaScript, takes full advantage of HTML5 and is executed within a web browser. The main functionalites of Radiogost include, but are not limited to life-cycle management, efficient communication using FIPA ACL [14], yellow-page service, interaction with agent in third-party multi-agent platforms.

ADiS [11] (*Agent-based Distributed Computing System*) is an agent-oriented system for distributed computing, designed for working in environments that are characterized by sudden, unexpected failures of computational nodes. ADiS brings efficient load-balancing and job-distribution architecture.

JADE [12,13] is a MAS written in Java and strongly adherent to FIPA standard.

Communication in JADE is based on asynchronous message delivery. Every agent has a mailbox (a queue meant for messages) where JADE runtime puts messages that an agent receives. When a message is delivered to the queue, the agent is notified, but the moment in which it will look and process it is determined by the developer. Messages in JADE are ACL messages.

JADE implements some of the protocols defined by FIPA: FIPA-Request, FIPA-Query, FIPA-Propose, iterative versions of FIPA-Request and FIPA-Query, Contract-Net and FIPA-Subscribe [14].

At first glance it may appear that JADE is similar to XJAF, however there are certain differences. XJAF uses JMS for message exchange, while JADE has its own system. The main difference between these two systems, however, lies in the fact that, when creating clusters, in JADE agents have to be manually divided between cluster nodes, while in XJAF it is done automatically. That is to say, in XJAF an agent is defined at the level of the computer cluster, and not at an individual node.

XJAF is the better solution for applications that have a large number of agents and/or need to have their fault-tolerance on a high level. On the other hand, JADE is the better solution since it consumes less resources and it is easier to use. In Sects. 4 and 5 we will demonstrate that XJAF-based solution can survive node failure, while JADE-based solution cannot.

3   XJAF Environment

XJAF (*EXtensible Java EE-based Agent Framework*) is a multi-agent system [4] based on Java platform, Enterprise Edition (Java EE). Most multi-agent systems that are based on Java don't take advantage of the benefits that Java EE offers; such as scalability, security, as well as resistance to software and hardware errors.

Basic features that MAS offers are life-cycle management, infrastructure for message exchange and subsystems that give the agents ability to use the resources at their disposal, as well as perform algorithms. In some cases, MAS offer a certain amount of security: mechanisms for agent coordination in distributed environments, as well as support for agent persistency and mobility. XJAF uses the following components of Java EE:

- Java Naming and Directory Interface (JNDI) used in agent directories and service implementation
- Java Message Service (JMS) enables communication
- Enterprise JavaBeans (EJB) used for agent and service implementation
- Java Serialization supports agent mobility and persistency.

Two important features that XJAF supports are: load-balancing and fault tolerance. Load-balancing guarantees that the agents will be distributed equally amongst the nodes of a computer cluster, meaning that no node will be overburdened in comparison to the others. The system also copies XJAF components and agents across the nodes, which guarantees that there will be no loss of data in case of an error.

Communication and coordination within the system are reliable. Namely, messages will be delivered to the agent, even if it isn't available at the moment the message is sent.

XJAF is defined as a set of loosely-based components that are dynamically accessed. In XJAF these components are called managers. Each manager is in charge of a specific part of the process. Managers can only be accessed through their interface, which means they can easily be adjusted to the needs of different applications. XJAF has the following managers, as shown on Fig. 1.

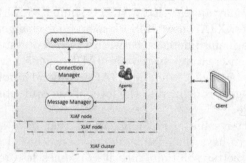

Fig. 1. The XJAF architecture

- AgentManager manages the agents life cycle
- MessageManager provides the basis for agent communication
- ConnectionManager used for maintaining networks of distributed XJAFs

3.1 Agents and AgentManager

AgentManager provides functionalities defined by FIPA [14]: registration, dereg-istration, modification and search. It maintains a list of registered agents.

XJAF agents are implemented as EJB (*Enterprise Java Bean*) components, which provides a certain resistance to software and hardware errors, and provides high availability for clients. Modern JavaEE application servers use the technique of EJB pooling to achieve high availability of EJB beans. A certain number of EJB instances is kept in the memory, so when a request for a new agent arrives, an agent instance is recycled from the pool. Of course, the number of instances in the pool can be changed in accordance to the number of requests. Agents can be represented as *stateful* or as *stateless* beans.

XJAF is deployed on the JBoss application server. JBoss server maintains a pool of threads that agents can use. Instead of assigning a thread to every agent, the threads are assigned as needed. The server will also try to minimize the number of threads when possible (e.g. if a thread is not used in a certain time period, it will be deleted). Also, if an agent is passive for a time, it will be removed from the run-time memory and stored in secondary memory (e.g. hard drive).

To support the agent mobility, AgentManager actually keeps two directories, one for local agents, and the other for remote agents. Every record in the direc-tory is kept in form of a pair (ID, reference). When an agent is created, a new pair is added to the local directory, and when its deactivated its pair is deleted. If an agent is moved to another node, the pair is deleted from the local directory and moved to the directory meant for remote agents where the pairs are in form of (ID, address). Agents within the system communicate based on their IDs, and not based on the references.

When a message is sent to an agent, AgentManager will first look in the local directory, and if it cant find its ID, only then will it look in the remote agent directory.

3.2 Message Exchange in XJAF

Message exchange between EJBs is done using JMS (Java Message Service). JMS is an API meant for asynchronous message exchange between loosely-based components.

JMS supports two basic means of communication:*point-to-point* and *publish-subscribe*. In point-to-point, a producer puts the message in the queue, and that message is consumed by a single consumer. In publish-subscribe, communication is done through topics. Namely, a producer posts a message in a topic, and all the consumers subscribed to the topic receive it. In XJAF, communication is point-to-point. Message delivery in JMS is shown on Fig. 2.

Messages are posted to the queue, and are consumed by MDBs (message-driven beans). MDBs are organized in a pool that can be extended if needed. MDB deliv-ers the message to a bean by calling its *onMessage* method. When the *onMessage* method is executed, the message is deleted from the queue. If an error occurs dur-ing the delivery, the process is repeated. In case of multiple unsuccessful deliveries,

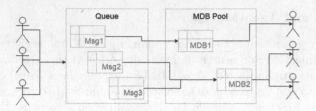

Fig. 2. Message Exchange in XJAF

the message is placed in the dead letter queue, at which point the developer can manually handle it.

Messages in XJAF are in ACL (Agent Communication Language specification by FIPA standard) [14] format.

3.3   XJAF on Clusters

A computer cluster is a network of nodes, which from a users point of view, function as a single system. Clusters enable *high availability* of deployed applications, which guarantees the client access to the application even in the case of a node failure, as well as timely response from the application regardless of the number of requests [15].

In case XJAF is executed on a cluster, organization of the cluster is as follows: one node in the cluster is the master node, while the rest are slave nodes. For controlling a node, JBoss *host controller* is used, while the master node itself can control the entire cluster using JBoss *domain controller*.

Processing power of a larger cluster is greater, but so is the possibility of an error with the larger number of nodes the possibility that one of them would fail rises.

The general advantages of a cluster are: failover and load-balancing. The first advantage is only applicable to EJBs - namely, whenever there is a change in a state of a stateful bean, its state is copied to all the nodes in the cluster, so that if there is an error on any of the nodes, its beans (whether agents or services) can easily continue their execution from another node.

There are two ways of state replication on a JBoss server: *replicated* and *distribution*.

In the replicated mode, the states are copied to all the nodes in the cluster, and are available from every node. However, this mode works well only when small clusters are in question (up to 10 nodes), because with the increase of number of nodes, the number of message exchanges needed for replication becomes too large.

In the distribution mode, the states are copied to a configurable number of nodes. To determine on which nodes the state needs to be copied to, a hash algorithm is used. While the number of nodes is configurable, it stands that with a larger number of copies the fault-tolerance is increased, but the performance decreases.

The advantage of load-balancing in multi-agent systems like XJAF is in the automatic distribution of agents across the nodes with the purpose of optimizing performances. If a bean is stateful, it is placed on a node, and all the method calls are performed on that node. In the case of a stateless bean, there can be multiple copies on different nodes, and on a method call one of the nodes performs it.

4 Fault-Tolerance Case Study - Contract Net Implementation

Contract Net protocol implementation will be used as a case study for demonstration of fault-tolerance in the XJAF agent middleware. XJAF middleware will be deployed in a cluster of JBoss application servers and all messages between agents, as well as agents themselves will survive random cluster node failure.

Contract Net protocol [14] is based on the business model of making a call for proposal and then assigning the task to a company that gave the best proposal. In the case of software agents, we have the *initiator* that makes the call for proposal, accepts the proposals of *participants*, analyses them, chooses the best one on offer and, finally, awards the task to the participant who gave the best

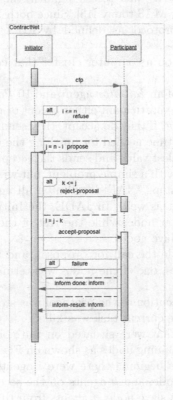

Fig. 3. Contract Net protocol

proposal (the best proposal can be the one that requires the least resources, or the one that offers the shortest time period for finishing the task).

The course of events in Contract Net protocol is shown on Fig. 3.

Initiator sends the CFP (Call For Proposals) to n participants. Based on the demands, the participants decide whether or not they will submit a proposal (*refuse* or *propose*). In total, there are n responses (for n participants), of which j are *propose* and the remaining n-j are *refuse*. After the time for proposals has passed (as defined in *reply-by* parameter of the CFP), initiator analyses the proposals and accepts the best proposal. More than one proposal can be accepted. The number of accepted proposals is marked with I. The remaining k proposals are refused.

The participant whose proposal was accepted performs the requested task and informs the initiator as to the results (it can be a *failure*, *inform-done* that lets the initiator know that the task is completed and *inform-result* that contains the results of the task, as well).

5   Results

The fault-tolerance of XJAF was tested against the JADE MAS, which is arguably the most popular MAS today [13]. Since both XJAF and JADE implement the Contract Net protocol as defined by FIPA, a simple example was created for testing purposes.

The example was run on a computer cluster that contained a master node and 2 slave nodes.

The example consists of an *Initiator* agent and 10 *Participants*. The initiator sends out the CFP, and the participant either accept or refuse the CFP based on a random number generator. If they do accept, they send the value of the random number as their proposed time needed for finishing the task. The initiator picks the proposal with the smallest value and sends the acceptance to the participant. The participant is tasked with a simple problem, but with an added sleep period of 5 seconds after every step. XJAF distributed all agents equally across the cluster (the Initiator agent as well). In JADE, the Initiator and 2 participants were situated on the master node, while the rest of the participants were split equally between the two slave nodes.

After the initiator made the call for proposal, and the participants started sending out their responses, one of the nodes was terminated. After this point, a difference was noted.

In JADE, once the execution of the protocol was stopped. The system itself continued running, but the current task was killed.

In XJAF, the agents that were situated on the node that was terminated were reassigned to the remaining nodes as shown on Fig. 4, which is possible due to state replication. In the begining there were i agents on the master node, j agents on the first slave node, and k agents on the second slave node. When a failure occured on the first slave node, j agents from that node were reassigned to the remaining nodes of the cluster. Since the messages are delivered to agents

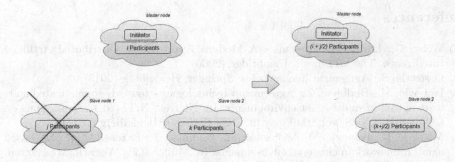

Fig. 4. The XJAF cluster with failed node

regardless of them being moved to another node [4], the execution of the protocol was not interrupted.

6   Conclusion and Future Work

Scalability and fault-tolerance are key features in distributed problem solving and multi-agent systems. Most state of the art frameworks support some features which enable scalability and fault-tolerance, but most of them rely on proprietary solutions. Contrary to them, XJAF agent middleware relies on industry-level of solutions for the scalability and fault-tolerance in clusters. XJAF is deployed on the JBoss application server which supports clustering, and this supports scalability and fault-tolerance for the instances in the cluster. This fact provides two advantages over proprietary solutions:

1. JBoss-implemented clustering is tested and implemented in multitude of industry solutions, which offers high level of scalability and availability for the nodes in the cluster, and
2. performance of the JBoss-based clusters is constantly improved in multiple projects, in time.

As a case study, we have implemented the Contract Net protocol which depends on intensive message exchange between agents. We have tested the implementation of the clustered version of Contract Net on two platforms: the JADE framework and the XJAF middleware. Since JADE-based solution does not automatically distribute agents in the cluster, failure of a single node in the cluster caused the failure of the complete implementation. On the other hand, XJAF-based solution was able to overcome the failure of any node(s) in the cluster, since the distribution of agents is done automatically, and agents that reside on the failed node are automatically restored on the existing nodes.

Since XJAF middleware is deployed on the JBoss application servers, and its clustering support is constantly enhanced, it can only benefit from those improvements, so in future we can expect even faster and more reliable clusters.

References

1. Weiss, G.: Multiagent Systems - A Modern Approach to Distributed Artificial Intelligence. The MIT press, Cambridge (2000)
2. Ossowski, S.: Agreement Technologies. Springer, Heidelberg (2013)
3. Ivanović, M., Budimac, Z.: Agreements technologies - towards sophisticated software agents in multi-agent environments. In: Nguyen, N.T. (ed.) Transactions on CCI XVIII. LNCS, vol. 9240, pp. 105–126. Springer, Heidelberg (2015)
4. Mitrović, D., Ivanović, M., Vidaković, M., Budimac, Z.: Extensible java EE-based agent framework in clustered environments. In: Müller, J.P., Weyrich, M., Bazzan, A.L.C. (eds.) MATES 2014. LNCS, vol. 8732, pp. 202–215. Springer, Heidelberg (2014)
5. Agent Developing Framework. http://adf.sourceforge.net/index.html
6. Voyager. http://www.recursionsw.com/
7. Voyager and Agent Platforms Comparison published by ObjectSpace Inc.for Voyager 1.0. http://www.cis.upenn.edu/bcpierce/courses/629/papers/unfiled/AgentPlatformsW97.PDF, September 1997
8. Cougaar. http://www.cougaar.org/
9. Alberola, J.M., Such, J.M., Botti, V., Espinosa, A., Garcia-Fornes, A.: A scalable multiagent platform for large systems. Comput. Sci. Inf. Syst. **10**(1), 51–77 (2013)
10. Mitrović, D., Ivanović, M., Budimac, Z., Vidaković, M.: Radigost: Interoperable web-based multi-agent platform. J. Syst. Softw. **90**, 167–178 (2014)
11. Mitrović, D., Ivanović, M., Geler, Z.: Agent-based distributed computing for dynamic networks. Inf. Technol. Control **43**(1), 88–97 (2014)
12. JADE. http://jade.tilab.com/
13. Bellifemine, F.L., Caire, G., Greenwood, D.: Developing Multi-agent Systems, vol. 7. Wiley, Chicheste (2007)
14. FIPA Standard Specification. http://www.fipa.org/repository/standardspecs.html
15. WildFly 8 high availability guide. https://docs.jboss.org/author/display/WFLY8/High+Availability+Guide

Agent-Based Support of a Virtual eLearning Space

Emil Doychev(✉), Asya Stoyanova-Doycheva(✉),
Stanimir Stoyanov(✉), and Vanya Ivanova(✉)

University of Plovdiv, Plovdiv, Bulgaria
{e.doychev,stani,vantod}@uni-plovdiv.bg,
astoyanova@uni-plovdiv.net

Abstract. This paper provides an overview and presents the architecture of a virtual space supporting eLearning. Various types of assistants are examined as well which are implemented as rational BDI agents supporting the operation of the space. Furthermore, development of the space as an IoT ecosystem is considered.

Keywords: Elearning · IoT · Virtual educational space · Assistants · Intelligent agents · BDI architecture

1 Introduction

In recent years, a Distributed eLearning Centre (DeLC) project was implemented in the Faculty of Mathematics and Informatics at the University of Plovdiv aiming at the development of an eLearning environment [1]. DeLC is implemented as a network infrastructure which consists of separate nodes, called eLearning Nodes. Each eLearning Node presents a real educational unit (laboratories, departments, faculties, colleges, and universities), which offers a complete or partial educational cycle. Each eLearning Node is an autonomous host of a set of electronic services.

DeLC suffers from the shortcomings of the widely used eLearning systems that ignore the physical world which they operate in. Observing the physical environment reveals opportunities for development of context-aware systems. The effective support of the learning process is many-sided and dependent on actions and events occurring in different places and at different times, e.g. attending lectures and seminars, self-studies, examinations, consultations. However, an analysis of the results of the learning process has to take into account all the various aspects and can make a connection between them. In this sense, we aim to transform DeLC into a new infrastructure, known as Virtual Educational Space (VES), where users, time, location, autonomy and context-awareness are first-class citizens, and which enables a uniform treatment and interpretation of information coming from both the virtual environment and the physical world [2].

This paper provides an overview and presents the architecture of VES. The rest of the paper is organized as follows: the Sect. 2 provides an overview of the virtual space, the Sect. 3 presents the assistants of the space, and the Sect. 4 presents the DeLC 2.0 portal as a special entry point of the space.

© Springer International Publishing Switzerland 2016
N.T. Nguyen et al. (Eds.): ICCCI 2016, Part II, LNAI 9876, pp. 35–44, 2016.
DOI: 10.1007/978-3-319-45246-3_4

2   Related Works

The broad usage of the Internet and its steady transformation into a network of objects [3], as well as the globalization of cyberspace, are a foundation for the rapid development of cyber-physical social systems which will lead to essential technological, economical and sociological consequences in the following years. The IoT paradigm began to be used in the field of education. In [4], a project is presented which aims to combine both the virtual and the physical environments providing a better learning experience to the students. A specific technical framework of a ubiquitous learning environment based on the Internet of Things (IoT) supports learners with increasing social skills. The environment integrates three layers: a perception layer, a network layer, and an application layer [5]. In this way, IoT-based learning can happen at any place, any time, with any people, and any content. The current situation of M-learning under the Internet of Things is discussed in [6]. In [7], a combination of IoT architecture and techniques of learning analytics is considered which can be used to record and conduct an analysis of the students' learning process and further enable learners and schools to obtain the feedback that they need and establish an effective lifelong learning environment. The paper [8] presents a Tempus project aimed at collaborating distant ELabs of different Maghrebian and European universities using IoT interoperability. In [9] is described the concept of Internet of Things and its role in the evolution of a SCORM-based eLearning application is demonstrated. In this case, the Content Aggregation Model of SCORM specifies a general framework that can be used in the learning process supported by the IoT standard.

3   Overview of VES

3.1   VES Features

The basic features of VES can be summarized as follows:

- *Autonomy*–VES is developing as an autonomous ecosystem delivering electronic educational services and teaching content that can be adapted to various educational institutions to support self-space learning, blended learning, and life-long learning. Individual users can create and use personalized subspaces.
- *Intelligence*–an intelligent space [10] can control what happens in it, interact with the components included, infer, make decisions and act in accordance with these decisions.
- *Context-awareness*–the context is all the information which can be used to characterize the situation of an identity where an identity can be a human, a place or an object which are viewed as significant for the interaction between a user and the system. According to the definition of a context, a system is context-aware if it uses contexts to deliver essential information or services important for the user's tasks [11, 12]. In our case, context-awareness means the ability of a system to find, identify and interpret changes in its environment and, depending on their nature, to undertake appropriate actions such as, for example, personalization or adaptation. Personalization

is the ability of the system to adapt to individual features, desires, intentions, and goals of the users. Adaptation is the system's ability to adapt to the remaining context features such as area of knowledge, school subjects, and types of devices used by the end-users.

- *Scenario-orientation*–from a user's point of view, the virtual educational space is an environment delivering separatee- learning services or completed educational scenarios accessible through both the DeLC 2.0 portal and the personal assistants. Scenarios are implemented by corresponding workflows rendering an account of the environment's state where it is possible to take into account various temporal attitudes of the educational process (e.g. duration, repetition, frequency, start, end) or events (planned or accidental) which can impede or alter the running of the current educational scenario. To deal with possible emergencies (such as an earthquake, flood or fire) emergency scenarios are defined to be executed with the highest priority.
- *Controlled access*–the virtual educational space is a controlled ecosystem which means that access to the resources of the space is only possible through the specialized supporting modules known as "entry points." The personal assistants operate as typical entry points while the DeLC 2.0 portal is a special entry point. A user has to be in possession of a personal assistant or to use the portal to be able to work in the space.

3.2   VES Architecture

The VES architecture contains different types of components (Fig. 1). In the next section will be considered assistants implemented as rational BDI-agents which play an important role in the space.

The basic functionality delivered by the space is deployed in both subspaces (D- and A-) that interact intensively during the execution of educational scenarios. D-Subspace is designed for direct support of the educational process providing the following three engines:

- *SCORM Engine (SEng)* – this engine delivers teaching content in the form of SCORM 2004 electronic packages to support students' self-study. The SCORM Engine integrates three separate modules (SCORM Player, SCORM Manager and SCORM Statistics) implemented according to the ADL's SCORM 2004 R4 specification [13]. The electronic packages are stored in a digital library that can be accessed by the students during their self-study. The SCORM Engine traces the progress of students actually working with the teaching material and the collected information (metadata) can be used by the TNB (Teacher's Note Book) for analysis and evaluation of the students' performance.
- *Test Engine (TEng)* – it provides all its functionalities in agreement with the QTI 2.1 standard [14] as a result of the communication between two base modules–a User Interaction Provisioning (UIP) and an Assessment Provisioning (AP) module. The UIP provides the sensing means of the system to the users' environment, and the AP is responsible for the analysis of the data received from all the eTesting system's sensors–not only the sensors targeting the user's environment (UIP) but also from

the ones referring to the inner VES space changes concerning the personalized learning state of the user (extracted from the SCORM engine).

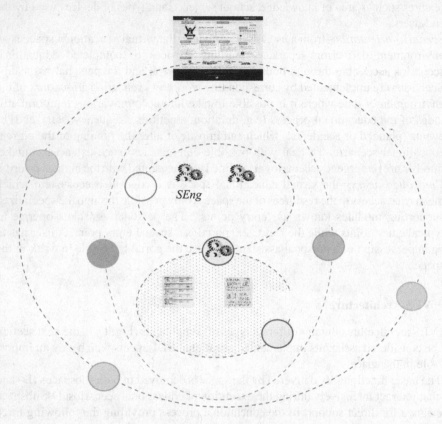

Fig. 1. VES Architecture

- *Event Engine (EEng)* – it is shared by both subspaces and implements the event model specified for the whole space. The events are used to create more complex structures, such as plans, schedules, and personal calendars. The interface also provides an event editor.

Furthermore, a digital library is managed in the D-Subspace where the teaching content is prepared mainly in accordance with the SCORM and QTI standards (other formats are also possible, e.g. .pdf, .ppt, .doc, etc.). In addition, the digital library provides a flexible security mechanism allowing the definition of cascading access rules per user, roles and/or role groups. There are three different access rights: to view, download and manage, each of which with several access levels.

The A-Subspace secures all activities related to the organization, control and documentation of the education process. In the administrated database is stored all the necessary information used for planning, organizing, protocoling and documenting the educational process.

3.3 VES as an IoT Ecosystem

VES is defined as an abstraction of the whole learning process decoupling it into its different aspects without interrupting the connectivity between them. Thus, VES is composed of components that handle the user interaction with the system and vice-versa. It is responsible for maintaining shared knowledge throughout its components as well as providing means for enabling the seamless communication between them via the Internet. In order to act as an IoT ecosystem, VES has to comply with the layers from the IoT stack [15]:

- The *sensing layer* is mainly composed of the SCORM Engine and Test Engine–they are sources of virtual sensor data and deliver information about student progress of self-study and examinations. In addition, various user interactive components such as personal assistants and the web application directly accessible to the users (DeLC 2.0) can provide virtual sensor data. Physical sensor data is supplied by the guards;
- The *data integration IoT stack layer* is implemented based on using specifications compliant with the content representation. In the space, two standards (SCORM 2004 and QTI 2.1) sharing a common metadata specification (LOM) are basically supported. The content is exchangeable both ways–towards the ecosystem by imports and out of the system by exports. Again, all the data exchange with the outer world is service-based adopting the REST concept for maximal platform independence;
- The *analytics of things IoT stack layer* is composed of a Grade Book and a Teacher's Notebook that receive sensor data directly provided by the SCORM and QTI engines. It could also be enriched with information as a result of the reasoning intelligent actions performed on a higher level over the straightforward statistics data. Thus, the analytics are personalized in a generalized way concerning the whole educational process not scoped to a certain educational aspect;
- The *cognitive actions layer* is implemented as a multi-agent system of rational agents having mind states–beliefs, desires, intentions. They gain and share common knowledge that has an impact on all the VES components' behaviors on different levels of granularity in order to complete the currently selected educational scenario.

3.4 Event Model of VES

The space is built as a completely decentralized system. To ensure the interoperability standards are used (e.g., SCORM 2004, QTI 2.1, Common Cartridge), together with appropriate ontologies. In addition, the interoperability of the VES IoT stack is enhanced by an event model implemented in the space. An event is presented as the triple (event_id, event_type, event_arguments). Three types of events are distinguished in the model:

- Basic events–date, hour, location;
- System events–generation/removing assistants, sending/receiving messages;
- Domain-dependent events–in our case events related to the educational process, e.g. lectures, tutorials, exams, self-study, consultations.

Furthermore, the events can have different attributes such as:

- Repeatability–single, periodic;
- Durability–discrete, continuous;
- Purpose–container, simple.

Dealing with the event in the model is formally presented with the help of various constructions proposed by a formalism known as Event Calculus (EC) [16].

4   The Agents of VES

The following three types of assistants are supported in the space (Fig. 1):

- Personal assistants (PAs);
- Operative assistants (OpAs);
- Guards.

4.1   Personal Assistants

PAs have to perform two main functions to provide the needed "entry points" of the space. Firstly, they operate as an interface between their owners and the space and, if necessary, carry out activities related to the personalization and adaptation. Secondly, they interact with other assistants in the space in order to start and control the execution of educational scenarios. The personal assistants will usually be deployed over the users' mobile devices.

The students' PA, known as LISSA (Learning Intelligent System for Student Assistance) is able to perform the following tasks:

- It monitors the student's subsequent curriculum and reminds them of upcoming exams, lectures, seminars, etc. The time for reminders of upcoming events depends on the type of event, the time needed for preparation and the student's current location thus taking into account the length of time needed to reach the location where the event will be taking place. In order to choose a specification in case of a conflict between two or more events, priorities are set which can be used during the deliberation process. After solving a conflict, the agent warns the student that it has taken a certain decision and consults with them to check whether they agree with it.
- It recommends useful bibliography to the student–upon a warning for an upcoming event LISSA will recommend learning materials that the student can use to study.
  - It automatically registers the student in the different events and according to the student's location it can determine whether they are present at a certain event (a lecture, seminar, exam, etc.) and will mark them as present.
  - Voice commands–besides the normal interaction with the agent through the touch screen, LISSA will allow voice commands. The voice commands are optimized so that there is no need for internet access to a server for their recognition, thus preserving battery life.

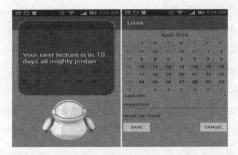

Fig. 2. LISSA

The early prototype of LISSA, implemented as a BDI agent, has a simplified design with an avatar on the main screen (Fig. 2). The mental attitudes of LISSA are events. Activities implemented as a personal calendar are a container event including various domain-dependent events which are under LISSA's control. Depending on the current assumptions (for date, time, location), the current intention is determined from the personal calendar. Depending on the type of the intention (event) a suitable plan can be activated. For example, when the agent wants to warn the student about an upcoming activity, it pronounces the message and synchronously displays it on the screen. The message is presented simultaneously visually and vocally so that the agent can be useful to visually impaired and hearing impaired students. The voice recognition allows calibration to easily recognize different voice types and accents. All announcement functionalities can be switched on and off.

4.2 Operative Assistants

The operative assistants are active components exhibiting a more complex architecture. An agent in itself is not a suitable software component for delivering business functionality. A service is a good decision for the functionality but it is static and cannot operate as a separate component in the space. For that reason corresponding service interfaces are implemented for the operative assistants. Since all functionalities that the VES components can use and expose while inhabiting the eLearning ecosystem are provided as services, the space is open for new components which can provide their capabilities as services as well regardless of the technologies used for their implementation.

Usually located on the server nodes of the space, the OAs support the execution of educational scenarios; therefore, they implement suitable interfaces to the available electronic services and data repositories. Operatives serve both subspaces. The two assistants operating in the A-Subspace are:

- *Grade book (GB)* – the student's grade-book stores and analyzes information on the success rate of students in all the studied courses. It is currently being developed in accordance with the Grade Book specification of the Common Cartridge standard [17].

- *Teacher's notebook (TNB)* – it is designed for the analysis of the success rate of students in a particular course of studies. In addition, it helps the teacher to organize his/her duties during the current education period.

Due to the inherent complexity, the TNB is realized as a multi-agent system, including the following agents:

- Agent One (A1) – Its main task is to analyze each test separately, and in tests with unsatisfactory results it sets the flag for further analysis of the test. As an output the agent calculates the average grade of the exam;
- Agent Two (A2) – It analyzes the sections of the marked tests for further analysis and provides summary results for specific sections of the exam tests. Finally, the agent sorts sections in an ascending order–from the section with the lowest results to the section with the highest ones. The agent also sets an additional flag if there is a need to analyze the individual questions in a particular section;
- Agent Three (A3) – It analyzes various questions in the marked sections for each test and provides summary results of the examination. The agent represents a list of questions that students answer incorrectly the most;
- Agent Four (A4) – It summarizes information received from the SCORM player–A4 checks the information about the activity of certain students in order to determine whether the learners have passed through the material and have done the tests in the SCORM player;
- Agent Five (A5) – This agent systematizes the information obtained from the analysis, summarizes the results and provides them to PAT and PAS respectively in an appropriate form.

4.3 Guards

The guards are special assistants which are responsible for the safety and efficient execution of the educational scenarios in the space. These are usually intelligent devices that react to various physical quantities in the environment such as smoke, temperature changes, and humidity. The guards represent the real world in the virtual one and act as an interface between both worlds in the space. In addition, the guards are responsible for activating emergency scenarios.

5   The DeLC 2.0 Portal

The DeLC 2.0 portal acts as an entry point in VES, which provides access to the resources of the space without using personal assitants. The architecture of the portal (Fig. 3) consists of an educational portal with a user interface and a server side. These units communicate with each other by HTTP, RESTful services and Web Sockets. The browser side application is built with HTML 5 and CSS 3 together with JQuery and Bootstrap, which delivers the responsive nature of the user interface.

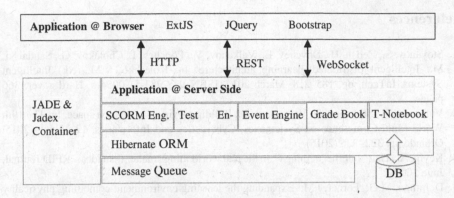

Fig. 3. The DeLC 2.0 Architecture

DeLC 2.0 provides the needed set of functionality for a modern web application: management of the static web content, security services, a control panel for management of all services, plugins and so on. As additional services the portal provides a Message Queue (MQ) and ORM layer. The MQ acts as a transport environment for asynchronous communication between the plugins and the rest of the components. The ORM layer is a built-in mechanism for transforming the database relational data into objects.

The portal implements interfaces to the D- and A-Subspaces by using plugins. Some of the implemented plugins are the SCORM Engine, Grade Book, Event Engine, Test Engine and Teacher Notebook. The interaction with the operative asisstants of the space is ensured by integration of the JADE [18] and Jadex [19] containers within the portal's server side. These containers are intended to provide an environment for running software agents which can communicate seamlessly with the operative assistants.

6   Conclusion

This paper provides an overview of the Virtual Educational Space implemented as an IoT ecosystem. The space is being implemented as a successor of an eLearning environment known as DeLC. The prototype implementation of VES is progressively extended and experimented in a real education process. In the last few years more than 2500 students have been educated using VES in more than 20 topics.

Acknowledgment. The authors wish to acknowledge the partial support of the NPD – Plovdiv University under Grant No. IT15-FMIIT-004 "Research in the domain of innovative ICT oriented towards business and education", 2015-16.

References

1. Stoyanov, S., Zedan, H., Doychev, E., Valkanov, V., Popchev, I., Cholakov, G., Sandalski, M.: Intelligent distributed elearning architecture. In: Koleshko, V.M. (ed.) Intelligent Systems, InTech, pp. 185–218, March 2012. ISBN 978-953-51-0054-6, Hard cover, 366 pages
2. Valkanov, V., Stoyanov, S., Valkanova, V.: Building a virtual education space. In: The 19th World Multi-Conference on Systematics, Cybernetics and Informatics, 12-15 July 2015, Orlando pp. 322–326 (2015)
3. Kevin, A.: That "Internet of things", in the real world things matter than ideas. RFID Journal, June 2009
4. Domingo, M.G., Forner, J.M.: Expanding the learning environment: combining physicality and virtuality. the internet of things for elearning. In: 10th IEEE International Conference on Advanced Learning Technologies, Tunisia, 730–731 (2010)
5. Xue, R., Wang, L., Chen, J.: Using the IOT to construct ubiquitous learning environment. In: Proceedings of the Second International Conference on Mechanic Automation and Control Engineering (MACE), Inner Mongolia, China, pp. 7878–7880 (2011). ISBN 978-1-4244-9436-1
6. Yang, B., Nie, X., Shi, H., Gan, W.: M-learning mode research based on internet of things. In: Artificial Intelligence, Management Science and Electronic Commerce (AIMSEC), Zhenzhou, China, pp. 5623–5627 (2011). ISBN 978-1-4577-0535-9
7. Cheng, H.-C., Liao, W.-W.: Establishing an lifelong learning environment using IoT and learning analytics. In: The 14th International Conference on Advanced Communication Technology (ICACT 2012), Phoenix Park, pp. 1178–1183 (2012). ISBN 978-89-5519-163-9
8. Lamri, M., Akrouf, S., Boubetra, A., Merabet, A., Selmani, L., Boubetra, D.: From local teaching to distant teaching through IoT interoperability. In: International Conference on Interactive Mobile Communication Technologies and Learning (IMCL), Thessaloniki, pp. 107–110 (2014). doi:10.1109/IMCTL.2014.7011115
9. Pau, V.C., Mihailescu, M.I.: Internet of things and its role in biometrics technologies and eLearning applications. In: 13th International Conference on Engineering and Modern Electric Systems (EMES), Oradea, pp. 177–180 (2015). ISBN 978-1-4799-7651-5
10. Liu, B., et al.: Intelligent spaces: an overview. In: IEEE International Conference on Vehicular Electronics and Safety, Beijing 13-15 December 2007. ISBN 978-1-4244-1266-2
11. Dey, A.K.: Understanding and using context. Pers. Ubiquitous Comput. J. 5(1), 4–7 (2001)
12. Dey, A.K., Abowd, G.D.: Towards a better understanding of context and context-awareness. In: Proceedings of the Workshop on the What, Who, Where, When and How of Context-Awareness. ACM Press, New York (2000)
13. SCORM 2004 Specification. http://adlnet.gov/adl-research/scorm/scorm-2004-4th-edition/
14. IMS Question & Test Interoperability Specification. https://www.imsglobal.org/question/index.html
15. Gramatova, K., Stoyanov, S., Doychev, E., Valkanov, V.: Integration of eTesting in an IoT eLearning ecosystem - virtual eLearning space, BCI 2015. ACM, Craiova (2015). ISBN 978-1-4503-3335-1/15/09, Art. 14
16. Mueller, E.T.: Commonsense Reasoning. An Event Calculus Based Approach. Elsevier, Amsterdam (2015)
17. IMS Common Cartridge Specification. https://www.imsglobal.org/cc/index.html
18. Bellifemine, F., Caire, G., Greenwood, D.: Developing Multi-Agent Systems with JADE. Wiley, New York (2007)
19. Jadex Active Components. https://www.activecomponents.org/bin/view/About/New+Home

Generalized Nets for Agent-Based Modeling

Galina Ilieva[✉] and Stanislava Klisarova

Plovdiv University "Paisii Hilendarski", Tsar Assen 24, 4000 Plovdiv, Bulgaria
galili@uni-plovdiv.bg, stanislava.klisarova@gmail.com

Abstract. The purpose of this paper is to evaluate the applicability of generalized nets for multi-agent systems modeling. Several models of interaction among intelligent agents as economic subjects were developed. The analysis that has been conducted shows that the proposed method is suitable for modeling and assessing information systems based on agents' technology.

Keywords: Multi-agents system modeling · Generalized nets · Economic model

1 Introduction

This paper is dedicated to applying generalized nets (GNs) for modeling economic interactions that involve participation of a large number of independent agents. The aim of the work is to enrich the tools for formal description of objects, processes and interactions in multi-agent systems (MAS).

Interest in the topic is inspired by recent changes in economics methodology, where a growing demand for alternative approaches to economic analysis is observed [3]. One option for achieving greater flexibility of economic models is employing multi-agent systems that simulate the behavior of a large number of individual participants, prominent in the literature as agent-based computational economics (ACE) [14].

ACE methods are suitable for the study of processes characterized by heterogeneity of the participants and a variety of interactions among them. These methods can successfully overcome some limitations of traditional economic models, such as requirements for the existence of equilibrium, assumptions about rationality of economic actors, etc. Modern MAS have various economic applications ranging from researching production processes [10] to modeling economic crises [12], financial markets [15] or automated auctions [8]. Despite the numerous advantages of ACE, however, research work on the specifics of building economic multi-agent systems is not plentiful. Enrichment of the tools for description, design and analysis of MAS would increase the quality of developed software and contribute to the further improvement to ACE modeling of economic objects, processes and phenomena.

The first section provides a brief overview of the basic ideas and functionality of some tools for describing MAS and emphasizes the advantages of generalized nets for modeling parallel processes in complex adaptive systems. The second section models the interaction of economic systems elements with the tools of generalized nets. The last section discusses the applicability of GNs in the design of multi-agent systems for

© Springer International Publishing Switzerland 2016
N.T. Nguyen et al. (Eds.): ICCCI 2016, Part II, LNAI 9876, pp. 45–55, 2016.
DOI: 10.1007/978-3-319-45246-3_5

economic simulations. The conclusion summarizes the main findings on the importance of the proposed method for researchers and analysts who deal with economic issues.

The main contribution of the work is in the construction of formal models of interaction among economic subjects (such as bank loan approval, stock trading, and online market). The created GNs information models reflect the mechanisms of interaction among participants and can be used in the analysis and prediction of economic processes as well as in the design of agent systems for their simulation.

2   Some Instruments for MAS Modeling

The topicality of the theme of this paper stems from issues surrounding the design of MAS. Applying the appropriate modeling instruments can significantly improve the quality of the development process and thus, the quality of the software system. Despite the significance of agent-based modeling, there is no standard protocol, language or tool to support the development of multi-agent applications.

In practice, different approaches are used to describe and present agent-based software. [16] provides a comprehensive overview of techniques for agent-oriented analysis and design, dividing them in two groups: those that extend or adapt existing object-oriented (OO) methodologies and those that adapt knowledge engineering or other techniques. The last decade saw the emergence of new instruments such as ODD, AML, etc.

The ODD protocol standardizes the three main blocks in the description of MAS – Overview, Design concept and Details. The basic idea behind the protocol is always structuring the information about an agent-based model in the same sequence. This sequence consists of several elements that can be grouped in the abovementioned blocks: purpose, state variables and scales, process overview and scheduling; design concepts; initialization, input, and submodels [6]. Although ODD model descriptions are understandable and complete, this protocol has several shortcomings and has been modified multiple times. For example, [7] adds more elements to the model. The advantages of ODD are its independence from subject area, operation system и programming language. Its main disadvantage is that it requires the properties and methods to be presented separately, which obstructs OO implementations.

AML is an extension of industrial standard UML and facilitates MAS specification in agent-oriented software engineering. This language is applied in all phases of MAS development and offers opportunities for project visualization through standard flowcharts types. Despite the fact that AML is sufficiently detailed and tangible enough, some aspects of agent-based systems such as concurrency, ontologies and mobility are not adequately supported. The capabilities of AML for visualizing the collaboration of several subjects is also limited [2].

The disadvantages of the abovementioned tools for design and technical description of MAS are:

- inability to design complex composite structures with a high degree of detail,
- problems in defining the dynamic behavior of the modeled objects,
- no temporal dimension in interactions, and others.

These restrictions hinder the adequate representation of contemporary MAS. A viable alternative are Petri nets (PNs), a mathematical modeling language and a well-known instrument for the description of distributed systems [9]. Generalized nets (GNs) as their name implies are a significant extension and generalization of the concept of PNs. GNs are currently present in various developments in the field of artificial intelligence. Two of their main applications can be found in [5, 11, 13] contains a comprehensive bibliographic database. As agent-based economics studies complex distributed systems concurrent actions of a set of interacting entities and their emergent behavior, GNs constitute a modern formalism that could be applied to MAS modeling.

The main advantages of generalized nets as a discrete tool for description of models of complex systems with many different and interacting components, often involved in parallel activities are:

- the conditions of predicate-transitive PNs are replaced by a predicate matrix which defines the rules for tokens' transfer to the different output places,
- in addition to discrete time in which processes in PNs take place, GNs contains an absolute time scale which can be used for keeping the time while a net is functioning,
- specific token's feature determines plausible predicate values,
- while going through transitions, the tokens receive new characteristics through a characteristic feature,
- practically no limitations to the characteristics of the tokens [1].

These features of generalized nets turn them into a potential tool for modeling dynamic processes within MAS with heterogeneous participants. Compared to well-known instruments for MAS modeling, GNs offer a simple and effective way of adequately modeling real processes. With GNs, designers are free to define different types of data and create models suitable for OO programming. Therefore, generalized nets are a possible alternative for overcoming the shortcomings of the abovementioned modeling languages and for expanding the applications of agent technology in economic research studies.

3   Generalized Net Models for MAS Simulation of Some Economic Processes

Through GNs, economic processes are described as a complex system of objects that interact and influence each other over time. Descriptions of two GN models of examples from the field of economics are discussed below.

Assume a business process of loan approvals in a financial institution is given. Loan candidates are represented by tokens with initial characteristics that include an agent's income, age, number of children, other debts, etc. The representation of this GN model (Fig. 1) contains only three transitions.

The form of the first transition is the following:

$$Z_1 = \langle \{l_1\}, \{l_2, l_3\}, r_1, M_1, \wedge(l_1) \rangle. \tag{1}$$

The token from place l_1 splits into two tokens α_2 and α_3, which take places l_2 and l_3, respectively. Token α_2 obtains characteristic "number of clients who receive the loan" at place l_2 and token α_3 obtains characteristic "number of clients who do not receive the loan" at place l_3. If we need any additional information, such as loan type, we can expand the respective token's characteristics.

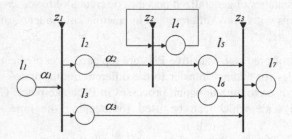

Fig. 1. A GN model for bank loan system

The transition condition r_1 is represented by the index matrix (IM):

$$r_1 = \begin{array}{c|cc} & l_2 & l_3 \\ \hline l_1 & W_{1,2} & W_{1,3} \end{array}$$

where $W_{1,2}$ and $W_{1,3}$ define the conditions for approving and declining a loan request.

The arc-capacity IM M_1 for the transition Z_1 can have the form:

$$M_1 = \begin{array}{c|cc} & l_2 & l_3 \\ \hline l_1 & n & n \end{array}$$

with possibility to receive n loan requests simultaneously. In this case we must set $c(l_1) = c(l_2) + c(l_3) = c(l_6) = c(l_7) = n$, where c gives the capacities of the places.

The last element of Z1 is the so-called transition type and it is an object having a form similar to a Boolean expression. It contains as variable the symbol that serves as label for the transition's input place, and the Boolean connective $\wedge$ determines that the place 11 must contain at least one token.

The second transition is the following:

$$Z_2 = \langle \{l_5, l_6\}, \{l_7\}, r_2, M_2, \vee(l_5, l_6) \rangle \tag{2}$$

Now,

$$r_2 = \frac{\begin{array}{c|c} & l_7 \\ \hline l_5 & W_{5,7} \\ l_6 & W_{6,7} \end{array}}{} \qquad M_2 = \frac{\begin{array}{c|c} & l_7 \\ \hline l_5 & n \\ l_6 & n \end{array}}{}$$

where $n \geq 1$.

This transition models the process of repaying the loan. If necessary, that repayment could be carried out by the loan guarantors (place l_6). In real life there are various other situations such as, for example, the borrower applying for refinancing.

The transition type $\vee(l_5, l_6)$ determines the following condition: there must be at least one token in the set of places l_5 and l_6.

Actions at Z_3 can be represented as follows:

$$Z_3 = \langle \{l_2, l_4\}, \{l_4, l_5\}, r_3, M_3, \vee(l_2, l_4) \rangle, \tag{3}$$

Where

$$r_3 = \frac{\begin{array}{c|cc} & l_4 & l_5 \\ \hline l_2 & W_{2,4} & W_{2,5} \\ l_4 & false & true \end{array}}{}$$

and predicate $W_{2,4}$ has characteristic "it is necessary to refinance the loan", $W_{2,5} = \neg\, W_{2,4}$. Token α_2 enters place l_4 or l_5 and obtains suitable characteristics. The form of IM M_3 is similar to the one above.

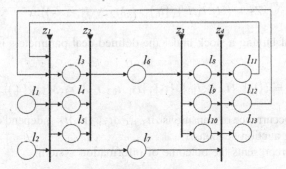

Fig. 2. A GN model of stock exchange system

In this first GN model, we discussed one of the ways to represent the process of lending a loan. All discussed situations can be broken down into further details.

Figure 2 presents a model of a part of MAS for stock exchange. Transition Z_1 models the process of receiving input orders to buy a particular amount of financial instruments and their distribution for further processing, e.g., market order (l_3), limit order (l_4) and market-to-limit order (l_5).

Here is what transition Z_1 looks like:

$$Z_1 = \langle \{l_1, l_2, l_{11}, l_{12}\}, \{l_3, l_4, l_5, l_7\}, r_1, M_1, \vee(l_1, l_2, l_{11}, l_{12}) \rangle. \tag{4}$$

The initial characteristics of the tokens from this part of MAS are: l_1 – reception of purchase orders; l_2 – presence of new information from the National Commerce

Register, which should be taken into consideration in the further implementation of the service; l_{11} – arrival of a purchase order, after the supervisory authorities denied participation for this buyer once; l_{12} – availability of information in the national database that needs to be compared with information in the submitted documents.

Matrices r_1 and M_1 have the following contents:

$$r_1 = \begin{array}{c|cccc} & l_3 & l_4 & l_5 & l_7 \\ \hline l_1 & W_{1,3} & W_{1,4} & W_{1,5} & false \\ l_2 & false & false & false & W_{2,7} \\ l_{11} & W_{11,3} & W_{11,4} & W_{11,5} & false \\ l_{12} & false & false & false & W_{12,7} \end{array}$$

$$M_1 = \begin{array}{c|cccc} & l_3 & l_4 & l_5 & l_7 \\ \hline l_1 & n & n & n & 0 \\ l_2 & 0 & 0 & 0 & n \\ l_{11} & n & n & n & 0 \\ l_{12} & 0 & 0 & 0 & n \end{array}$$

Transition Z_2 models the process of structuring the received applications as market orders (l_3), limit orders (l_4) and market-to-limit orders (l_5):

$$Z_2 = \langle \{l_3, l_4, l_5\}, \{l_6\}, r_2, M_2, \vee(l_3, l_4, l_5) \rangle. \tag{5}$$

The process of buying a stock under the defined deal parameters is modeled with transition Z_3:

$$Z_3 = \langle \{l_6\}, \{l_8, l_9, l_{10}\}, \{t_3\}, \{t_{31}, t_{32}, t_{33}\}, r_3, M_3, \vee(l_6) \rangle. \tag{6}$$

The times of occurrence of transitions $\Delta t_{3\text{-}31}$, $\Delta t_{3\text{-}32}$ и $\Delta t_{3\text{-}33}$ depend on the length of the corresponding auction sessions.

Transition Z_4 represents the outcome of information system:

$$Z_4 = \langle \{l_7, l_8, l_9, l_{10}\}, \{l_{11}, l_{12}, l_{13}\}, r_4, M_4, \vee(l_7, l_8, l_9, l_{10}) \rangle. \tag{7}$$

For this transition IM have the following contents:

$$r_4 = \begin{array}{c|ccc} & l_{11} & l_{12} & l_{13} \\ \hline l_7 & false & W_{7,12} & W_{7,13} \\ l_8 & W_{8,11} & W_{8,12} & W_{8,13} \\ l_9 & W_{9,11} & W_{9,12} & W_{9,13} \\ l_{10} & W_{10,11} & W_{10,12} & W_{10,13} \end{array} \qquad M_4 = \begin{array}{c|ccc} & l_{11} & l_{12} & l_{13} \\ \hline l_7 & 0 & n & n \\ l_8 & n & n & n \\ l_9 & n & n & n \\ l_{10} & n & n & n \end{array}$$

The presented model reflects only a small part of the actions included in the procedures of an agent-based exchange information system.

4 Generalized Net Model of Electronic Market Simulation

Both examples above demonstrate the possibilities for real use of GN in agent-based economic modeling. The first one represents a bank activity in a credit department whose basic processes are described. This model can be used for predicting different situations that may arise at the department (delaying payments, modifying credit terms and conditions, refinancing, etc.).

The second model visualizes purchases at a stock exchange. This model can be expanded to present a market where every agent, buyer or dealer, could pursue their own strategy for buying or selling. Both models could also be improved to provide mechanisms for reporting the impact of external effects on interest rates, stock prices and more.

The main advantages of both GN models from Sect. 3 are:

- describing and modeling of real-life parallel processes,
- simulating of processes,
- monitoring and controlling of processes.

However, these two examples do not demonstrate the essence of applying a GN model to an agent-based system, which is the representation of the behaviors and interactions of many agents, each of which has unique characteristics. For demonstrating these advantages of GNs, we will build a model of MAS for many-to-many negotiation (Fig. 3).

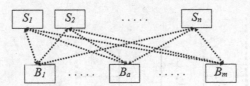

Fig. 3. Many-to-many negotiation

In order to describe the interactions between buyer B_i, $i = 1, ..., m$ and seller S_j, $j = 1, ..., n$ [4] proposes a model of bilateral agents' negotiation, where each agent's behavior depends on a scoring function and the remaining time that an agent has. The corresponding GN model of buyer Ba behavior is shown on Fig. 4.

The first transition Z_{1a} has $n + 1$ input places, labeled $l_{1,a}, l_{2,a}, ..., l_{n,a}$ – receiving sellers' offers (the notation used here is $l_{i,j}$, where i is buyer number, and j is seller number) and l_{ago} means that the buyer is ready to receive the next offer (i.e. it has finished processing the previous one). There is only one output–l_a and it is the offer that is currently being processed.

$$r_{Z1a} = \begin{array}{c|c} & l_a \\ \hline l_{1,a} & W_1 \\ l_{2,a} & W_2 \\ \cdots & \cdots \\ l_{n,a} & W_n \\ l_{ago} & false \end{array} \qquad M_{Z1a} = \begin{array}{c|c} & l_a \\ \hline l_{1,a} & 1 \\ l_{2,a} & 1 \\ \cdots & \cdots \\ l_{n,a} & 1 \\ l_{ago} & 0 \end{array}$$

Transition Z_{1a} chooses an offer to process provided that processing the previous offer has been completed and place l_{ago} contains a token:

$$Z_{1a} = \langle \{l_{1,a}, l_{2,a}, \ldots, l_{n,a}, l_{ago}\}, \{l_a\}, r_{Z1a}, M_{Z1a}, \wedge\big(l_{ago}, \vee(l_{1,a}, l_{2,a}, \ldots, l_{n,a})\big)\rangle. \quad (8)$$

The transition condition r_{Z1a} and arc-capacity M_{Z1a} are represented by IMs, where $W_1, \ldots, W_n$ are predicates, through which the next offer for processing is determined. It choice may depend on negotiation history with a certain buyer, offer conditions, etc. As only one offer can be processed at a time, it is implied that only one of the predicates $W_1, \ldots, W_n$ can be assigned a value of *true* in case a certain choice is made. A value of *false* assigned to transfer predicate $l_{ago} - l_a$ means that the token of l_{ago} could not be assigned to l_a.

The second transition is the following:

$$Z_{2a} = \langle \{l_{ainit}, l_a, l_{atm}\}, \{l_{a\alpha}, l_{a\beta}, l_{a\gamma}\}, \{0, 0, t_{prop}\}, r_{Z2a}, M_{Z2a}, \vee(l_{ainit}, l_a, l_{atm})\rangle \quad (9)$$

and it models the process of assessing the currently processed offer and making a decision. The offer l_a chosen in transition Z_{1a} serves as input. Output could be $l_{a\alpha}$, $l_{a\beta}$

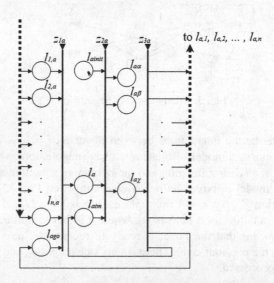

Fig. 4. GN model of buyer agent in electronic market MAS

and $l_{a\gamma}$, which correspond respectively to decisions to withdrawing from negotiations due to time running out (timeout); accepting the processed offer and making a deal; preparing a counteroffer. Additional input for this transition may also be provided: l_{ainit} – for initializing an agent's activity and making an opening offer and l_{atm} – for forceful activation of a transition in case of prolonged lack of new offers.

Now,

$$
r_{Z2a} = \begin{array}{c|ccc} & l_{a\alpha} & l_{a\beta} & l_{a\gamma} \\ \hline l_{ainit} & false & false & Init \\ l_a & F_1 & F_2 & F_3 \\ l_{atm} & T_1 & T_2 & T_3 \end{array} \qquad M_{Z2a} = \begin{array}{c|ccc} & l_{a\alpha} & l_{a\beta} & l_{a\gamma} \\ \hline l_{ainit} & 0 & 0 & 1 \\ l_a & 1 & 1 & 1 \\ l_{atm} & 1 & 1 & 1 \end{array}
$$

where $Init$ is a predicate for making an opening offer, $F_1, \ldots, F_3$ and $T_1, \ldots, T_3$ are predicates for transitions to the respective output places. The two $false$ predicates in r_{Z2a} guarantee that, during initialization no timeout or making a deal may occur. The third element of a transition's definition gives the length of the transfer to output states (places) where t_{prop} indicates the time needed for preparing a new offer (transfer to $l_{a\gamma}$).

The last transition Z_{3a} is as follows:

$$
Z_{3a} = \left\langle \{l_{a\gamma}\}, \{l_{a,1}, l_{a,2}, \ldots, l_{a,n}, l_{ago}, l_{atm}\}, \{0, 0, \ldots, 0, 0, t_{timeout}\}, r_{Z3a}, M_{Z3a}, \wedge(l_{a\gamma}) \right\rangle.
$$

$$(10)$$

The last transition directs the prepared specific offer ($l_{a\gamma}$) to the respective seller S_1, $S_2, \ldots, S_n$. Moreover, activating this transition ensures unconditional (predicate $true$ in r_{Z3a}) tokens for l_{ago} and l_{atm}, which are necessary for allowing transition Z_{1a} and activating transition Z_{2a} in case of timeout, respectively. Transfer time of token to l_{atm} is given by the parameter $t_{timeout}$ (Eq. 10). This guarantees periodic self-activation of this transition when there are no new offers.

The IMs are:

$$
r_{Z3a} = \begin{array}{c|cccccc} & l_{a,1} & l_{a,2} & \ldots & l_{a,n} & l_{ago} & l_{atm} \\ \hline l_{a\gamma} & Y_1 & Y_2 & \ldots & Y_n & true & true \end{array} \qquad M_{Z3a} = \begin{array}{c|cccccc} & l_{a,1} & l_{a,2} & \ldots & l_{a,n} & l_{ago} & l_{atm} \\ \hline l_{a\gamma} & 1 & 1 & \ldots & 1 & 1 & 1 \end{array}
$$

where $Y_1, \ldots, Y_n$ are predicates for directing the generated offer to the respective seller. Only one of predicates $Y_1, \ldots, Y_n$ may be assigned a value of $true$ when activating the transition.

A GN model of agent-buyer in MAS for simulating an e-market is very similar.

The model that was described earlier demonstrates modeling time-related characteristics via a GN. For example, to measure the time needed for generating new offers in different bidding strategies, the t_{prop} characteristic is used, and for ensuring activity when there are no offers, $t_{timeout}$ is set.

The three GN models that were presented demonstrate some of the advantages of GN. For example, detailed and individualized description of the mechanism of interaction among participants facilitates the software realization of MAS via an OO programming language. Agents' behaviors and interactions can be described in detail and time-related characteristics can also be added.

5 Conclusion

In this paper we show that generalized nets are detailed, comprehensive and tangible enough to be a useful tool for building multi-agent information systems. A comparative analysis has been conducted on instruments for multi-agent modeling. The advantages of GNs as a generalization of Petri nets variations were listed. Simplified generalized nets which model some business processes were developed. Employing GNs, a new model of agent-participant (buyer or seller) in MAS for electronic market simulation was built. The analysis that was performed shows that GNs are a suitable instrument for preliminary modeling of basic parts of large scale ACE applications before their software implementation. Due to their features GNs might form a significant contribution to the effort of bringing about widespread adoption of intelligent agents across various areas of applications, such as economic research. Future work will include integrating the developed models in complex models in MAS for electronic commerce.

Acknowledgement. This work was partially sponsored by the Scientific Research Fund of the Plovdiv University "Paisii Hilendarski" as part of project SR15 FESS 019/24.04.2015.

References

1. Atanassov, K.: On Generalized Nets Theory. "Prof. M. Drinov" Academic Publishing House, Sofia (2007)
2. Chervenka, R.: Modeling multi-agent systems with AML. In: Essaaidi, M., Ganzha, M., Paprzycki, M. (eds.) Software Agents, Agent Systems and their Applications, vol. 32, pp. 9–27. IOS Press (2012)
3. Colander, D., Follmer, H., Haas, A., Goldberg, M., Juselius, M., Kirman, A., Lux, T., Sloth, B.: The Financial Crisis and the Systemic Failure of Academic Economics. Kiel Working Papers 1489, Kiel Institute for the World Economy (2008)
4. Faratin, P., Sierra, C., Jennings, N.: Negotiation decision functions for autonomous agents. Int. J. Robot. Auton. Agents **24**(3), 159–182 (1998)
5. GNs Online Resources Web Address. http://ifigenia.org/wiki/Category:Publications
6. Grimm, V., Berger, U., DeAngelisc, D., Polhilld, J., Giskee, J., Railsback, S.: The ODD protocol: a review and first update. Ecol. Model. **221**(23), 2760–2768 (2010). doi:10.1016/j.ecolmodel.2010.08.019
7. Grimm, V., Berger, U., et al.: A standard protocol for describing individual-based and agent-based models. Ecol. Model. **198**(1–2), 115–126 (2006). doi:10.1016/j.ecolmodel.2006.04.023
8. Ilieva, G.: A fuzzy approach for bidding strategy selection. Cybern. Inf. Technol. **12**(1), 61–69 (2012)

9. Murata, T.: Petri Nets: properties, analysis and applications. Proc. IEEE **77**(4), 541–580 (1989)
10. Ostrosi, E., Fougères, A.J.: Optimization of product configuration assisted by fuzzy agents. Int. J. Interact. Des. Manuf. **5**(1), 29–44 (2011)
11. Roeva, O., Pencheva, T., Shannon, A., Atanassov, K.: Generalized nets in artificial intelligence. Generalized Nets and Genetic Algorithms, vol. 7. "Prof. M. Drinov" Academic Publishing House, Sofia (2013)
12. Setterfield, M., Gibson, B.: Real and financial crises: a multi-agent approach. Working Paper No 1309 (2013)
13. Sotirov, S., Atanassov, K.: Generalized Nets in Artificial Intelligence. Generalized Nets and Supervised Neural Networks, vol. 6. "Prof. M. Drinov" Academic Publishing House, Sofia (2012)
14. Tesfatsion, L.: Agent-based computational economics: modeling economies as complex adaptive systems. Inf. Sci. **149**(4), 262–269 (2003)
15. Thurner, S., Farmer, D., Geanakoplos, J.: Leverage causes fat tails and clustered volatility. Quant. Financ. **12**(5), 695–707 (2012)
16. Wooldridge, M.: An Introduction to MultiAgent Systems. Wiley, Chichester (2002)

Social Media and the Web of Linked Data

Networking Readers: Using Semantic and Geographical Links to Enhance e-Books Reading Experience

Dan Cristea[1,2(✉)], Ionuț Pistol[1], Daniela Gîfu[1(✉)], and Daniel Anechitei[1]

[1] Faculty of Computer Science, "Alexandru Ioan Cuza" University of Iaşi, Iaşi, Romania
{dcristea,ipistol,daniela.gifu,daniel.anechitei}@info.uaic.ro
[2] Institute for Theoretical Computer Science, Romanian Academy - Iaşi branch, Iaşi, Romania

Abstract. This paper describes how a system currently developed can be used to connect readers of enhanced e-books both to each other, to web resources and to real world locations and events. A set of Natural Language Processing resources are used to annotate relevant e-books and a framework is developed using the original text and the annotated metadata to detect and display semantic connections within the text and from text to relevant web data. This system can be further enhanced to detect connections between users using common reading interests and habits, their location in relation to locations found in text, and their reading and real world localisation history. Users could also be able to share collected data (text and web references, video and audio recordings, other interested readers) improving an individual reader experience and helping to establish a community around a particular e-book or a real life location with literary significance.

Keywords: Social networks · Semantic links · NLP · Geographic data · e-Books

1 Introduction

Building communities of users around technologies can enhance the ways users' benefits from the implemented functionalities and can increase both the frequency and the duration of usage of said technology. E-books have become a common way of experiencing written content, the most significant contributing factor being the increasing usage of mobile devices both for communication and for access to documents.

This paper aims to show how a currently developed system designed to add semantic links within texts and between text and outside resources (web pages and maps) can be further developed to build communities of users around particular E-books using discovered links (mostly of geographical and social nature).

Employing text mining techniques to improve social networks is not a novel idea. Ever since the emergence of large networks of user contributed data (mostly text), researchers have described methods of using that text to extract additional data about the network and its users [1, 2]. Two areas in which Natural Language Processing technologies have been particularly successful in enhancing a social network user experience have been e-learning [3, 4] and consumer reviews [5, 6]. In both cases, links between users are established based on their consumed and produced data and in both cases this is shown to significantly increase users satisfaction and user retention.

© Springer International Publishing Switzerland 2016
N.T. Nguyen et al. (Eds.): ICCCI 2016, Part II, LNAI 9876, pp. 59–67, 2016.
DOI: 10.1007/978-3-319-45246-3_6

The novelty of our approach is in applying these techniques in a richer text environment, using heavily annotated texts (with both syntactic and semantic metadata) especially including real-world geographical references. Quite often a reader feels the need to supplement the knowledge on certain places or notorious people that are mentioned in the book she/he reads by searching the web or visualising places on geographical atlases (in digital or classical printed forms). MappingBooks (Fig. 1) is a technology that facilitates these searches by pre-computing links outside the book text itself.

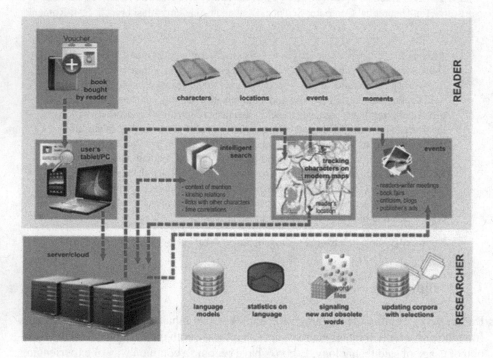

Fig. 1. MappingBooks in a bird's eye view

Installed on mobile devices, and accessible by readers on the basis of subscriptions, the technology will enrich the texts with contextual links intended to enhance the reading satisfaction, offering supplementary and interactive guide beyond the text itself and making the reading entertaining and pleasant.

When comparing the proposed functionalities with those offered by other social communities built around reading hobbits (such as Goodreads[1]), our approach offers a closer connection to the actual content of the books, basing most detected connections between users on the existing metadata previously added to the books, as described in Sect. 4 of this paper.

The paper has the following structure. In Sect. 2 we briefly describe the way documents are processed in MappingBooks, detailing the annotations the technology adds

---

[1] https://www.goodreads.com/.

to the raw text. Section 3 shows what types of semantic relations are currently added to texts and what further efforts could be made in this regard. Section 4 describes how users can be connected using both the viewed text content and their activity. Finally, Sect. 5 summarises the proposed development efforts and shows how they could further benefit other systems.

2   Text Mining in MappingBooks

A MappedBook, the technological output of the MappingBooks (MB) ongoing project is an online connected book that facilitates creation of social networks based on reading preferences. Mentions of locations and other entity names, which the book contain, are automatically identified and put in correlation with geographical artifacts (like maps, coordinates, layouts) and the web. As such, the user (considered to be reading the book while connected to the application) will be directed, at appropriate moments, towards significant events happening in the real world in locations mentioned in the book, which are reflected in the virtual world. Moreover, sensible to the instantaneous location of the user, the system can locate her/his position in connection with geo-mentions contained in the book, creating thus a more intimate relationship between the text and its readers.

Hypermaps [7] composed of background base layers with continuous cover (mainly represented by raster data) or overlay layers, as discrete, raster and vector data, will be associated to geonames in text, thus offering additional multimedia information, mainly through pop-up windows and hyperlinks.

To support these functionalities, the base text is automatically processed as follows. First, the base text is extracted from the original document (usually a PDF file which includes images or other graphical artifacts). This step is performed by using first the iText library, then by manually correcting the resulted text (fixing diacritics and hyphen segmented words, removing page numbers, image captions and other additional elements). All text extraction errors are also fixed, the quality of the resulted text being extremely important for increasing the accuracy of the further automated processing.

The corrected text is then passed through a series of annotators: POS tagger [8], NP-Chunking [9], NER (Name Entity Recognizer) [10] and RARE (Robust Anaphora Resolution Engine) [11]. The result is a stand-off annotated XML document which is then used as input for the following semantic processing.

3   Linking Entities in and Out of Text

Entities are uniquely identifiable physical/abstract things. Entity detection (i.e. finding a string denoting an entity) and understanding (linking the string to an entry in a knowledge repository where the entity is described) takes us a long way towards understanding the text itself, and provides additional knowledge to the reader. The detection of geographical entities belonging to 15 classes is done by a Named Entity Recogniser [13].

External repositories, such as Wikipedia and Open Linked Data, which combine numerous, rather stable sources of knowledge, are used to link the mentions outside the book. This process is illustrated in Figs. 2 and 3.

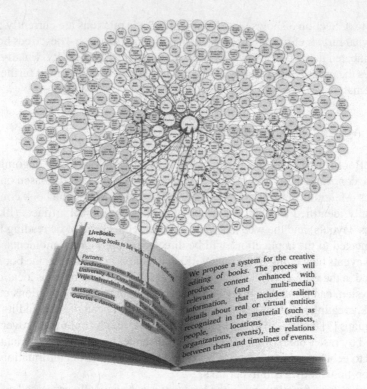

LiveBooks:
Bringing books to life with creative editing

Partners:
Fondazione Bruno Kessler
University A.I. Cuza, Iasi, Romania
Vrije Universiteit Amsterdam
ArtSoft Consult
Guerini e Associati

We propose a system for the creative editing of books. The process will produce content enhanced with relevant (and multi-media) information, that includes salient details about real or virtual entities recognized in the material (such as people, locations, artifacts, organizations, events), the relations between them and timelines of events.

Fig. 2. Linking entities to Open Linked Data.

Our aim is not only to find additional information, but interesting additional information. The notion of interestingness, used in the fields of knowledge discovery and text summarization, contrasts the notion of importance. Important things may be common knowledge, while something interesting is idiosyncratic, specific to a certain reader, unexpected and previously unknown. The following reader profiling sources are used in our project: age, sex, profession, nationality, city and region address, musical, reading and other cultural preferences, hobbies, etc. Profiling the user means filling in a vector of characteristics (by using different sources agreed by the user to be connected to the application, the least liked being direct acquisition at sign up). Then, for each piece of online linked information a similar vector is filled in, by using contextual sources (as given by headers, origin of sites, etc. and using bag of words and tf*idf measures). Then, vectors of acquired pieces are matched against the user's vector and only the best ranked are retained.

Entity mentions in a text are said to be coreferent when they refer to the same entity. Mentions may take different syntactical forms in a text, but the most common ones are noun phrases built around proper nouns (named entities), common nouns or pronouns. While reading a book, the reader continuously deciphers coreferential mentions, most of the time without a conscious effort. Coreferential (and in general, anaphoric) relations give a text cohesiveness, allowing a reader to connect entities between them and connect events through their participating entities, to build the picture the author intended. The set

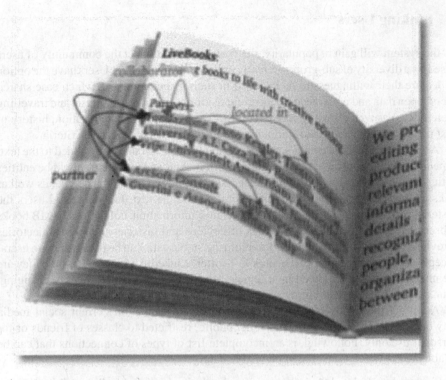

Fig. 3. Entities and the links between them.

of relations that link mentions referring to the same entity over a whole text can take the form of a chain or a tree, with links always directed from the current mention, called anaphor, to coreferent mentions that appeared before them in the text – antecedents [12]. Thus, to each entity that appears in a book corresponds a chain or a tree of coreferential links, with the root usually anchored in the first mention of the corresponding entity. We call this structure the coreference chain. Coreference chains are important for building and synthesizing information from a book. Extracted information about the entity (e.g. images, URL to Wikipedia or Open Linked Data) will be shared by all nodes in the chain. It is also important to distinguish shared information from information particular to a specific mention. Temporal or geographical coordinates, for example, may not be shared– the same entities may appear in different locations at different times. Such information must remain linked only to the specific mention to which it applies, based on the context.

It makes explicit the participants in events by detecting the entities to which they refer, thus preparing the text for the event analysis phase. It is also crucial in establishing connections between events that share participants, and it may help detect relationships between entities [13]. Relations between different types of entities may include: people are located in specific places, events occur at specific points in time, mutual positioning of locations in space, spatial distances and directions, etc., as exemplified in Fig. 2. In MB there are identified 17 types of semantic relation between entities [14], adding to the outside text relations mentioned above.

4   Linking Users

As the system will gain in popularity, the possibility to connect the community of users based on a diversity of sub-group preferences becomes interesting. Users have the option to declare their willingness to be included in such a community, in which case shared users' open data and preferences (education, readings, sensibilities, music and travelling preferences) as well as instantaneous contexts (as is the immediate location, history of past travels or the intention to start a journey) can be used as selection criteria.

As described in the previous sections, the annotation automatically added to the text, depending on the content, could be extremely rich. Among them there should be entities with geographical real-world significance, institutions and notorious people, as well as links between these entities both inside and outside the text. For registered users the system keeps personal data, such as identification information, collections of MB books subscripted for, instantaneous locations of devices and histories of users' trajectories (previous locations). Part of this information, the stable data, can be obtained, with users' accepts, from social networks (Facebook, Twitter, LinkedIn, Google +) where they are members, other data are typed in at registration, while the volatile, quickly changing, data, such as location and journeys is collected during daily or sporadic use.

Adopting the trendy typology of accessibility of data from current social media networks, here too the user's data can be: public, restricted to classes of friends or for personal use only. Following, is an incomplete list of types of connections that can be established between users, based on visible data:

- If a user has declared the information on books "subscripted for" as visible, then the system can form the user's community **current co-readers of B**. A user is in this community if MB actively changes information about B with her/him. Users of this community would thus be able to share hot impressions on the reading B, supplement their reading lists with similar suggestions; invite people to visit their personal cultural forums, etc.
- The above community can be enlarged if the necessity to read the book B at the very current moment is removed. A user is in the **co-readers of B** community, even if not online, if MB knows about her/him to have been changing data about this book with her/him at any time now or in the past.
- If a user has declared as visible the information on "instantaneous location", then the system can form the current **co-proximity of L** community. A user is in this community if MB knows about her/him that is currently online and is physically located in that location or in proximity of it. This type of information can create very interesting links between readers, based on personal impressions, photos, shared real-time descriptions.
- Again, the above condition can be relaxed if the demand to be in the location L at this very moment is removed. A user is in the **co-proximity of L** community, even if she/he is not currently online, if MB knows about her/him that has been at any time, now or in the past, in that location or in proximity of it.
- If the location is enlarged to include points assigned to a journey, communities of people sharing a route (track) can be formed. The community represents a

generalisation of the MB **co-proximity of** L community, known as **co-track of T**. A user is in this community, even if she/he is not currently online, if MB knows about her/him that has been at any time, now or in the past, in any of the points belonging to track T or in proximity of it. Also as above, the subcommunity of people sharing the same track at the current moment, **current co-track of T**, can be formed.

- Any combination of the above species of communities can be dynamically formed on request by the user. As such, a user can contact very constrained communities such as current **co-readers of B** AND **current co-proximity of L** down to much less constrained, such as **co-readers of B** AND **co-track of T**. Thus, links can be established between people that have common reading preferences and have pursuit similar travelling experiences on places inspired by these lectures.

All described connections are unilaterally established from the initiating user (the active user). They dont require any action/confirmation from the other users involved, as they dont make available to the original user any sensitive data (such as personal info, search/reading history, travel history). All user data shared is only relevant in the local context it is made available.

Since the original MappingBooks system is designed as a client-server application, considerations have to be made with respect to the usage of the limited memory and processing resources available on the device, the client. The features described in this section are under development, but we believe that adding social networking capabilities is feasible, without a major impact on processing or memory overload of the client-server application. This is because all relevant data on users should be kept on the server, as specific data paired with the users' IDs. Then, after receiving from a user the type of community asked for, they are formed by filtering conditions expressed on a common database and only results will be communicated back to users. Actually, following classic algorithms used in artificial intelligence that minimise the matches between data and patterns [15], a fixed but large number of predefined communities can be updated permanently and their retrieval made instant.

5   Conclusions

At the present moment MappingBooks offers a basic mobile app serving as proof-of-concept for part of the functionalities described above. As an example of relevant types of e-books the app currently offers an annotated geography manual (described in [13, 14]). The included annotations are initially added automatically (at the surface morpho-syntactic level), and further enhanced with manual annotation of entities and semantic relations.

The proposed functionalities are well within the technological capabilities of current mobile devices. The MappingBooks system and its connection facilities addresses a diversity of possible users: from the passionate readers, people enjoying to read books everywhere, to occasional readers, those reading only during travelling or in vacations, from youngsters, school children and students to retired people, from adventures, those often on route to people travelling only in their minds, who have never stepped out of their town. Linking entities identified in text to Open Linked Data will project the book into a huge semantic space, which may contain snippets of information that would make useful additions to the book content. Moreover, finding people with similar reading preferences and easily establishing contacts with them will be enjoyable and rewarding. Novels, biographies, books with historical or geographical subjects, class manuals and travel guides are only some examples of styles that are lend of being transposed in the MB technology.

Publishing houses could be the principal beneficiaries of the MappingBooks technology, as it could generate increased book sales over time if correctly mastered from the point of view of business models. A system of bonuses may also bring advantages to publishing houses in partnership with local administration or tourism agencies, as for instance, one that would challenge the readers to hit as many of the books places, visiting all locations or getting through all mentioned routes. A way in which providers of tourism services could be informed on the user's travel interests and particular locations and routes associated with a popular new book can also be easily imagined.

With respect to forming communities, the MB interface can be extended to include lists of "achievements" in connection with a text. Thus, users can be automatically upgraded for "connoisseurs" levels with respects to, for instance, tourist objectives mentioned in text.

It is easy to imagine other ways to form communities rooted in lectures, as, for instance, selections that intersect common readings and attended places with levels of friendship reported by other social media, like Facebook or Twitter. Events and entities mentioned in a book can be associated with a real-world location and a particular time of the year (or of the day). If a history is available for users' locations, a reader can identify users who visited that location and/or witnessed that event at the relevant time of the year/day.

However, there remains the problem of finding a few needles in numerous haystacks, and putting them together into a coherent whole, or otherwise the reader will soon be suffocated by the amount of useless information made available. Even if the text may contain clues that can be used as constraints about what relevant information is (such as the time frame where the entity is mentioned, the location and the general context), we are not yet totally clear about the right way to filter the linked information.

Acknowledgements. The work reported in this paper was achieved with the support of the PN-II-PT-PCCA-2013-4-1878 Partnership PCCA 2013 grant "MappingBooks - Intră în carte!", having as partners UAIC, SIVECO and „Ștefan Cel Mare" University of Suceava. We address our thanks to Vivi Năstase for relevant ideas and realisation of Figs. 2 and 3.

References

1. Aggarwal, C.: Text mining in social networks. In: Aggarwal, C. (ed.) Social Network Data Analytics, 2nd edn, pp. 353–374. Springer, Heidelberg (2011)
2. Irfana, R., Kinga, C.K., Gragesa D., Ewena S., Khana, S.U., Madania, S.A., Kolodzieja, J., Wanga, L., Chena, D,, Rayesa, A, Tziritasa, N., Xua, C., Zomayaa, A.Y., Alzahrania, A.S., and L, H.: A Survey on Text Mining in Social Networks. The Knowledge Engineering Review. Cambridge University Press, Cambridge, pp. 1–24 (2015)

3. Romero, C., Ventura, S.: Data mining in education. Wiley Interdisc. Rev. Data Min. Knowl. Disc. **3**(1), 12–27 (2013)
4. Hung, J.L., Zhang, K.: Examining mobile learning trends 2003–2008: a categorical meta-trend analysis using text mining techniques. J. Comput. High. Educ. **24**(1), 1–17 (2012)
5. Cambria, E., Schuller, B., Xia, Y., Havasi, C.: New avenues in opinion mining and sentiment analysis. IEEE Intell. Syst. **2**, 15–21 (2013)
6. Netzer, O., Feldman, R., Goldenberg, J., Fresko, M.: Mine your own business: market-structure surveillance through text mining. Mark. Sci. **31**(3), 521–543 (2012)
7. Kraak, M.-J., Rico, V.D.: Principles of hypermaps. Comput. Geosci. **23**(4), 457–464 (1997)
8. Simionescu, R.: UAIC Romanian Part of Speech Tagger, resource on nlptools.info.uaic.ro, "Alexandru Ioan Cuza" University of Iaşi (2011)
9. Simionescu, R.: Romanian deep noun phrase chunking using graphical grammar studio. In: Moruz, M.A., Cristea, D., Tufiş, D., Iftene, A., Teodorescu, H.N. (eds.) Proceedings of the 8th International Conference Linguistic Resources and Tools for Processing of the Romanian Language, pp. 135–143 (2012)
10. Gîfu, D., Vasilache, G.: A language independent named entity recognition system. In: Colhon, M., Iftene, A., Barbu Mititelu, V., Cristea, D., Tufiş, D. (eds.) Proceedings of ConsILR-2014, pp. 181–188, "Alexandru Ioan Cuza" University Publishing House, Iaşi (2014)
11. Ignat, E.: RARE-UAIC (Robust Anaphora Resolution Engine), open-resource on META-SHARE, "Alexandru Ioan Cuza" University of Iaşi (2011)
12. Cristea, D., Postolache, O.: Anaphora resolution: framework, creation of resources, and evaluation. In: Proceedings of the Fifth International Conference Formal Approaches to South Slavic and Balkan Languages, FASSBL-2006, 18–20 October, Sofia, Bulgaria (2006)
13. Cristea, D., Gîfu, D., Pistol, I., Sfirnaciuc, D., Niculita, M.: A mixed approach in recognising geographical entities in texts. In: Trandabat, D., Gîfu, D. (eds.) EUROLAN 2015. CCIS, vol. 588, pp. 49–63. Springer, Heidelberg (2016). doi:10.1007/978-3-319-32942-0_4
14. Gîfu, D., Pistol, I., Cristea, D.: Annotation conventions for geographical relations. In: Proceedings of the 11th International Conference Linguistic Resources and Tools for Processing the Romanian Language, ConsILR-2015, pp. 67–78 (2015)
15. Waterman, D.A., Hayes-Roth, F. (eds.): Pattern-Directed Inference Systems. Academic press, New York (2014)

Detecting Satire in Italian Political Commentaries

Rodolfo Delmonte[✉] and Michele Stingo

Department of Language Studies and Department of Computer Science,
Ca' Bembo 1075 – Ca' Foscari University, 30123 Venezia, Italy
delmont@unive.it, stingomichele@gmail.com

Abstract. This paper presents computational work to detect satire/sarcasm in long commentaries on Italian politics. It uses the lexica extracted from the manual annotation based on Appraisal Theory, of some 30 K word texts. The underlying hypothesis is that using this framework it is possible to precisely pinpoint ironic content through the deep semantic analysis of evaluative judgement and appreciation. The paper presents the manual annotation phase realized on 112 texts by two well-known Italian journalists. After a first experimentation phase based on the lexica extracted from the xml output files, we proceeded to retag lexical entries dividing them up into two subclasses: figurative and literal meaning. Finally more fine-grained Appraisal features have been derived and more experiments have been carried out and compared to results obtained by a lean sentiment analysis. The final output is produced from held out texts to verify the usefulness of the lexica and the Appraisal theory in detecting ironic content.

Keywords: Semantic annotation · Pragmatic annotation · Appraisal theory · Automatic irony detection · Literal vs nonliteral language

1 Introduction

We present work carried out on journalistic political commentaries in two Italian newspapers, by two well-known Italian journalists, Maria Novella Oppo, a woman, and Michele Serra, a man[1]. Political commentaries published on a daily basis consists of short texts not exceeding 400 words each. Sixty-four texts come from Michele Serra's series titled "L'Amaca", published daily on the newspaper "La Repubblica" between 2013 and 2014; usually the targeted subjects are politicians, bad social habits and in general every trendy current event. Forty-nine texts come from Maria Novella Oppo's series titled "Fronte del video", published daily on the newspaper "L'Unità" in a previous span of time, from 2011 to 2012; the targeted subjects are usually politicians and televised political talk shows.

 The two journalists have been chosen for specific reasons: Oppo is a master in highly cutting and caustic writing, Serra is less so. Both are humorous, both use sophisticated rhetorical devices in building the overall logical structure of the underlying satiric network of connections. Oppo borders sarcasm, Serra never does so. Oppo's texts are

---

[1] Permission to republish excerpts from their articles has been granted personally by the authors.

© Springer International Publishing Switzerland 2016
N.T. Nguyen et al. (Eds.): ICCCI 2016, Part II, LNAI 9876, pp. 68–77, 2016.
DOI: 10.1007/978-3-319-45246-3_7

slightly longer than Serra's. In order to focus on the specific features connotating political satire, manual annotation has been carried out on the 112 texts using at first a reduced
version of the Appraisal Framework [1]. Following the annotation activity, a typological
classification has been produced for all the entries contained in the automatically
collected lexica (one for each author) composed of the annotated items/phrases (see also
[2, 3, 4]). The classification has been carried out using three linguistic traits: namely
idiomatic, metaphorical – these two being figurative uses – and none for the rest. This
has been done in order to set apart figurative uses the author chose for a specific item/
phrase from nonfigurative ones(see [5]). All the annotations have been done by the
second author and counterchecked by the first author.

2 Satire and the Appraisal Framework

The decision of adopting Appraisal Theory (hence APTH) is based on the fact that
previous approaches to detect irony – a word we will use to refer to satire/sarcasm – in
texts have failed to explain the phenomenon. Computational research on the topic has
been based on the use of shallow features to train statistical model with the hope that
when optimized for a particular task, they would come up with a reasonably acceptable
performance. However, they would not explain the reason why a particular Twitter
snippet or short Facebook text has been evaluated as containing satiric/sarcastic expressions. Except perhaps for features based on text exterior appearance, i.e. use of specific
emoticons, use of exaggerations, use of unusually long orthographic forms, etc. which
however is not applicable to the political satire texts [6]. These latter texts are long texts,
from 200 to 400 words long and do not compare with previous experiments.

In the majority of the cases, the other common approach used to detect irony is based
on polarity detection. So-called Sentiment Analysis is in fact an indiscriminate labeling
of texts either on a lexicon basis or on a supervised feature basis where in both cases, it
is just a binary decision that has to be taken. This is again not explanatory of the
phenomenon and will not help in understanding what it is that causes humorous reactions
to the reading of an ironic piece of text. It certainly is of no help in deciding which
phrases, clauses or just multiwords or simply words, contribute to create the ironic
meaning (see [7, 8]).

By adopting the Appraisal analysis, we intended not only to describe but also to
compute with some specificity the linguistic regularities which constitute the evaluative
styles or keys of political journalistic texts. The theory put forward by White & Martin
[1] (hence M&W) makes available an extended number of semantically and pragmati-
cally motivated annotation schemes that can be applied to any text. In particular, one
preliminary hypothesis would be being able to ascertain whether the text under analysis
is just a simple report, a report with criticism, a report with criticism and condemnation.
In the book by M&W there's a neat distinction between these three types of voices:
'reporter voice', 'correspondent voice' and 'commentator voice'. Since the commentator voice has the possibility to condemn, criticize and report at the same time, and
since we assume that satire, and even more, sarcasm have a strong component made of
social moral sanction, this is our option and our first hypothesis.

In APTH, the evaluative field called Attitude is organized into three subclasses, Affect, Appreciation and Judgement, and it is just the latter one that contains subcategories that fit our hypothesis. We are referring first of all to Judgement which alone can allow social moral sanction, and to its subdivision into two subfields, Social Esteem and Social Sanction. In particular, whereas Social Esteem extends from Admiration/Admire vs Criticism/Criticise, Social Sanction deals with Praise vs Condemn etc.[2] So in our texts we are dealing with the "commentator voice", which may consist of authorial social sanction, plus authorial directives (proposals), in addition to criticism. The second hypothesis is that both commentators are characterized by a high number of Judgements and possibily, negative ones. Then we also hypothesized that there should be an important difference between the two corpora, Oppo's being the one with the highest number. This hypothesis has been borne out by the results of the annotation as can be seen in the distribution of categories in the tables presented below.

There are three possible strategies writers can use to produce humorous effects: the superiority presumption, [9], relief presumption ([10, 11]) and incongruity presumption [12]. The first speculative contribution was proposed by [13], further revised by [14, 15, 16] as a general theory of verbal humour. The hypothesis we will now formulated is based on the contribution that our new annotation traits can bring to the detection task. The superiority presumption assumes that the object of the ironic process be sanctioned, so here we refer to the Judgement Social Sanction/Esteem Negative classified items of our lexicon. The relief presumption could be based again on the use of the previous features in addition to Positively marked features. The relief is given by laughter, i.e. by humorous meaning which generates positive energy. This physic energy is built anytime we need to suppress negative feelings in our psyche and every time we release this energy, by virtue of jokes related to taboos and cultural values induced by society (namely when we suppress the mental censorship mechanism), we experience laughter and a psychological benefit is reached. This may be obtained by the use of figurative language, i.e. the use of a word/phrase/expression with the opposite meaning it usually has. Finally the incongruity presumption can be again achieved by combining Positive and Negative Judgement/Appreciation features with strong socially related nuances. As to the satiric discourse we rather deem the incongruity presumption [16] to be more adequate to explain the humurous mechanism. In particular, at the heart of this approach there is an opposition between two dimensions, and in order for a text to be processed as humorous – in addition to the opposition feature, the dimensions have to share a common part, so that it is possible a shift from one dimension to another. First of all we present general data about the annotations (Table 1):

Table 1. General data about the corpus

	NoSents	No.Toks	No.Annots
Oppo	514	14350	1651
Serra	561	14641	1849

---

[2] As reported in M&W p. 52.

When we collapse polarity in the two main categories we obtain the picture reported in Table 2. below. As can be noted, differences in total occurrences of Negative Judgements are very high and Oppo has the highest. Also Positive Judgements shows a majority of cases annotated for Oppo's texts.

Table 2. Annotations split by polarity

Writers	JudgNegat	JudgPost	ApprNegat	ApprPost
Serra	577	216	678	385
Oppo	824	260	442	188

On the contrary, in the Appreciation class differences are all in favour of Serra, both for Negative and Positive polarity values. Finally, we can see that Oppo's commentaries are based mainly on Judgement categories and their polarity is for the majority of cases Negatively marked. Also Appreciation has a strong Negative bias as can be gathered from Table 2. On the contrary, Serra's commentaries are more based on Appreciation and polarity is almost identically biased.

3 Experiments to Validate Nonliteral Language

The first approach to better understand the semantic/pragmatic features of our texts has been that of automatically deriving a lexicon from the annotated texts and then proceed to some further investigation. We extracted some 3500 annotations overall, one third has been identified as belonging to figurative language, that is idiomatic expressions and similes, metaphors and metonymies. The remaining 2/3, i.e. 2300 has been assigned to the neutral category NONE. However this classification was not satisfactory and so we started detecting literal from nonliteral expressions at first using automatic procedures. We produced a lexicon of Appraisal Categories related to lexical entries as they are listed in the book by M&W. We came up with some 500 entries which we then used to retag the 3500 lexical entries. We wrote a simple script that took each lexical entry, produced the lemmata for every semantic word, and then tried to match it with the Appraisal lexicon. The results have been very poor and we only managed to cover 10 % of all entries. So we decided to manually retag the 2300 neutral entries dividing them up into three subcategories: a. Literal meaning – whenever the appraisal category coincided with the literal meaning of the entry; b. Nonliteral meaning – whenever the appraisal category was not related to the literal meaning associated to the entry; c. Semantically hard to compute literal meaning – whenever the meaning of the entry required some compositional analysis to recover the literal meaning and there was not a one-to-one correspondence between the entry and the appraisal category. We ended up by reclassifying 16 % of the 2300 None as belonging to category b, i.e. 244 new entries as nonliteral; and another 22.96 % as semantically hard, i.e. 528 entries. The new organization of the two lexica is now as follows (Table 3):

Now proportions are reversed and literal language covers only 43 % of all lexical entries. As to appraisal classification, lexicon values repeat the opposition we found in counting annotations in texts: Oppo's lexicon has a majority of Negative Judgements,

Table 3. Semantic subdivision of lexical entries

	None	Idiom-atic	Figur-ative	Non-literal	Sem_hard	Totals
Oppo	711	201	422	153	187	1674
Serra	816	143	449	91	341	1840
Totals	1527	344	871	244	528	3514

Serra's lexicon has a majority of Negative Appreciations. Serra's Positive Appreciations are almost the double of Oppo's, whereas Positive Judgements are comparable (Table 4):

Table 4. Subdivision of lexica with fine grained Judgement subclasses

	Judgmt. Negat.	Judgmt. Posit.	Apprec. Negat.	Apprec. Posit.	Negative Esteem	Negative Sanction
Oppo	*742*	275	396	181	375	363
Serra	554	214	*618*	343	275	274
Totals	1296	489	1014	524	650	637

In order to comply with our interpretation of commentators' role, we expected then to have an internal subdivision of Negative Judgement showing a high percentage of Negative Sanction. So we proceeded in the reclassification of all Judgement lexical entries into the new two subcategories, Sanction and Esteem. All these values are referred to the types listed in the new lexica and they only represent potential new automatic annotations which however need to be tested on the corpus. The subdivision of Negative Judgements between Sanction and Esteeem is strongly in favour of Oppo with slight differences in distribution between the two classes.

3.1   Computing Nonliteral Language

We will now delve into the experimental part of the work which is strictly related to the fine-grained classification and the subdivision of lexical entries into Literal and Figurative language, which should allow better performances as far as irony detection is concerned (see [17, 18, 19]). We then set up our algorithm for irony detection with the following instructions:

```
 -   SEARCH inside a sentence all annotations
         - of type Judgement_Sanction_Negative
         - or Judgement_Esteem_Negative
         - or Appreciation_Negative
         - or Emotion_Negative
 - ELSE none found
     END
         AND
         - together with annotations with the
                 opposite polarity, Positive
             EITHER
             - belonging to LITERAL type
             - belonging to NONLITERAL type
             output=TRUE
         ELSE
         output=FALSE
         END
```

where True indicates a possible condition for irony detection and False the opposite. The combination of all the different parameters has given as a result six different outputs which confirm all the hypothesis we put forward in the previous section.

In Fig. 1 below main differences can be found when Figurative language is used and Negative features are involved. In particular When SocialSanction_Negative and Social-Esteem_Negative Figurative annotations are used together with any Negative annotations in the same sentence, Oppo's texts show a great jump up when compared to Serra's – that's the orange cylinders. On the contrary, when Only Positive Figurative annotations are used together with any Positive annotation in the same sentence, we see that Serra's values are higher, light green. Using all Negatives Figurative annotations with Negatives again favours Oppo's texts – light blue, whereas Negatives Figurative with Positive annotations favours Serra's – light red.

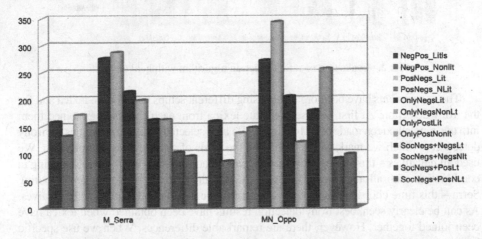

Fig. 1. 12 experiments with new lexica (Color figure online)

This distribution of the data confirms our previous hypothesis: Oppo's text are more close to sarcasm, while Serra's text are less so and just satiric. Oppo's appraisal configuration for best irony detection requires the presence of Negatively marked Judgements, socially biased, and also with a preference for literal meaning. On the contrary, Serra's texts are characterized by the preference of purely Positively marked words/phrases with a strong bias for nonliterality.

3.2 Experimenting with New Texts

We now report results obtained with held out texts for the two journalists. We ran our automatic annotation algorithm based on the lexica created from the manual annotation and further modified, on 20 texts, ten for each author, to verify whether the setup we derived from our previous analysis is directly applicable to any new text or not. Oppo's texts contain 118 sentences, Serra's texts contain 96 sentences. Oppo's texts have been automatically assigned 100 annotations; Serra's texts, only 66. From Fig. 2 below we have some confirmations but also some new data.

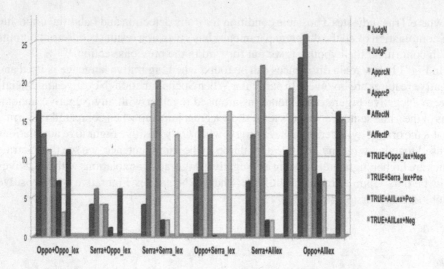

Fig. 2. Evaluation of satire detection algorithm with held out texts

The experiments have been organized using different setups both for the lexica and for the irony detection. At first, we used separate lexica from each corpus, then joined them into one single lexicon made of 3514 entries. We also selected different strategies for irony detection – which we mark with TRUE – on the basis of our previous computation. We used all negatives – this strategy favouring Oppo – choosing those with literal meaning in combination with all negatives. Then we selected all positives –this strategy favouring Serra – this time choosing those with nonliteral meaning in combination with positives. As can be clearly seen, best irony detection results have been obtained when lexica have been joined together. However, there are remarkable differences. When we use specific lexica we see important improvements in the number of annotations, in particular in the case of Serra's texts. With Oppo's texts, we get more TRUE detection cases when Serra's lexicon is used compared with Oppo's lexicon. Remember that when we use Serra's lexicon, we also modify our strategy for irony detection to Positive + Nonliteral. Generally speaking, however, it is always Oppo's texts and lexica that produce the highest number of Judgements Negative. Strangely enough, Oppo's texts are also characterized by a great number of Judgement Positive, in fact the highest number. Then, contrary to expectations, TRUE decisions in Serra's texts are determined by the Positive strategy which obtains higher results than the Negative one. In the case of Oppo, we see a slightly higher number of TRUE when the Positive strategy is applied.

So it would seem that this experiment does only partially confirm our hypotheses. However we need to consider that the lexica produced from previous manual analysis do not cover completely the new texts in that the number of automatic annotations obtained is only a small percentage: 100/118, 66/96, i.e. not even one per sentence. On the contrary, in the previous manual work, we had an average of 3.2 annotation per sentence.

To improve recall, we then collected all lexical items contained in the book by M&W and we used them with the lexicons with shallow analysis as before, and we labeled them all as having literal meaning in association with each appraisal category. Results are reported in Fig. 3 below. First of all, number of automatic annotations now increased to 103 for Serra, and 146 for Oppo: still not comparable to manual annotations but certainly much better than before - we are now halfway from the target of 3.2 annotations per sentence. Coming now to the new automatic classification of test texts, Appraisal categories are divided up as follows, where N = negative and P = positive, Af = affect, Ap = appreciation, Jg = judgement, and Sct = Sanction, Est = Esteem:

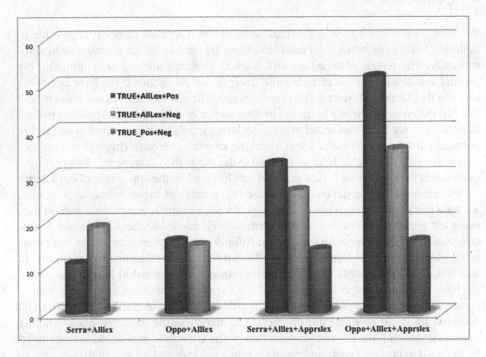

Fig. 3. Irony detection using augmented lexica

In this figure we present final results for irony detection using appraisal theory by simply checking three possible combinations of polarity values: only Positives, marked PP, only Negatives, marked NN and then Positives and Negatives marked NP.

As can be noticed, best results are NN combinations and as before, they are higher in Oppo's texts. Then come the PP combinations and finally the NP which are however much lower. In this case, Oppo's True cases are over 50, which when compared to number of sentences makes almost 50 % of them. In the case of Serra's True they only reach 36 sentences, which is a much lower percentage when compared to number of sentences, only 37.5 %.

Negative Esteem seems to be used a lot more than Sanction which is however used in the opposite manner, more Positive evaluations than negative ones. Here we must remind that we have decided to treat all new lexical entries derived from M&W as semantically literal, but we have seen from previous analysis that this may only be true for 40 % of all data (Table 5).

Table 5. Classification of test texts into Appraisal categories

Authors	Af N	Af P	Ap N	Ap P	Jg N	Jg P	Sct.	Est.	Sct N	Sct. P	Est. N	Est. P
Oppo	9	11	21	27	27	40	15	22	7	8	20	2
Serra	3	3	21	27	12	20	9	12	3	6	9	3
Total	12	14	42	54	39	60	24	34	10	14	29	5

4   Conclusion

We have shown that by using the framework of the Appraisal theory it is possible to highlight features of ironic texts and to use these features to detect satire/sarcasm automatically. The results obtained are still work in progress and we are continuing the manual annotation work to include more fine-grained distinctions. We have been able to show that Oppo and Serra stylistic devices are different in a significant manner, and that this difference is clearly borne out by the categories derived from Appraisal theory. In particular, we have succeeded in showing how Oppo's texts constitute more cutting political comments than Serra's text, speaking in general terms. This stylistic characteristic is strictly derivable from and related to the use in their comments of more Judgement rather than Appraisal lexical material for Oppo, while the opposite applies to Serra.

Future work will be devoted to increase the number of experiments. In particular, we want to try to show correlations existing between automatic and manual annotations, using test texts where however manual verification is needed to check how many nonliteral uses have been done with the specific Attitude related categories. Annotating texts using M&W theoretical framework is hard and it requires specific linguistic training. In addition, classifying political commentaries requires a lot of world knowledge due to the habit of commentators to refer to real life events and use them as a comparison to comment on the current political issue. This aspect could be covered by accessing LOD data and by using ground truth description to match satiric distorted ones. Another important element that has not yet been part of the automatic evaluation is constituted by the need to corefer events and people, again a difficult task to accomplish.

References

1. Martin, J., White, P.R.: Language of Evaluation, Appraisal in English. Palgrave/Macmillan, London/New York (2005)
2. Taboada, M., Grieve, J.: Analyzing appraisal automatically. In: Proceedings of the AAAI Spring Symposium on Exploring Attitude and Affect in Text: Theories and Applications, pp. 158–161. AAAI Press (2004)
3. Fletcher, J., Patrick, J.: Evaluating the utility of appraisal hierarchies as a method for sentiment classification. Proceedings of the Australasian Language Technology Workshop, pp. 134–142, Sydney (2005)
4. Khoo, C., Nourbakhsh, A., Na, J.: Sentiment analysis of online news text: a case study of appraisal theory. Online Inf. Rev. 36(6), 858–878 (2012)

5. Sarmento, L., Carvalho, P., Silva, M., de Oliveira, E.: Automatic creation of a reference corpus for political opinion mining in user-generated content. In: Proceedings of the 1st international CIKM workshop on Topic-sentiment analysis for mass opinion, pp. 29–36. ACM, Hong Kong (2009)
6. Carvalho, P., Sarmento, L., Silva, M., de Oliveira, E.: Clues for detecting irony in user-generated contents: oh...!! it's so easy;-). In: Proceeding of the 1st International CIKM Workshop on Topic-Sentiment Analysis for Mass Opinion, pp. 53–56. ACM, Hong Kong (2009)
7. Reyes, A., Rosso, P.: Mining subjective knowledge from customer reviews: a specific case of irony detection. In: WASSA 2011 Proceedings of the 2nd Workshop on Computational Approaches to Subjectivity and Sentiment Analysis, pp. 118–124. Association for Computational Linguistics, Stroudsburg (2011)
8. Özdemir, C., Bergler, S.: CLaC-SentiPipe: SemEval2015 Subtasks 10 B, E, and Task 11. In: Proceedings of the 9th International Workshop on Semantic Evaluation (SemEval 2015), pp. 479–485. Association for Computational Linguistics, Denver (2015)
9. Gruner, C.: The Game of Humor: A Comprehensive Theory of Why We Laugh. Transaction Publishers, New Brunswick (1997)
10. Freud, S.: Der Witz und seine Beziehung zum Unbewußten. Franz Deuticke, Leipzig (1905)
11. Minsky, M.: Jokes and their Relation to the Cognitive Unconscious. In: Vaina, L., Hintikka, J. (eds.) Cognitive Constraints on Communication: Representations and Processes. D. Reidel, Dordrecht (1981)
12. Koestler, A.: The Act of Creation. Hutchinson & Co, London (1964)
13. Raskin, V.: Semantic Mechanisms of Humor. D. Reidel, Dordrecht (1985)
14. Attardo, S., Raskin, V.: Script theory revis(it)ed: joke similarity. HUMOR: Int. J. Humor Res. 4(3/4), 293–347 (1991)
15. Attardo, S.: Linguistic Theories of Humor. Mouton de Gruyter, Berlin (1994)
16. Bosco, C., Patti, V., Bolioli, A.: Developing corpora for sentiment analysis: the case of irony and Senti-TUT (extended abstract). In: Yang, Q., Wooldridge, M. (eds.) Proceeding of 24th International Joint Conference on Artificial Intelligence, IJCAI 2015, pp. 4158–4162. AAAI Press, Buenos Aires (2015)
17. Birke, J., Sarkar, A.: Active learning for the identification of nonliteral language. In: FigLanguages 2007 Proceedings of the Workshop on Computational Approaches to Figurative Language, pp. 21–28. Omnipress Inc, Rochester (2007)
18. Turney, D., Neuman, Y., Assaf, D., Cohen, Y.: Literal and metaphorical sense identification through concrete and abstract context. In: EMNLP 2011 Proceedings of the Conference on Empirical Methods in Natural Language Processing, pp. 680–690. Association for Computational Linguistics, Edinburgh (2011)
19. Hernandez Farias, D., Sulis, E., Patti, V., Ruffo, G., Bosco, C.: ValenTo: sentiment analysis of figurative language tweets with irony and sarcasm. In: Proceedings of the 9th International Workshop on Semantic Evaluation (SemEval 2015), pp. 694–698. Association for Computational Linguistics, Denver (2015)

Semantic Diversification of Text Search Results

Andrei Micu and Adrian Iftene[(✉)]

Faculty of Computer Science, Alexandru Ioan Cuza University,
Berthelot 16, 700483 Iasi, Romania
{andrei.micu,adiftene}@info.uaic.ro

Abstract. Search engines are getting faster and more feature-rich year by year, striving to bring their users the information they need as fast as possible. Bringing relevant information to the user in an effortless manner is no easy task. The search feature set is where search engines compete to win their users and it usually describes in what manner a search engine may be different from others. One of the most challenging features in a search engine is to diversify the search results in a way that each result has different meaning or different content from others. The goal is to free the user from the burden of separating redundant results. What is redundant for the user is the key challenge of this feature and may have different meanings depending on the format of the information. For text searches, user's input is commonly used for diversification of results [1]. This input may include information like the topic of search, previous searches or supplementary parameters asked by the engine. This paper describes a different approach on text diversification, based on text semantics analysis and combined with clustering algorithms. The aim is to explore how similarities from the semantic point of view can be used to eliminate redundant texts.

Keywords: Information retrieval · Diversification · Clustering

1 Introduction

When it comes to diversifying text results the common element in each technique is the intent to cluster the results in an attempt to optimize an objective function. The selected text features, similarity function, clustering algorithm, as well as the objective function, are influenced by the use cases that must be covered. The most natural way to approach a diversification problem is to start from the requirements and to experiment with the other variables.

The most natural way to approach a diversification problem is to start from the requirements and to experiment with combinations of the components enumerated above. Other solutions ensemble more than one technique to produce results that are closer to the objective than each of the techniques alone.

A comprehensive overview on the formal definitions for search results diversification was described in an article written by two researchers from the University of Ioannina, Greece [2]. The article describes diversity definitions based on content, novelty, coverage and combinations of the three. It also contains algorithms and heuristics that can be used to achieve diversification of text search results.

© Springer International Publishing Switzerland 2016
N.T. Nguyen et al. (Eds.): ICCCI 2016, Part II, LNAI 9876, pp. 78–88, 2016.
DOI: 10.1007/978-3-319-45246-3_8

In the area of diversification based on user's input, a group of researchers at the Autonomous University of Madrid studied a set of intent-oriented diversification algorithms [3] in an attempt to build a diversification framework out of them. The set of algorithms include xQuAD, IA-Select, PIA-Select, PxQuAD and others. When introducing user's input as a random variable, the researchers discovered an increase of 3 %–11 % in terms accuracy values, and between 3 %–8 % in terms of diversity values.

Diversification of text search results at the term level was studied by two researchers at the University of Massachusetts [4]. For identifying relevant topic terms inside a text, they proposed an algorithm called DSPApprox. Using algorithms like PM-2 and xQuAD adapted to work with the terms inside a text search result, they obtained results comparable with the results of commercial search engines. In our group, we do diversification in an image retrieval system based on textual and visual processing [5, 6].

The semantic diversification approach described in this paper is an attempt to explore the advantages of text semantics when comparing similarities between them. It uses the entities and categories found in text documents resulted from a search to eliminate redundancy among the results. Since a statement can be formulated in multiple ways, which mean the same thing, the plain text may not be an accurate indicator that the search results represent something different for the user. The entities and categories, however, are the same whenever the formulation of a statement changes.

This approach is one step further from the term-level diversification presented in the state of the art section. The difference between terms and entities is the fact that a single entity can be represented by multiple terms, a combination of terms, or even the context in which the terms appear. Another difference is the fact that the semantic diversification also considers the categories (or topics) in which the text is already placed.

Because the semantic diversification problem can be modeled differently, depending on the nature of the elements it works with, diversification based on entities and categories will have their own algorithms that fit best. This paper describes a formulation of the problem from this point of view, a proposed solution based on hierarchical clustering algorithms and a demonstration on how such a solution can perform.

2   Problem Formulation

To state the problem that the semantic diversification of text search results tries to solve, there are a couple of terms that need to be defined:

- *Text search result* (or *text* for short): any document in plain text format that can be a result of a search from a search engine;
- *Entity*: something that exists and can be recognized by a user as an element that has meaning;
- *Category*: a label that associates a text search result to a set of text search results which share similar meanings.

An entity is not necessarily a word found inside the text. An entity can be contained by a text through a synonym or it can be implied by the meaning. For example, "The Netherlands national football team" can also appear inside a text like "Holland national soccer team", or it can be implied by the phrase "the clockwork orange".

In this paper we are not going to discuss the manner in which entities and categories are extracted from a text. Instead, we are going to focus on the part that differentiates the text search results based on the entities that they contain. Given this approach, we are going to assume that each text already has an associated set of entities and categories and the texts must be returned to the user grouped using the associated sets. The groups must contain texts that are as similar as possible by meaning and each group of texts must be as different as possible by meaning. Then it is up to the search engine how to present the information (pick up a document from each group and present them to the user, show the documents in groups and let the user choose, etc.). Before defining the problem formally, we must also define a distance function that will be an indicator of how different two texts are. From the semantic point of view, defining how different two texts are is not an easy task, but for a starting point we can consider some simple functions like the following:

Let $S = \{T_1, T_2, \ldots, T_n\}$ be the set of texts that need to be grouped. Each text T_i contains a set of entities and categories: $T_i = \{t_1, t_2, \ldots, t_n\}$. We define:

$$cd : S \times S \to \mathbb{N}, \ cd(T_i, T_j) = |(T_i \cup T_j) \backslash (T_i \cap T_j)| \text{(count distance)} \quad (1)$$

$$pd : S \times S \to [0, 100], pd(T_i, T_j) = \frac{cd(T_i, T_j) \times 100}{|T_i| + |T_j|} \text{ (percentage distance)} \quad (2)$$

Basically, the count distance measures the number of different distinct entities and categories between two texts, while the percentage distance measures the percentage of different distinct entities and categories from the total number of them in the two texts. Now that we have a distance function we can formally define our problem:

Given $S = \{T_1, T_2, \ldots, T_n\}$, a set of texts, and d, a distance function (cd or pd), find $P = \{S_1, S_2, \ldots, S_m\}$, a partition of S, such that, $\forall S_i = \{T_1, T_2, \ldots, T_p\} \in P$ and

$$\forall T_j, T_k \in S_i, P minimizes \frac{\sum_{j,k} d(T_j, T_k)}{|S_i|} \quad (3)$$

The problem above would easily be solved if m would be equal to $|S|$, i.e. if we return all the text search results as they appear in the input, each one of them being a group in itself. However, this would not be the result we need because the purpose of the solution is to diversify the results to eliminate redundancy up to a certain level of similarity. Depending on what the user requires in the end, certain restrictions may be added to the problem. An example restriction can be a fixed m, i.e. restricting the number of groups to a fixed value such that a user will receive exactly m results. Other useful restrictions will be discussed later in this paper.

3 Proposed Solution

The solution proposed in this paper is based on agglomerative hierarchical clustering algorithms [7, 8] and the previously defined distance functions. Because clustering algorithms work well with distance functions and a function to score the output, we are going to apply the most appropriate ones to obtain the groups. There are several reasons why hierarchical clustering is the most appropriate for this problem:

- The number of clusters can be controlled efficiently and flexibly by using any stop condition (which is checked at each step);
- It is not restricted to numerical values (like EM, for instance);
- It does not require other structures than the data (for example, it does not require centroids like K-Means does, which would not make sense in our use case).

The distance functions we previously defined and generically named as d can be used for calculating the intra-cluster distances, which is useful for rating the solution of the algorithm. The inter-cluster distance can be also obtained by using these distance functions. Some basic inter-cluster distances can be adapted as follows:

Let $S_1, S_2 \in 2^S, S_1 = \{T_1, T_2, \ldots, T_n\}$ and $S_2 = \{T'_1, T'_2, \ldots, T'_m\}$. We define:

$$sl : 2^S \times 2^S \to \mathbb{N}, \ sl(S_1, S_2) = \min\left(d\left(T_i, T'_j\right)\right), T_i \in S_1, T'_j \in S_2 \text{ (single − link)} \quad (4)$$

$$sl : 2^S \times 2^S \to \mathbb{N}, \ sl(S_1, S_2) = \min\left(d\left(T_i, T'_j\right)\right), T_i \in S_1, T'_j \in S_2 \text{ (complete − link)} \quad (5)$$

$$al : 2^S \times 2^S \to \mathbb{R}, \ al(S_1, S_2) = \frac{\sum d\left(T_i, T'_j\right)}{|S_1| \times |S_2|}, T_i \in S_1, T'_j \in S_2 \text{ (average − link)} \quad (6)$$

We will also note with d the inter-cluster distance, which can be any of the three above. Rating the output of this algorithm can be done using metrics that involve the clusters. One such metric can be the average distance inside a cluster, defined using the intra-cluster distance function:

Let $P = \{S_1, S_2, \ldots, S_m\}$ be a partition of S $(P \subset 2^S)$. We define:

$$ad : P \to \mathbb{R}, \ ad(P) = \frac{\sum d\left(S_i, S_j\right)}{C^2_{|P|}} \text{ (average intra − cluster distance)} \quad (7)$$

4 Implementation

In order to prove the utility of the semantic text diversification solution, we created a system which implements the data structures and algorithms previously described and it can be used to perform experiments on a demo dataset. This system is currently hosted in this GitHub repository: https://github.com/andrei-micu/semantic-text-diversification.

The final result intended for the experiments is to determine the best combinations of intra-cluster distance function, inter-cluster distance function and stop condition that suit the requirements of the user. Such requirements can be:

- Return only 20 results per search;
- Group the results that contain similar entities (let's say less than 3 entities/categories differ);
- Return only results that differ by more than 50 % in terms of entities and categories.

The dataset on which the experiments are performed is a collection of news descriptions from RSS news feeds of main worldwide newspapers. The reason why this set of data was chosen for experimenting is the fact that they are likely to contain news about the same event, but written in different ways. Such cases are perfect for demonstrating the ability of the semantic text diversification to differentiate new from redundant information.

4.1 Architecture

The system consists of a processing pipeline that fetches the texts from the RSS news feeds as input and outputs them grouped in clusters, along with metrics like average intra-cluster distance and time spent for clustering. Each experiment that is run through the pipeline has parameters like intra-cluster distance function, inter-cluster distance function and stop condition.

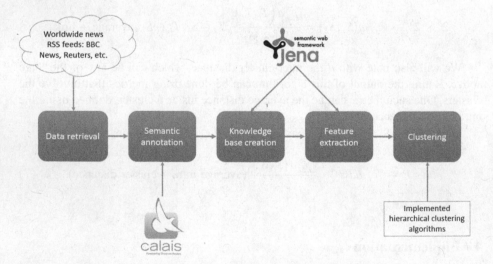

Fig. 1. Overview of the process which is involved in an experiment

Figure 1 depicts an overview of the experiment process, along with the technologies used for accomplishing each task. The first three steps are preparatory and can be run independently from the others, as their purpose is to build the dataset gradually. These steps can be run daily to fetch data from the RSS feeds and to store it on the local machine.

The other two steps are responsible with the actual grouping of the texts. These steps, which have the dataset and the parameters (intra-cluster distance function, inter-cluster distance function and stop criterion) as input define an experiment. The output of these steps are the grouped texts, but also information like the time spent clustering, the number of clusters, the intra-cluster average distance of the results and the variance of these distances.

4.2   Data Retrieval

The data retrieval component is responsible with fetching data from the public RSS feeds of well-known worldwide newspapers. The choice for these particular newspapers is purely based on the availability and licensing, which does not restrict the usage for academic purposes. The data is retrieved using ROME Tools[1], which is a library for easily fetching, parsing and manipulating RSS feed data.

The title and description of each news entry is stored on the local machine along with a unique ID. The ID is bound to the specific news entry in all the later steps and is crucial for managing them along the processing pipeline.

4.3   Semantic Annotation

The following component in the processing pipeline is the semantic annotator. This annotator takes every news entry along with its title and transforms them into RDF structured data. This prepares the text for feature extraction by marking the entities and categories that it contains.

To perform the annotations, the component uses a free service called Open Calais[2] that converts unstructured data into RDF structured data containing entities, categories and relations between them.

The choice to offload this work to a service is based on the fact that, as we previously mentioned, my work is focused on clustering the results and not the preprocessing of data. Additionally, the license of Open Calais permits the usage for academic purposes. Another advantage of it is that the output format of the structured data can be read by semantic frameworks and it can be further processed by them for feature extraction.

---

[1] ROME Tools - http://rometools.github.io/rome/.

[2] Open Calais - How Does Calais Work? - http://www.opencalais.com/about.

4.4   Knowledge Base Creation

Converting plain text to structured data would not be of great use if there would not be tools to work with it. To load the structured data into memory as a model, the system uses Apache Jena[3], a framework for manipulating this kind of information. In this step each news entry will get an associated knowledge graph with the entities and categories that are present in the text.

Jena creates knowledge graphs called "Models" from each news entry that was converted by Open Calais. These knowledge graphs are stored on the local machine as a checkpoint in the process. Loading the RDF annotated data in a Jena Model enables working with tools for the feature extraction step, like SPARQL.

4.5   Feature Extraction

The feature extraction is also a preparatory step, but it is run in the experiment phase because the impact on running time is insignificant. Another reason why this step is performed at experiment-time is the fact that it strips information from the RDF data in the output.

Feature extraction is performed by querying the knowledge base for information that is relevant to the similarity function. Because Apache Jena supports querying the Model with SPARQL, this step applies a query to extract the entities and categories using Jena's API.

In the Open Calais output we associated the "SocialTag" entries with the entity concept and the "cat" entries with the category concept. Because each of the two similarity functions require the categories and entities, the feature extraction is performed through the following SPARQL query:

```
PREFIX calais-pred: <http://s.opencalais.com/1/pred/>

SELECT ?result
WHERE {
 OPTIONAL { ?subject calais-pred:name ?result }
 OPTIONAL { ?subject calais-pred:categoryName ?result }
 FILTER regex(str(?subject), "/SocialTag/|/cat/")
}
```

4.6   Clustering

The clustering component is the implementation of the agglomerative hierarchical clustering algorithm. Currently, it implements the *cd* and *pd* intra-cluster distance functions, the *sl*, *cl* and al inter-cluster distance functions and stop criteria based on the number of clusters and the maximum inter-cluster distance.

---

[3] Apache Jena - https://jena.apache.org/.

The intra-cluster distance functions are implemented in the EntryDistance enumeration and the inter-cluster distance functions are implemented in the ClusterDistance enumeration. The StopCriterion class encapsulates all the parameters that must be checked for stopping the clustering algorithm.

5 Experiments and Comparisons

The following part of this paper is composed of several experiments performed with the system we created, along with comparisons and conclusions based on the results. The experiments are made on a dataset consisting of around 2.000 news entries fetched from the RSS news feeds over a timespan of one month. The news entries can be found on the GitHub repository where the implementation is stored.

All combinations of intra-cluster distance functions, inter-cluster distance functions and stop criteria were tested and compared based on the time spent by the whole clustering process, the intra-cluster average distance and the variance of these distances. It is important to mention that between the count distance (cd) and the percentage distance (pd) there is a difference between the value intervals: cd has a value interval of $[0, +\infty]$ (which denotes the number of different entities/categories) and pd has a value interval of $[0, 100]$ (which denotes the percentage of different entities/categories).

Table 1 describes the experiments for the use case where exactly 10 results or groups must be returned. Similar to the classic implementation of the hierarchical clustering algorithms, the single-link and the average-link distances tend to bring most of the texts into one cluster, leaving the other clusters with only entries that have duplicates. Complete-link, one the other hand, groups the results more even, in the sense that the news entries are distributed almost equally between the 10 clusters. This fact is more evident in the percentage distance case, where the variance is significantly lower.

Table 1. Comparison of the results for the 10 clusters restriction

Distance function	sl (single-link)	cl (complete-link)	al (average-link)
cd (count distance)	Time spent: 229 s Avg. distance: 1.98 Variance: 23.72	Time spent: 324 s Avg. distance: 14.93 Variance: 23.81	Time spent: 260 s Avg. distance: 3.38 Variance: 37.13
pd (percentage distance)	Time spent: 245 s Avg. distance: 9.79 Variance: 958.66	Time spent: 325 s Avg. distance: 95.23 Variance: 1.48	Time spent: 262 s Avg. distance: 9.79 Variance: 958.70

Another difference between the three distances is the fact that complete-link requires around 50 % more time than the others. This is due to the fact that the other two distances group the majority of the entries in a single cluster and that reduces the number of comparisons when calculating the inter-cluster distances.

Although the results do not show great differences between the count distance and the percentage distance in this experiment, the latter one has the most impact combined with the complete-link distance. It appears that this combination distributes the results the best, leading to a significantly lower variance between intra-cluster distances of 1.48 %.

Table 2 describes the same combinations of distances, but with a different stop criterion. The criterion makes the algorithm stop when the minimum inter-cluster distance found at one step passes a certain threshold. This also guarantees that the intra-cluster average distance is below the certain threshold. These threshold are: "3 different entities/categories" for the count distance and "20 % of instances are different" for the percentage distance.

Table 2. Comparison of the results for the inter-cluster maximum distance restriction

Distance function	sl (single-link)	cl (complete-link)	al (average-link)
cd (count distance) (stop when cd <= 3)	Clusters: 1375 Time spent: 96 s Avg. distance: 0.08 Variance: 0.16	Clusters: 1477 Time spent: 140 s Avg. distance: 0.09 Variance: 0.14	Clusters: 1443 Time spent: 96 s Avg. distance: 0.09 Variance: 0.15
pd (percentage distance) (stop when pd <= 20 %)	Clusters: 1559 Time spent: 74 s Avg. distance: 0.27 Variance: 4.03	Clusters: 1575 Time spent: 112 s Avg. distance: 0.24 Variance: 3.13	Clusters: 1567 Time spent: 71 s Avg. distance: 0.26 Variance: 3.47

These stop criteria based on the maximum different instances are fitting well for use cases where the focus is not on the number of results, but the actual removal of duplicate texts. For low thresholds like the ones in Table 2, the groups are numerous and contain only highly similar texts.

Similar to the previous experiments with fixed cluster number, the results highlight complete-link's need for around 50 % more processing time. The other similarity is the fact that complete-link produces groups of more evenly distrib-uted texts, leading to better average intra-cluster distance and variance.

As a conclusion from the experiments, the complete-link distance creates clusters that are more evenly distributed and is a better fit for both cases, even if the numbers sometimes show otherwise. Also, the percentage distance may give the algorithm a better distributing behavior, mostly when the number of resulted clusters is fixed to a small value.

6   Conclusions

Diversification of text search results using text semantics is a topic that shows great potential in several areas of computer science. Depending on the use case, there can be more than one solution for diversifying text considering its semantics, leaving many possibilities to customize the behavior and to fine-tune it by both the developers and the users. These means of identifying and eliminating redundant information has the potential to replace or augment other techniques that are currently applied by search engines.

The solution we proposed has fulfilled its purpose: it demonstrated that text semantics have a powerful impact on text diversification and the usage can lead to results that are helpful for the end user. The implementation and experiments are a proof that the theory inside the solution can be translated into real-world systems and that it is practical in terms of resources like memory and time.

6.1 Future Work

Future improvements on the current solution can be done by extending the current system with more distance functions and stop criteria. These elements can be changed to best fit a particular use case, which can increase the accuracy of the results for targeted users.

A yet unexplored inter-cluster distance function is the average intra-cluster distance of the two clusters union, named "average union link" and formally defined as below:

Let $S_1, S_2 \in 2^S$, $S_1 = \{T_1, T_2, \ldots, T_n\}$ and $S_2 = \{T'_1, T'_2, \ldots, T'_m\}$. We define:

$$aul : 2^S \times 2^S \rightarrow \mathbb{R}, \ aul(S_1, S_2) = \frac{\sum d\left(T_i, T'_j\right)}{C^2_{|P|}}, T_i, T'_j \in S_1 \cup S_2 \ (\text{average} - \text{union} - \text{link}) \quad (8)$$

This distance function will result in a direct optimization of the rating function presented in the solution as "average intra-cluster distance", since this is actually an adaptation of it at inter-cluster level. The algorithm ran with this distance should guarantee the minimization of the rating function because at each step it selects the pair of clusters which can form a new cluster that has the minimum value for the rating function.

Acknowledgments. This work is supported by the PRIVATESKY project (Experimental development in public-private partnership for creating native Cloud platform with advanced features for data protection), from POC 2014-2020, Action 1.2.3, Partnerships for knowledge transfer.

References

1. Welch, M.J., Cho, J., Olston, C.: Search result diversity for informational queries. In: Twentieth International World Wide Web Conference, Hyderabad, India, March (2011)
2. Marina, D., Evaggelia, P.: Search result diversification. SIGMOD Rec. **39**(1), 41–47 (2010). ACM, New York, NY, USA
3. Vallet, D., Castell, P.: Personalized diversification of search result. In: SIGIR 2012, August 12–16, Portland, Oregon, USA (2012)
4. Van, D., Bruce, C.W.: Term Level Search Result Diversification, SIGIR'13, July 28–August 1. ACM, Dublin, Ireland (2013)
5. Iftene, A., Alboaie, L.: Diversification in an image retrieval system based on text and image processing. Comput. Sci. J. Moldova **22**(66), 339–348 (2014)

6. Iftene, A., Alboaie, L.: Diversification in an image retrieval system, IMCS-50. In: The Third Conference of Mathematical Society of the Republic of Moldova Dedicated to the 50th Anniversary of the Foundation of the Institute of Mathematics and Computer Science, pp. 521–524, August 19–23, Chisinau, Republic of Moldova (2014)
7. Manning, C., Schutze, H.: Foundations of Statistical NLP ch. 14. MIT Press, Cambridge (2002)
8. Mitchell, T.: Machine Learning ch. 6.12. McGRAW Hill, Boston (1997)

Computational Swarm Intelligence

Ant Colony System
with a Restart Procedure for TSP

Rafał Skinderowicz[✉]

Intitute of Computer Science, University of Silesia,
Będzińska 39, 41-205 Sosnowiec, Poland
rafal.skinderowicz@us.edu.pl

Abstract. Ant Colony Optimization has proven to be an efficient optimization technique for solving difficult optimization problems. Nonetheless, the convergence of the ACO can still be prohibitively slow. We investigate how the recently proposed Restart Procedure (RP) can be used to
improve convergence of the Ant Colony System (ACS) algorithm, which
is among the most often applied algorithms from the ACO family. In
particular, we present a series of computational experiments to answer
the question about how the values of the RP-related parameters influence the convergence of the ACS combined with the RP (ACS-RP). We
also show that the ACS-RP achieves significantly better results than the
standard ACS within the same computational budget.

Keywords: Ant Colony System · Restart procedure · Travelling
salesman problem

1 Introduction

Metaheuristic optimization algorithms (MOAs) are often a useful tool if one tries
to solve a difficult optimization problem for which no efficient exact algorithm
exists [9]. The main drawback of the MOAs is often their slow convergence.
Even if an algorithm theoretically converges to the optimum, the expected time
to reach the optimal solution can be prohibitive. There are many ways to improve
the algorithm convergence or shorten the computation time. One, often applied,
concept is a *restart procedure*. It consists in restarting the MOA after a certain
criterion is met. In consequence, we obtain a number of, possibly different, solutions from which one with the highest quality becomes the final solution to the
problem. Restarting an algorithm after a fixed number of iterations is one of the
simplest restart criteria.

Recently Carvelli proposed a generic Restart Procedure (RP) [3]. He conducted a theoretical analysis of the conditions necessary for the RP to improve
the convergence of the underlying MOA. Specifically, he focused on finding the
optimal value of the restart time given the available computation time. The optimal restart time is such that corresponds to the minimum *failure probability*, i.e.
probability of not finding the optimum.

© Springer International Publishing Switzerland 2016
N.T. Nguyen et al. (Eds.): ICCCI 2016, Part II, LNAI 9876, pp. 91–101, 2016.
DOI: 10.1007/978-3-319-45246-3_9

In this work we investigate how the Restart Procedure (RP) by Carvelli [3] can be used to improve the convergence of the Ant Colony System (ACS). First, we conduct a number of experiments to investigate how the values of the RP-related parameters impact the performance of the algorithm. Next, we evaluate the performance of the ACS with the RP (ACS-RP in short) relative to the performance of the standard ACS. The computations are performed on a set of TSP instances from the well known TSPLIB repository. It is worth noting that we are mainly interested in the performance of the algorithm when the computational budget is rather small, i.e. we do not require it to converge to the optimum. To make the work more thorough we consider both the basic ACS and the ACS paired with an efficient local search heuristic (2-opt).

The structure of the article is as follows. In Sect. 2 a brief summary of the related work is given. Section 3 contains a brief description of the ACS algorithm, while Sect. 4 presents the Restart Procedure in more details. The experiments conducted are described in Sect. 5 and a short summary is given in Sect. 6.

2   Related Work

The idea of restarting computations to improve the convergence of various heuristic algorithms solving $\mathcal{NP}$-hard problems is not new and is an integral part of a more broad family of *multi-start methods* [11]. The main goal of the multi-start methods is to increase diversification of the search process to overcome local optimality [11]. The multi-start methods can be divided into *memory-based* and *memory-less* groups. Tabu Search [7] and the Path-relinking [8] are example methods from the first group. Simulated Annealing and the Greedy Randomized Adaptive Search Procedure (GRASP) [6] can be seen as example methods from the second group. A detailed overview of the multi-start methods can be found in [9,11].

The idea of restarting computations in case of the ACO algorithms usually takes a form of pheromone resetting. If all pheromone trails are reset the approach can be classified as the memory-less multi-start method. Otherwise, if only some portion of the pheromone trails is reset, it can be classified as the memory-based multi-start approach. A decision when to reset the pheromone values is another important factor. In the most simple case the pheromone is reset after a predefined number of iterations. The more complex ideas involve detection of the algorithm's stagnation, e.g. based on the diversity of generated solutions or on the relative differences between the pheromone trails. For example, Oliveira et al. [12] applied the restart procedure to periodically reset the pheromone values in the Population-based ACO. The experiments conducted on a set of TSP and Quadratic Assignment instances showed that the pheromone resetting significantly improved results of the algorithm. Blum et al. [2] proposed Hyper-cube Framework for the ACO in which the pheromone trails are reset to initial values if the algorithm convergence is detected. The resulting algorithm was competitive with the state-of-the-art metaheuristics for the Unconstrained Binary Quadratic Programming Problem. A similar idea was used by Blum in the Beam-ACO algorithm [1]. The pheromone trails are reset if the algorithm convergence is detected based on the relative differences between the pheromone values. Guntsch et al. [10] applied the ACO to solve the Dynamic TSP problem and compared the relative performance of a simple restart strategy with two other strategies involving a partial reinitialization of the pheromone values. The former proved to be efficient especially when the dynamic changes to the TSP instance considered were relatively big, while the latter was efficient in case of relatively small changes to the problem.

3   Ant Colony System

The Ant Colony System belongs to a family of nature inspired metahueristic algorithms. More specifically, it is an improved version of the Ant System algorithm inspired by the foraging behavior of certain species of ants [4]. Usually the problem tackled is modeled using a complete, weighted graph $G = (V, A)$, where V is a set of nodes and $A = \{(i, j)|(i, j) \in V, i \neq j\}$ is set of edges connecting the nodes. In the context of the TSP graph G is undirected and the set of nodes V corresponds to the set of n cities, while the edges represent roads between the consecutive pairs of cities with the edge weights equal to the distances between the cities. In a single iteration of the ACS each of artificial ants constructs a complete solution to the problem. An ant starts at a random node and in each of the subsequent steps extends its partial solution with an unvisited node. An ant selects one of the unvisited nodes based on a *pseudo-random proportional rule* [4]. The rule takes into account external information about the problem, which in case of the TSP is simply the distance between cities, and the values of the *artificial pheromone trails* deposited by the ants during the algorithm execution. The pheromone trails comprise a *pheromone memory* which allows the algorithm to learn. In the ACS there are two pheromone memory update rules – local and global. The former involves evaporation of a small amount of pheromone from an edge's trail each time it is traversed by an ant. The latter is applied after all the ants have completed their solutions and involves increasing the pheromone levels on the trails corresponding to the edges belonging to the best so far solution. For a more detailed overview of the ACS please refer to [4,5].

4   Restart Procedure

In the present work we focus on the Restart Procedure proposed by Carvelli [3]. The main idea of the RP is as follows. Given a fixed computation time t one has two options. The MOA can be executed *once* using all the available time or it can be *restarted* every $T < t$ time units (or iterations, assuming a single iteration takes one time unit). Whether the later option is preferable depends on the so-called *failure probability* of the MOA which is a non-increasing function of the number of iterations defined as:

$$p(t) = P(X(t) \notin X^*),\tag{1}$$

where $f : S \rightarrow \mathbf{R}$ is the function to be minimized (maximized), $X(t)$ is a stochastic process corresponding to the state of the MOA at the time t, and X^* as set of optimal solutions in the set of all feasible solutions S. If the failure probability is big enough, the resulting failure probability of the MOA with the RP decreases to zero geometrically with the number of restarts, i.e. it equals $p(T)^{\lfloor \frac{t}{T} \rfloor}$. On the one hand, one wants to maximize the number of restarts, $\lfloor \frac{t}{T} \rfloor$, on the other hand, the duration T of a single MOA run (restart) should be long enough to make $p(T)$ small. Carvelli [3] showed that the Restart Procedure could be *optimized* by choosing for the restart time a value σ that minimizes the function $g(t) = p(t)^{\frac{1}{t}}$. In practice, however, the failure probability, $p(t)$, may not be known, for example if the optima of are not known. Carvelli proposed a procedure of calculating $\hat{\sigma}$, i.e. an estimation of σ that converges to σ as the number of the procedure steps grows.

```
1   for  i ← 1 to r₀ do
2   |      execute algorithm    𝒜ᵢ until T₀
3   end
4   save  Y_{A₀}
5   compute  σ₀ from Y₀
6   for  k ← 1 to X do
7   |      if  σ_{k−1} > λ · T_{k−1} then
8   |      |      T_k ← f_T(T_{k−1})
9   |      |      r_k ← r_{k−1}
10  |      |      for  replication  i = 1 to r_k do
11  |      |      |      continue the execution of   𝒜ᵢ until T_k
12  |      |      |      save  𝒜ᵢ(T_k)
13  |      |      end
14  |      else
15  |      |      r_k ← f_r(r_{k−1})
16  |      |      T_k ← T_{k−1}
17  |      |      for  replication  i = r_{k−1} + 1 to r_k do
18  |      |      |      execute  𝒜ᵢ until T_k
19  |      |      |      save  𝒜ᵢ(T_k)
20  |      |      end
21  |      end
22  |      save  Y_{A_k}
23  |      compute  σ_k from Y_k
24  end
```

Fig. 1. Pseudocode of the Restart Procedure

The general idea of the RP is as follows. We start with an initial number r_0 of the MOA replications (instances) and based on the current progress we calculate an estimation of the optimum restart time σ. Based on the estimation, we either allow the current MOA replications to continue computations or start a number of new replications and run them until the current replication length. Figure 1 shows the pseudo-code of the RP. It starts with the execution of r_0 initial *replications* of the underlying MOA, denoted by $\mathcal{A}$, for T_0 iterations each (lines 1–3). In line 5 $\hat{\sigma}_0$ is computed from matrix Y_{A_0}, where i-th row of the matrix corresponds to a sequence of T_0 best-so-far solutions found by i-th replication of the MOA in the first T_0 iterations. In lines 6–24 the main loop of the RP is performed until a predefined stopping criteria is met. In the main loop one of the two cases is performed based on the estimation $\hat{\sigma}_{k-1}$ of the optimum restart time, σ. The first one is taken (line 7) if the current estimation of the optimum restart time $\hat{\sigma}_{k-1}$ is close to the current replication length T_{k-1}. This could be explained as if we were *not sure* if our estimation of the restart time is good enough. In such case we increase the number of iterations per replication (line 8) to a new value T_k and allow each replication to continue computations until T_k. In the second case (lines 14–20) the estimation of the optimum restart time seems appropriate and a number of new MOA instances (replications) is started and executed until the current replication length $T_k = T_{k-1}$. After each step (loop iteration) a new estimation $\hat{\sigma}_k$ is computed (line 23) from matrix Y_{A_k}, which accumulates the progress of all r_k replications of the MOA.

The estimation $\hat{\sigma}_k$ of the optimum restart time (lines 5 and 23 in Fig. 1) is equal to *the smallest value of the time (iteration)* when the current global best solution was found by as many of the current replications as possible. Two important issues remain. The first one concerns the calculation of the new replication length according to the function f_T (line 8 in Fig. 1), while the second one concerns increasing the number of replications according to the function f_r (line 15 in Fig. 1). Carvelli gave the necessary requirements for the proper f_T and f_r (for the details please refer to [3]) and provided the example functions, as follows. The replication length is increased according to $T_{k+1} = f(T_k) = q(T_k) \cdot T_k$ where

$$q(T_k) = c_2 + C \cdot \sqrt{\frac{|\bar{Y}(r_k, T_k) - \bar{Y}(r_k, \hat{\sigma}_k)|}{(\bar{Y}(r_k, T_k) + \bar{Y}(r_k, \hat{\sigma}_k))/2}}, \qquad (2)$$

where c_2 and C are constants larger than one and zero, respectively, $\bar{Y}(r, s) = \frac{1}{r}\sum_{i=1}^{r} Y_i(s)$ and $Y_i(s)$ is the value of the best solution found by the i-th replication of the MOA until time s. In [3] the number of replications was increased according to $r_{k+1} = f_r(r_k) = c_1 \cdot r_k$, where $c_1 > 1$. In our work we use a simpler yet still valid formula: $r_{k+1} = f_r(r_k) = r_k + 1$, which is more suitable if the computation budget is tight.

5   Experiments

The ACS-RP requires setting a few additional parameters. The aim of the first part of the experiments was to better understand how the performance of the ACS-RP changes depending on the values of the RP related parameters. The second part of the experiments was focused on assessing the performance of the ACS-RP relative to the ACS. A series of computational experiments was conducted on a set of the TSP instances from the TSPLIB repository. Both algorithms were run without and with the local search heuristic applied (2-opt). It was dictated by the fact that the use of the LS often significantly improves the convergence of the ACO algorithms and is often used in practice [5].

The algorithms were implemented in C++ and run on a machine with Intel Xeon E5-2680v3 2.5 GHz CPU. The values of the ACS parameters were chosen based on preliminary experiments: number of ants $m = 20$, $\beta = 3$, $q_0 = (n -$

$20)/n$, where n is the size of the problem, size of the candidates list $cl = 32$, global pheromone update coefficient $\rho = 0.1$, local pheromone update coefficient $\alpha = 0.01$. Initial pheromone level was set to $(n \cdot L_{nn})^{-1}$, where L_{nn} is the length of the solution produced by the nearest neighbor greedy heuristic. For each combination of the parameter values the computations were repeated 30 times.

5.1 Initial Number of Replications

In an ideal case the optimum restart time needed to maximize the probability of finding the optimum solution would be known. Unfortunately in case of difficult problems, such as the TSP, such knowledge cannot be provided *a priori*. The usefulness of the RP stems from the fact that there is no need to know the optimum restart time but it is approximated during the computations. The algorithm starts with the initial number r_0 of the algorithm runs (replications) which are allowed to conduct T_0 iterations each. From the practical point of view, the initial number of algorithm runs r_0 should be as small as possible, especially if the computation budget is relatively small.

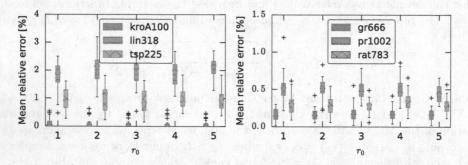

Fig. 2. Boxplot of the mean relative solution error vs the initial number of replications, r_0. The plot on the left shows the results for the ACS-RP with no local search used, while the plot on the right presents the results for the ACS-RP with 2-opt local search heuristic applied.

In order to observe how the ACS-RP algorithm performance (in terms of the solutions quality) changes relative to the initial number of replications r_0, we run the algorithm with r_0 varying from 1 to 5 ($T_0 = 100$ in all the cases) for a total of 6 TSP instances: 3 smaller and 3 larger. The smaller were: *kroA100*, *tsp225* and *lin318* for which no LS used and the number of constructed solutions was equal to $2000 \cdot n$. The larger instances were: *gr666*, *rat783* and *pr1002* for which the algorithm was run with the 2-opt LS and the total number of constructed solutions was $500 \cdot n$. Figure 2 shows box plots of the mean solution error (relative to an optimum) vs r_0. As can be seen, there were significant differences in the solution quality between the problem instances, the larger the problem size, the worse the results. This is expected as the problem complexity (in terms of the total number of feasible solutions) increases non-linearly with the number of nodes n. Also, please note that the use of the LS greatly improved the quality of solutions for the larger instances, despite the lower total number of constructed solutions. Nevertheless, in both cases the differences in the quality depending on the number of initial replications were small. In fact, there were only two statistically significant differences between the results for the various r_0 values, according to the non-parametric, two-tailed Mann-Whitney U test with $\alpha = 0.05$. In the first case, the results for the ACS-RP with $r_0 = 5$ for the *lin318* instance were better than the results for $r_0 = 1$. In the second case, the results for the ACS-RP with the 2-opt LS and with $r_0 = 3$ for the *rat783* instance were better than the results for $r_0 = 4$.

Based on the results, the ACS-RP can be run with r_0 equal to 1 but in the following experiments we chose $r_0 = 2$ to force the computations to differ from the ACS even if the number of iterations is small. In fact, if the computation budget is small there should be no benefit of using the ACS-RP over the standard ACS.

5.2 λ and c_2 Parameters

The most important parameters that govern the RP are λ and c_2. To measure the effect those parameters have on the quality of solutions the ACS-RP algorithm was run with several combinations of λ and c_2 values. Namely, $\lambda \in \{0.4, 0.45, 0.5, 0.55, 0.6, 0.65, 0.7, 0.75, 0.8\}$ and $c_2 \in \{1.1, 1.2, 1.3, 1.4, 1.5\}$, what makes a total of 45 different pairs of values. The rest of the RP specific parameters were set as follows: $c_1 - 1.5$ based on the suggestions in [3], number of initial algorithm runs $r_0 = 2$, number of initial iterations $T_0 = 100$. The values of the ACS specific parameters were set based on the preliminary experiments: $m = 20$ (the number of ants), $q_0 = (n - 20)/n$, $\beta = 3$, $c_1 = 1.5$, $\rho = 0.1$ (global pheromone evaporation rate), $\phi = 0.01$ (local pheromone evaporation rate), $cl = 32$ (size of candidate set). For each set of the parameters values the computations were repeated 30 times.

A total of four TSP instances were used: *kroA100*, *tsp225*, *gr666* and *rat783*. For the first two instances the algorithm was run with a limit of 5000 iterations, while the remaining two were run with the limit set to 3000 but a local search heuristic (2-opt) was used to improve the solutions generated by ants.

Figure 3 shows how c_2 and λ influenced the number of replications used during the ACS-RP run. As can be seen, the strongest effect had λ parameter which directly influences the decision whether to increase the current length of replication (line 7 in Fig. 1) or increase the number of replications. The closer to 1 is the value of λ, the greater is the chance that the current replication length will be increased instead of starting new replications, hence the lower number of replications observed. Influence of c_2 parameter stems from Eq. 2 which is used to calculate a new replication length.

The more important is the question how the values of c_2 and λ affect the quality of the generated solutions. To answer the question, the results obtained for a given pair of values were compared with the results for every other pair

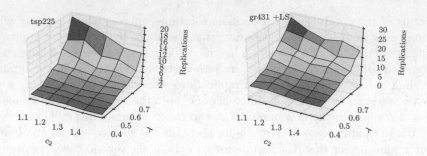

Fig. 3. Boxplots of the mean number of replications vs λ and c_2.

of the parameter values using the non-parametric Wilcoxon rank-sum test (at significance level of $\alpha = 0.05$). There were 45 different combinations of values tested, hence the maximum number of times a single pair could be better than the rest was 44 for a single TSP instance, and a total of $4 \cdot 44$ for all four instances considered. The summary is presented in Table 1. Clearly, there is no single combination of values that is significantly better than all the others. Generally, the results obtained for $0.6 \leq \lambda \leq 0.7$ and $c_2 \geq 1.3$ were more often significantly better than the other combinations. Based on those observations in the subsequent experiments we set $\lambda = 0.7$ and $c_2 = 1.4$.

Table 1. Summary of a statistical comparison between the results of the ACS-RP with various combinations of (c_2, λ) values for a set of four TSP instances. The first (resp. second) number in a cell equals to the total number of times the results were significantly better (resp. worse) than the results for the other pairs of parameter values according to the non-parametric Wilcoxon rank-sum test ($\alpha = 0.05$).

c_2\λ	0.4	0.45	0.5	0.55	0.6	0.65	0.7	0.75	0.8
1.1	11/13	16/25	7/12	24/1	18/6	19/2	27/12	20/20	26/71
1.2	17/24	6/32	26/3	14/11	12/23	34/1	17/18	23/13	5/54
1.3	16/21	12/47	11/11	13/17	50/1	8/8	17/2	27/28	16/62
1.4	12/66	12/41	6/26	10/8	10/4	36/10	38/0	21/1	23/26
1.5	15/13	25/11	22/37	17/11	13/13	29/1	23/1	21/4	11/25

5.3 Algorithms Comparison

To assess how the Restart Procedure influences the convergence of the algorithm we compared the results of the ACS with the results of the ACS-RP over a set of the TSP instances divided into two subsets. For the first subset, containing the instances *kroA100, tsp225, lin318, gr431, d493, d657, gr666, rat783* and *u1060*, the algorithms were run without a local search. In case of the second subset, containing the instances *gr431, d493, d657, gr666, rat783, pr1002, u1060, fl1400* and *u2152*, the local search (2-opt) heuristic was applied to improve the solutions

Table 2. Results for the ACS and ACS-RP. Values in bold refer to a significant difference to the other algorithm according to the non-parametric Wilcoxon rank-sum test (at significance level of $\alpha = 0.05$).

Problem	ACS			ACS-RP		
	Mean sol. value	Mean error [%]	Calc. time [s]	Mean sol. value	Mean error [%]	Calc. time [s]
kroA100	21463.97	0.86	3.25	**21343.10**	**0.29**	3.13
tsp225	3962.87	1.20	15.35	3959.17	1.10	15.48
lin318	43023.17	2.37	31.90	42922.13	2.13	30.94
gr431	177860.80	3.76	57.66	177418.80	3.50	61.62
d493	37542.70	7.26	82.79	**37054.80**	**5.86**	83.33
d657	53710.80	9.81	171.30	**52709.23**	**7.76**	163.03
gr666	310277.63	5.41	162.97	310348.90	5.43	164.52
rat783	**9123.20**	**3.60**	220.72	9185.23	4.31	250.74
u1060	241888.90	7.94	467.72	241999.60	7.99	482.86

generated by the ants. In fact, the ACS is a general purpose metaheuristic and its performance, although good, can still be significantly improved if a dedicated LS heuristic is applied.

Tables 2 and 3 show results of the algorithms comparison. There are two important facts worth noticing. The first one is that the local search significantly improved the quality of the results. It is consistent with the previous research [5]. The second one is the advantage of the ACS-RP over the standard ACS in almost

Table 3. Results for the ACS and ACS-RP with the 2-opt local search heuristic applied. Values in bold refer to a significant difference to the other algorithm.

Problem	ACS			ACS-RP		
	Mean value	Mean error [%]	Calc. time [s]	Mean sol. value	Mean error [%]	Calc. time [s]
gr431	171777.57	0.21	95.64	**171434.60**	**0.01**	92.31
d493	35109.37	0.31	131.21	**35059.03**	**0.16**	123.63
d657	49063.33	0.31	288.20	**49004.83**	**0.19**	239.00
gr666	295326.17	0.33	236.95	**294744.53**	**0.13**	222.10
rat783	8842.53	0.41	321.28	**8828.20**	**0.25**	321.53
pr1002	260681.50	0.63	533.09	**260227.77**	**0.46**	546.16
u1060	225130.83	0.46	685.33	224962.73	0.39	654.79
fl1400	20201.77	0.37	1251.29	20192.83	0.33	1288.59
u2152	64566.30	0.49	2432.08	**64438.80**	**0.29**	2418.00

all cases. In fact, the mean error for the ACS-RP was smaller than the error for the ACS for the six smaller TSP instances and only slightly above for the three largest instances. It is an expected result because the increase in the problem difficulty .could not be compensated by the linear growth of the computational budget, i.e. the number of generated solutions, equal to $1000 \cdot n$, where n is the size of the problem (number of cities). The results for the ACS-RP with the LS were better in all the cases despite the larger size (up to 2152 cities). The use of the LS, even as basic as the 2-opt heuristic, greatly improved the ACS convergence, hence the computational budget was enough for the ACS-RP to gain advantage from the multiple replications of the ACS. Although, similarly to the previous case, the difference between the results diminished as the size of the problem grew.

6   Conclusions

The experiments conducted confirm that the Restart Procedure is an efficient metaheuristic that can be successfully applied to improve the convergence of the ACS. In almost all cases the results obtained were significantly better relative to the standard ACS, both with and without the local search applied. However, for the RP to improve the results the computational budget (time) has to be large enough to allow the RP to find a good estimate of the optimal restart time. The efficiency of the RP depends also strongly on the values of λ and c_2 parameters which directly influence the calculation of the restart time (equal to the replication length).

It is worth noting that although the RP has little impact on the computation time it increases the memory complexity of the whole algorithm because all the replications of the underlying MOA have to be kept in memory. It may become a problem for many of the ACO-related algorithms which use a pheromone matrix whose size in case of the TSP is at order of $O(n^2)$. A possible solution could be to use the ACS with a selective pheromone memory whose size could be made much smaller (even linear) without sacrificing the quality of results [13].

Acknowledgments. This research was supported in part by PL-Grid Infrastructure.

References

1. Blum, C.: Beam-aco–hybridizing ant colony optimization with beam search: an application to open shop scheduling. Comput. Oper. Res. **32**(6), 1565–1591 (2005)
2. Blum, C., Dorigo, M.: The hyper-cube framework for ant colony optimization. IEEE Trans. Syst. Man Cybern. Part B: Cybern. **34**(2), 1161–1172 (2004)
3. Carvelli, L.: Improving convergence of combinatorial optimization meta-heuristic algorithms. Ph.D. thesis, Sapienza Universita di Roma, Facolta di Scienze Matematiche Fisiche e Naturali (2013)
4. Dorigo, M., Gambardella, L.M.: Ant colony system: a cooperative learning approach to the traveling salesman problem. IEEE Trans. Evol. Comput. **1**(1), 53–66 (1997)

5. Dorigo, M., Stützle, T.: Ant Colony Optimization. MIT Press, Cambridge (2004)
6. Feo, T.A., Resende, M.G.: Greedy randomized adaptive search procedures. J. Global Optim. **6**(2), 109–133 (1995)
7. Glover, F., Laguna, M.: Tabu Search. Springer, New York (2013)
8. Glover, F., Laguna, M., Martí, R.: Fundamentals of scatter search and path relinking. Control Cybern. **29**(3), 653–684 (2000)
9. Glover, F.W., Kochenberger, G.A.: Handbook of Metaheuristics, vol. 57. Springer, New York (2006)
10. Guntsch, M., Middendorf, M.: Pheromone modification strategies for ant algorithms applied to dynamic TSP. In: Boers, E.J.W., Gottlieb, J., Lanzi, P.L., Smith, R.E., Cagnoni, S., Hart, E., Raidl, G.R., Tijink, H. (eds.) EvoWorkshops 2001. LNCS, vol. 2037, p. 213. Springer, Heidelberg (2001)
11. Martí, R., Resende, M.G.C., Ribeiro, C.C.: Multi-start methods for combinatorial optimization. Eur. J. Oper. Res. **226**(1), 1–8 (2013)
12. Oliveira, S.M., Hussin, M.S., Stützle, T., Roli, A., Dorigo, M.: A detailed analysis of the population-based ant colony optimization algorithm for the TSP and the QAP. In: Krasnogor, N., Lanzi, P.L. (eds.) GECCO 2011, Companion Material Proceedings, Dublin, Ireland, 12–16 July, pp. 13–14. ACM (2011)
13. R. Skinderowicz. The GPU-based parallel ant colony system. J. Parallel Distrib. Comput. (2016, in press)

Differential Cryptanalysis of FEAL4 Using Evolutionary Algorithm

Kamil Dworak[1,2]([✉]) and Urszula Boryczka[1]

[1] University of Silesia, Sosnowiec, Poland
{kamil.dworak,urszula.boryczka}@us.edu.pl
[2] Future Processing, Gliwice, Poland
kdworak@future-processing.com

Abstract. This paper presents a differential cryptanalysis attack on the Fast Data Encipherment Algorithm (*FEAL4*) reduced to four rounds, using an Evolutionary Algorithm (*EA*). The main purpose of the developed attack is to find six subkeys of the encryption algorithm, which will be used to decipher the captured ciphertext. Furthermore, an additional heuristic negation operator was introduced to improve local search of *EA*. The algorithm is based on a chosen-plaintext attack. In order to improve an effectiveness, the attack uses the differential cryptanalysis techniques. The results of the developed algorithm were compared against a corresponding Hill Climbing (*HC*), Simple Evolutionary Algorithm (*SEA*) and Brute Force (*BF*) attacks.

Keywords: Differential cryptanalysis · Evolutionary algorithm · FEAL4 · Cryptography · Hill climbing

1 Introduction

A high level of security has become a standard for every IT system [1]. Security norms should be characterized by appropriate integrity, availability and confidentiality of information. It is not only associated with the data storage, but also with its processing and computation [2]. Over time, special encryption algorithms, such as the Advanced Encryption Standard (*AES*), were developed. They fully satisfy mentioned above security requirements. Cryptography does not aim to obscure information itself from unauthorised access, but process it to such form, that it is only readable to the sender and the proper recipient [3].

The application of cryptography is closely associated with the concept of cryptanalysis. It involves on using appropriate mathematical methods to demonstrate whether given cipher or cryptographic system afford sufficient protection [4]. It usually consists in guessing a valid decryption key or a set of subkeys which can be used to decipher the intercepted ciphertext.

Increasing complexity of modern ciphers require advanced cryptanalysis. It takes more processor capacity, which significantly extend their memory consumption and performance. Computational intelligence (*CI*) is becoming more

© Springer International Publishing Switzerland 2016
N.T. Nguyen et al. (Eds.): ICCCI 2016, Part II, LNAI 9876, pp. 102–112, 2016.
DOI: 10.1007/978-3-319-45246-3_10

popular as an optimization tool in recent years. It based on nature and artificial intelligence (AI) methodologies. These kind of metaheuristics are addressed to various optimization problems, also in the computer security. Over the past few years, there have been many publications, which concern the application of diverse evolutionary methods to optimize currently applied ciphers. Many of them are used in cryptanalysis such as EA and GA [5–8], Particle Swarm Optimization (PSO) [9,10], Ant Colony Optimization (ACO) [11,12] and Simulated Annealing (SA) [13]. The current advances in the analysis of symmetric block ciphers cryptanalysis have been summarized of many thoroughly surveys and reviews [14,15]. The interest of evolutionary computation techniques, in the field of cryptology, is becoming more popular, although there are many problems that should be addressed.

The attack proposed in this paper uses the most popular metaheuristic optimization algorithm used in cryptology – EA. This type of algorithm operates on a finite set of individuals, known as the population [16]. Each individual is characterised by a certain numerical value called fitness. It defines the quality of the solution and it is calculated on the basis of a special fitness function (F_f), appropriately selected for the given problem [16]. EAs are nature inspired and they use operations such as natural selection, reproduction and mutation. More detailed information about EAs can be found in [16,17].

The proposed attack tries to find the six 32-bit subkeys which are used to decipher an original ciphertext intercepted earlier. The algorithm is based on a chosen-plaintext attack, which assumes that the cryptanalyst has access to the encryption algorithm. It is helpful to preview and analyse the ciphertexts generated by him on an ongoing basis [4]. The attack applies some differential cryptanalysis methods. In order to achieve better EA functional quality, an additional heuristic negation operator was introduced.

The next section of this paper contains the formulation of the problem. The third section describes basic concepts of differential cryptanalysis for $FEAL4$ cipher. The fourth section presents the proposed negation EA attack (NEA). The fifth section contains experimental study and comparison to the HC, SEA and BF attacks. Last section concludes the paper and highlights the direction of future work.

2   Fast Data Encipherment Algorithm

The $FEAL4$ cipher was developed by Shimizu and Miyaguchi in 1987 [18]. At first the algorithm was defined as $FEAL$, but later it was expanded by extra rounds and the original name has been changed to $FEAL4$. The authors were looking to create a stronger, faster and simpler cipher than the Data Encryption Standard (DES) used at that time [19]. Each round of the algorithm was to be stronger. $FEAL4$ does not use any permutations or tables, it is only restricted to simple bit operations such as exclusive disjunctions (xor) and cycle shifts [19]. After a few years it was discovered that the algorithm is not so secure after all. Only the version with the number of rounds increased to 16 or 32 is on par with

DES's security [19]. The *FEAL4* cipher has a significant impact for developing differential (*DC*) and linear cryptanalysis (*LC*) [20].

The presented encryption algorithm is a four round symmetric block cipher, which operates on a 64-bit blocks of data and uses a 64-bit key [18]. This paper describes a modified version of the *FEAL4* [20]. To increase problem complexity, the key will be divided into six 32-bit subkeys. The original version uses twelve 16-bit subkeys. The full description of the modified encryption algorithm is presented in [20]. It will not be necessary to discover the original 64-bit key, obtaining all subkeys will be sufficient to decrypt given ciphertext. More information about subkeys and their generation is described in [18]. The whole encryption process of the modified *FEAL4* algorithm is presented on Fig. 1 [18].

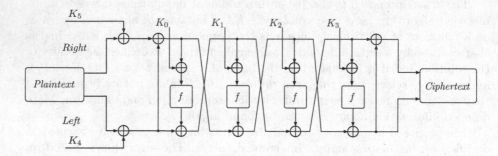

Fig. 1. *FEAL4*'s encryption algorithm.

At first the 64-bit block of plaintext is split into two 32-bit parts. The left-hand side part is xored with subkey K_4, the right one with K_5. Then the left and right parts are xored together to create a new right-hand side part. The newly generated fragment, with the left part, go through four cycles (rounds) of the encryption algorithm. In each cycle, the right-hand part is xored with 32 bits of the round subkey and subjected to the round function f. Once processed, the result is xored with the left part of data block. At the end of the cycle, the left and right parts are swapped. After all four cycles, the last right part is xored with the left one and later concatenated with this left part. A 64-bit block of ciphertext is generated [20].

2.1 Round Function f

The *FEAL4*'s round function f is presented on Fig. 2. A 32-bit portion of data is passed as input. Data block is split into four 8-bit parts (x_0, x_1, x_2, x_3). They are xored and passed to G_0 and G_1 functions as shown on Fig. 2 [18].

Functions G_0 and G_1 have been defined with the following expression:

$$G_x(a,b) = (a + b + x(mod256)) \lll 2, \tag{1}$$

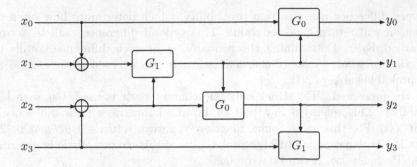

Fig. 2. *FEAL4*'s round function f.

where the $\lll$ operator signifies a cyclical shift to the left. It is possible to determine:

$$y_0 = G_1(x_0 \oplus x_1, x_2 \oplus x_3), \tag{2}$$

$$y_1 = G_0(x_0, y_1), \tag{3}$$

$$y_2 = G_0(y_1, x_2 \oplus x_3), \tag{4}$$

$$y_3 = G_1(y_2, x_3). \tag{5}$$

Concatenation of the above values (y_0, y_1, y_2, y_3) yields a 32-bit value for round f function.

3 *FEAL4*'s Differential Cryptanalysis

DC was developed by Biham and Szamir in 1990 [21]. It is directed for symmetric block ciphers attacks. Next to *LC*, it is used as a basic tool for finding loopholes in modern encryption algorithms and cryptographics systems [20]. *DC* compares pairs of ciphertexts, generated from encryption process of plaintexts pairs which differ in a certain particular manner [21]. For the *FEAL4* cipher, the difference between plaintexts is determined using the simple xor operation. Both plaintexts are encrypted using the same key and the differences between them are analysed in subsequent cycles of the encryption algorithm [19]. It does not matter what texts are used, they can be generated randomly. They have to be related by a given difference. Unfortunately, according to a cipher, differential cryptanalysis requires large collection of chosen plaintext pairs to be effective.

Encryption algorithms contain some non-linear elements. In the described cipher case, it refers to round function f. It is not possible to find any formula or pattern which would predict the next value of that function. It is worth pointing out, that a function of this type should not be able to generate any pseudo-random values, because it would be irreproducible, which means that it would not be possible to decipher generated ciphertext.

Every difference has a certain probability, which determines how often the f function will return expected value. This type of differences will be termed as characteristics. Determining the probability for each difference entails the generation of whole tables of characteristics. It can be used to determine the most probable subkeys [21].

In the presented *NEA* attack only one characteristic is used, but with high probability. This refers to any pair of input plaintexts with a difference of *0x80800000*. For this value round function f always returns *0x02000000* [20]. A detailed explanation about high probability of this characteristic is described in [20] (Problem 28). It can be wrote as:

$$Y = Y_0 \oplus Y_1 = 0x80800000, \tag{6}$$

$$Z = f(Y_0) \oplus f(Y_1) = 0x02000000, \tag{7}$$

where Y_0 and Y_1 are two blocks of data, which are passed to the round function f, Y is the difference between blocks Y_0 and Y_1, whereas Z constitutes the difference between the function values for each block.

4   Proposed *NGE* Attack

When the characteristic with an appropriately probability is found, a differential cryptanalysis process may be performed. Figure 3 depicts the encryption algorithm once again, but with differences between data blocks on every step of the cipher [20]. The plaintexts can be generated randomly, but with properly difference between them:

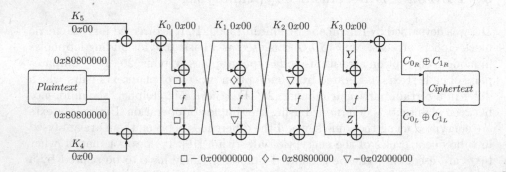

Fig. 3. *FEAL4*'s differential analysis.

The attack uses differences between blocks, that is why it is necessary to perform the xor on subkeys too. Their value will always be *0x00*. Predicting the result of each operation is possible up to the third round of the algorithm, when *0x02000000* is passed as input for the round function. It is not possible to clearly determine the value of the f for this difference. This problem can be overcome approaching from the end of the algorithm, analysing generated ciphertexts.

Analyzing pairs of obtained ciphertexts allows to determine:

$$Y_0 = C_{0_L} \oplus C_{0_R}, \tag{8}$$

$$Y_1 = C_{1_L} \oplus C_{1_R}, \tag{9}$$

$$Z = (C_{0_L} \oplus C_{1_L}) \oplus 0x02000000. \tag{10}$$

The first of the subkeys K_3, can be easily determined from the expressions presented above. Usage of a typical brute force algorithm can require checking of all 2^{32} possible subkeys, which constitutes approximately 4 294 967 296 combinations. In this situation EA can be used as a great optimization tool.

An individual, hereinafter referred to as a chromosome, represents one 32-bit subkey inside an entire population of potential solutions. Initially the population comprises N randomly generated subkeys. With subsequent iterations and the genetic operators, population will evaluate in order to improve the quality of solutions. The function, adjusted to the proposed cryptanalysis attack, is defined as follows:

$$F_f = \sum_{i=0}^{n} H((f(k \oplus Y_{0_i}) \oplus f(k \oplus Y_{1_i})), Z_i), \tag{11}$$

where H is the Hamming distance, f is the round function, k is the currently tested subkey under evaluation, n is the number of generated ciphertext and plaintext pairs. F_f determines the number of differences between the value obtained from the round function and the known Z difference determined on the basis of ciphertexts. The lower value of F_f corresponds to the higher-quality individual.

When the initial population is generated and all individuals are assessed, the evolution process begins. This process comprises three genetic operators, such as selection, crossover and mutation. In the proposed attack, to improving local searches, an additional heuristic negation operator was introduced. This operator is activated after crossover and mutation. The execution of all genetic operations leads to a new population of chromosomes, which is assessed anew. This process is repeated until a certain number of iterations is reached or a satisfactory result is obtained. Additionally, it was decided to save the best fit individual in a population.

In NEA attack, it was decided to use tournament selection, which aims to select two parents for the crossover process. Chromosomes are selected randomly for each tournament, out of which the leader becomes the first parent. The remaining individuals are returned to the population. This process is repeated in order to select next candidate for crossover.

A single point crossover operator is applied to selected parents. The crossover point is randomly choosen in the 1 to 31 range. The chromosomes are cut at that point and they exchange their genetic material between each other. The new offspring is subject to the mutation operator. It randomly selects one gene of the chromosome, which is xored with the corresponding bit from the second individual. The heuristic negation operator is activated last. It entails negating each bit of the chromosome, with a certain probability P_n, and remembers the most favourable variant.

A full cryptanalysis attack activates EA four times, in order to discover the K_0 – K_3 subkeys according to Fig. 3. The remaining two K_4 and K_5 subkeys are determined using the simple xor operation. The algorithm reduces the set of possible solutions and later, using detailed local search (heuristic negation operator) tries to find a valid subkey for each algorithm's round.

5   Experimental Results

All algorithms were implemented using the C++ programming language and were executed on a computer equipped with an Intel i7 processor clocked at 2.1 GHz. All plaintexts were generated randomly. The maximum number of iterations (generations) for all attacks was set to 70. The proposed NEA attack was compared to hill climbing (HC), a simple - without any additional operators - evolutionary algorithm (SEA) and the brute force (BF) approaches. When mentioned HC attack finds a better solution it goes to the next iteration and starts local searching again from the beginning of the subkey. For each compared algorithm a population consisted of 30 chromosomes. For NEA and SEA attacks the crossover probability P_c was set to 0.8, the mutation probability P_m to 0.02. The additional heuristic negation operator, used in SEA, has its own probability $P_n = 0.35$. The population leader was saved and was moved to the offspring population in each generation. For each run 30 pairs of plaintexts were generated. Also the number of checked keys have been counted for every experiment. Subkeys were generated randomly.

Table 1. F_f values for EA with additional heuristic negation operator (NEA)

ID	Minimum	Median	Average	Maximum	Standard deviation	Checked subkeys	Correct subkey	Iteration
1	0	120	169.63	375	131.12	28086	2837813671	53
2	0	169	190.37	436	146.46	21526	169015406	37
3	0	303	244.4	394	139.48	9756	2627704252	18
4	18	273.5	238.3	410	142.76	38294	1562400545	70
5	0	173	179.17	371	123.49	16290	4037332048	31
6	0	87	167.4	409	133.65	22256	596189050	40
7	0	123	163.47	411	124.06	30476	3773544206	58
8	0	122	177.3	413	143.54	17724	2262932401	34
9	0	132	186.97	404	141.54	14826	931539171	27
10	14	81	155.97	374	126.57	36630	4132119337	70

Table 1 presents results of *NEA* algorithm. It contains statistics with information about the valid decryption subkey, total number of checked keys during algorithm execution and iteration where the best chromosome was found. Presented experiments were selected randomly. 80 % of tested subkeys where successfully broken with *NEA* attack. In the 4th and 10th experiment we can see algorithm convergence. The attack was not able to leave the local optimum.

Table 2. F_f values for simple *EA* attack (*SEA*)

ID	Minimum	Median	Average	Maximum	Standard deviation	Checked subkeys	Correct subkey	Iteration
1	59	310	302.97	423	86.1	2100	2837813671	70
2	219	349	336.67	435	55.37	2100	169015406	70
3	80	294	284.63	407	95.88	2100	2627704252	70
4	32	318	301.57	426	109.23	2100	1562400545	70
5	159	328	301.77	445	104.88	2100	4037332048	70
6	113	351	315.57	425	106.05	2100	596189050	70
7	50	297	278.43	411	111.92	2100	3773544206	70
8	94	321	288.03	391	97.15	2100	2262932401	70
9	185	350	327.5	434	76.32	2100	931539171	70
10	41	343	309.8	411	102.939	2100	4132119337	70

Table 2 juxtaposes results of *SEA* attack. Simple *EA* did not find any valid subkey. In each experiment 2100 subkeys where checked. *SEA* algorithm does not work well without the additional heuristic negation operator.

Table 3. F_f values for hill climbing attack (*HC*)

ID	Minimum	Median	Average	Maximum	Standard deviation	Checked subkeys	Correct subkey	Iteration
1	67	148	182.23	315	72.76	120336	2837813671	70
2	99	194	213.7	362	83.78	121934	169015406	70
3	43	226	214.33	337	72.45	122622	2627704252	70
4	83	258	237.73	331	75.04	121618	1562400545	70
5	133	275	258.37	310	54.54	122564	4037332048	70
6	47	199	203.77	321	72.28	121926	596189050	70
7	20	197	202.2	321	86.8	121632	3773544206	70
8	85	219	219.3	384	87.46	122106	2262932401	70
9	104	183	204.87	344	67.76	121514	931539171	70
10	35	212	212.9	368	88.26	123158	4132119337	70

Results presented in Table 3 refer to HC algorithm. Similarly as it was in the SEA attack, the HC algorithm was not able to find any valid subkey. In each experiment it checked around 120000 subkeys. Values of the F_f are much better than in SEA algorithm but still inadequate to find the perfect subkey. The brute force attack (BF) always starts from 0 and checks all possibilities until it find a valid solution or reaches the last 2^{32}th subkey.

Comparing all presented algorithms and their results can be helpful to easily notice, the proposed NEA algorithm with additional heuristic negation operator was the most effective one. The total number of checked chromosomes was not higher than 40000 which is 0.001 % of all possible subkeys. High values of the standard deviation are caused by small set of plaintexts and ciphertexts pairs differences. 30 pairs were sufficient to find valid subkeys.

The proposed attack, in each algorithm's round, finds three different subkeys with value of F_f equal 0. Usage of all these subkeys allows to generate next three solutions for next rounds. Following this line the presented NEA algorithm is able to find a set of solutions which consisted of 81 possible combinations of subkeys. All of these combinations can be used to decrypt the captured ciphertext.

All implemented algorithms can be tested on an online web platform on the following website: http://heucrypt.azurewebsites.net/.

6   Conclusions and Future Work

This paper presents differential cryptanalysis attack extended by EA with an additional heuristic negation operator, directed to ciphertexts encrypted by the $FEAL4$ cipher. The main purpose of the exploiting algorithm is to find six 32-bit subkeys useful to decrypt a full encrypted message. The proposed attack was compared with brute force (BF), hill climbing (HC) and simple EA (SEA) - without any additional genetic operator or steps - approaches. The presented attack turned out to be the best from all algorithms. The NEA checks a much smaller number of possible subkeys in contrast to a simple brute force exhaustive attack or HC algorithms. Developed attack looks for one subkey per execution, so the avalanche effect, common in symmetric block ciphers, is eliminated. The other subkeys are obtained in next cipher cycles.

Also the proposed algorithm was able to find 81 combinations of subkeys for each ciphertext. Every obtained set of six subkeys was tested and can be used to decrypt the proper ciphertext.

The differential cryptanalysis evolutionary attack should be tested on more complex block ciphers like $FEAL8$, $FEAL-NX$ or the well known DES, where differential cryptanalysis processes are more advanced. Also we aim at designing any adaptive techniques used to modify EA parameters during execution. Other metaheuristic alternatives can be used in this optimization problem such as Particle Swarm Optimization (PSO) or Differential Evolution (DE). Also another computational learning technique like Machine Learning (ML) can provide some interesting results.

References

1. Kenan, K.: Cryptography in the Databases. The Last Line of Defense. Addison Wesley Publishing Company, New York (2005)
2. Stallings, W.: Cryptography and Network Security: Principles and Practice, 5th edn. Pearson, New York (2011)
3. Stinson, D.R.: Cryptography: Theory and Practice. CRC Press Inc., Boca Raton (1995)
4. Pieprzyk, J., Hardjono, T., Seberry, J.: Fundamentals of Computer Security. CRC Press Inc., Boca Raton (2003)
5. Boryczka, U., Dworak, K.: Genetic transformation techniques in cryptanalysis. In: Nguyen, N.T., Attachoo, B., Trawiński, B., Somboonviwat, K. (eds.) ACIIDS 2014, Part II. LNCS, vol. 8398, pp. 147–156. Springer, Heidelberg (2014)
6. Song, J., Zhang, H., Meng, Q., Wang, Z.: Cryptanalysis of four-round DES based on genetic algorithm. In: Proceedings of IEEE International Conference on Wireless Communications, Network and Mobile Computing, pp. 2326–2329. IEEE (2007)
7. Bhasin, H., Hameed, K.A.: Cryptanalysis using soft computing techniques. J. Comput. Sci. Appl. **3**(2), 52–55 (2015)
8. Garg, P., Varshney, S., Bhardwaj, M.: Cryptanalysis of simplified data encryption standard using genetic algorithm. Am. J. Net. Comm. **4**, 32–36 (2015)
9. Dworak, K., Boryczka, U.: Cryptanalysis of SDES using modified version of binary particle swarm optimization. In: Núñez, M., Nguyen, N.T., Camacho, D., Trawinski, B. (eds.) ICCCI 2015. LNCS, vol. 9330, pp. 159–168. Springer, Heidelberg (2015). doi:10.1007/978-3-319-24306-1_16
10. Abd-Elmonim, W.G., Ghali, N.I., Hassanien, A.E., Abraham, A.: Known-plaintext attack of DES-16 using particle swarm optimization. In: Proceedings of Third World Congress on Nature and Biologically Inspired Computing (NaBIC), pp. 12–16. IEEE (2011)
11. Russell, M., Clark, J.A., Stepney, S.: Using ants to attack a classical cipher. In: Cantú-Paz, E., et al. (eds.) GECCO 2003. LNCS, vol. 2723, pp. 146–147. Springer, Heidelberg (2003)
12. Mekhaznia, T., Menai, M.E.B.: Cryptanalysis of classical ciphers with ant algorithms. Int. J. Metaheuristics **3**, 175–198 (2014)
13. Garg, P.: Genetic algorithms, tabu search and simulated annealing: a comparasion between three approached for the cryptanalysis of transposition cipher. IJNSA **1**(1), 34–52 (2009)
14. Dewu, X., Wei, C.: A survey on cryptanalysis of block ciphers. In: 2010 International Conference on Computer Application and System Modeling (ICCASM), vol. 8, pp. 218–220 (2010)
15. Laskari, E.C., Meletiou, G.C., Stamatiou, Y.C., Vrahatis, M.N.: Evolutionary computation based cryptanalysis: a first study. Nonlinear Anal. **63**, e823–e830 (2005)
16. Michalewicz, Z.: Genetic Algorithms + Data Structures = Evolution Programs, 3rd edn. Springer, London (1996)
17. Goldberg, D.E.: Genetic Algorithms in Search. Optimization and Machine Learning. Addison-Wesley Longman Publishing, Boston (1989)
18. Shimizu, A., Miyaguchi, S.: Fast data encipherment algorithm FEAL. In: Price, W.L., Chaum, D. (eds.) EUROCRYPT 1987. LNCS, vol. 304, pp. 267–278. Springer, Heidelberg (1988)

19. Schneier, B.: Applied Cryptography: Protocols, Algorithms, and Source Code in C. Wiley, New York (1996)
20. Stamp, M., Low, R.M.: Applied Cryptanalysis. Breaking Ciphers in the Real World. Wiley, New York (2007)
21. Biham, E., Shamir, A.: Differential cryptanalysis of DES-like cryptosystems. J. Cryptology **4**(1), 3–72 (1991)

Differential Evolution in a Recommendation System Based on Collaborative Filtering

Urszula Boryczka and Michał Bałchanowski[(✉)]

Institute of Computer Science, Unversity of Silesia, Bedzinska. 39,
41-200 Sosnowiec, Poland
{urszula.boryczka,michal.balchanowski}@us.edu.pl

Abstract. Recommendation systems have become an integral part
of e-Commerce websites, as they facilitate the user's decision-making.
To improve the performance of these systems, new techniques are proposed. One of them is the use of the heuristic algorithm, which learn the
user's preferences and provide tailored suggestions. In this article the
application of the Differential Evolution algorithm (DE), with a view to
creating neighborhood in a Recommendation System, based on the collaborative filtering technique, will be presented. To this end a modified
Euclidean metric, which (taking into consideration additional weights
found by DE) generates the closest neighborhood for an active user, is
used. The results of the experiment are compared with the linear measure
of similarity Pearson's correlation.

Keywords: Recommendation systems · Collaborative filtering ·
Differential evolution

1 Introduction

Nowadays we experience a rising trend within the area of obtaining and analyzing information. Customers who browse an online shop website are not capable
of familiarizing themselves with all offered products, and new products are added
on a daily basis. Taking into consideration this congestion of information and
problems it breeds, new recommendation systems were proposed. Naturally, the
aim of these recommendations is not only suggesting products that the potential
customers might find interesting, but also increasing the sales in a given shop.
These recommendations have already been successfully introduced on such websites as: Amazon or eBay, and help users to take a decision [1].

However, together with the development of these systems and a will to
improve their quality, it became more and more clear that making recommendations exclusively on the basis of users votes made out for given items may prove
insufficient. It is related to the fact that other interrelations may occur between
system users and while generating neighborhood (selecting similar users) it is
also worth to take into consideration other attributes (for instance age, gender
or occupation). The problem is that for some users certain attributes may be of
a greater importance than for others.

© Springer International Publishing Switzerland 2016
N.T. Nguyen et al. (Eds.): ICCCI 2016, Part II, LNAI 9876, pp. 113–122, 2016.
DOI: 10.1007/978-3-319-45246-3_11

Therefore, our aim was to create a system that would define the similarity between users, not only on the basis of users votes made out for a given item, but also through the application of a modified measure of similarity that would additionally allow for the weights of particular attributes for a given user during the process of creating recommendations. To this end the differential evolution algorithm was implemented, the aim of which (within the process of searching for a solution on training data) was to adjust these weights for an active user of the system. Subsequently, these were used on a test set to create neighborhood and generate recommendations, and their results were compared with an approach using the Pearson's correlation coefficient. It is important to emphasize that an active user of the system is a user for whom the recommendations are generated and who in this article is defined as Active User A.

The use of heuristic algorithm in this paper is dictated by desire to improve the speed of finding the appropriate weights, which then can be used to generate a better recommendation for active user. This idea has been taken from [2,3] where Genetic Algorithm (GA) and Particle Swarm Optimization (PSO) has been used for this purpose. GA mimics the process of natural selection and was invented by John Holland in the 1960s [4] and PSO is algorithm that is inspired by the social behaviour of bird flocking [5]. In this article [6] we can find usage of Genetic and Memetic algorithms for building user based clustering models. Also heuristic graph-based recommendation systems have been tested in [7], where authors used Ant Colony Optimization (ACO) [8] for student courses recommendation.

The remaining part of the paper is organized as follows: the next section presents basic information about collaborative filtering. The third section shows how neighborhood can be found and presents types of similarity measures in Recommendation Systems. The fourth section provides elementary information about differential evolution (individuals, population and operators). The fifth section presents the fitness function and how it is calculated, whereas the sixth section shows how a recommendation process works. Section number seven presents experiments and results. The final, eight section concludes our research work and possible future work is discussed.

2   Collaborative Filtering

Recommendation systems usually produce recommendations throught collaborative or content-based filtering. Content-based filtering recommends items based on keywords or attributes used to describe items and collaborative filtering (user-based) is suggesting new items to a defined user, based on opinions of different users that are similar to him. The origin of this method is a simple observation which assumes that the probability that people who have similar tastes or preferences will take a liking to similar items is higher. Algorithms applied within this method can be divided into two main categories:

- memory-based - Within this approach an entire set of users, who have a similar record of votes, is defined on the entire data set, by means of statistical techniques. Subsequently, after creating neighborhood for an active user, preferences of neighbors are connected, with a view to generating recommendations,

- model-based - Within this approach, on the basis of data exploration techniques and machine learning, an appropriate pattern is created on training data, which, afterward, is conveyed onto real data, with a view to defining a proper prediction. Majority of models are based on classification or clustering [9].

Collaborative filtering algorithms usually describe the relationship between the user and the item as a vote matrix of $m \times n$ dimensions, in which every row corresponds with user's vote for a given item. Apart from the very votes, it can include additional information, both about the user and the items he rated (Table 1).

Table 1. Utility matrix user/item completed with additional attributes for the users.

Additional user features	User/item	Item 1	Item 2	Item 3	Item 4
gender: M, age: 24	user 1	3		4	3
gender: F, age: 17	user 2		5		3
gender: M, age: 44	user 3	2	3	2	2
gender: F, age: 54	user 4		2	1	

3 Neighborhood and Similarity Measure in Recommendation Systems

One of the crucial elements of collaborative filtering is finding "neighborhood", so a group of like-minded users. To this end very often various statistical methods are applied, which define the degree of similarity between given users. The elementary similarity measure, used within this technique, is Pearson's correlation coefficient, which defines the level of linear relationship between random variables. In our case we will analyze the interrelations between users and therefore we will apply the following formula:

$$w_{u,v} = \frac{\sum_{i \in I}(r_{u,i} - \bar{r}_u)(r_{v,i} - \bar{r}_v)}{\sqrt{\sum_{i \in I}(r_{u,i} - \bar{r}_u)^2}\sqrt{\sum_{i \in I}(r_{v,i} - \bar{r}_v)^2}}, \tag{1}$$

where $r_{u,i}$ denotes the rate i-of this item by the user u, and $\bar{r}_u$ is the mean for collaboratively rated items by the user u.

Subsequently, on the basis of the calculated correlation, we can define the size of neighborhood, so a certain subset of users, who are most similar to the active user. It is possible to apply a value threshold here, above which users have to be to get included in the neighborhood.

As part of the next stage predictions about particular items i are calculated for active user A. To this end weighted sum of all votes for a given item is calculated, according to the formula:

$$P_{A,i} = \bar{r}_A + \frac{\sum_{u \in U} (r_{u,i} - \bar{r}_u) w_{A,u}}{\sum_{u \in U} |w_{A,u}|}, \tag{2}$$

where $r_{u,i}$ is the vote of user u for item i, $\bar{r}_A$ and $\bar{r}_u$ are the vote mean for user A and user u from collaboratively rated items. $w_{A,u}$ is the correlation coefficient between user A and u.

A different means of defining similarity between users is calculating the Euclidean distance, according to the following formula:

$$Euclidean(A, j) = \sqrt{\sum_{i=1}^{n} (x_{A,i} - x_{j,i})^2}, \tag{3}$$

where A is an active user for whom we generate recommendations, j defines the user who has the same items as user A, $x_{A,i}$, $x_{j,i}$ are values (votes) for a particular item made out by user A and user j, i is the same item users A and j have rated, while n defines the number of the same products the users have.

However, we can also add additional features to this computation. The Weighted Euclidean distance used in this paper includes additional user features f and their weights w_f.

$$WeightedEuclidean(A, j) = \sqrt{\sum_{i=1}^{n} \sum_{f=1}^{z} w_f (v_{A,i,f} - v_{j,i,f})^2}, \tag{4}$$

where A is an active user for whom we generate recommendations, j defines the user who has the same items as user A, n defines the number of items both users have, z is a total number of features, w_f is the weight of feature f for user A and $v_{A,i,f}$ is the value of feature f on the item i for user A. Before this calculation is done the profiles values need to be normalized and lie between 0 and 1. Also the sum of all weights for z features should be equal to 1.

Before calculating the correlation between the two users it is also crucial to pay attention to the number of shared items that are taken into consideration while calculating it. It is connected with the fact that, e.g. if two users rated the same two movies the same way and it is the only two movie they both rated, then such correlation in Pearson's measure will equal 1. Therefore, it is a very strong correlation between these users, which, however, can prove untrue and very likely lead to poor recommendations. In order to prevent similar situations (lessen the probability of their occurrence), the MCI (Min Common Item) parameter was introduced, which defines the minimum number of items both users rated, which, in turn, is needed to calculate the correlation. Users who do not have the required number of shared items with an active user are, therefore, excluded while creating neighborhood.

4 Differential Evolution

Differential Evolution is an evolution technique introduced by K. Price and R. Storn [10]. It was prepared, first and foremost, for the optimization of continuous functions, therefore, individual is a certain vector of real numbers. The number of elements in the vector equals the number of dimensions, which is, at the same time, the potential answer. The individual can be presented in the following way:

$$x_i = \{x_{i,1}, x_{i,2}, ..., x_{i,j}\}, j = 1, 2, ..., g, \tag{5}$$

where j is the number of the independent variable of the vector x_i and g is the number of dimensions.

Similarly as in the evolutionary algorithms, at the beginning we initialize a certain initial population, which, after some time, owing to the application of operators, is changed. The population comprises NP (number of individuals in population) vectors x_i and can be defined in the following way [11]:

$$S = \{x_1, x_2, ..., x_m\}, m = 1, 2, ..., NP, \tag{6}$$

4.1   Operators in DE Approach

In the Differential Evolution algorithm we distinguish two basic operators:

- Mutation - For every individual from the S population, a new individual is created and added to the mating pool V. Mutation is the main operator here and is conducted prior to crossover. Creating a new individual can be expressed by the means of the following formula:

$$v_i = x_{r1} + F(x_{r2} - x_{r3}), \tag{7}$$

where $r1, r2, r3$ are three randomly generated identifiers of individuals from the S population, and parameter F is an amplification factor, where $0 \le F \le 1$. This mutation strategy is called DE/rand/1/z.
- Crossover - Then using the crossover operator a new u_i individual is created, which is created through the connection of the genotype of parent x_i from population S, and an individual created as a result of the application of the mutation operator v_i from population V. What is also applied here is the CR parameter, which is introduced by the system user and is responsible for the crossover probability, which is a number generated form the $[0, 1)$ open interval. $Rand(j)$ is a generated random number also from the $[0, 1)$ interval. This process can be expressed according to the following formula:

$$u_{i,j} = \begin{cases} v_{i,j} & if(rand(j) \le CR) \\ x_{i,j} & otherwise \end{cases} \tag{8}$$

Then the value of adjusting every individual x_i from the parents' population S is compared with the created individual u_i from the offspring population U. If the adjustment of the individual u_i is higher than x_i, u_i substitutes individual x_i in the population, otherwise individual u_i is rejected.

5   Fitness Function

Fitness function will define how "good" is a given individual, selected by the means of the DE technique. It is essential to be able to manage the evolution process. In order to calculate it, we have to analyze to what extent the votes of the recommended items differ from these really chosen by a given user, therefore, we apply here the training set, where we will have such a possibility. It is also important to bear in mind that for every new set of weights (for every individual), a new neighborhood is generated, using the entire training set. Owing to that it is not a trivial process and it consumes relatively much time and processing power.

Subsequently, with a view to calculating the predicted vote of active user A for item i, on the basis of the generated neighborhood, we have to apply the following formula, suggested in [12]:

$$PredictVote(A, i) = \bar{V}_a + k \sum_{j=1}^{n} Euclidean(a, j)(v_{j,i} - \bar{v}_j), \qquad (9)$$

where v_a is the mean vote for active user A, $\bar{v}_j$ is the mean vote for active user j, $v_{j,i}$ is actual vote for user j on item i, k is a normalizing factor such that the sum of the Euclidean distances is equal to 1 and n is the size of the neighborhood.

Having the votes predicted by the algorithm and actual votes, which the user made out for a given item, we can calculate the fitness function, which will be the mean squared error RMSE, according to the following formula:

$$RMSE = \sqrt{\frac{1}{\theta} \sum_{(u,i) \in \theta} (\widehat{R_{u,i}} - R_{u,i})^2}, \qquad (10)$$

where θ is a set of pairs of users and items for which predictions were made, $\widehat{R_{u,i}}$ and $R_{u,i}$ are respectively the predicted and actual value of the item vote i for user u.

6   Recommendation Process

To achieve an appropriate recommendations for a given active user A, the method of collaborative filtering is applied, the aim of which is to build profiles and find the nearest neighborhood. In order to initiate the process of recommendation, first of all, we have to build up profiles for every user in the system. A profile will be defined as a set of all movies i, which were rated by a given user j.

The next step consists in defining neighborhood for an active user. It is a very important stage of creating recommendations, since the degree to which the user will like the recommended items depends on the quality of the created neighborhood. In order to do that, we have to, by means of a chosen similarity measure, define the correlation between system users and an active user. It can be achieved through computing Pearson's correlation. However, it takes into account exclusively information about voting by users for a given item, completely omitting additional interrelations that can occur between them. In order to improve the quality of the generated recommendations it was decided to make use of a modified Euclidean distance, expressed by the means of formula (4), which additionally takes into account different features of users, such as age or gender, while defining correlation. However, when making use of such an approach we have to define with what degree of importance will a given feature be taken into consideration. To this end weights w_f, which will define it, have to be indicated. It is not a trivial task and hence the differential evolution technique is applied. An individual comprises weights w_f for all features f in the system as shown in Fig. 1.

Fig. 1. Individual in differential evolution as set of weights.

After defining all correlations between active user A and other users j of the system, the nearest neighbors have to be selected. It can be achieved using the kNN classifier and assuming a certain amount of nearest users, the aim of which will be to narrow down the neighborhood. The purpose of it is to improve the quality of generated recommendations.

Subsequently, on the basis of the generated neighborhood, generating recommendations and predicted votes of active user A for item i takes place, according to the formula (9). Having the predicted vote and the actual vote from the training set, we can compare the quality of these recommendations, using formula (10). Owing to that, we are able to calculate the fitness function, which is indispensible to rate the individual in population, described in Sect. 5. The aim of it is to fine tune weights w_f in such a way so as to obtain the best quality of recommendations.

During the next step we select the individual with the highest fitness function (set of weights) created in the process of generating neighborhood on the training set. Then these weights are used to generate recommendations on a test set, the quality of which is also defined by the means of formula (10). The obtained results are compared with the algorithm generating recommendations, on the basis of neighborhood defined with the help of Pearson's correlation coefficient, according to formulas (1) and (2).

7   Experiments and Results

The experiments were made possible owing to a database downloaded from http://grouplens.org/ [13]. Every user in this database rated at least 20 movies and the incomplete data was removed. Our Recommendation System called "DE Recommender" uses 22 features from this data set: movie rating, age, gender, occupation and 18 movie genre frequencies: action, adventure, animation, children, comedy, crime, documentary, drama, fantasy, film-noir, horror, musical, mystery, romance, sci-fi, thriller, war, western.

Then the entire data set is used as training set, on which the differential evolution technique defines the appropriate weights for an active user. Movie items that

Table 2. Parameters of the differential evolution algorithm

Population	50
Number of Iterations	200
Crossover Probability	0.9
Amplification Factor F	0.6

Table 3. Results achieved by DE Recommender with different neighbourhood sizes

ExperimentId	UserId	MCI	NS	Paerson	Euclidean	WeightedEuclidean
1	1	12	10	23 %	23 %	**27 %**
2	2	5	10	**42 %**	34 %	38 %
3	3	9	10	41 %	41 %	**55 %**
4	4	10	10	**38 %**	29 %	36 %
5	5	4	10	23 %	37 %	**47 %**
6	6	12	10	23 %	19 %	**24 %**
7	7	13	10	29 %	35 %	**36 %**
8	8	5	10	49 %	51 %	**52 %**
9	1	12	25	27 %	25 %	**28 %**
10	2	5	25	34 %	43 %	**45 %**
11	3	9	25	45 %	47 %	**56 %**
12	4	10	25	29 %	30 %	**32 %**
13	5	4	25	40 %	42 %	**43 %**
14	6	12	25	24 %	22 %	**25 %**
15	7	13	25	35 %	33 %	**40 %**
16	8	5	25	53 %	49 %	**60 %**
17	1	12	50	26 %	**28 %**	27 %
18	2	5	50	40 %	42 %	**43 %**
19	3	9	50	50 %	50 %	**55 %**
20	4	10	50	29 %	**31 %**	28 %
21	5	4	50	35 %	**37 %**	30 %
22	6	12	50	21 %	27 %	**28 %**
23	7	13	50	41 %	37 %	**42 %**
24	8	5	50	51 %	51 %	51 %

the active user has seen are split into two datasets: a training set (1/3) and a test set (2/3).

Three different methods of comparing similarity between items were applied. As part of the first one the Pearson Coefficient was used, in the second the Euclidean distance and finally in the third one the weighted Euclidean distance. The results were presented in the form of tables, included below. Three types of experiments were conducted with a view to comparing the quality of the generated recommendations. During the experiments eight users, who rated at least 20 movies, were selected and then recommendations were generated for them by choosing 10, 25 and 50 best correlated users. MCI parameter described in Sect. 3 is set to 20 % of all movie items rated by active user. Differential evolution algorithm parameters are shown in Table 2.

In Table 3 UserId is the Id of an active user, MCI is the minimum number of common movies between an active user and the potential neighbor (other user), NS is the size of neighbourhood. Paerson, Euclidean and WeightedEuclidean is the percentage of right prediction for Pearson correlation, Euclidean distance and Weighted Euclidean distance respectively between the predicted recommendations generated by our system and the actual recommendation from a test set for a specific item (movie).

Analyzing the results presented in Table 3, it can be easily noticed that for the neighbourhood of size 25, all the weighted Euclidean measure rendered better results, both for its non-modified version and the Pearson Coefficient. After decreasing the size of neighbourhood, this measure did not achieve similarly good results. Increasing the size of neghbourhood to 50, which, however, did not improve the quality of the generated recommendations by the modified Euclidean distance (Fig. 2).

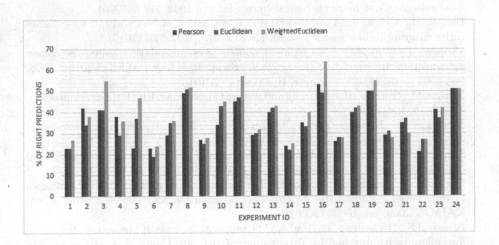

Fig. 2. Results achieved by DE Recommender

8   Conclusion and Future Work

The experiments and their analysis conducted in the previous point show that it is possible to successfully apply the differential evolution technique in recommendation systems based on collaborative filtering of users. However, particular attention has to be given to the size of neighbourhood, on the basis of which the list of recommendation items is created. The experiments show that for particular sizes, a decline of the quality of generated recommendations using the modified correlation measure, in compare with other measures, is likely to occur. Owing to that, in the future we will conduct a more precise analysis of this type of cases. We will also increse set of users (statistical sample), and we will take a better look into which weights are more important then others in a proper neighborhood creation. Besides that we will also aim at improving the quality of the recommendations proposed by our system.

References

1. Schafer, B.J., Konstan, J., Riedl, J.: Recommender systems in E-Commerce. In: Proceedings of the 1st ACM Conference on Electronic Commerce, pp. 158–166 (1999)
2. Ujjin, S., Bentley, P.J.: Learning user preferences using evolution. In: 4th Asia-Pacific Conference on Simulated Evolution and Learning, Singapore (2002)
3. Ujjin, S., Bentley, P.J.: Particle swarm optimization recommender system. In: IEEE International Conference on Evolutionary Computation, pp. 124–131 (2003)
4. Eiben, A.E., Smith, J.E.: Introduction to Evolutionary Computing. Springer, Berlin (2003)
5. Kennedy, J., Eberhart, R.: Particle swarm optimization. In: Proceedings of IEEE International Conference on Neural Networks, pp. 1942–1948 (1995)
6. Banati, H., Mehta, S.: A multi-perspective evaluation of MA and GA for collaborative filtering recommender system. IJCSIT **2**(5), 102–122 (2010)
7. Sobecki, J., Tomczak, J.M.: Student courses recommendation using ant colony optimization. In: Nguyen, N.T., Le, M.T., Świątek, J. (eds.) ACIIDS 2010. LNCS, vol. 5991, pp. 124–133. Springer, Heidelberg (2010)
8. Dorigo, M.: Optimization learning and natural algorithms. Ph.D. thesis, Politecnico di Milano, Italy (1992)
9. Xiaoyuan, S., Khoshgoftaar, T.: A survey of collaborative filtering techniques. Advances in Artificial Intelligence archive (2009)
10. Storn, R., Price, K.: Differential evolution a simple and efficient heuristic for global optimization over continuous spaces. J. Global Optim. **11**, 341–359 (1997)
11. Boryczka, U., Juszczuk, P., Kosowicz, L.: A comparative study of various strategies in differential evolution. In: Evolutionary Computing and Global Optimization, KAEiOG 2009, pp. 19–26 (2009)
12. Breese, J.S., Heckerman, D., Kadie, C.: Empirical analysis of predictive algorithms for collaborative filtering. In: Proceedings of the 14th Conference on Uncertainty in AI, pp. 43–52 (1998)
13. Maxwell Harper, F., Konstan, J.A.: The MovieLens datasets: history and context. ACM Trans. Interact. Intell. Syst. (TiiS) **5**(4), 19 (2015). Article 19

Using Genetic Algorithm to Aesthetic Patterns Design

Grzegorz T. Machnik, Miłosław Chodacki$^{(\boxtimes)}$, and Wiesław Kotarski

Institute of Computer Science, Silesian University in Katowice, Katowice, Poland
{grzegorz.machnik,miloslaw.chodacki,wieslaw.kotarski}@us.edu.pl

Abstract. This article presents possibilities of using a genetic algorithm as a method of artificial intelligence which is able to generate fractal structures with value of beauty. Fractal structure can be used as a utility model. The main problem is to define measure of beauty parameter. Using aesthetic measure algorithm can automatically, without human interaction, create nice looking various fractal structures.

1 Introduction

In this paper we present some usage of genetic algorithm to design aesthetic patterns. At the beginning of the twentieth century, through the Art Deco mainstream, designers tried to combine functionality and visual arts to increase aesthetics feeling of objects [1]. Nowadays consumers are looking for objects that not only fulfill the required technical characteristics but also present an extraordinary attention to details and unique design. That direction in software engineering is known as User Experience [2]. The aim – nice graphical patterns – can be obtained using e.g. fractal structures (Fig. 1).

Fractals are self–similar objects that have non–integer dimensions, fine and detailed structures and are generated by simple recurrence formulas. Fractals and orbits of dynamical systems are interesting in visualization meaning and in this paper they are objectives of research.

The problem is how to control genetic evolution to obtain beautiful objects. The first issue is connected with a possibility of beauty measurement and there is a question how to create fitness function that makes evaluation possible. What kind of geometrical parameters which describe fractal or dynamical system object should be used to create appropriate evaluation? How can we establish a proper relations among the parameters to define right evaluation? A lot of designers use a few kind of symmetry, golden ratio and other geometrical parameters to measure aesthetic value of objects [3,4].

Also the evaluation of beauty must satisfy a human sense of aesthetics. This creates another problem related to human perception and the well know Latin sentence "De gustibus non est dispudanum" (There is no accounting for taste). To solve these problems we applied an approach based on statistics. Using questionnaire techniques and analyze answers of a group of responding people we could extract an important geometrical features from presented objects and establish their relations to obtain a proper beauty evaluation.

N.T. Nguyen et al. (Eds.): ICCCI 2016, Part II, LNAI 9876, pp. 123–132, 2016.
DOI: 10.1007/978-3-319-45246-3_12

Fig. 1. Fractals on tiles, napkin and cup as decorative theme.

Earlier, in [5] a genetic algorithm was used in creation of artistic fractal images by ArtiE-Fract software. ArtiE-Fract is not a fully automatic computer program but it is designed to a human-computer interaction that makes possible to control evolutionary process depending on human evaluation. Fractals were also used for example to design beautiful jewelry [6] and on the other hands to create fractal antennas for mobile devices [7] and in many other kinds of applications.

The fitness function controls genetic evolution. Generally, it should increase with the number of generations. But its local fluctuations can occur. To measure aesthetics we propose to extract from fractal objects some features, associated to beauty evaluation, like consistency, egdes shapes and fractal dimension.

The paper is organized as follows. In Sect. 2 basic information on dynamical systems are presented. In Sect. 3 we show a regressive model of an aesthetics evaluation, and in Sect. 4 the genetic algorithm that uses the fitness function introduced in Sect. 3 is described. Next, in Sect. 5 some results of genetic searching for an aesthetic object are presented, and finally Sect. 6 concludes the paper.

2   Dynamical System

In this section we focus on dynamic systems. We start from some definitions:

Definition 1. *A set X is a metric space (X, d) with a distance function $d : X \times X \to \mathbb{R}$ that satisfies well-known properties for metric function.*

Definition 2. *A dynamical system is a transformation $t : X \to X$, where $t = \underbrace{t \circ \cdots \circ t(x)}_{n \ times} = t^{\circ n}(x)$.*

Definition 3. *Let $t : X \to X$ is the dynamical system on X. An orbit of a point $x \in X$ is a sequence $\{x_n\}_{n=0}^{\infty} = \{x_0, t(x_0), t^2(x_0), \dots\}$. For $n > 0$ a value of x_n can be written as $x_n = t(x_{n-1}) = t^n(x_0)$.*

Not all of dynamical system orbits have interesting geometrical shapes from aesthetics point of view. We focus on transformations which produce nontrivial and aesthetics patterns. Some nice looking visualizations of dynamical systems one can find in [8]. In this paper we use Euclidean metric space $X = \mathbb{R}^2$ and focus on transformations listed bellow:

1. Gumowski-Mira [9]:

$$x_n = y_{n-1} + \alpha(1 - 0.05y_{n-1}^2)y_{n-1} + f(x_{n-1}),$$
$$y_n = -x_{n-1} + f(x_n),$$
(1)

where

$$f(x) = \mu x + \frac{2(1 - \mu)x^2}{1 + x^2} \text{ and } \mu, \alpha \in \mathbb{R} .$$
(2)

2. Hopalong [10]:

$$x_n = y_{n-1} - sgn(x)\sqrt{|bx_n - c|},$$
$$y_n = a - x_{n-1},$$
(3)

where $a, b, c \in \mathbb{R}$ and $sgn(x)$ is defined as follows:

$$sgn(x) = \begin{cases} -1 & \text{if } x < 0 \\ 0 & \text{if } x = 0 \\ 1 & \text{if } x > 0 . \end{cases}$$
(4)

3. Quadrup Two transformation [11]:

$$x_n = y_{n-1} - sgn(x_{n-1})\sin(\ln(|bx_{n-1} - c|))\arctan(cx_{n-1} - b)^2,$$
$$y_n = a - x_{n-1},$$
(5)

where $a, b, c \in \mathbb{R}$.

Some visualization examples of orbits described above are shown in Fig. 2. We did research on many kinds of function (2) in Gumowski-Mira system [12].

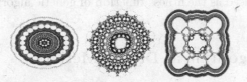

Fig. 2. The examples of orbits (from the left): Gumowski-Mira, Hopalong, Quadrup Two.

Proper modifications of formula's parameters can lead to a variety of results. Performed experiments showed that random choice of variables' values leads to chaotic and generally non–interesting structures. Non-interesting means that patterns are similar to regular scattered grains of sand in a given space without any recognizable structure or points leading to small individual areas. Finding proper variables' values without using any heuristic algorithm can be time-consuming. Further it is not specified which pattern will be acceptable. And at the end it is impossible to present the best solution. Also the problem can have multiple correct solutions. The solution to the problem should create base of variety interesting fractal structures. The final decision on the selection will still belong to the user. But the algorithm excludes non-interesting and uncomplicated in comparison to the others structures.

3   Model and Evaluation Criteria

The process of aesthetic design creation is a very complicated one, especially due to the difficulty of measuring of patterns taste. It can be described as some optimization problem in which one is looking for such $x^* \in V(p)$ that maximizes the following formula:

$$f(x^*) = \max_{x \in V(p)} f(x), \tag{6}$$

where $V(p)$ is a multidimensional vector of parameters.

The evaluation function (fitness function in evolutionary domain) that determines the effect of parameters related to the measure of aesthetics was initially constructed on the base of a survey conducted among 100 person group of respondents. The survey contained 30 patterns with parameters such as: symmetry, consistency and complexity (Fig. 3). Each of the respondents described a subjective assessment by taking into account the above aesthetic parameters. The WEB survey was designed by G. T. Machnik [13]. The results were subjected to statistical analysis determined the impact of various parameters on the value of the assessment. The obtained multidimensional regression function is presented in (7).

The statistical results showed that the fractal dimension of geometrical object is the most important evaluation parameter for respondents.

$$Fitness(x \in V(p)) = 117.21 + 124.99x_1 - 5.34x_2 - 1.95x_3, \tag{7}$$

where x_1 is a fractal dimension, x_2 is a symmetry and x_3 is edge smooth values. This function was used as the fitness function of genetic algorithm that we used in our research.

4   Genetic Algorithm as Design Method

Genetic algorithms are used to solve optimization problems, which cannot be described analytically or their description is too complicated or even unknown. Genetic algorithm is a heuristic approach, which uses a simplified imitation of the natural evolution. Although genetic algorithms are not well grounded in theory, their application in many fields of science and technology, as well as social and economic issues, indicate the validity of their use [14–16].

As already mentioned above, rating aesthetics of visual properties of geometric patterns is difficult to define precisely. Although one can use certain measures that are based on, among others: the golden ratio, symmetry, consistency and fractal dimension. Rating aesthetics is also dependent on the perception of the observer. Hence the justification for the use of genetic algorithms as a method of generating aesthetic designs, for genetic algorithms do not require the delivery of explicit knowledge on the problem being solved [14].

This algorithm also has some useful features, such as: the ability to deliver multiple point solutions, and so the lack of concentration of solutions around a certain class of geometric patterns. The algorithm mimics natural evolutionary

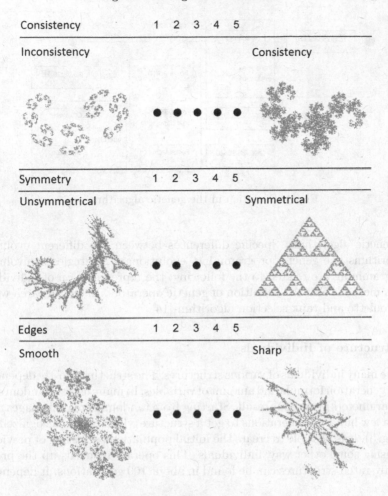

Fig. 3. A few examplary images of Survey.

processes, and therefore there exists the possibility of self-control calculations in such a way that a solution better adapted to a greater extent affects the entire population of solutions (Selective Pressure).

Genetic algorithm directs the search in the space of feasible solutions by environmental evaluation of the fitness function of each solution (Individual). The course of the algorithm is presented in Fig. 4.

In the initial stage of the genetic algorithm essentially random P_0 population base is created. The population is further assessed by the environment. Based on the adaptation of individuals their reproduction to temporary populations T_i is made. Then from T_i using genetic operators crossover and mutation with some probabilities the descendant population (Offsprings) O_i is created. Next evaluation of the newly established offspring population P_{i+1} takes place iteratively until the stopping criteria are fulfilled. This evolutionary process is common

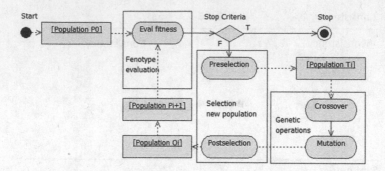

Fig. 4. Evolution in the genetic algorithm.

to all genetic algorithms. Specific differences between the different evolutionary algorithms like genetic programming, evolutionary strategies or evolutionary programming are related to the following: the representation of individuals, selection methods and the definition of genetic operators. In our research we use simple roulette and rank selection algorithm [14].

4.1   Structure of Individuals

There are many individuals of various structures. The structure mostly depends on the used generation formula and amount of variables. In many cases a random selection of parameters gives poor result. Starting from randomly created images needs at least a few hundred generations to get a structure which can be recognized by a man (Fig. 5). It is possible to create the initial population consisting of previously found, using some other way, individuals. This operation speeds up the process and highly-rated structures can be found in about 100 generations. It depends on

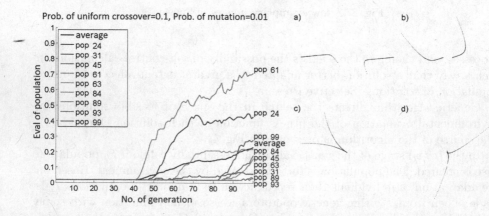

Fig. 5. Random initial population with low rate. Fitness value is (a) 0.46 (b) 0.42 (c) 0.38 (d) 0.34.

Fig. 6. From the top: Gumowski-Mira, Hopalong, QuadrupTwo system visualizations.

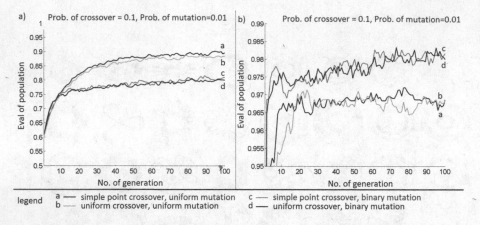

legend   a —— simple point crossover, uniform mutation   c —— simple point crossover, binary mutation
          b —— uniform crossover, uniform mutation   d —— uniform crossover, binary mutation

Fig. 7. Effects of Crossover and Mutation on the (a) population and (b) best individual.

generation formula whether 100 is enough amount to create a large variety of images. A diverse collection of fractal structures can be increased by reviewing the result of each generation, focusing not only on the final population.

4.2   Genetic Operators

There are basically two genetic operators: crossover and mutation. Every fractal structure such as Gumowski-Mira has a set of parameters that determine its visual structure. It is always the starting point (x, y) and, depending on the type of dynamic system, two, three or more parameters (a, b, etc. according to type of dynamic system: (1), (3) or (5)). Parameters are grouped in the binary chromosome and are modified using genetic operators. In crossover operation structures are matched in pairs. Next follows the exchange of binary form of parameters between selected individuals. Using this method algorithm creates two new structures (offsprings) from two individuals (parents). In the next step part of structure's parameters, which are selected using probability, are binary modified. Operator switches 0 value to 1 and opposite.

5   Results and Discussion

The experiment starts with population consisting of 100 individuals with low rating. Algorithm stops after 100 generations (stop condition). Probability of binary crossover was set at 0.1 and binary mutation – 0.01. Result of the experiment are presented in Fig. 6. The article presents part of the research devoted to the issue. Genetic algorithm works with binary and uniform crossover and mutation as well. Each run of the algorithm, started with the same structures, generated various result. In Fig. 7 effects of Crossover and Mutation on the fitness of whole population and the best individual are presented. As one can see uniform mutation with any kind of crossover significantly increases value of fitness function. Genetic algorithm can successfuly produce aesthetic fractals despite of the random population with individuals rated very low (Fig. 5).

6 Conclusions

It is difficult to define exact structure which is searched for. The advantage of the genetic algorithm is to modify fractal structure only on the basis of fitness function and it is not required to enrich algorithm with additional knowledge. The algorithm has been tested on more than 30 various fractal generation formulas. For every tested formula, end rating of individuals was higher than with initial structures. The algorithm can be successfully extended to additional formulas.

We observed a limitation of the algorithm used. Namely, the experiments showed that the genetic algorithm finds automatically nicely looking and well-evaluated fractal patterns rather among similar members of a class of shapes determined by a specific dynamic system. That effect following [17] can be explained by the usage of fixed parametric representation of the problem considered in the paper. To enrich the variety of generated patterns we plan to mix different dynamic systems which is equivalent to the next stage - extensible parametric representation [17] of our problem. Also in future we want to find automatically the formulas of dynamical systems producing nicely looking patterns. The both mentioned improvements of the genetic algorithm may help to solve the sameness and innovation problem of generated fractal patterns.

References

1. Bayer, P.: Art Deco Architecture: Design, Decoration and Detail from Twenties and Thirties, 1st edn. Harry N. Abrams, New York (1992)
2. Laurel, B., Mountford, J.S.: The Art of Human-Computer Interface Design. Addison-Wesley Longman Publishing Co., Inc., Boston (1990)
3. Huntley, H.E.: The Divine Proportion. Dover, New York (1970)
4. Field, M.: Designer chaos. Comput. Aided Des. **33**, 349–365 (2001)
5. Lutton, E., Costa, A.: ArtiE-Fract, used by Anabela Costa, visual Artist. Computational Aesthetics in Graphics Visualization, and Imaging. The Eurographics Association (2012)
6. Wannarumon, S., Bohez, E.L.J.: A new aesthetic evolutionary approach for jewelry design. Comput. Aided Design Appl. **3**(1–4), 385–394 (2006)
7. Pourahmadazar, J., Nourinia, J., Shirzad, H.: Multiband ring fractal monopole antenna for mobile devices. IEEE Antennas Wirel. Propag. Lett. **9**, 863–866 (2010)
8. Gdawiec, K., Kotarski, W., Lisowska, A.: Automatic generation of aesthetic patterns with the use of dynamical systems. In: Bebis, G., et al. (eds.) ISVC 2011, Part II. LNCS, vol. 6939, pp. 691–700. Springer, Heidelberg (2011)
9. Gumowski, I., Mira, C.: Recurrences and Discrete Dynamic Systems. Springer, New York (1980)
10. Martin, B.: Graphic potential of recursive functions. In: Computers in Art: Design and Animation, pp. 109–129. Springer, New York (1989)
11. Peters, M.: HOP - Fractals in Motion. http://www.mpeters.de/mpeweb/hop/

12. Böhm, J.: Dynamic System on Various Platforms. An Excursion from Environment and Tourism to Strange Attractors, Austrian Center for Didactics of Computer Algebra and DUG (2012)
13. Machnik, G.T.: Google Docs Service as Statistical Research Automation Tool in Aspect of Fractal Patterns Perception. Internet in the Information Society, Computer Systems Architecture and Security, Dąbrowa Górnicza, pp. 151–158 (2013)
14. Goldberg, D.E.: Genetic Algorithms in Search, Optimization, and Machine Learning. Addison-Wesley Professional (1989)
15. Michalewicz, Z.: Genetic Algorithms + Data Structures = Evolution Programs. Springer (1996)
16. Ashlock, D., Jamieson, B.: Evolutionary exploration of complex fractals. In: Hingston, F.P., Barone, C.L., Michalewicz, Z. (eds.) Design by Evolution. Advances in Evolutionary Design. Natural Computing Series. Springer, Heidelberg (2008)
17. Galanter, P.: The problem with evolutionary art Is... In: Di Chio, C., Brabazon, A., Di Caro, G.A., Ebner, M., Farooq, M., Fink, A., Grahl, J., Greenfield, G., Machado, P., O'Neill, M., Tarantino, E., Urquhart, N. (eds.) EvoApplications 2010, Part II. LNCS, vol. 6025, pp. 321–330. Springer, Heidelberg (2010)

Ambient Networks

A Survey of ADAS Technologies for the Future Perspective of Sensor Fusion

Adam Ziebinski[1(✉)], Rafal Cupek[1], Hueseyin Erdogan[2], and Sonja Waechter[2]

[1] Institute of Informatics, Silesian University of Technology, Gliwice, Poland
{Adam.Ziebinski,Rafal.Cupek}@polsl.pl
[2] Conti Temic microelectronic GmbH, Ingolstadt, Germany
{Hueseyin.Erdogan,Sonja.Wachter}@continental-corporation.com

Abstract. Traffic has become more complex in recent years and therefore the expectations that are placed on automobiles have also risen sharply. Support for drivers and the protection of the occupants of vehicles and other persons involved in road traffic have become essential. Rapid technical developments and innovative advances in recent years have enabled the development of plenty of Advanced Driver Assistance Systems that are based on different working principles such as radar, lidar or camera techniques. Some systems only warn the drivers via a visual, audible or haptical signal of a danger. Other systems are used to actively engage in the control of a vehicle in emergency situations. Although technical development is already quite mature, there are still many development opportunities for improving road safety. The further development of current applications and the creation of new applications that are based on sensor fusion are essential for the future. A short summary of capabilities of ADAS systems and selected ADAS modules was presented in this paper. The review was selected toward the future perspective of sensors fusion applied on the autonomous mobile platform.

Keywords: ADAS · Radar sensor · Lidar sensor · Camera sensor · Sensor fusion

1 Introduction

The expectations that are placed on automobiles have changed continuously in recent years and have risen sharply. Nowadays, in particular, safety plays an enormous role, in addition to the performance and the comfort of a vehicle. The protection of a vehicle's occupants and other persons involved in road traffic has become essential. In today's traffic, a variety of manoeuvers have to be performed by a driver, which can easily lead to an excessive demand. These manoeuvers range from navigation, to driving and stabilizing a vehicle. With the expansion of infrastructures and ever-evolving technology, the technical possibilities in the field of automobile production have also improved continuously. To this end, vehicles are equipped with numerous electronic components that provide support for a driver. However, the application of these systems differs. Some systems only warn a driver via a visual, audible or haptical signal of a danger. Other systems are used to actively engage in the control of a vehicle in emergency situations, which is intended to avoid accidents that result from mistakes and carelessness in

© Springer International Publishing Switzerland 2016
N.T. Nguyen et al. (Eds.): ICCCI 2016, Part II, LNAI 9876, pp. 135–146, 2016.
DOI: 10.1007/978-3-319-45246-3_13

complex traffic. Furthermore, most of these systems contribute to more fuel-efficient and relaxing driving. These complex components can be divided according to their intended use as follows [1–5]:

- Human Machine Interfaces form the basis for communication between humans and machines. Passing information by screening them on a display is much more visual, and therefore, it provides a driver with a better understanding. For example, navigation tools are optimised by such visualisations. Furthermore, the outputs of a request about a vehicle's status are more transparent. Regulating the heating, ventilation and radio station setting is clear, and therefore, easier. For this reason, a driver is less distracted and can concentrate more on driving the car. These components mainly increase the level of comfort during driving while at the same time. They also increase security by reducing distractions from the road traffic [6].
- Safety Telematics Systems are solutions for processing information via telecommunications that enable a car-to-car or car-to-environment communication. These solutions can be grouped as systems for intelligent driving. They provide more comfortable and relaxed driving and help to handle confusing situations in traffic, for example, via navigation and the announcement of traffic jams and road closures. A driver can, therefore, focus more on the traffic situation and does not have to search for the right route. One application area is, for example, the avoidance of traffic jams or the use of navigation systems that take over or assist a driver in the task of navigation and target acquisition. Although at first sight this is a comfort-enhancing system, it also contributes to security because a driver can focus more on the management responsibilities by not having to navigate [7].
- Vehicle Surrounding Sensors are used to prevent accidents by monitoring a vehicle's environment. Through various sensors, a car's environment is checked for obstacles or other cars that are not always visible to a driver. The sensors warn a driver of a potential collision in different ways or work together with active sensors to actively support a driver in critical or dangerous situations. The known technical benefits that are realised through the use of these sensors are, for example, parking assistance or lane change assistance [6].
- Another group, Active Safety Sensors, actively support a driver in dangerous situations. They take control of a vehicle, for example via Electronic Stability Control or Brake Assist. These sensors extend the impact of the actions of a driver and therefore reduce the negative effects of a slow human response. Thus, traffic accidents can be reduced [8].
- Passive Safety Systems act immediately after an accident has already happened in order to reduce the consequences of the accident and prevent passengers and pedestrians from worse injuries. The most well-known example is the airbag. An example of pedestrian protection is a pressure sensor in the bumper that can reliably detect accidents involving people and activate safety systems in a vehicle's body to lessen the injuries [6].

Presented above, in-vehicle elements together with roadway sensors and vehicular communication create a new perspective for intelligent transportation systems [9]. However, full and comprehensive analysis of all components and technologies that leading the way in the realization of intelligent transportation systems is out of the scope of this paper. Instead the authors focus on selected application examples (in Sect. 2) and sensors used in contemporary Advanced Driver Assistance Systems (in Sect. 3). Presented level of details should be sufficient to compare the areas of application for different ADAS solutions and sensors. This analysis does not cover whole range of possible options but only shows selected representative examples of given technologies in order to find the proper coverage for the sensor fusion discussed in Sect. 4. The final conclusions are presented in Sect. 5.

2   Various Applications of Adas

The aim of Advanced Driver Assistance Systems (ADAS) is to reduce the consequences of an accident, to prevent traffic accidents and in the near future to facilitate fully autonomous driving. The development of driver assistance systems began with Anti-lock Braking System (ABS) introduced into a serial production in the late 70 s of the twentieth century. The comprehensive description of the evolution of Driver Assistance Systems as well as the expected development directions can be found in [10] The main steps on the development path in driver assistance systems can be classified as: proprioceptive sensors, exteroceptive sensors, and sensor networks. The proprioceptive sensors were able to detect and respond to danger situation by analysing the behaviour of the vehicle. The exteroceptive sensors like ultrasonic, radar, lidar, infrared and vision sensors are able to respond on an earlier stage and to predict possible dangers. Further improvements are expected by application of multisensory platforms and traffic sensor networks. The aim of this section is to make a survey of the capabilities of contemporary exteroceptive sensors and based on them ADAS systems.

ADAS do not act autonomously but provide additional information about the traffic situation to support a driver and assist him in implementing critical actions. The synchronization of a driver's actions and the information from the environment and furthermore the recognition of the current situation and the possible vehicle manoeuvers is essential for the efficient performance of the various applications of ADAS [10]. Some ADAS application examples are described below.

Blind Spot Detection (BSD) monitors the area next to a vehicle. The function of a Blind Spot Detection System is to warn a driver with a visual such as a sign in the side-view mirror [6] or with an audible signal, when there are objects in the blind spots. The aim of this system is to avoid potential accidents, especially during lane change manoeuvers in heavy traffic [11].

The Rear Cross Traffic Alert (RCTA) can help to avoid accidents when reversing out of a parking space, which can often lead to serious accidents with pedestrians or cyclists that involve personal injuries. For this function, the environment behind a vehicle is monitored and checked for objects. In of the event an object is detected in the reverse driving direction, a driver receives an audible and a visible warning [6].

The Intelligent Headlamp Control (IHC) regulates the lights of a vehicle automatically according to the environmental conditions. This application optimizes changes between full-beam and dipped headlights during night-time drives. Driving at night or through tunnels is therefore more comfortable and safer. Moreover, drivers in oncoming vehicles are no longer blinded by a vehicle's lights [6].

Another safety application is the Traffic Sign Assist (TSA). This system automatically recognises traffic signs (also the signs of different countries) and can process the information they contain. Therefore, a driver is able to receive important information such as the legal speed limits or the actual priority rules. As a result of providing such information, the traffic sign assist enables more relaxed and also safe driving [11].

The Lane Departure Warning (LDW) scans the sides of the road and detects when a vehicle is leaving the lane or the road. By controlling the steering movement, the system is able to evaluate whether the lane change is intentional. The system warns a driver that the lanes have been changed inadvertently with a visual or a haptical warning such as steering wheel vibrations. Traffic accidents that are caused by vehicles leaving the road or collisions with passing or parked cars can be reduced [6].

The Emergency Brake Assist (EBA) enhances driving safety via active braking support and by automatically braking in dangerous situations. Rear-end collisions can therefore be avoided entirely. Furthermore, the consequences of accidents are reduced because of the reduction in the impact speed and impact energy. The emergency brake assist is also a possible interface for pre-crash applications and restraint systems or pedestrian protection [11].

The Adaptive Cruise Control with Stop & Go functions (ACC + S&G) controls the distance to the vehicle in front, even in stop-and-go situations. It either warns a driver or actively slows a vehicle's speed if the relative distance becomes too small. This application supports a driver, particularly in congested traffic and tailback situations. One consequence is more comfortable and stress-free driving with the flow of traffic. Furthermore, safety is enhanced due to a pre-defined distance and warning when emergency braking is needed [11].

In addition to the above-mentioned many new applications [10] are being developed and optimized continuously to enhance the safety of passengers [12] and pedestrians or animals [13, 14] and also to provide more comfortable and economical driving.

3 Advanced Driver Assistance Systems

Advanced Driver Assistance Systems have become indispensable in today's vehicles. Due to the increasing need for mobility, traffic has become more and more complex and therefore has become a greater challenge for all road users. ADAS are essential in order to avoid accidents and any concomitant injuries or possible fatalities [15]. Furthermore, they provide solutions for comfortable, economical and intelligent driving. In the last few years, more and more different types of complex control units have been developed and integrated into vehicles [9, 16]. These systems differ in their operating principles and application areas. ADAS use surrounding sensors such as radar, infrared, video or ultrasound to monitor and analyses a vehicle's environment.

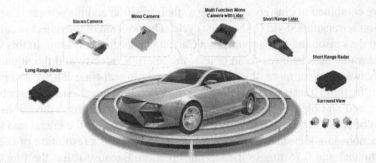

Fig. 1. ADAS and their vehicle position [6]

Various companies such as Bosch, Continental, Delphi Automotive, Freescale, Texas Instruments and many other suppliers provide different types of ADAS solutions for End-User Applications. They offer microcontrollers, microprocessors or complete sensors for ADAS. Freescale, for example, creates embedded processing solutions for the automotive industry and provides solutions with basic-rear-, smart-rear- and surround-view-cameras as well as a 77 GHz radar system [17]. With ADAS application processors, Texas Instruments enables Original-Equipment and Design Manufacturers a fully integrated mixed processor solution. The processors are highly integrated, programmable platforms whose benefits are the combination of high processing performance, low power usage, smaller footprint and a flexible and robust operating system. It enables embedded automobile technology by including a front camera, rear camera, surround view, radar and fusion on a single, heterogeneous and scalable architecture [18]. Other suppliers such as Continental or Bosch [19] provide complete sensors for ADAS that are also based, for example, on a camera or radar. Although the various ADAS solutions of the several suppliers differ in their construction, the main working principle, the mounting placement in the car and their application for ADAS are similar. Due to the usage of the equipment from Continental in AutoUniMo project [20], a major review was prepared based on examples of their ADAS products (Fig. 1) [6].

3.1 Radar Sensor

A Short Range Radar (SRR) detects objects and can measure their relative distance and velocity. This radar sensor is integrated at the four corners behind the bumpers on a car. Two sensors work for the forward detection of an object and two for the reverse detection. The sensor can be mounted in a vertical mounting window of 0.5 m up to 1.2 m from the ground. The sensor is based on Pulse Compression Modulation, which means that pulsed electromagnetic waves at the speed of light are emitted by the sensor. The operating frequency is 24 GHz. The emitted waves are reflected when they hit an object. The reflection is dependent on the material, size and shape of the object. Due to special antennas, the electromagnetic waves can be bundled in a particular direction. Thus, it is possible to determine the exact angular coordinates of the reflecting object. By measuring the time between the transmission and reflection, the distance and speed of the object can be detected in a field view from ± 75° up to 50 m [21]. These reflected radar signals are combined in clusters according to their position and movement. The cluster information is transmitted via CAN every cycle. The position of the object is calculated relative to the sensor and the output is in a Cartesian coordinate system. In this way the sensor measures the distances to an object, the relative velocity and the angular relationship between two or more objects simultaneously in real-time. The sensor provides two values for the distance and two values for the velocity. The relative distance between the sensor and an object is measured in the longitudinal and lateral directions with a resolution of up to 0.1 m. Additionally, the velocity is given for the lateral and longitudinal directions in a measurement range of up to 35 m/s and a resolution of 0.2 m/s in the longitudinal and 0.25 m/s in the lateral direction. Because it has the functions of object detection and distance measuring, the sensor is used for several safety applications such as Blind Spot Detection and Rear Cross Traffic Alert. Different sensor types for several measurement ranges are available for the radar systems. Long-range radar is designed for long-distance applications, for example, Emergency Brake Assist and Adaptive Cruise Control [21]. With an operating frequency of 77 GHz, the sensor has a measurement range of up to 250 m.

3.2 Multi Function Camera

The Multi-Function Camera (MFC) [6] is mounted in the vicinity of the rearview mirror behind the windscreen at a height above the ground of 1-1,45 m with a max. lateral shift of $\pm 0,1$ m. The optical path can be cleaned by wipers. The camera is available in a mono or stereo version. Due to the perspective differences between the two camera images, the stereo camera can detect objects in front of the car and measure their distance within a range of between 20-30 m. The redundancy of the second camera also provides a higher degree of reliability. The camera can furthermore detect whether and where an object is moving. A grayscale image is captured and processed using the camera. The camera operates with a monochrome CMOS image sensor. This provides 40 ms an image with a resolution of 640×496 pixels, a pixel size of $7,5 \times 7$ microns and a colour depth of 8 bits. A detection range of approximately 44° is achieved at a focal length of 6 mm. Plausibility algorithms and filters are used to gather further information on the environment. Visible edges in the video image, for example, can represent vehicle contours, a lane marking or headlights from other cars. A typical processing frequency is 12 Hz.

With these characteristics the camera can be used for different advanced driver assistance functions such as Intelligent Headlamp Control [21]. For this application, the camera provides information about the angle and distance to the next light point. The sensor is also used for the Lane Departure Warning and therefore measures the distance to the left and right edges of the roadway and the allowable distance before an alarm is triggered. This function is possible up to a road width of 2,5–4,6 m and at a distance of up to 7,5–90 m. Another application of the camera is the Traffic Sign Assist. The camera sensor provides information about the current speed limit, gives feedback about no-passing bans and warns when a street is entered in the wrong driving direction. All of the system outputs are transmitted via a CAN bus [6, 21].

3.3   Surround View Camera

ADAS are continually being developed in order to improve their efficiency and to increase their applications. The Multi-Function Camera that has already been described is a very useful sensor and is used in a large number of safety applications although its field of view is limited. An essential prerequisite for autonomous driving, however, is knowledge of all relevant information from the vehicle's entire environment. An innovative system that fulfils these requirements is the Surround View Camera [6]. A Surround View Camera provides a complete three-dimensional 360° view via a "fisheye" camera and also provides the possibility for augmented reality. Therefore, the surround view sensor can be used for several applications. One of these functions is Blind Spot detection, which enables cyclists and pedestrians to be easily tracked. The camera provides stress-free vehicle maneuverability. Parking by a driver as well as automated parking is especially easy and trouble free. Moreover, the camera ensures safer lane changes and overtaking.

Furthermore, the surround view camera is a very small sensor and can therefore be built in a very space-saving manner such as the ceiling of a car or on the side-view mirrors. Size is another challenge in development and production. Sensors have to be smaller, while still integrating new and complex technology that has to be of a high quality and reliable. The dimensions and tolerance deviations have become smaller, which requires ever more precise manufacturing [1, 6].

3.4   Lidar

A Short Range Lidar (SRL) sensor [6] is installed in the area of the rear-view mirror of a vehicle and is useful for a measurement range of between 1 and 10 m. A Lidar sensor is used for the emergency brake assist. The distance to objects in front of a vehicle, in proportion to the vehicle's speed, can be measured with an SRL sensor. The sensor can distinguish between obstacles for which a vehicle needs to stop such as pedestrians who are crossing a street and those where a vehicle need not stop such as rising emissions, road damage etc. These parameters, including the force of braking, differ for each customer. Infrared sensors operate by transmitting energy from a laser diode. The reflected energy is focused onto a detector consisting of an array of pixels. The measured data is then processed using various signal-processing methods. They can operate both night and day. Their disadvantages are their sensitivity to inclement weather conditions and ambient light. This sensor uses three independent channels. A field of view of 27° in horizontal and 11° in vertical direction is achieved via the three laser pulses [21]. By measuring the time between the emitted and received pulses, the relative distance to the reflecting object can be determined with an accuracy of $\pm$ 0.1 m. The desired results about the relative distance and speed are given in the CAN frames. One value for the distance and one value for the speed are measured by each of the three independent laser beams. Thus, the sensor provides three distance frames as well as three speed frames. All of the frames have the same structure. The measured data are communicated via a High-Speed-CAN bus [21].

3.5   Multi Function Lidar

A Multi-Function Lidar sensor [6] is a combination of an infrared and camera sensor. It can be used for the same applications as a camera and lidar sensor, e.g. Emergency Brake Assist, Lane Departure Warning, Intelligent Headlamp Control or for Traffic Sign Assist but it achieves this with a much smaller range. Multi-Function Lidar has a compact One Box Design with a colour CMOS imager of 1024 x 605 pixels and a 905 nm infrared laser. A combination of several different sensors is necessary to enable the maximum benefits and an efficient safety function. Space-saving sensors are also required by automobile manufacturers. New innovative systems show that this technical trend is currently being pursued. One step in the further development of ADAS is the integration of several different techniques into one sensor. The Multi-Function Lidar is a smart sensor that has the advantages of a lidar as well as those from a camera sensor in a space-saving manner [6, 11, 22].

4   Sensor Fusion

The different ADAS all have their advantages and disadvantages and can be used for different functions and in different application areas (Fig. 2). A camera, for example, of course has a better performance in object-detection. It has not just the possibility to detect an object but the fact that it can also provide specific information about what kind of object has been detected. A camera also has a better resolution in laterally and at elevations. However, a camera does not perform very well at night or in bad weather conditions. A safety sensor is supposed to work in all conditions. In such cases, radar and lidar perform much better. These two sensor types also have a reliable principle to measure the distance and velocity to an object. The camera is only able to measure distance in a stereo version. Each sensor is suitable for different applications and use cases. The aim of this section is to define how to join specified in Sect. 3 ADAS devices and take additional advantage by sensors fusion.

A good solution is the merging of data from multiple environment detecting sensors in order to improve the efficiency of environmental detection. One possibility is to use sensors that combine several operating principles such as the Multi-Function Lidar sensor, which was already introduced. This sensor combines the benefits of a camera and of an infrared sensor. A single working principle provides additional information. For example, due to the higher lateral resolution of the camera, the cross-location and cross-speed can be determined more precisely. The complementary information helps to determine whether an emergency manoeuver has to be carried out.

Another example is image-based approach for fast and robust vehicle tracking from a moving platform based on a stereo-vision camera system. Position, orientation, and full motion state, including velocity, acceleration, and yaw rate of a detected vehicle, are estimated from a tracked rigid 3-D point cloud. This point cloud represents a 3-D object model and is computed by analyzing image sequences in both space and time, i.e., by fusion of stereo vision and tracked image features [23, 24].

The risk of false triggering can be reduced by the possibility of a more accurate calculation. ADAS use information from a vehicle's environment. Individual solutions calculate information independently and are partially redundant. With a central environment modelling, synergy effects can be used thus resulting in an economically better technical feasibility. Moreover, the increased performance of the systems can be achieved via the fusion of the information from different environmental sensors [25] which could be also used for indoor location, identification of objects and navigation system [26, 27]. A basic question for the design process is what type of hardware and software architecture is required for such a sensor data fusion. As yet, no unified architecture has been determined for the automobile industry. The targeted development of a unified architecture and algorithms is desirable.

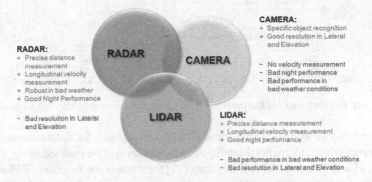

Fig. 2. Technological comparison of different advanced driver assistance systems

A fusion of several sensors with different working principles such as radar, infrared and camera are possible using an embedded system [28, 29]. For autonomous driving is required preparation a functional architecture which allow to control usage signals from all ADAS [30]. The use of multiple sensor systems to realize several active safety applications implies a number of other problems associated with merger signals [25]. In this case problem occurs with the assessment of reliability information provided by each of the sensors. While in the case of a single sensor the information about result is used, a merger between several may be overlapping and possibly with conflicting information requires additional output information form devices providing component information [25]. This problem causes, that in the case of a merger standard devices and algorithms must be developed with functionality to allowing for a reliability assessment of the information provided. Such solutions are realized in the AutoUniMo project. The project deals with such new challenges for intelligent driving. A concept for a mobile platform is being developed including the implementation of real-time CAN communication between several different ADAS in order to create autonomous and intelligent driving. The ADAS that are used in an actual vehicle application were chosen for this project. A part of described above sensors will be implemented on a mobile platform which could be used additionally in indoor environment. In order to enhance the number of advanced control units in a vehicle, the development of efficient solutions for the combination of these different units is essential. The system will be controlled using a Raspberry Pi (RPI), which will receive and process the measurement frames from the connected sensors. The RPI will be connected to the electronic sensors and ADAS

modules by several interfaces (I2C, SPI, UART, CAN). According to technical requirements ADAS sensors will be connected by four separate CAN buses. These four CAN
buses will allow two SRR on the front, two SRR on the back, SRL and MFC to be
connected. The requirements for the preparation of communication with the ADAS
modules using CAN buses are presented in Fig. 3.

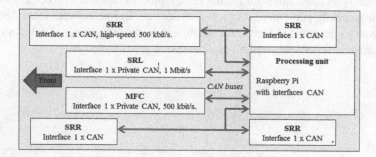

Fig. 3. The requirements for the preparation of communication with the ADAS modules

The capability of working in real-time is the essential basis for autonomous driving.
Fast data processing and artificial intelligence functions will be realised using a RPI or
partially in a Field Programmable Gate Array [31] for some of the functions. RPI solution
allows for fast prototyping. For RPI many common applications are available. They are
useful in rapid implementation of the new features: remote monitoring of the mobile
platform using WiFi; viewing, processing and analysis of measurements; communication with ADAS modules; etc. RPI solution on mobile platform with connected ADAS
modules will allow in the future to develop and implement new computational intelligence applications for ADAS and to find new solutions for sensor fusion.

5   Conclusions

Traffic has become more complex in recent years. Humans alone are sometimes no
longer able to cope with the traffic situation. Increased mobility and therefore an
increasing number of road users, confusing traffic and traffic sign signage, time pressure
and the desire for continuous availability, have made support for drivers a necessity.
Rapid technical developments and innovative advances in recent years have enabled the
development of Control Units to enhance the safety of passengers and pedestrians. Many
ADAS are on the market and are no longer a luxury item for upscale vehicles but are
now standard equipment in low-cost cars. These control units are based on different
working principles such as radar, lidar or camera techniques. All of these sensors have
their advantages and disadvantages, which means that the most efficient application is
using a combination of the different types of sensors. The technical developments are
already quite mature but there are still many development opportunities for improving
road safety. The further development and the creation of new applications are essential
for the future. The design of the hardware and software architecture of the sensor data
fusion is a determining factor. By the CAN interface, the Raspberry Pi can communicate

with ADAS modules. So, now it is possible to use standard ADAS solutions not only for normal vehicle but also in control mobile platform with use simple processing units.

Acknowledgements. This work was supported by the European Union through the FP7-PEOPLE-2013-IAPP AutoUniMo project "Automotive Production Engineering Unified Perspective based on Data Mining Methods and Virtual Factory Model" (grant agreement no: 612207) and research work financed from funds for science for years: 2016-2017 allocated to an international co-financed project.

References

1. Winner, H., Hakuli, S., Wolf, G.: Handbuch Fahrerassistenzsysteme. Spriger Vieweg, Wiesbaden (2012)
2. Brookhuis, K.A., de Waard, D., Janssen, W.H.: Behavioural impacts of Advanced Driver Assistance Systems–an overview. TNO Human Factors Soesterberg; The Netherlands
3. Piao, J., McDonald, M.: Advanced driver assistance systems from autonomous to cooperative approach. Trans. Rev. **28**, 659–684 (2008)
4. Schneider, J.H.: Modellierung und Erkennung von Fahrsituationen und Fahrmanövern für sicherheitsrelevante Fahrerassistenzsysteme. Fakultät für Elektrotechnik und Informationstechnik, TU Chemnitz (2009)
5. Bertozzi, M., Broggi, A., Carletti, M., Fascioli, A., Graf, T., Grisleri, P., Meinecke, M.: IR pedestrian detection for advanced driver assistance systems. In: Michaelis, B., Krell, G. (eds.) DAGM 2003. LNCS, vol. 2781, pp. 582–590. Springer, Heidelberg (2003)
6. Continental Automotive, 'Advanced Driver Assistance Systems'. http://www.continental-automotive.com/www/automotive_de_en/themes/passenger_cars/chassis_safety/adas/
7. Boodlal, L., Chiang, K.-H.: Study of the Impact of a Telematics System on Safe and Fuel-efficient Driving in Trucks. U.S. Department of Transportation Federal Motor Carrier Safety Administration Office of Analysis, Research and Technology (2014)
8. Keller, C.G., Dang, T., Fritz, H., Joos, A., Rabe, C., Gavrila, D.M.: IEEE Xplore abstract - active pedestrian safety by automatic braking and evasive steering. IEEE Trans. Intell. Transp. Syst. **12**, 1292–1304 (2011)
9. Tewolde, G.S.: Sensor and network technology for intelligent transportation systems. Presented at the May (2012)
10. Bengler, K., Dietmayer, K., Farber, B., Maurer, M., Stiller, C., Winner, H.: Three decades of driver assistance systems: review and future perspectives. IEEE Intell. Trans. Syst. Mag. **6**, 6–22 (2014)
11. Vollrath, M., Briest, S., Schiessl, C., Drewes, K., Becker, U.: Ableitung von Anforderungen an Fahrerassistenzsysteme aus Sicht der Verkehrssicherheit. Berichte der Bundesanstalt für Straßenwesen. Bergisch Gladbach: Wirtschaftsverlag NW (2006)
12. Fildes, B., Keall, M., Thomas, P., Parkkari, K., Pennisi, L., Tingvall, C.: Evaluation of the benefits of vehicle safety technology: The MUNDS study. Accid. Anal. Prev. **55**, 274–281 (2013)
13. David, K., Flach, A.: CAR-2-X and pedestrian safety. IEEE Veh. Technol. Mag. **5**, 70–76 (2010)
14. Horter, M.H., Stiller, C., Koelen, C.: A hardware and software framework for automotive intelligent lighting. Presented at the June (2009)
15. Hegeman, G., Brookhuis, K., Hoogendoorn, S.: Opportunities of advanced driver assistance systems towards overtaking. Eur. J. Trans. Infrastruct. Res. EJTIR **5**(4), 281 (2005)

16. Lu, M., Wevers, K., Heijden, R.V.D.: Technical feasibility of advanced driver assistance systems (ADAS) for road traffic safety. Trans. Planning Technol. **28**, 167–187 (2005)

17. NXP - Automotive Radar Millimeter-Wave Technology. http://www.nxp.com/pages/automotive-radar-millimeter-wave-technology:AUTRMWT

18. TDA2x - Texas Instruments Wiki. http://processors.wiki.ti.com/index.php/TDA2x

19. Bosch Mobility Solutions. http://www.bosch-mobility-solutions.com/en/

20. AutoUniMo: FP7-PEOPLE-2013-IAPP AutoUniMo project "Automotive Production Engineering Unified Perspective based on Data Mining Methods and Virtual Factory Model" (grant agreement no: 612207). http://autounimo.aei.polsl.pl/

21. Continental Industrial Sensors-Willkommen bei Industrial Sensors. http://www.conti-online.com/www/industrial_sensors_de_de/

22. Kaempchen, N., Dietmayer, K.C.J.: Fusion of laserscanner and video for ADAS. IEEE Trans. Intell. Transp. Syst. TITS **16**(5), 1–12 (2015)

23. Błachuta, M., Czyba, R., Janusz, W., Szafrański, G.: Data fusion algorithm for the altitude and vertical speed estimation of the VTOL platform. J. Intell. Rob. Syst. **74**, 413–420 (2014)

24. Budzan, S., Kasprzyk, J.: Fusion of 3D laser scanner and depth images for obstacle recognition in mobile applications. Opt. Lasers Eng. **77**, 230–240 (2016)

25. Sandblom, F., Sorstedt, J.: Sensor data fusion for multiple configurations. In: Presented at the June (2014)

26. Grzechca, D., Wrobel, T., Bielecki, P.: Indoor location and idetification of objects with video survillance system and WiFi module. In: Presented at the September (2014)

27. Tokarz, K., Czekalski, P., Sieczkowski, W.: Integration of ultrasonic and inertial methods in indoor navigation system. Theor. Appl. Inform. **26**, 107–117 (2015)

28. Pamuła, D., Ziębiński, A.: Securing video stream captured in real time. Przegląd Elektrotechniczny. R. **86**(9), 167–169 (2010)

29. Ziebinski, A., Swierc, S.: Soft core processor generated based on the machine code of the application. J. Circ. Syst. Comput. **25**, 1650029 (2016)

30. Behere, S., Törngren, M.: A functional architecture for autonomous driving. In: Presented at the Proceedings of the First International Workshop on Automotive Software Architecture (2015)

31. Cupek, R., Ziebinski, A., Franek, M.: FPGA based OPC UA embedded industrial data server implementation. J. Circ. Syst. Comp. **22**, 18 (2013)

Modified Random Forest Algorithm for Wi–Fi Indoor Localization System

Rafał Górak[✉] and Marcin Luckner

Faculty of Mathematics and Information Science, Warsaw University of Technology,
ul. Koszykowa 75, 00–662 Warszawa, Poland
{R.Gorak,M.Luckner}@mini.pw.edu.pl

Abstract. The paper presents a modification of Random Forest approach to the indoor localization problem. The localization solution is
based on RSS (Received Signal Strength) from multiple sources of Wi–Fi
signal. We analyze two localization models. The first one is built using
a straightforward application of a random forest method. The second
model is a combination of localization models built for each Access Point
from the building's network using similar technique (Random Forests)
as for the first model. The modification proposed in the second model
gives us a substantial accuracy improvement when compared to the first
model. We test also the solution against a network malfunction when
some Access Points are turned off as the malfunction immunity is another
important feature of the presented localization solution.

1 Introduction

The problem of building an effective Indoor Positioning System (IPS) is very
complex. Firstly, there is no GPS signal accessible inside the buildings. Secondly,
a greater accuracy is usually required indoor than outdoor where even a small
localization error may lead to a completely different localization results. For
example 10 m error for the outdoor positioning is acceptable while the same
localization error inside the building may place you in a completely different
area of the building. In our paper we describe the solution that is based on
the existing Wi–Fi infrastructure of a building. This solution will relay on the
fingerprint approach.

The fingerprint approach requires to collect measurements (fingerprints) of
RSS (Received Signal Strength) from various Access Points in the known locations. The measurement points should approximately cover the whole building.
We describe the details of data collection in Sect. 2. The collected data of RSS
vectors gathered in the known localizations constitutes a training set(*radio map*)
which allows us to create a localization model. In Sect. 3 we formally describe
what we mean by the localization model as well as we introduce some standard
error notions that will allow us to evaluate and compare models.

The research is supported by the National Centre for Research and Development,
grant No PBS2/B3/24/2014, application No 208921.

© Springer International Publishing Switzerland 2016
N.T. Nguyen et al. (Eds.): ICCCI 2016, Part II, LNAI 9876, pp. 147–157, 2016.
DOI: 10.1007/978-3-319-45246-3_14

In [2] authors compared different methods of creating such a localization model for a given radio map. The paper divides the methods into two categories: deterministic and probabilistic. The deterministic methods are based on distances between RSS vectors while probabilistic ones use the estimation of the probability that we are in a given localization while a particular RSS vector was measured. Another technique is used in [6] which is a multilayer perceptron and [5] provides the analysis k Nearest Neighbours algorithm.

In Sect. 3.1 we explain a construction of localization model based on a straightforward application of the classical Random Forest method as described in [1]. Then in Sect. 3.2 we present a modification of the previous approach. The comparison of both methods is in Sect. 4. The obtained results indicate a substantial accuracy improvement after the suggested modification of the classical Random Forest approach.

When comparing the localization methods in Sect. 4 we will also consider long term changes in accuracy of the localization solutions. We analyze data collected between years 2012–2014 and check if there is any decrease in accuracy. This is very important issue from a perspective of future implementations, since building a new radio map as well as updating it is quite expensive and time consuming process. Hence, it is very important to obtain a solution that works well also in longer periods of time.

In Sect. 4.1 we compare the accuracy of the proposed Random Forest methods with the approach that uses a multilayer perceptron presented in [6]. This comparison is very valuable since the results presented in [6] consider the same building and partially the same data (for example the training data are the same).

Another important issue while comparing the localization models is its immunity to the malfunction of a Wi–Fi infrastructure. In Sect. 4.2 we briefly discuss this problem by considering the situation when some Access Points are turned off. Once again the modified Random Forest method performs better than the classical Random Forest approach. The final conclusions are presented in Sect. 5.

All the tests were held inside the building of Faculty of Mathematics and Information Science of Technical University of Warsaw. This is a 6 floor building of irregular shape that fits inside the rectangle of dimensions 50 m by 70 m. In order to build and test the model we used the data gathered in five different series each one covering the same area of the building. The data was collected using the Android application installed on mobile phones. For more details about the application see [8].

2   Data Collection

Let us first formally describe the structure of the data and the concept of fingerprinting. For the precision sake we introduce the following definitions:

Definition 1. *(i) $\mathcal{AP}$ is the set of all Access Points used for the localization model.*

(ii) $\mathcal{F} = \mathbf{R}^2 \times \mathbf{Z} \times \mathbf{R} \times \mathbf{R}^n$ is the space of all possible measurements (fingerprints) where $n = \sharp \mathcal{AP}$.

For $f \in \mathcal{F}$

(a) first two coordinates $f_1[m]$, $f_2[m]$ denote a horizontal position of a place where measurement was taken and $f_3 \in \mathbf{Z}$ is the floor number;

(b) $f_4[s]$ is the time of the measurement;

(c) $f_k[dBm]$, where $4 \leq k \leq n + 4$, is the RSS from kth source from $\mathcal{AP}$. If there is no signal from kth transceiver station than $f_{k+4} = noSignal$, where noSignal is a special unique value.

(iii) We call a set of fingerprints $S \subset \mathcal{F}$ a measuring series. Usually S is collected during one or a few consecutive days in a particular building.

(iv) $\mathcal{L} = (\mathcal{L}_x, \mathcal{L}_y, \mathcal{L}_f) : \mathcal{F} \mapsto \mathbf{R}^2 \times \mathbf{Z}$ is the projection onto the first three coordinates of the set $\mathcal{F}$ and $\pi : \mathcal{F} \mapsto \mathbf{R}^n$ is the projection onto the last n coordinates. In other words $\mathcal{L}(f)$ provides us with the location of a fingerprint f while $\pi(f)$ is the RSS vector associated with f (measurement).

(v) For $v = (v_1, ..., v_n) \in \mathbf{R}^n$ we denote $supp(v) = \{k : v_k \neq noSignal\}$ which is just the set of visible Access Points for the measurement v.

Let us denote by S_i, where $i \in \{1, 2, 3, 4, 5\}$, the ith measuring series. The series S_1, S_2, S_3 were collected in 2012 within three months period and data sets S_4, S_5 were collected during two series in 2014 within two months period. All series were collected on different days. In each of five series the location fingerprints create mostly a 1.5×1.5 m grid. Only when it was impossible due to the building's structure (walls or different obstacles) the grid was slightly sparser. It should be mentioned that for every i the measuring series S_i covers the same areas of the building. There were 40 fingerprints taken in every measuring point. Table 1 shows the number of fingerprints in each series. Hence the number of measuring points (elements of the grid) is around 1100–1200 for each series.

Table 1. Number of fingerprints in each series.

i	1	2	3	4	5
$\sharp S_i$	47960	47960	43680	43680	46760

Creating a model we considered only the academic Wi–Fi infrastructure that consists of 46 APs. In our paper n will always denote the number of Access Points in the academic network that is the number $\sharp \mathcal{AP} = 46$. At this point it is worth mentioning that we observed far more than 46 Access Points inside the building but we decided not to take them into account while building a model. It is because we controlled only the academic net and we can guarantee no changes in its infrastructure (location of APs, device changes etc.) between years 2012 and 2014. In our considerations we define $noSignal[dBm]$ as the minimal signal strength ever recorded minus 2 (our devices reports only odd values). The minimal value that was ever reported was -115 hence we set $noSignal = -117$.

3   The Localization Models

The localization model is a function $\widehat{\mathcal{L}} : \mathbf{R}^n \mapsto \mathbf{R}^2 \times \mathbf{Z}$ such that given a RSS vector $v \in \mathbf{R}^n$, $\widehat{\mathcal{L}}(v)$ predicts a localization of the point where the measurement v was taken. However, before we describe some particular examples of the localization models let us introduce the error functions:

Definition 2. *Let S be a measuring series. For an element $s \in S$ we introduce the following notions:*
horizontal error

$$\mathcal{E}_h(\widehat{\mathcal{L}}, s) = \sqrt{(\hat{x} - x)^2 + (\hat{y} - y)^2}$$

and the floor error:

$$\mathcal{E}_f(\widehat{\mathcal{L}}, s) = |\hat{f} - f|,$$

where $\widehat{\mathcal{L}}(\pi(s)) = (\hat{x}, \hat{y}, \hat{f})$ and $\mathcal{L}(s) = (x, y, f)$.

In order to evaluate how well $\widehat{\mathcal{L}}$ predicts a localization for a given testing series S we introduce the following:

Definition 3. *For a testing series S and the localization model $\widehat{\mathcal{L}}$ let us define:*

(i) Mean horizontal error

$$\mu\mathcal{E}_h(\widehat{\mathcal{L}}, S) = mean\{\mathcal{E}_h(\widehat{\mathcal{L}}, s) : \ s \in S\};$$

(ii) Median horizontal error

$$m\mathcal{E}_h(\widehat{\mathcal{L}}, S) = median\{\mathcal{E}_h(\widehat{\mathcal{L}}, s) : \ s \in S\};$$

(iii) Percentile horizontal error

$$p_{90}\mathcal{E}_h(\widehat{\mathcal{L}}, S) = Perc_{90}\{\mathcal{E}_h(\widehat{\mathcal{L}}, s) : \ s \in S\};$$

(iv) Classification error for floor's prediction

$$\varepsilon_f(\widehat{\mathcal{L}}, S) = \frac{\#\{s \in S : \ \mathcal{E}_f(\widehat{\mathcal{L}}, s) \neq 0\}}{\#S};$$

When it is clear from the context which model we test we omit the name of the model, i.e. we use the notation: $\mu\mathcal{E}_h(S)$, $m\mathcal{E}_h(S)$, $p_{90}\mathcal{E}_h(S)$, $\varepsilon_f(S)$.

Obviously the main goal in the localization problem is to make $\mu\mathcal{E}_h, m\mathcal{E}_h, p_{90}\mathcal{E}_h$ and ε_f as small as possible.

3.1   Random Forest Localization Model

Let us describe the localization model $\widehat{\mathcal{L}}_{RF}$, which is built using Random Forest algorithm. First we create $\widehat{\mathcal{L}}_x$, $\widehat{\mathcal{L}}_y$ and $\widehat{\mathcal{L}}_f$ by applying a Random Forest algorithm described in [1] where the training set is $\pi(S_1)$ with responses $\mathcal{L}_x(S_1)$, $\mathcal{L}_y(S_1)$ and $\mathcal{L}_f(S_1)$, respectively. Obviously for creating $\widehat{\mathcal{L}}_x$ and $\widehat{\mathcal{L}}_y$ we grow regression trees and for $\widehat{\mathcal{L}}_f$ the decision trees are grown. We select the number of grown trees to be 30 as we checked that growing more does not improve the accuracy of the localization algorithm. Finally we may define $\widehat{\mathcal{L}}_{RF}(v) = (\widehat{\mathcal{L}}_x(v), \widehat{\mathcal{L}}_y(v), \widehat{\mathcal{L}}_f(v))$. For $v_0 = (noSignal, noSignal, ..., noSignal)$ we set $\widehat{\mathcal{L}}_{RF}(v_0) = NaN$. However, it appears that the Access Point coverage is such that $v_0 \notin \pi(S_i)$ for all $i \in \{1, 2, ..., 5\}$.

3.2   The Modified Random Forest Localization Model

In this part we describe a modification of the straightforward application of Random Forests described above. The idea behind the modification is to create a localization model $\widehat{\mathcal{L}}^a$ for every Access Point $a \in \mathcal{AP}$ that predicts the localization only when there is a signal from Access Point a. In order to create $\widehat{\mathcal{L}}^a$ we apply the same Random Forest technique as in the previous section but using only this part of training data set for which the signal from AP a was reported. By the appropriate combination of models $\widehat{\mathcal{L}}^a$ we obtain the final result, that is the modified localization model $\widehat{\mathcal{L}}_{mRF}$. The following part of this section formally describes this construction.

For every $a \in \mathcal{AP}$ let us define the set of all fingerprints from the training series S_1 for which we have signal from Access Point a i.e. $S^a = \{s \in S_1 : a \in supp(s)\}$. Then for every $a \in \mathcal{AP}$ we create $\widehat{\mathcal{L}}_x^a$, $\widehat{\mathcal{L}}_y^a$ and $\widehat{\mathcal{L}}_f^a$ by applying Random Forest algorithm in a similar manner as in Sect. 3.1 where the training set is $\pi(S^a)$ with responses $\mathcal{L}_x(S^a)$, $\mathcal{L}_y(S^a)$ and $\mathcal{L}_f(S^a)$, respectively. Once again we select the number of grown trees to be 30 as greater values do not improve the accuracy. Finally for every $v \neq v_0 = (noSignal, noSignal, ..., noSignal) \in \mathbf{R}^n$ we define $\widehat{\mathcal{L}}_{mRF}(v) = (\hat{x}, \hat{y}, \hat{f})$ where

$$\hat{x} = \text{mean}\{\widehat{\mathcal{L}}_x^a(v) : a \in supp(v)\},$$
$$\hat{y} = \text{mean}\{\widehat{\mathcal{L}}_y^a(v) : a \in supp(v)\},$$
$$\hat{f} = \text{mode}\{\widehat{\mathcal{L}}_f^a(v) : a \in supp(v)\}.$$

For $v_0 = (noSignal, noSignal, ..., noSignal)$ we set $\widehat{\mathcal{L}}_{mRF}(v_0) = NaN$. The idea behind the construction is simple. The Fig. 1 shows four Access Points a_1, a_2, a_3 and a_4. We also marked a position where the measurement $v \in \mathbf{R}^n$ (RSS vector) was taken. The circles around Access Points shows the range of each Access Point that is where the terminal reports a signal from that Access Point.

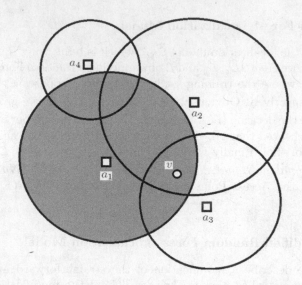

Fig. 1. Access Points $a_1,...,a_4$ and the position where the measurement $v \in \mathbf{R}^n$ was taken

In order to calculate the position where v was taken, we apply models created for a_1 a_2 and a_3 but not for a_4. In other words $\widehat{\mathcal{L}}_{mRF}(v) = (\hat{x}, \hat{y}, \hat{f})$ where

$$\hat{x} = (\widehat{\mathcal{L}}_x^{a_1}(v) + \widehat{\mathcal{L}}_x^{a_2}(v) + \widehat{\mathcal{L}}_x^{a_3}(v))/3$$
$$\hat{y} = (\widehat{\mathcal{L}}_y^{a_1}(v) + \widehat{\mathcal{L}}_y^{a_2}(v) + \widehat{\mathcal{L}}_x^{a_3}(v))/3,$$
$$\hat{f} = \mathrm{mode}\{\widehat{\mathcal{L}}_f^{a_1}(v), \widehat{\mathcal{L}}_f^{a_2}(v), \widehat{\mathcal{L}}_f^{a_3}(v)\}.$$

In order to create models $\widehat{\mathcal{L}}_x^{a_1}$, $\widehat{\mathcal{L}}_y^{a_1}$ and $\widehat{\mathcal{L}}_f^{a_1}$ we apply Random Forest method in a similar manner as in Sect. 3.1 but we use only the fingerprints from the training data that where taken in the range of Access Point a_1 (the shaded region in Fig. 1). We do the similar for all Access Points inside the building (for all $a \in \mathcal{AP}$).

4   Comparison of the Models

Let us see how the models $\widehat{\mathcal{L}}_{RF}$ and $\widehat{\mathcal{L}}_{mRF}$ perform with respect to the measures introduced in Definition 3 for all the series S_i for $i \in \{1, ..., 5\}$. When looking at all the tables below, please have in mind that S_1 was used for training purposes.

One can see that for all the testing series the model $\widehat{\mathcal{L}}_{mRF}$ performs better than $\widehat{\mathcal{L}}_{RF}$. We can also observe that we obtain the worst results for the testing series S_4 and S_5 that were collected about two years after the training series S_1. There can be several reasons. One of them is that in series S_1, S_2, S_3 different mobile phones

Table 2. Performance of $\widehat{\mathcal{L}}_{RF}$ model that was built using the classical Random Forest approach

i	$\varepsilon_f(S_i)$	$\mu\mathcal{E}_h(S_i)$	$m\mathcal{E}_h(S_i)$	$p_{90}\mathcal{E}_h(S_i)$
1	0.0094	1.32	0.18	3.41
2	0.0284	3.66	2.89	6.95
3	0.0264	3.55	2.95	6.80
4	0.0849	4.36	3.72	8.34
5	0.0769	4.53	3.76	8.53

Table 3. Performance of $\widehat{\mathcal{L}}_{mRF}$ model that was built using the modified Random Forest approach

i	$\varepsilon_f(S_i)$	$\mu\mathcal{E}_h(S_i)$	$m\mathcal{E}_h(S_i)$	$p_{90}\mathcal{E}_h(S_i)$
1	0.0087	0.57	0.11	1.55
2	0.0291	3.34	2.72	6.36
3	0.0222	3.29	2.82	6.10
4	0.0822	4.11	3.52	8.03
5	0.0711	4.30	3.63	8.23

were used than 2 years later during collection of series S_4 and S_5. For the discussion on the difference of RSS vector reported by different models of mobile phones see [7]. Another reason is that some minor changes could occur inside the building. As we mentioned before, we can guarantee that the academic infrastructure was unchanged (i.e. Access Points and their locations). However we had no control over such changes as furniture replacements or rooms rearrangements, although we have not observed any major changes during this two years period. Perhaps it will be illustrative if we look at the Fig. 2. This are the graphs of functions $\tau_i(k) = \frac{\sharp\{s \in S_i : \sharp supp(s)=k\}}{\sharp S_i}$ for $i \in \{1, 2, ..., 5\}$. The value of $\tau_i(k)$ shows the rate of all fingerprints in S_i for which there were k Access Points visible. We can see that the characteristics of Access Points coverage has changed between years 2012 and 2014 and it influenced the performance of both localization models $\widehat{\mathcal{L}}_{RF}$ and $\widehat{\mathcal{L}}_{mRF}$ (see Tables 2 and 3 once again).

4.1   Comparison with Other Works

Although there are many indoor localization solutions based on Wi–Fi signals, the comparison of their performance is very difficult. It is because the testing conditions depend very much on the building where the tests are performed. The construction of a building or the density of the WI–Fi network may affect the accuracy. For example the accuracy of Ekahau Real-Time Location System is claimed to be 3 m on average when using Wi–Fi solution only, while independent

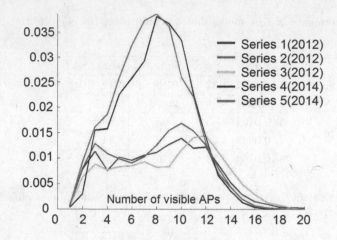

Fig. 2. Graphs of τ_i functions for all series

tests described in [3] show that it is in fact $7\,m$ on average. The best method to avoid such problems it to compare different methods using the same data.

In [6] authors create a localization model using a multlilayer perceptron technique, in the same building, using the same data as series S_1, S_2, and S_3. The authors use S_1 as the training set, S_2 as the validation set and finally S_3 is used for testing purposes. They introduce a different horizontal accuracy measures $|e_X|$ and $|e_Y|$ what is the difference between the coordinate and its prediction. In our notation this would be $|e_X|(s) = |\hat{x} - x|$ and $|e_Y|(s) = |\hat{y} - y|$ where x and y are the true coordinates where the fingerprint s was taken and $\hat{x}$, $\hat{y}$ the predicted coordinates by the model. In order to asses the model they use 90th percentile of both $|e_X|$ and $|e_Y|$. The best results that authors obtained in [6] are $5.18\,m$ for 90th percentile of $|e_X|$ and $5.82\,m$ for 90th percentile of $|e_Y|$ for the testing series S_3. When testing $\widehat{\mathcal{L}}_{mRF}$ on series S_3 we obtain $4.11\,\text{m}$ for 90th percentile of $|e_X|$ and $4.72\,\text{m}$ for 90th percentile of $|e_Y|$. It is worth mentioning that in another very recent paper [9] authors built different localization model using optimization of the Random Forest model as presented in Sect. 3.1, by the Particle Swarm Optimization algorithm. Moreover authors used the same data set for learning purposes as in this paper. However, the testing procedure as well as the testing data set are different. It would be interesting to see the comparison of gains we obtain using the modification of the Random Forest method presented in this paper and the optimization using Particle Swarm Optimization algorithm as presented in [9].

4.2   Immunity to the System Malfunction

It is a common situation that some Access Points may malfunction, can be removed or replaced having different MAC address and hence being "invisible" for the localization model. Most probably this is a situation that influences the

Table 4. Performance of $\widehat{\mathcal{L}}_{RF}$ model that was built using the classical Random Forest approach after turning off 3 Access Points

i	$\varepsilon_f(S_i)$	$\mu\mathcal{E}_h(S_i)$	$m\mathcal{E}_h(S_i)$	$p_{90}\mathcal{E}_h(S_i)$
1	0.0094	1.32	0.18	3.41
2	0.0565	4.80	3.51	9.99
3	0.0581	4.36	3.43	8.73
4	0.1101	4.88	4.07	9.23
5	0.1089	5.09	4.21	9.60

Table 5. Performance of $\widehat{\mathcal{L}}_{mRF}$ model that was built using the modified Random Forest approach after turning off 3 Access Points

i	$\varepsilon_f(S_i)$	$\mu\mathcal{E}_h(S_i)$	$m\mathcal{E}_h(S_i)$	$p_{90}\mathcal{E}_h(S_i)$
1	0.0087	0.57	0.11	1.55
2	0.0487	3.80	3.11	7.36
3	0.0436	3.74	3.16	7.08
4	0.1002	4.34	3.77	8.29
5	0.0933	4.58	3.88	8.79

performance of localization models. Indeed, Tables 4 and 5 when compared to Tables 2 and 3, show the decrease of accuracy after turning off 3 Access Points (the same APs for both $\widehat{\mathcal{L}}_{RF}$ and $\widehat{\mathcal{L}}_{mRF}$).

We can see that this time the localization model $\widehat{\mathcal{L}}_{mRF}$ performed much better compared to $\widehat{\mathcal{L}}_{RF}$. It is especially important when we look at the gross errors that is when we compare $p_{90}\mathcal{E}_h$ measure for both models. The difference can reach up to 2.5 m (for testing series S_2). We should point out that after turning off the chosen 3 Access Points, there were only 52 measurements in all series where no Access Point was detected that is $\widehat{\mathcal{L}}_{RF}$ and $\widehat{\mathcal{L}}_{mRF}$ did not provide a localization (NaN value). We do not provide a deeper analysis of the malfunction immunity of both models. We decided to choose particular 3 APs to illustrate a situation that is typical for both models. In other words we have run the experiment choosing many times 3 random APs and turning them off in all the series $S_2...S_5$. In all cases model $\widehat{\mathcal{L}}_{mRF}$ performed much better. The extended analysis of malfunction immunity of the classical Random Forest model $\widehat{\mathcal{L}}_{RF}$ is presented in [4] while the paper containing the same extended analysis for the modified $\widehat{\mathcal{L}}_{mRF}$ is in preparation.

5   Conclusions

This work shows the method of modifying the Random Forest approach in building the localization model $\widehat{\mathcal{L}}_{mRF}$. This way we were able to improve the localization accuracy over the accuracy of our reference model $\widehat{\mathcal{L}}_{RF}$ that was build by straightforward application of Random Forest technique. The horizontal accuracy gains are about 5 % to 9 % when looking at the mean horizontal error. It was confirmed by four independent tests (series S_2 to S_5). The vertical accuracy (floor's detection) remained almost the same for both models.

The comparison of $\widehat{\mathcal{L}}_{mRF}$ model with the model built using multilayer perceptron as described in [6] also revealed that the $\widehat{\mathcal{L}}_{mRF}$ performs much better when considering horizontal accuracy and almost the same when comparing the vertical accuracy (floor's detection). We also checked that the model $\widehat{\mathcal{L}}_{mRF}$ performed much better than $\widehat{\mathcal{L}}_{RF}$ when some Access Points were turned off during the testing procedure. For instance, after turning off 3 Access Points for the testing series S_2, the loss was about 30 % for $\widehat{\mathcal{L}}_{RF}$ when looking at the mean horizontal error while the same loss for the modified model $\widehat{\mathcal{L}}_{mRF}$ was about 14 %. The advantage of using $\widehat{\mathcal{L}}_{mRF}$ is even more evident when we look at the gross errors that is when we compare $p_{90}\mathcal{E}_h$ measure. Once again in case floor's detection both models behave similarly.

It is worth mentioning that this modification can be applied to different localization models, not only to the ones based on Random Forests. For example a similar modification of kNN method could be considered by constructing $\widehat{\mathcal{L}}^a$ in Sect. 3.2 using kNN algorithm. We also believe that a similar technique can be applied when dealing with problems different than indoor localization. The only condition is that we deal with the sparse data (where *noSignal* value is the most common value in the sparse data) and one has a reasonable number of features. Our results suggest that one may count on some improvements especially for regression problems.

References

1. Breiman, L.: Random forests. Mach. Learn. **45**(1), 5–32 (2001)
2. Dawes, B., Chin, K.W.: A comparison of deterministic and probabilistic methods for indoor localization. J. Syst. Softw. **84**, 442–451 (2011)
3. Gallagher, T., Tan, Y.K., Li, B., Dempster, A.G.: Trials of commercial wi-fi positioning systems for indoor and urban canyons. In: IGNSS Symposium on GPS/GNSS, Gold Coast, Australia (2009)
4. Górak, R., Luckner, M.: Malfunction immune Wi–Fi localisation method. In: Núñez, M., Nguyen, N.T., Camacho, D., Trawinski, B. (eds.) ICCCI 2015. LNCS, vol. 9329, pp. 328–337. Springer, Heidelberg (2015). doi:10.1007/978-3-319-24069-5_31
5. Grzenda, M.: On the prediction of floor identification credibility in RSS-based positioning techniques. In: Ali, M., Bosse, T., Hindriks, K.V., Hoogendoorn, M., Jonker, C.M., Treur, J. (eds.) IEA/AIE 2013. LNCS, vol. 7906, pp. 610–619. Springer, Heidelberg (2013)
6. Karwowski, J., Okulewicz, M., Legierski, J.: Application of particle swarm optimization algorithm to neural network training process in the localization of the mobile terminal. In: Iliadis, L., Papadopoulos, H., Jayne, C. (eds.) EANN 2013, Part I. CCIS, vol. 383, pp. 122–131. Springer, Heidelberg (2013)
7. Kjargaard, M.B.: Indoor location fingerprinting with heterogeneous clients. Pervasive Mob. Comput. **7**, 31–43 (2011)

8. Korbel, P., Wawrzyniak, P., Grabowski, S., Krasinska, D.: LocFusion API - programming interface for accurate multi-source mobile terminal positioning. In: Federated Conference on Computer Science and Information Systems (FedCSIS), pp. 819–823, September 2013

9. Okulewicz, M., Bodzon, D., Kozak, M., Piwowarski, M., Tenderenda, P.: Indoor localization of a moving mobile terminal by an enhanced particle filter method. In: Rutkowski, L., Korytkowski, M., Scherer, R., Tadeusiewicz, R., Zadeh, L.A., Zurada, J.M. (eds.) ICAISC 2016. LNCS, vol. 9693, pp. 512–522. Springer, Heidelberg (2016). doi:10.1007/978-3-319-39384-1_45

Smart Underwater Positioning System and Simultaneous Communication

Boguslaw Szlachetko[(✉)] and Michal Lower

Faculty of Electronics, Wroclaw University of Science and Technology,
ul. Wyb. Wyspianskiego 27, 50-370 Wroclaw, Poland
{Boguslaw.Szlachetko,Michal.Lower}@pwr.edu.pl

Abstract. This paper touches a problem of development of effective technology of underwater positioning, which allows control position of more than one object i.e. submarines, remotely operated underwater robots or divers. The paper presents a solution which bases on the use of a modulated ultrasound waveform instead of bunch of square-shaped pulses utilized by a traditional positioning systems. The proposed system can be seen as smart because it not only solves a positioning problem but also delivers communication link between a supervising ship and remotely controlled objects. Some preliminary results of experiments with real underwater robots are also presented. The resultant relative positional accuracy was about 1 m and the data rate can be estimated about 1 kb/s.

Keywords: Underwater localization · Underwater robot positioning · ROV · Ultrasound distance measurement · Ultra short baseline system

1 Introduction

Positioning of underwater vessels and a Remotely Operated underwater Vehicle (ROV) as well as underwater communication is a specific task because of the environment impact [4]. The water attenuates radio waves very effectively, thus the direct using of traditional technologies like the Global Positioning System (GPS) or radars, which are commonly used in free space, are not possible. In opposite to radio waves an ultrasound waves can be applied in underwater scenario with much higher effectiveness [2]. Acoustic waves propagate five times quicker in a water than in an air. In spite of dispersion and diffraction effects the range of 300 m using not so powerful 40 kHz transducers is possible to reach [3].

Because of technical limitation of ROV position determination the Inertial Navigation Systems (INS) using the Inertial Measurement Units (IMU) are commonly used to support dead reckoning navigation task [6,7,9,12]. Such methodology is subject to cumulative errors, so inaccuracy of position determination increases in time. An error of position estimation based on IMU is integrated constantly, therefore only other independent position measurements can introduces a new information, which resets such errors. Those other underwater measurements can rely on measure of distance from reference point placed on the

© Springer International Publishing Switzerland 2016
N.T. Nguyen et al. (Eds.): ICCCI 2016, Part II, LNAI 9876, pp. 158–167, 2016.
DOI: 10.1007/978-3-319-45246-3_15

water surface or on the bottom of the reservoir. Unquestionable advantage of reference points placed on the bottom of the reservoir is geographical placement stability and lack of impact of water movement and waves on its position. However, such a placement of reference points complicates communication with them and puts greater demand on design related to water pressure influence on devices which realize function of reference points. Placement of such devices on the water surface simplify by far their design, mechanical construction and maintenance. However, their placement stability depends heavily on weather condition and specificity of a basin [5]. Thus, in such a solution a smart sensor fusion algorithm which joins position measurement based on reference points with dead reckoning calculations based on IMU measurement is often utilized [7].

Position determination systems based on measurement of a distance from reference points with the use of ultrasound waves are commonly used nowadays, however, such systems utilize very simple signaling schemes. Usually a bunch of several short square impulses are generated by an ultrasound transducers in periodic manner [2,8]. In our work an application of modulated ultrasound signals is proposed, which allows to develop smart distance measurement system. Each bunch of impulses is replaced by a modulated waveform, so some important information which can simplify measurement methodology, is carried by the waveform during each period. Such a solution significantly widens functional capabilities of a whole measurement system.

2 Conventional Ultrasound Methodology of Position Measurement

For the dive depth exceeding 100 m and in cases of broad space of work of submarines or ROVs the Long Base-Line (LBL) systems are used. Baseline means the distance between the transponders (aka reference points). In such systems a measured distance sometimes reaches a few kilometers. For such a long distance a problem of rising of an absolute error is observed, when one utilizes an ultrasound methodology. Thus, in order to lowering the absolute error of the measurement of a distance (and in consequence a position), multiple number of transponders support these measurements. In some papers a using of seabed transponders is reporter [10,11]. But in other papers a system which is composed of several transponders placed on the water surface is considered. A position of transponders is adjusted by GPS coordinate measurement. Such systems are called GPS intelligent buoys (GIB) [1].

In some cases baseline do not cross 100 m. Therefore, such systems are called Short Base-Line (SBL) acoustic positioning systems. The principle of measurement is similar like in LBL systems, but in SBL some transponders are often placed on body of a ship which performs or supervises an underwater explorations done by ROVs. Such placement gives quite stable reference points for calculations of a distance and a position relative to the ship's body, because of rigid arrangement of reference points. Moreover, using a GPS system mounted on the ship, calculation of an absolute coordinate of reference points is possible. Thus, this allows to determine absolute position of underwater ROVs.

However, the most frequent case of using acoustics positioning technology is called Ultra Short BaseLine (USBL), due to the fact that usually a small ship, with limited technical possibility of placing transponders is used to supervise ROVs. In such cases a baseline is very short - usually does not cross several meters, and simultaneously it is multiple time shorter than a distance to measure. In this type of systems obtaining a good enough localization estimation based on reference points only arises many problems. For example - this situation can be observed when working depth of a ROV is a dozen times bigger than a length of a baseline. For this reason the USBL technology is usually supported by additional measurement systems. Calculated USBL coordinates are combined with dead reckoning calculations e.g. a ROV speed and direction of its movement is taken from IMU while actual depth is calculated on a base of pressure measurement. Moreover, if transponders are equipped with the receivers which can determine a direction of arrival of ultrasound waves an USBL system can be reduced to only one such a transponder called base-point.

3   Proposed Positioning System

3.1   Smart Ultrasound Transponder

Regardless of the problem of baseline's length there exist a second one factor, which limits conventional ultrasound positioning systems. Nearly all of known systems utilize simple signaling scheme - namely a bunch of short ultrasound impulses. Such bunches are repeated with constant period. Therefore, several effects have to be accounted. First, each bunch is identical, which creates a problem with using more than one transponders. Second, it is hard task to differentiate the bunch impinging receiver directly (through a straight line between a transponder and a receiver) from reflected one. A common solution of the second problem rely on setting up a long period of repetition of consecutive bunches. The other solution consists in generating bunches with different number of impulses in it i.e. each consecutive bunch contains the increasing number of impulses.

Our method relies on the use of modern modulation technology. First, we decided to invert a role of transponders. Usually they are placed in a stable location while a ROV with a receiver mounted on it moves around. In our solution receivers are mounted on a ship (in a stable location joined to the ship's body) whereas a transponder is joined to the ROV's body and moves together with it. Second, a transponder generates an ultrasound waveform through modulation of a carrier signal. The carrier signal produces main waveform whereas a modulating signal transports some information. The modulation allows to carry on communications between ROV and receivers. Currently, our system utilizes short waveform thus it allows to send information about start time of a transmission, but nothing prevents to expand amount of informations. A BPSK modulation scheme is used by our transponder, where carrier signal is a waveform of a square shape. Further details are beyond of the scope of this paper and will be published in the near future.

A principle of distance measurement, which make use of modulated ultrasound waveforms, is quite simple. A transponder triggers a transmission sending an information about it's start time. A receiver detects a beginning of reception and stores it in memory. Then, received data are demodulated, which allows to recover information sent by the transponder. The time difference between time of reception and start time of a transmission gives a value t called Time-of-Flight (ToF). For the known speed of ultrasound waveform in a water v_s one can calculate a distance as follows $r - v_s t$. Of course an assumption has to be made that a time base on both the transponder and the receiver is the same. It means that timers which are used on both devices are synchronized.

3.2 Position Calculation

General scheme of the proposed system, its arrangement and reference points placement are shown in Fig. 1.

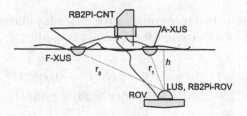

Fig. 1. Arrangement of an underwater position determination system.

Our USBL positioning system is composed of:

- ultrasound receiver placed on a stern called A-XUS; additionally equipped with a GPS receiver,
- ultrasound receiver placed on a bow called F-XUS; additionally equipped with a GPS receiver,
- ultrasound transmitter placed on the ROV, which works as an illuminator LUS,
- ROV with on-board computer RPi-ROV equipped with an accelerometer, gyroscopes, magnetometers and a hydrostatic pressure sensor,
- central processing unit RPi-CNT placed on a supervising ship, the USBL positioning system works on it.

It is worth noting that usually reference point is a device broadcasting ultrasound waves at known a-priori position. In our configuration it is a smart device which can receive and broadcast ultrasound waves. Moreover, it has the possibility to analyze modulated ultrasound waves.

Direct quantity measured with the use of ultrasound waves flowing through the water is ToF evaluated in seconds. This delay is measured with the use of an embedded timer of STM32F4 microcontroller. A frequency of the timer's count equals 10 kHz, which means that delay can be measured with a precision of 10 µs. In effect, this precision results in accuracy of distance measurement of about 1.5 cm. Because of much larger ROVs size this accuracy is sufficient. It is worthwhile add that 32-bit timer is used, so for frequency of 10 kHz an overflow of the timer would occur after nearly 12 h of continuous work. In case of using frequency of 100 kHz, accuracy of distance measurement reaches theoretical value of 2 mm, but unfortunately the timer overflow will occur after only 70 min, which could result in some calculation problems if the ROV is designed to operate longer than one hour.

The designated distance r (in [m]) is a function of the speed v_s (in [m/s]) and time delay t (in [s]). The speed of ultrasound waves in water depends on several factors. Three of them have significant impact on the speed:

T – a temperature of water evaluated in [°C],
P – a hydrostatic pressure of water evaluated in [bar] and
s – a salinity of water evaluated in [%].

Accordingly, the speed of an ultrasound wave depends indirectly on a depth. To calculate this speed Medwins empirical formula is used:

$$v_s = 1449.2 + 4.6T - 0.055T^2 + 0.00029T^3$$
$$+ (1.34 - 0.010T)(s - 35) + 0.158P \tag{1}$$

According to Fig. 1 our system makes simultaneously two distance measurements:

r_1 – a distance from the ROV to the vessel stern (LUS → F-XUS)
r_2 – a distance from the ROV to the vessel bow (LUS → A-XUS)

Additionally a system measures a hydrostatic pressure and calculates a dive depth of submersion. Next, a data fusion algorithm evaluates a 3D position of ROV. This algorithm collects data from:

– hydrostatic pressure sensor,
– r_1 and r_2 measurements,
– IMU sensors installed on ROV,
– GPS coordinates of both reference points (A-XUS, F-XUS),
– physical distance L between A-XUS and F-XUS.

The outline of communication in the proposed USBL positioning system is presented in the Fig. 2. In this outline an ultrasound signal connection is also pointed out. Note that all timers utilized on LUS, A-XUS, F-XUS have to be synchronized before our system starts to calculate ROV's position.

First, a ROV position in coordinate system connected to the supervising ship is calculated. Next, base on a GPS measurement taken on both bow and

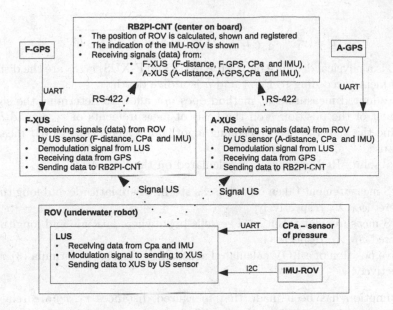

Fig. 2. The outline of communication and ultrasound signaling in the proposed USBL positioning system.

stern, geographical coordinates are calculated. A ROV position relative to the supervising ship is determined in (x, y, h) coordinate system. Axes OX and OY are perpendicular to each other and parallel to a water's surface, while a third axis OH is perpendicular to a water's surface and directed down. A start of coordinating system is located in a central point of a ship's stern according to the Fig. 3.

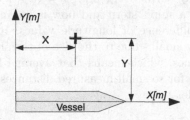

Fig. 3. Coordinates system relative to the supervising vessel.

A dive depth (in [m]) can be easily estimated based on hydrostatic pressure taken from sensor mounted on ROV according to the following formula:

$$h = 10P \tag{2}$$

where P is a measured pressure in [bar]. Next a ROV's position can be calculated according to a system of formulas:

$$x = (r_1^2 - r_2^2 + L^2)/(2L), \quad y = \sqrt{r_1^2 - x^2 - h^2} \tag{3}$$

where L is a physical distance between A-XUS and F-XUS, r_1, r_2 are the distance measurements according to Fig. 1 and h is a dive depth.

The adopted measurement method does not allow to determine the sign of component of the position y on the basis of measurements of r_1, r_2 and h. To determine the sign the dead reckoning algorithm which utilizes IMU measurement is used.

An absolute ROV's position is calculated on the basis of:

- a GPS measurement taken on a vessel's stern, where latitude and longitude is gathered ϕ_A, λ_A respectively,
- a GPS measurement taken on a vessel's bow, where latitude and longitude is gathered ϕ_F, λ_F respectively,
- relative position of a ROV calculated according to the set of formula (3) x, y, h respectively.

Assumption has been made that measured distances r_1, r_2 are relatively short, not longer than 100 m. It simplified the proportional correction coefficient during procedure of recalculation of angular longitude to metric values. Hence, formulas for calculation of ROV's position, it means their latitude ϕ_{xy} and longitude λ_{xy}, can be derived as follows:

$$\phi_{xy} = \phi_A + \frac{1}{L}(y\lambda_L \cos \hat{\phi} + x\phi_L) \tag{4}$$

$$\lambda_{xy} = \lambda_A + \frac{1}{L}\left(x\lambda_L - y\frac{\phi_L}{\cos \hat{\phi}}\right) \tag{5}$$

where $\phi_L = \phi_F - \phi_A$, $\lambda_L = \lambda_F - \lambda_A$ and $\hat{\phi} = (\phi_F + \phi_A)/2$. A difference of latitude values taken on a ship's stern and bow is an angular value. The same situation is in case of difference of longitude values. It is worth noting that these differences are very small - due to the size of the ship. Thus, formulas (4) and (5) are simplified ones, which means that average latitude is used at the whole line segment. But, for so small measured distances an error made by this simplification is negligible.

3.3   Error Analysis

In goal of determining an accuracy of obtained ROV's position formulas of error calculation are provided. An exact differential method has been used. Total error is calculated as a radius of a ball surrounding ROV's location. A value of total error is given in [m]. Also r_1, r_2 values are given in [m]. Note, that latitude and longitude values are given as angle, therefore, one should recalculates it using the mean earth radius $R_E = 6371$ km.

$$\Delta x = \frac{r_1}{L}\Delta r_1 + \frac{r_2}{L}\Delta r_2 + \frac{1}{2}\Delta L + \left|\frac{r_1^2 - r_2^2}{2L^2}\right|\Delta L \qquad (6)$$

$$\begin{cases} y \neq 0 : \Delta y = \dfrac{r_1}{|y|}\Delta r_1 + \dfrac{h}{|y|}\Delta h + \dfrac{|x|}{|y|}\Delta x \\ y = 0 \wedge h > 0 \wedge x > 0 : \Delta y = \sqrt{2r_1\Delta r_1 - 2h\Delta h - 2x\Delta x + \Delta^2 r_1 - \Delta^2 x - \Delta^2 h} \\ y = 0 \wedge h > 0 \wedge x = 0 : \Delta y = \sqrt{2r_1\Delta r_1 - 2h\Delta h + \Delta^2 r_1 - \Delta^2 h} \\ y = 0 \wedge h = 0 \wedge x > 0 : \Delta y = \sqrt{2r_1\Delta r_1 - 2x\Delta x + \Delta^2 r_1 - \Delta^2 x} \\ y = 0 \wedge r_1 = 0 : \Delta y = \Delta r_1 \end{cases} \qquad (7)$$

$$\Delta\phi_{xy} = \Delta\phi_A + R_E\left(\left|\frac{x\phi_L - y\lambda_L cos\hat{\phi}}{L^2}\right|\Delta L + \left|\frac{\lambda_L cos\hat{\phi}}{L}\right|\Delta y + \left|\frac{\phi_L}{L}\right|\Delta x\right)$$
$$+ \left|\frac{x}{L}\right|\Delta\phi_L + \left|\frac{ycos\hat{\phi}}{L}\right|\Delta\lambda_\Delta + \left|\frac{y\lambda_L sin\hat{\phi}}{L}\right|\Delta\hat{\phi} \qquad (8)$$

$$\Delta\lambda_{xy} = R_E\left(\frac{1}{L^2}\left|\frac{y\phi_L}{cos\hat{\phi}} - x\lambda_L\right|\Delta L + \left|\frac{\lambda_L}{L}\right|\Delta x + \left|\frac{\phi_L}{Lcos\hat{\phi}}\right|\Delta y\right)$$
$$+ \left|\frac{x}{L}\right|\Delta\phi_L + \left|\frac{y}{Lcos\hat{\phi}}\right|\Delta\lambda_L + \left|\frac{y\phi_L sin\hat{\phi}}{Lcos^2\hat{\phi}}\right|\Delta\hat{\phi} + \Delta\lambda_A \qquad (9)$$

where a maximal error of latitude difference calculation is $\Delta\phi_L = \Delta\phi_F + \Delta\phi_A$, a maximal error of longitude difference calculation is $\Delta\lambda_L = \Delta\lambda_F + \Delta\lambda_A$ and a maximal error of average latitude calculation equals $\Delta\hat{\phi} = (\Delta\phi_F + \Delta\phi_A)/2$.

4   Results

The evaluation of our system has been done on a small lake near the Wroclaw. Because of the lake's size our test is focused on errors of estimation of a ROV's position. Two experiments were provided. In the first one a ROV was trying to keep its position close to a ship's bow and on submersion depth of approximately 2.5 m. In the second one a ROV was trying to keep its position far from a ship nearly between the bow and the stern on the same submersion depth. In both cases measurements have been collected during a period of one minute.

In the Fig. 4 results of experiments are presented. Every point represents one measurement from the set. Positions of ship's bow and stern are calculated thanks to the formula (3), but a note should be done that the GPS measurement taken as a first one at the stern is placed in point $(0,0)$. All consecutive measurements (i.e. stern, bow and ROV positions) are calculated in relation to this position. As can be seen, variation of a position of the stern and the bow

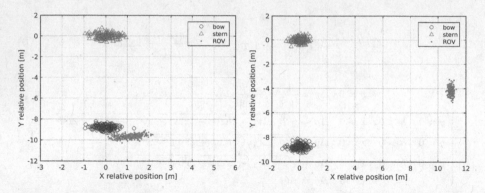

Fig. 4. Two experiments of estimation of a ROV's position in relation to a supervising ship's position. Positions of a stern and a bow is calculated base on GPS measurement. The position of ROV is computed according to (3).

is bounded to the circle with diameter of about two meters, which is coherent with horizontal and vertical accuracy reported by GPS chip during its work. A second observation can be done that for the case when the ROV is close to a reference point (the bow or the stern), errors of position estimation of the ROV are bigger than in case when both distances to reference points are similar.

5 Conclusions

The proposed system has been developed as USBL positioning one, but its capabilities are much wider. To reach the goal our system utilizes a moving ultrasound transmitter and two stable receivers, where both a transmitter and a receiver are completely redesigned regarding conventional solutions. It was done by using modulated ultrasound waveforms with BPSK modulation instead of simple bunches of square-shaped pulses.

Our system allows to realize two working scenario. In the first simple solution, a ROV works as remotely controlled submarine. The ROV is equipped with a transmitter only and receivers are mounted on a ship's stern and bow. In this arrangement a communication between the ROV and a supervising ship take place in one direction only. This allows to determine a ROV's position, but the ROV do not know it. In the second solution a communication can take place in two direction namely from a ROV to a supervising ship and from a ship to a ROV. Thus, a ROV can receive information about its current position. In a consequence a ROV can operate autonomously.

Conventional positioning systems use a simple identical bunches of ultrasound pulses as measurement signals. Due to simplicity of such signals a possibility of miss discrimination is very probable. So, a USBL system is able to work with only one ROV, usually. Our proposal overcomes this limitation. Modulated signals allows to distinguish many ROVs sharing the same basin. Moreover, our system can be used to organize communication link between all of them. It is an important factor because currently in industrial application a communication between a ship and a ROV is made through the cable, which limits operational capabilities of a ROV. Currently our system offers link speed of 1 kb/s, which is not so big, but enough to send some important information e.g. about submersion depth of a ROV or a diver.

References

1. Alcocer, A., Oliveira, P., Pascoall, A.: Underwater acoustics positioning systems based on buoys with GPS. In: Proceedings of the Eighth European Conference on Underwater Acoustics, ECUA, 12–15 June 2006, Carvoeiro, Portugal (2006)
2. Catipovic, J.A.: Performance limitations in underwater acoustic telemetry. IEEE J. Oceanic Eng. **15**(3), 205–216 (1990)
3. Chitre, M., Shahabudeen, S., Stojanovic, M.: Underwater acoustic communications and networking: recent advances and future challenges. Mar. Technol. Soc. J. **42**(1), 103–116 (2008)
4. Christ, R.D., Wernli, S.R.L.: Underwater acoustics and positioning. In: Christ, R.D., Wernli, S.R.L. (eds.) The ROV Manual, pp. 81–124. Butterworth-Heinemann, Oxford (2007). Chapter 4
5. Curcio, J.A., McGillivary, P.A., Fall, K., Maffei, A., Schwehr, K., Twiggs, B., Kitts, C., Ballou, P.: Self-positioning smart buoys, the "un-buoy" solution: logistic considerations using autonomous surface craft technology and improved communications infrastructure. In: OCEANS 2006, pp. 1–5, September 2006
6. Hegrenas, O., Berglund, E., Hallingstad, O.: Model-aided inertial navigation for underwater vehicles. In: IEEE International Conference on Robotics and Automation, ICRA 2008, pp. 1069–1076, May 2008
7. Karras, G.C., Kyriakopoulos, K.J.: Localization of an underwater vehicle using an imu and a laser-based vision system. In: Mediterranean Conference on Control Automation, MED 2007, pp. 1–6, June 2007
8. Kilfoyle, D.B., Baggeroer, A.B.: The state of the art in underwater acoustic telemetry. IEEE J. Oceanic Eng. **25**(1), 4–27 (2000)
9. Menon, M., Dixon, T., Tena, I.: Resolving subsea navigation, tracking and positioning issues by utilising smart ROV control system software. In: 2013 MTS/IEEE on OCEANS, Bergen, pp. 1–8, June 2013
10. Vickery, K.: Acoustic positioning systems. A practical overview of current systems. In: Proceedings of the 1998 Workshop on Autonomous Underwater Vehicles, AUV 1998, pp. 5–17, August 1998
11. Willemenot, E., Morvan, P.Y., Pelletier, H., Hoof, A.: Subsea positioning by merging inertial and acoustic technologies. In: OCEANS 2009, EUROPE, pp. 1–8, May 2009
12. Yoo, T., Hong, S.K., Ryuh, Y.: Fuzzy logic based 2d position estimation for small robotic fish using low cost mems accelerometer. In: 2011 15th International Conference on Advanced Robotics (ICAR), pp. 175–179, June 2011

Hierarchical Data Aggregation Based Routing for Wireless Sensor Networks

Soumyabrata Saha[1($\boxtimes$)], Rituparna Chaki[2], and Nabendu Chaki[3]

[1] Department of Information Technology,
JIS College of Engineering, West Bengal, India
som.brata@gmail.com
[2] A.K.Choudhury School of Information Technology,
University of Calcutta, West Bengal, India
rituchaki@gmail.com
[3] Department of Computer Science and Engineering,
University of Calcutta, West Bengal, India
nchaki@gmail.com

Abstract. Designing routing protocols to serve real time disaster management, is a challenging task involving optimum energy utilization along with reduced delay in event reporting. In this paper, we have focused on the data aggregation techniques which alleviate the constraints of sensor networks and put forward as an essential paradigm for routing in Wireless Sensor Networks. We have presented a hierarchical data aggregation based routing algorithm. Extensive analysis and simulation shows that proposed algorithm outperforms other existing approaches.

Keywords: Wireless sensor networks · Routing · Hierarchical · Data aggregation

1 Introduction

The characteristics of WSNs and its application requirements have direct impact on the design issues in terms of network performance and capabilities. Despite the tremendous and numerous advantages like distributed localized computing, wide area coverage, extreme environment area monitoring, WSNs stance various challenges to the research community. The current trend of utilizing the powers of wireless sensor networks for disaster management and other real time decision making systems have opened up new research domains. The most important challenge remains to deliver the information from the source of occurrence to the reporting station at the quickest possible manner. The routing of such information must ensure uniform energy dissipation across the network, quickly converge irrespective of the network node density, and be flexible in terms of the routing framework and the route computation metric. The main aim of hierarchical routing is to efficiently maintain the energy consumption of sensor nodes by involving them in multi-hop communication within a particular region. Clustering is a network management technique, which creates a hierarchical structure over a flat network and provides scalability and robustness for that network. In a clustered network, as the

© Springer International Publishing Switzerland 2016
N.T. Nguyen et al. (Eds.): ICCCI 2016, Part II, LNAI 9876, pp. 168–179, 2016.
DOI: 10.1007/978-3-319-45246-3_16

correlation of data among neighbor sensor nodes is very high, it has been observed that use of data aggregation is beneficial for energy efficiency.

Data aggregation is the process of aggregating data from multiple sources. It is used to eliminate redundant data transmission and provide fused information to the destination node and communication cost decreases and network lifetime also increases. Researchers have presented several aggregation techniques which are used to combine data coming from different nodes and most of them are related to specific sensor applications.

This paper presents a novel hierarchical data aggregation based energy efficient routing protocol for WSNs. The logic is then evaluated using MATLAB and a comparative analysis against the performance of well-known cluster based aggregation techniques in WSNs.

The rest of the paper is organized as follows. Different data aggregation techniques have also been presented in the Sect. 2. In Sect. 3, we have presented a hierarchical data aggregation based energy efficient routing technique for WSNs. Result analysis has been presented in Sect. 4. In Sect. 5, this paper concludes and identifies some of the future directions with open research issues.

2 Related Works

In this section, we have presented a study on different hierarchical energy efficient routing for WSNs.

The well appreciated LEACH [1] protocol uses cluster based hierarchical model with random rotation of the cluster heads to evenly distribute the energy load among sensor nodes. There have been many extensions of the basic LEACH protocol [10–12], aiming to balance network lifetime vs. delay. Some have used temporally and spatially co-related nodes for [16] cluster formation. A hybrid structure has also been used [18] for combining both tree structure and cluster structure in a bid to ensure maximum utilization of network energy which in turns provides longer lifetime and low network delay. CBR Mobile-WSN [17] is a query base, round free clustering protocol which is used to handle packet loss and reduces the energy consumption and ensures efficient bandwidth utilization.

Improving network lifetime has been one of the most important goals of researchers as seen in countless works such as [6–8, 13–15]. Query driven routing has been preferred by scientists in order to reach the site of eventuality in fast and simple manner, as seen in [24]. Weight based mechanisms have been used for effective high energy efficient data deliveries in WSN, as evident from [25, 28].

Clustering techniques have to provide low overhead of cluster head rotation as well as optimal traffic distribution among cluster heads while keeping network connectivity and coverage. In clustering networks, the imbalanced energy consumption among nodes is the key factor considered to be affecting network lifetime. None of the above algorithms focused on the data aggregation technique to minimize redundant data transmission for achieving energy minimization.

In [3, 6] authors presented different aggregation techniques and scheduling [4] aim for high aggregation without incurring the overhead of structure approaches In [5], the position of the sensor nodes and number of the child node in the aggregation tree determines the aggregation timeout and ensuring maximum data processing time to maximize the opportunity for data aggregation. PEGASIS [2] is a chain based routing that are used collaborative techniques to increase the lifetime of each node and reduces the overhead of dynamic cluster formation in [1] and decreases the number of data transmission volume through the chain of data aggregation.

Different techniques have been used for data aggregation [9, 19, 20, 22] to reduce energy consumption, minimize the energy consumption cost and communication overheads. Aggregation methods are grouped into three classes depending on the level of sensed data as; data level aggregation, feature level aggregation and decision level aggregation. Based on the aggregation strategy, it has divided into three types: in-network query type, data compression type and representative type.

In [19] presented a simple, least time, energy efficient routing protocol with one level data aggregation that increased life time for the network. A structure freer-time data aggregation protocol [23] have considered temporal and spatial convergence of packets and improved network performance. An attribute aware data aggregation technique consisting of a packet driven timing algorithm and a potential based dynamic routing has been presented in [26, 27]. A data density correlation degree clustering method has used for measuring the correlation between a sensor node's and its neighbour's data. [30] Uses a collision free minimum latency aggregation scheduling algorithm for tree based WSNs to achieve an optimal time slot assignment.

Different correlation metrics [20, 21, 29] have used for measuring the correlation of the collected sensor node's data as, Pearson correlation coefficient, Euclidean score etc. Different researches have presented different filtering mechanism [3], as threshold filter, deviation filter, semantic filter, location filter for reducing redundant data and extracting useful information.

Energy efficiency is unanimously considered as one of the core design issue and in order to improve the deficiencies of aforementioned schemes, the challenge is to develop a routing protocol that can meet these conflicting requirements while minimizing compromise. In the next section we propose a hierarchical data aggregation based energy efficient routing for wireless sensor networks.

3   Proposed Routing Protocol

The main goal of the proposed scheme is to apply data aggregation technique on the clustered network and evenly dispense the energy usage among all nodes within the network. This helps to ensure that there are no overly utilized sensors nodes which might run out of energy before the others.

Assumptions: All sensor nodes in the network are immobile, homogeneous and initially all are assigned the same energy and each node knows its initial energy; Each sensor node has a unique identifier; The transmission range of each node is same; The data collections have been done based on the locality and the collection rate of each sensor node is different and unknown to each other.

Network Model: Here we have considered in an example network, where sensor nodes $\{s_1, s_2, \ldots s_n\}$ are randomly distributed and divided into 'i' clusters $\{c_1, c_2, \ldots\ldots, c_i\}$. Each cluster has one cluster head $\{Ch_i\}$ and the 'n' no of cluster members $\{Cm_{in}\}$. Cluster members sense the event of interest in the specified time interval and generate data packets $\{q_1(t), q_2(t), \ldots\ldots, q_n(t)\}$ of fixed size, where the collected information is redundant and few are distinct. Cluster members perform the aggregation technique at node level and send the aggregated data packets to its cluster head. Cluster head receives aggregated data packets from its member nodes, where one cluster member's collected data may be same or distinct comparing with other neighbour cluster member's collected data. The head level aggregation technique is defined as: $\left[f\{d_{Aggr}(Ch_i)\} = \sum_{i=1}^{n} C_i * q_i(t)\right]$. Cluster head transfers the aggregated data packets to its next hop neighbour. Next hop cluster head receives data from its neighbour cluster heads and performs the role of aggregator to communicate the aggregated data packets to the sink node. In the following subsection we are going to define about different proposed aggregation functions.

Aggregation Function Definitions: Let, X_i and Y_j are two variables representing temporal or spatial correlation between number of packets generated by the 'n' cluster members in a cluster, provided that $i = 1 \ldots p$ and $j = 1 \ldots k$, where 'p' & 'k' are finite integer. Depending on the characteristics of sensed data we have categorized aggregation techniques into different cases, those are discussed below:

Case1: Different Sensor Nodes Sense Different Data. In this case the final aggregation value would be Summation of all data receive from all participating nodes. This can be defined as:

$$f\{C_i(fn_s)\} = \sum_{i=1}^{p} X_i \; for \; \forall(X_i), \; where \, i = 1 \ldots p \tag{1}$$

Case 2: Different Sensor Nodes Sense Same Data. In this case the spatial correlations between the data packets are as average or duplicate. This can be defined as:

$$f\{C_i(fn_a)\} = 1/k \sum_{j=1}^{k} Y_j \; for \; \forall(Y_j), \; where \, j = 1 \ldots k \tag{2}$$

Case 3: Different Sensor Nodes Sense Different Value for Same Attribute. In this case different sensor nodes generate different packets containing different value for same attribute then the maximum value from the given set of data is:

$$f\{C_i(fn_{mx})\} = f(S_1, \ldots S_n) = max|S_i|, \; where \; i = 1 \ldots n \tag{3}$$

Minimum value is:

$$f\{C_i(fn_{mn})\} = f(S_1, \ldots S_n) = min|S_i|, \; where \, i = 1 \ldots n \tag{4}$$

Case 4: Nodes Require Median Value of Collected Data. In this case the median function would apply on the collected data and can be defined as:

$$f\{C_i(fn_{mdn})\} = f(S_1, \ldots S_n) = \sum\nolimits_{i=1}^{n} S_r, \text{ where } r = (n+1)/2 \tag{5}$$

The contributions of our work including the distinctive properties are mentioned here with. In this proposed work, data aggregation is done at node level and as well in the cluster head level to eliminate redundant data. As data aggregation has performed both at node level and cluster head level, sink node wouldn't receive any redundant data, and as well as less numbers of packets are transferred in the network.

In this scheme, we have considered the technique [25, 28] for weight based cluster head selection mechanism. The proposed scheme is divided in two models. Section 3.1 is used for cluster formation and tree set up phase. Data aggregation phase has been described in Sect. 3.2 (Table 1).

Table 1. Data dictionary

Variable	Description	Variable	Description
Ch_i	Cluster head	W_{Ch_i}	Weight factor of cluster head
Cm_{in}	Cluster member	Ch_Es	Residual energy of cluster head
Msg_Id	Message Id	$f\{C_i(fn_s)\}$	Sum of data packets
Sn_i	Node Id	$f\{C_i(fn_a)\}$	Average of data packets
E_{S_i}	Residual energy of node S_i	$f\{C_i(fn_{mx})\}$	Maximum value of data packet
ΔTh_{Chs}	Threshold value	$f\{C_i(fn_{mn})\}$	Minimum value of data packet
Val_i^n	Data in bucket	$f\{C_i(fn_{mdn})\}$	Median value of data packet
Cnt_i	Number of data in the bucket	$q_i(t)$	Data packet

3.1 Cluster Formation and Tree Setup Phase

The objective of this phase is to construct clusters of almost same sizes. After cluster head selection process [25, 28] has been completed, cluster head broadcast a CH_ADV_MSG {Ch_i, Msg_Id, W_{Ch_i}, Ch_Es, TTL} for the network nodes. Accordingly, nodes send a reply CM_RPLY_MSG {Ch_i, Sn_i, Nd_lc, W_{Ch_i}, Msg_Id, E_{S_i}, TTL} to join as the cluster member of that cluster. Depends upon the equation $\sum_{i=1}^{n} W_{S_i} \leq (W_{Ch_i})$, each cluster head would add maximum number of sensor nodes to its cluster and sends CHM_FRM_MSG {Ch_i, Cm_{in}, W_{Ch_i}, Ch_Es, E_{S_i}, TTL} to the cluster member of the corresponding cluster head and cluster formation has been executed. Sink node act as root of the tree with its own level to 0. Cluster head belong to transmission range of sink node add himself as the child of sink node and assign its own level incremented by one. The above process is applicable for all the remaining clusters in the network and tree formation has been executed.

Algorithm 1: Algorithm for Cluster Formation and Tree Set Up

Ch_i broadcast a CH_ADV_MSG for the rest of the nodes

For each node in the network:

   If any node is already another cluster head or cluster member

        Then discards the previous message

   Else

        Check the $Max(W_{s_i}, P_{s_i}, E_{s_i})$ and $Min(dist(Sn_i, d_{Ch_i})$ and it sends a reply CM_RPLY_MSG to appropriate Ch_i

        If $\sum_{i=1}^{n} W_{s_i} \leq (W_{Ch_i})$ is TRUE

            Then Ch_i sends CHM_FRM_MSG and corresponding node add as Cm_{in} of corresponding Ch_i

        Else

            Then it discards the previous message

        Ch_i maintains the location information of each member and used for establishing the route to each cluster member

Set the level of sink node to 0

Sink node sends RT_MSG for its neighbour cluster heads.

If $\{dist (Sn_i, d_{Ch_i}) < Th_{dist} \}$ is TRUE

        Then Ch_i copies all the information to its corresponding MSG_HS_TBL and performs operations.

            Increase its own level by one

            Reply with SNC_MSG and becomes the child of the sender node.

            Forward RT_MSG for its next hop neighbour nodes.

If any Ch_i receives the RT_MSG from more than one sender

        Then it will join as a child with $Max(W_{Ch_i})$ and $Min(dist(Sn_i, d_{Ch_i})$

Else

        If any Ch_i receives RT_MSG more than one sender with same level

            Then ignore RT_MSG

        Else Continue

END

By using the above procedure sensor node can join as a cluster member to the cluster head and cluster head add itself as the child node of the sink node and tree formation has been executed.

3.2 Data Aggregation Phase

In this section we have presented an aggregation based technique for the clustered WSN to send the aggregate data to the sink by using data aggregation functions. Cluster member Cm_{in} collects raw data in time instance t_i and stores the data $[x_1, x_2, \ldots x_k]$ in the specified bucket, where the cache size is same i.e. K. The bucket consists of node id, minimum and maximum values of collected data $X_i = \left[\dfrac{\sum_{n=1}^{Cnt_i} Val_i^n}{Cnt_i} \right]$. After specified time instances Cm_{in} collects another data $\left[x'_1, x'_2, \ldots, x'_k \right]$ and store them in the specified bucket with the existing data and calculates x'_i for that new data. We have calculated the adaptive threshold value by using the following equation as; $\left\{ \Delta Th_{val} = \sqrt{\frac{1}{n} \sum_{i=1}^{n} (x_i - \mu)^2} \right\}$ and the inter node data similarity/dissimilarity is measured using the Euclidian Score.

Algorithm 2: Algorithm for Data Aggregation Procedure

Each Cm_{in} calculates $X_i = \left[\frac{\sum_{n=1}^{Cnt_i} Val_i^n}{Cnt_i} \right]$ for a specified time t_i

After time interval T_n, Cm_{in} repeats above step for another collected data set and find X'_i

Using Euclidian Score, calculate data similarity/dissimilarity of different collected data set

   If $|X_i - X'_i| < \Delta Th_{val}$

   Then Cm_{in} selects X_i for performing aggregation, otherwise selects X'_i for aggregation

Repeat the steps for each member node in the network:

   According to decision, Cm_{in} selects data set for applying aggregation techniques and stores in the bucket, $[d_{Aggr}(Cm_{in}) = f\{C_i(fn_s)\} \oplus f\{C_i(fn_a)\} \oplus f\{C_i(fn_{mx})\} \oplus f\{C_i(fn_{mm})\} \oplus f\{C_i(fn_{mdn})\}]$

Cm_{in} sends aggregate data to its Ch_i

Ch_i receives data from its different cluster members, i.e. $[d_s(Ch_i) = \{\sum_{i=1,n=1}^{p} d_{Aggr}(Cm_{in})\}]$

Ch_i apply $[d_{Aggr} = f\{H\{f(d_s(Ch_i)) \oplus f_{(mm)}(d_s(Ch_i)) \oplus f_{\{th \oplus dv \oplus sm\}}(d_s(Ch_i))\}, Cm_{in}\}]$ and aggregated data will be delivered to next hop node

   If next hop is sink node, Then Ch_i can directly communicate with sink and aggregated data $\{d_{Aggr}\}$ will be delivered to sink node

   Else

   If next hop is next level neighbour cluster head, Then Ch_i follows the same process and aggregated data $\{d_{Aggr} = \sum_{i=1,n=1}^{p} d(Ch(Cm)_{in})_i\}$ will be delivered to next neighbour node.

END

Using the above methods data aggregation has been executed.

4   Result Analysis

The simulation model consists of a network model that has a number of wireless nodes, which represents the entire network to be simulated. We have implemented different sets of simulations using the well-known tool MATLAB to evaluate the performance of our proposed HDARP and compare with existing algorithms [1, 2, 21, 24, 25, 28] (Table 2).

Table 2. Simulation parameters

Variable	Description	Variable	Description
Channel	Wireless channel	Sink location	(50, 150)
Propagation	Two ray ground	Initial energy	2 J
Antenna	Omni antenna	d_0	50 m
Number of nodes	0–50	EP_{elec}	50 nJ/bit
Node deployment	Random	EP_{amp}	10 pJ/bit/m^2
WSNs field	(0, 0)–(200, 200) m	E_{aggr}	5 nJ/bit/signal
ε_{cg}	0.0012 pJ/bit/m^4	ε_{fs}	10 pJ/bit/m^2

In this section the performance metrics namely, network lifetime, number of dead nodes in each round, throughput, average energy consumptions have been considered. In Fig. 1, we have depicted a relation between numbers of nodes involved in cluster formations vs. required energy. From this Fig. 1, it has been observed that in the proposed HDAR scheme, node number 40 onwards if more sensor nodes are involved in cluster formation process but almost same energy consumption has been required up to certain threshold and in all situations HDAR requires less energy consumption compared with other existing hierarchical cluster based routing algorithms [1, 21, 24, 25, 28].

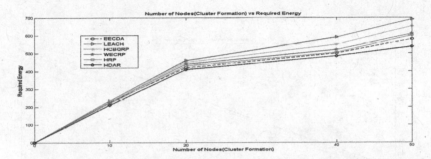

Fig. 1. Number of nodes (cluster formation) vs. required energy

Network Lifetime: It is defined as the time, until when all sensor nodes in the network die out of their energy and it depends upon the average energy spent. We have taken different necessary key factors into consideration in our proposed approaches as; joining of cluster members with the appropriate cluster head for cluster formation and tree setup phase and reconstruction operation balances the load of the whole network. Proposed data aggregation methods help to minimize the redundant data transmission and less energy consumption has required and all these aspects contribute to prolong the network lifetime. From the Fig. 2, we have observed that in the proposed HDAR scheme the network lifetime is greater than the existing approaches [1, 21, 24, 25, 28].

Throughput: It is defined as the total number of message send or received in per time unit. In this proposed HDAR scheme, throughput is measured in terms of data delivery in per time unit by varying the number of sensor node. Here cluster members have applied data aggregation technique on the collected data and transmitted to the cluster head, which has also performed data aggregation mechanism and send the aggregated data to sink.

From Fig. 3, we have observed that HDAR achieves better throughput comparative with [1, 21, 24, 25, 28]. In our proposed protocol cluster head executes its responsibility up to the specified threshold value, if the residual energy of any cluster head goes below the specified threshold, then that cluster head is released from its responsibilities and another sensor node would select as the new cluster head.

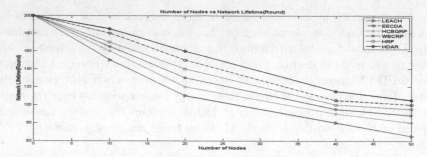

Fig. 2. Node number vs. network life time (round)

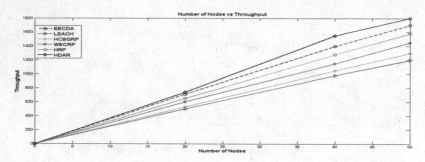

Fig. 3. Node number vs. throughput

In HDAR technique clustering, re-clustering, data aggregation methods and communication phases have been designed with care such that carefully that less numbers of sensor nodes are dead due to lack of energy. According to Fig. 4, as less numbers of dead nodes are present in HDAR and it has mentioned that less energy has been dissipated in the new proposed scheme comparative with the existing approaches [1, 21, 24, 25, 28].

Average Energy Consumption: The following parameters are involved for average energy consumption as; total numbers of nodes are present in the network; energy consumed for single transmission by each node; total number of transmissions by all sensor nodes; cluster head selection and cluster formation; data aggregation by cluster members and as well as cluster heads.

In [1, 2, 24, 25, 28], existing cluster/chain based routing protocols don't consider data aggregation approach while making the communication with the sink node, whereas in [21], author considered data aggregation technique along with the cluster formation and communication technique. From the Fig. 5, we have observed that our proposed protocol HDAR includes both the cluster formation, re-clustering mechanism along with data aggregation technique that have been performed by both cluster members and cluster head.

As a result, number of packet transmission from source to sink has been decreased which helps to minimize the average energy consumption for proposed HDAR protocol

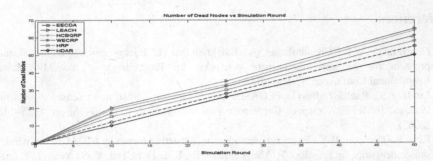

Fig. 4. No. of dead node vs. simulation round

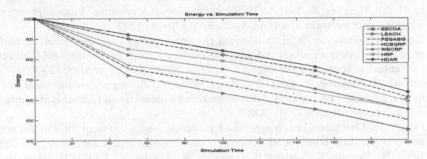

Fig. 5. Energy vs. simulation time

in comparing with other existing approaches as LEACH [1], PEGASIS[2], EECDA [21], HCBQRP [24], WECRP [25], and HRP [28].

5 Conclusions

The significant performance improvement has observed for proposed algorithm in terms of energy, network lifetime, throughput, average packet delivery ratio. As the algorithm is very simple and no complex computation are there and it provides less overheads and less complexity. The total energy consumption in the network is low and the network lifetime time has also increased. As our study reveals, it is not possible that a routing algorithm is suitable for all scenarios. Although many routing protocols have proposed in WSNs, many issues still exist and there are still many challenges that need to be solved in sensor networks. The future vision of WSNs is to embed numerous distributed devices to monitor and interact with physical world phenomena, and to exploit spatially and temporally dense sensing and actuation capabilities of those sensing devices.

References

1. Heinzelman, W.R., Chandrakasan, A., Balakrishnan, H.: Energy efficient communication protocol for wireless micro-sensor networks. In: Proceedings of the 34th Hawaii International Conference System (2000)
2. Lindsey S., Raghavendra C.: PEGASIS: power-efficient gathering in sensor information systems. In: IEEE Aerospace Conference Proceedings, vol. 3, no. 9–16, pp. 1125–1130 (2002)
3. Kim, K.-T., Youn, H.Y.: Energy-driven adaptive clustering hierarchy (EDACH) for wireless sensor networks. In: Enokido, T., Yan, L., Xiao, B., Kim, D.Y., Dai, Y.-S., Yang, L.T. (eds.) EUC-WS 2005. LNCS, vol. 3823, pp. 1098–1107. Springer, Heidelberg (2005)
4. Fasolo, E., Rossi, M., Widmer, J., Zorzi, M.: In-network aggregation techniques for wireless sensor networks: a survey. IEEE Wirel. Commun. **14**(2), 70–87 (2007)
5. Li, H., Yu, H., Yang, B., Liu, A.: Timing control for delay constrained data aggregation in wireless sensor networks. Commun. Syst. **20**(7), 875–887 (2007)
6. Ziaoyan, C., Zhao, L.: BCEE: a balanced clustering, energy-efficient hierarchical routing protocol in wireless sensor networks. In: Proceedings of the IEEE International Conference on Network Infrastructure and Digital Content (ICNIDC 2009), pp. 26–30 (2009)
7. Xuxing, D., Fangfang, X., Qing, W.: Energy balanced clustering with master/slave method for wireless sensor networks. In: Ninth International Conference on Electronic Measurement and Instruments (ICEMI 2009) (2009)
8. Fang, L.L., Chengchew, L.: Weight-based clustering routing protocol for wireless sensor networks. In: IEEE International Symposium on IT in Medicine & Education 2009 (2009)
9. Jiman, H., Joongjin, K., Sangjun, L., Dongseop, K., Sangho, Y.: T-LEACH: the method of threshold based cluster head replacement for wireless sensor networks. Inf. Syst. Front. **11**, 513–521 (2009)
10. Omer, F.M., Basit, D.A., Asadullah, S.G.: MR-LEACH: multi-hop routing with low energy adaptive clustering hierarchy. In: Fourth International Conference on Sensor Technologies and Applications (2010)
11. Abdulsalam, M., Kamellayla, M.: W-LEACH: weighted low energy adaptive clustering hierarchy aggregation algorithm for data streams in wireless sensor networks. In: IEEE International Conference on Data Mining Workshops, pp. 1–8 (2010)
12. Li, H., Jianghong, S., Yang, Q., Zhang, D.: An energy efficient hierarchical routing protocol for long range transmission in wireless sensor networks. In: 2nd International Conference on Education and Computer (2010)
13. Liu, Y., Xiong, N., Zhao, Y., Vasilakos, A.V., Gao, J., Jia, Y.: Multilayer clustering routing algorithm for wireless vehicular sensor network. IET Commun. **4**(7), 810–816 (2010)
14. Khattak, A.U., Shah, G.A., Ahsan, M.: Two tier cluster based routing protocol for wireless sensor networks. In: IEEE/IFIP 8th International Conference on Embedded and Ubiquitous Computing (2010)
15. Ying, T., Yang, O.: A novel chain cluster based routing protocol for mobile wireless sensor networks. In: 6th IEEE International Conference on Wireless Communications Networking and Mobile Computing (2010)
16. Misra, S., Thomasinous, P.D.: A simple, least-time, and energy-efficient routing protocol with one-level data aggregation for wireless sensor networks. J. Syst. Softw. **83**(5), 852–860 (2010)
17. Kumar, D., Trilok, C.A., Patel, R.B.: Multi-hop communication routing protocol for heterogeneous wireless sensor networks. Int. J. Inf. Technol. Commun. Convergence **1**(2), 130–145 (2011)

18. Wang, C., Jiang, C., Tang, S., Li, X.: Select cast: scalable data aggregation scheme in wireless sensor networks. IEEE Trans. Parallel Distrib. Syst. **23**(10), 1958–1969 (2011)
19. Jingsha, H., Yang, M., Xuguang, S.: Research on secure data aggregation in wireless sensor networks based on clustering method. In: IEEE International Conference on International Technology and Applications (2011)
20. Tharini, C., Ranjan, P.V.: An energy efficient spatial correlation based data gathering algorithm for wireless sensor networks. Int. J. Distrib. Parallel Syst. **2**(3), 16–24 (2011)
21. Kumar, D., Aseri, T.C., Patel, R.B.: EECDA: energy efficient clustering and data aggregation protocol for heterogeneous wireless sensor networks. Int. J. Comput. Commun. Control **VI**(1), 113–124 (2011)
22. Yousefi, H., Yeganeh, M.H., Alinaghipour, N., Movaghar, A.: A structure free real time data aggregation in wireless sensor networks. J. Comput. Commun. **35**(9), 1132–1140 (2012)
23. Kim, J.J., Shin, I., Zhang, Y.S., Kim, D.O., Han, K.J.: Aggregate queries in wireless sensor networks. Int. J. Distrib. Sensor Netw. **2012** (2012)
24. Saha, S., Chaki, R.: Hierarchical cluster based query-driven routing protocol for wireless sensor networks. In: Satapathy, S.C., Avadhani, P.S., Abraham, A. (eds.) Proceedings of the International Conference on Information Systems Design and Intelligent Applications 2012 (INDIA 2012) held in Visakhapatnam, India, January 2012. AISC, vol. 132, pp. 657–667. Springer, Heidelberg (2012)
25. Saha, S., Chaki, R.: Weighted energy efficient cluster based routing for wireless sensor networks. In: Cortesi, A., Chaki, N., Saeed, K., Wierzchoń, S. (eds.) CISIM 2012. LNCS, vol. 7564, pp. 361–373. Springer, Heidelberg (2012)
26. Ren, F., Zhang, J., Wu, Y., He, T., Chen, C., Lin, C.: Attribute-aware data aggregation using potential-based dynamic routing in wireless sensor networks. IEEE Trans. Parallel Distrib. Syst. **24**(5), 881–892 (2013)
27. Yuan, F., Zhan, Y., Wang, Y.: Data density correlation degree clustering method for data aggregation in WSN. Sensors J. **14**(4), 1089–1098 (2014)
28. Saha, S.B., Chaki, R.: Hierarchical routing protocol for wireless sensor network. In: Proceedings of 8th International Multi Conference on Information Processing (2014)
29. Yousefi, H., Malekimajd, M., Ashouri, M., Movaghar, A.: Fast aggregation scheduling in wireless sensor networks. IEEE Trans. Wirel. Commun. **14**(6), 3402–3414 (2015)
30. Tang, D., Li, T., Ren, J., Wu, J.: Cost-aware secure routing protocol design for wireless sensor networks. Trans. Parallel Distrib. Syst. **26**(4), 960–973 (2015)

Towards Cloud-Based Data Warehouse as a Service for Big Data Analytics

Hichem Dabbèchi[1]([⊠]), Ahlem Nabli[1,2]([⊠]), and Lotfi Bouzguenda[1]([⊠])

[1] MIRACL Laboratory, University of Sfax, Route de Tunis, Km 10, BP 242 3021,
Sakeit Ezzit, Sfax, Tunisia
dabb.hich@gmail.com, ahlem.nabli@fsegs.rnu.tn,
lotfi.bouzguenda@isimsf.rnu.tn
[2] Faculty of Computer Science and Information Technology, Al-Baha University,
Al-Baha, Kingdom of Saudi Arabia

Abstract. Business intelligence (BI) solutions help managers to make decisions. Big Data and Cloud Computing are both the most important thechnologies that offer new opportunities for business intelligence and data analytics systems. However, traditional data warehouse must be revised to provide business intelligence services based on cloud computing from big data sources. This data in these systems is collected from a variety of sources and stored in various types. Consequently, they need a high performance information technology infrastructure that provides superior computational efficiency and storage capacity. One possible way to deal with new data warehouse architecture design is the use of cloud computing paradigm. This latter offers useful methods, platforms and services that manage in an efficient way this massive data during the processing, computing, storage, and analyzing steps. In this paper, we propose a new cloud based data warehouse architecture for big data analytics perspective. More precisely, we detail the proposed layers such as data warehouse infrastructure, platform and analytics software as a service for supporting big data analytics.

Keywords: Data warehouse as a service · Big data · Cloud computing · NoSQL database · Big data analytics

1 Introduction

Big data analytics systems cover several private and public service areas such as information retrieval engines, social medias, electronic-commerce sites and multimedia, as well as a variety of scientific research areas such as business intelligence, bioinformatics, environment and so on... [10]. From design point of view, big data are characterized by very large data volume and velocity, highly variety in data types and sources, and stringent requirements of data veracity [5,14,17].

In the era of big data, the complexity, diversity of big data analytics systems and the emergence of new systems give rise to new challenges in how to explore and analyze over big data to benefit from structured and unstructured data and

© Springer International Publishing Switzerland 2016
N.T. Nguyen et al. (Eds.): ICCCI 2016, Part II, LNAI 9876, pp. 180–189, 2016.
DOI: 10.1007/978-3-319-45246-3_17

extract values and knowledge from it [9]. Cloud computing seems to be a convenient solution for managing and processing big data repositories [20], thanks to its innovative characteristics. It provides access to large amounts of computing power by aggregating resources and offering a single system view. Cloud computing is becoming a powerful architecture to perform large-scale and complex computing, and has revolutionized the way that computing infrastructure is abstracted and used [13]. In addition, an important aim of these technologies is to deliver computing as a solution for handling big data such as large-scale, multi-media and high dimensional data sets.

There is growing demand for information systems able to assist companies in their decision-making processes. To support these important decisions that companies, politicians, and institutions have to make, it has become necessary to analyze large amounts of data in order to obtain up-to-data and relevant knowledge [2,3,7].

The problem addressed in this paper is the following: how to redefine traditional data warehouse architecture to provide business intelligence services based on cloud computing from big data sources?

The rest of the paper is organized as follows. In Sect. 2 we discuss some related works, in Sect. 3 we introduce and detail our cloud data warehouse architecture layers, and finally, we draw the conclusions and future work.

2   Related Works

Big data analytics is a new area. Without the emergence of new information and communication technologies, there would be no big data phenomenon. Data will be more extreme in the future (e.g. with the four Vs) and new techniques needed, making it possible to analyze this data [16]. New data warehouse and database technologies introduced to address this problem [8,18]. In this section we discuss some related works in big data analysis in order to introduce the study framework of our approach.

The amount of data has been increasing and data set analyzing become more competitive. The challenge is not only to collect and manage vast volume and different type of data, but also to extract meaningful value from it [4]. Also needed for managers and analysts with an excellent insight of how big data can be applied. There are a number of solutions and actual researchs trends related to big data analytics systems. At present, the companies who have occupied big data analytics are EMC, IBM, Microsoft, Amazon, Google, Snaplogic, Oracle, SAP, etc., which mainly provide big data storage and analytics services. In the following, we provide an overview on the most significant of them. For example, The Ophidia project [11] is a research effort on big data analytics facing scientific data analysis challenges in the climate change domain. It provides parallel (server-side) data analysis, an internal storage model and a hierarchical data organization to manage large amount of multidimensional scientific data. EMC offers for enterprises big data storage and analysis services. Greenplum is data storage and analysis tool set of EMC, which consists of three parts:

Greenplum Database, Greenplum HD and Isilon Greenplum database manages, storages and analyzes PB-level data. Greenplum HD is the commercial branch of Hadoop, it allows user to use Hadoop for Big Data Analytics without considering the complexity of Hadoop versions. Isilon clustered storage is a scale-out Network Attached Storage (NAS) platform, it can support storing 15 PB data in a single file system and easy to manage. Amazon provides independent big data analysis services though AWS Marketplace. Microsoft is also provide big data analysis service through Windows Azure MarketPlace. Google offers Google BigQuery to support big data analysis. SnapLogic provides enterprises for big data processing service solutions which help them to obtain value by analyzing both business data and external data.

In conclusion, we believe that companies will need to develop a physical architecture that will not be constrained by a single platform like the RDBMS. That next-generation data warehouse will be able to consist of a numerous technologies that will be extremely flexible and scalable. Therefore, it is necessary to have a single infrastructure which provides common functionality of big data management, and flexible enough to handle different types of big data and big data analysis tasks. In the next section, we will give an overview of our Cloud Data Warehouse architecture that addresses the previous drawbacks.

3   A Novel Cloud Data Warehouse Architecture

A cloud data warehouse architecture is one that is built upon a large number of low cost computers to meet the needs of storing and computing big data in BI applications. It is a new way to get insight from big data; it is also a new form of service-oriented decision support systems. Cloud data warehouse architecture encapsulates various big data storage, management, and analytics techniques into services and the users just concern on what they want and get the service whenever and wherever to store, search, analyze and visualize the data. The end users of the architecture means all outside users that would use the cloud data warehouse, it contains administrators, developers and consumers. The service consumer can be divided into normal user and professional user.

Using service-oriented decision support systems (DSS in cloud) is one of the major trends for many organizations in hopes of becoming more agile [1, 21, 22]. In this section, after defining our contribution, we propose a conceptual framework for DSS in cloud named "Cloud Data Warehouse Architecture".

Analogous to the cloud architecture, the cloud data warehouse architecture is divided into three layers shown vertically. These can be roughly categorized by level of abstraction from infrastructure to analytics software, as visualized in Fig. 1. Cloud data warehouse architecture provides different levels of services, including infrastructure level, platform level and analytics software level. These services can be easily used and integrated into other systems. By encapsulating the complex details, cloud data warehouse architecture offers great opportunities to create new business values.

In the following sections we will detail the three layers of our proposed architecture.

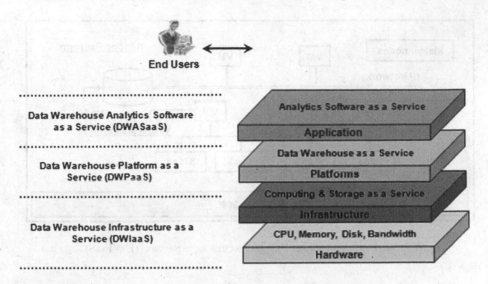

Fig. 1. Cloud Data Warehouse Architecture Layers.

3.1 Data Warehouse Infrastructure as a Service (DWIaaS)

Starting from the bottom layer, any Cloud Data Warehouse infrastructure provides necessary computing and storage capacity for big data. Data warehouse infrastructure as a service can leverage Infrastructure as a Service (IaaS) in cloud computing, including Storage as a Service (SaaS) and Computing as a Service (CaaS), to store and process the massive data. Big data impose significant challenges to the traditional infrastructure due to its characteristics. These big data need to be supported by a new type of Infrastructure tailored for big data which must have the performance to provide fast data access and process to satisfy users just in neededs times. One of the challenges of designing Data Warehouse Infrastructure as a Service is the requirement to support many different data types, not only the existing data types but also the new types that are emerged. It is noted that the distributed file systems and NoSQL DBMS are part of this layer to support unstructured data and performe parellel processing.

As shown in Fig. 2, the data warehouse infrastructure as a service (DWIaaS) is defined through two sub-layers, namely the hardware resources layer and the infrastructure resources layer. The first layer offers low level abstractions of physical devices that can be a server or a simple machine and the virtualization management tool used to deploy large numbers of virtual machines (VMs) on hardware. Each VMs will be used as a computing and storage node. There are two types of nodes, the master node and the slave node. The infrastructure resources layer is the software part deployed in VMs to provide Hadoop as a service (HDaaS). HDaaS is the deployment of the open source hadoop framework [12]. It is composed through two services, namely computing as a service (CaaS) and storage as a service (STaaS). CaaS is the implementation of MapReduce paradigm [15] for the distribution of data processing in each VM. STaaS is the data storage service. In this

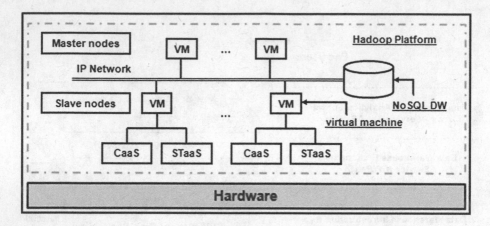

Fig. 2. Data Warehouse Infrastructure as a Service (DWIaaS).

service, we choose to use two storage techniques such as the distributed data storage systems and the NoSQL data stores [19]. The Hadoop Distributed File System (HDFS) [6] was used as a distributed data storage systems to store Big Data from various sources in each VM to process them later with CaaS. HDFS is a distributed file system designed to run on commodity hardware. It is highly fault-tolerant and is designed to be deployed on low-cost hardware. The NoSQL data stores represent another storage technique that uses the NoSQL DBMS to store the data warehouse. NoSQL database, also called Not Only SQL, is useful for very large sets of distributed data. It is especially useful when an enterprise needs to access and analyze massive amounts of unstructured data or data that stored remotely on multiple virtual servers in the cloud as in our case.

3.2   Data Warehouse Platform as a Service (DWPaaS)

The next layer up in the tiered Cloud Data Warehouse Architecture model is Data Warehouse Platform as a Service. Here the service provider puts not only a data management infrastructure in place but also the execution environment for data processing applications and scripts called Data Warehouse as a Service. The NoSQL data stores and distributed data are queried using query languages form the platform layer of cloud data warehouse architecture. This layer provides the logical model for the data warehouse stored in the NoSQL databases. The Data Warehouse as a Service allows users to build analytic applications on top of large data sets. Efficient integration of different data sources and combined data from different sources to provide end users with an unified view of these data is the challenges for the data warehouse as a service. How to make efficient data warehouse as a service with the 4 V characteristics of big data is become a new research direction for business intelligence.

As shown in Fig. 3, the data warehouse platform as a service (DWPaaS) is defined through six main services: DW Design as a Service (DWDaaS),

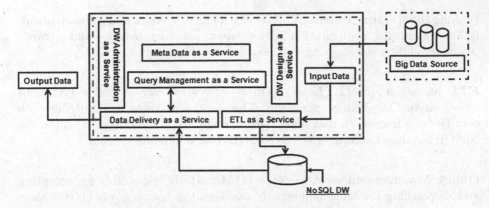

Fig. 3. Data Warehouse Platform as a Service (DWPaaS).

Meta Data as a Service (MDaaS), DW Administration as a Service (DWAaaS), ETL as a Service (ETLaaS), Query Management as a Service (QMaaS) and Data Delivery as a Service (DDaaS).

DW Design as a Service (DWDaaS). Is the service responsible of data warehouse design and modeling. The purpose of design service is to design back and front office of services according to the needs of customers and the skills/capabilities of service providers. These services offer an on-demand data warehouse design in order to ensure a powerful toolbox to help end users deal with internal challenges, to design and implement a data warehouse model in NoSQL databases.

Meta Data as a Service (MDaaS). Allows metadata management to facilitate information sharing and exchange between all services. This service maintains metainformation about the data warehouse, the system configuration and its status, available data sources, registered end users, and the available nodes. This includes the following: connection parameters such as NoSQL database location and credentials, metadata such as data sets contained in the cluster, replica locations and data partitioning properties. It stores all the information about the tables, their partitions, the schemas, the columns and their types, the table locations etc. This information can be queried or modified using a web service interface and as result it can be called from clients in different web service. Therefore, all other services of DWPaaS interact with the MDaaS.

DW Administration as a Service (DWAaaS). Offers a web-based tool for administrators to manage end users accounts, to customize services configuration and to report some information on platform usage and performance. DWAaaS provides installation, management and support for physical and virtual servers running the operating systems. This service includes systems and software installation, monitoring, configuration, management, patching, backups and recovery planning, and troubleshooting and responding to incidents.

ETL as a Service (ETLaaS). Offers an ad-hoc way to define ETL jobs based on the MapReduce paradigm. This service executes a MapReduce job over Hadoop framework that reads data stored in HDFS and insert them in the NoSQL databases according to the NoSQL data warehouse schema.

Query Management as a Service (QMaaS). Is responsible for accepting and dispatching incoming requests. It manages the life cycle of the end users queries during compilation, optimization and execution phases. Current data warehouse were designed for the scenario where sources are fixed and hard to change. Wich is not the case in the cloud, resources are available on-demand and can be changed at any time, something a traditional data warehouse can't do. QMaaS is the service in charge of executing end users queries. The metadata stored in the MDaaS is used by QMaaS to generate the query execution plan.

Data Delivery as a Service (DDaaS). Is responsible for delivering data from the data storage systems to the top architecture layer (DWASaaS). This data can be used and processed by the data warehouse analytics software as a service to execute scripts and queries and as result it can generate reports, visualizations, and dashboards.

3.3   Data Warehouse Analytics Software as a Service (DWASaaS)

Finally, the data warehouse analytics software as a service layer contains big data analytics services. It is the process of studying large amounts of data of various types to uncover hidden patterns, unknown correlations and other useful information. The big data analytics algorithms are complex and far beyond the reach of most organization's IT capabilities. Therefore, more and more organizations turn to data warehouse analytics software as a service to obtain the business intelligence service that turns their unstructured data into an enhanced asset. Cloud data warehouse architecture exploits massive amounts of structured and unstructured data to deliver real-time and intelligent results. In contrast to that, the end users of an Analytics Software as a Service (ASaaS) offering would be more familiar with interacting with an analytics platform on a higher abstraction level, that is, they would typically execute scripts and queries that data scientists or programmers developed from them, to generate reports, visualizations, and dashboards.

As shown in Fig. 4, various big data analytics approaches can be implemented and encapsulated into services such as:

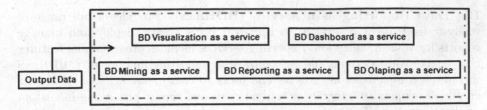

Fig. 4. Data Warehouse Analytics Software as a Service (DWASaaS).

Big Data Visualization as a Service (BDVaaS). Is the presentation of data in a pictorial or graphical format. For centuries, people have depended on visual representations such as charts and maps to understand information more easily and quickly. As more and more data is collected and analyzed, decision makers at all levels welcome data visualization software as a service that enables them to see analytical results presented visually, find relevance among the millions of variables, communicate concepts and hypotheses to others, and even predict the future. BDVaaS conveys information in a universal manner and makes it simple to share ideas with others.

Big Data Mining as a Service (BDMaaS). Includes extracting and analyzing huge amounts of data to discover models from big data. Extracting information from big data takes two major forms: prediction and description. It is tough to know what the data shows with classical methods. BDMaaS lets end users design, create, and visualize data mining models that are constructed from big data by using a wide variety of data mining algorithms.

Big Data Olaping as a Service (BDOLAPaaS). Is an online analytical processing service used to analyze the big data and make sense of information possibly spread out across multiple NoSQL databases, or in the Hadoop Distributed File System. It supports OLAP by letting end users design, create, and manage multidimensional structures that contain data aggregated from big data. BDOLAPaaS is computer processing that enables end users to easily and selectively extract and view big data from different points of view. This service provides a number of OLAP data cube operations allowing querying and analysis of the big data.

Big Data Dashboard as a Service (BDDaaS). Is a user interface, showing a graphical presentation of the current status and historical trends of an organization's key performance indicators to enable instantaneous and informed decisions to be made at a glance. It is a whole new kind of dashboard software who is interactive, creative, analytical, customizable and very powerful graphics. It lets end users access and consolidate data in a few clicks, then visualize that information in interactive dashboard.

Big Data Reporting as a Service (BDRaaS). Provides a full range of ready-to-use tolls and services to help end users create, deploy and manage reports for your organization. Reporting services includes programming features that enable you to extend and customize your reporting functionality. BDRaaS includes a complete set of tools for end users to create, manage and APIs that enable developers to integrate or extend data and report processing in custom applications. With BDRaaS, end users can create interactive, tabular, graphical or free-form reports from NoSQL database.

By this way, users will be able to interact with web-based analytics services easily without worrying about the data storage, management, and analyzing procedures.

4 Conclusion and Future Work

This paper has studied in depth some existing approaches devoted to the big data analytics systems design. Although these propositions are powerful, they suffer, however, of the taking into account the different format of data and the analytic perspective as defined previously. More precisely, this paper has:

– defended the use of cloud computing technology to deal with big data ware-house designing and development issues;
– proposed our cloud based data warehouse architecture considering the three complementary levels services, infrastructure, platform and analytics software.
– detailed the **D**ata **W**arehouse **I**nfrastructure, **P**latform and **A**nalytics **S**oftware as a **S**ervice (respectively called DWIaaS, DWPaaS and DWASaaS).

As future work, we plan to implement our proposed architecture and tested it in industrial case study and within a scalable context. We have already started with the establishment of the infrastructure of the data warehouse using some free tools. More precisely we have used the Hadoop platform and we have tested the MapReduce paradigm and HDFS. The next step is to develop services proposed in the data warehouse platform layer.

References

1. Big data-as-a-service: a market and technology perspective. Technical report, EMC Solution Group (2012)
2. Agrawal, D., Das, S., El Abbadi, A.: Big data and cloud computing: current state and future opportunities. In: Proceedings of the 14th International Conference on Extending Database Technology, pp. 530–533. ACM (2011)
3. Aloisioa, G., Fiorea, S., Foster, I., Williams, D.: Scientific big data analytics challenges at large scale. In: Proceedings of Big Data and Extreme-scale Computing (BDEC) (2013)
4. Bakshi, K.: Considerations for big data: architecture and approach. In: Aerospace Conference, 2012 IEEE, pp. 1–7. IEEE (2012)

5. Bhatia, A., Vaswani, G.: Big data-a review. In: IEEE International Journal of Engineering Sciences & Research Technology IJESRT (2013)
6. Borthakur, D.: The hadoop distributed file system: architecture and design. Hadoop Project Website **11**(2007), 21 (2007)
7. Chaudhuri, S.: What next?: a half-dozen data management research goals for big data and the cloud. In: Proceedings of the 31st symposium on Principles of Database Systems, pp. 1–4. ACM (2012)
8. Chaudhuri, S., Dayal, U., Narasayya, V.: An overview of business intelligence technology. Commun. ACM **54**(8), 88–98 (2011)
9. Cuzzocrea, A., Bellatreche, L., Song, I.: Data warehousing and OLAP over big data: current challenges and future research directions. In: Proceedings of the Sixteenth International Workshop on Data Warehousing and OLAP, DOLAP 2013, San Francisco, CA, USA, 28 October 2013, pp. 67–70 (2013)
10. Cuzzocrea, A., Song, I.Y., Davis, K.C.: Analytics over large-scale multidimensional data: the big data revolution! In: Proceedings of the ACM 14th International Workshop on Data Warehousing and OLAP, pp. 101–104. ACM (2011)
11. Fiore, S., DAnca, A., Palazzo, C., Foster, I., Williams, D.N., Aloisio, G.: Ophidia:toward big data analytics for escience. Procedia Comput. Sci. **18** (2013)
12. Hadoop, A.: Hadoop (2009)
13. Ji, C., Li, Y., Qiu, W., Awada, U., Li, K.: Big data processing in cloud computing environments. In: 12th International Symposium on Pervasive Systems, Algorithms and Networks (ISPAN), pp. 17–23. IEEE (2012)
14. Kataria, M., Mittal, M.P.: Big data: a review. Int. J. Comput. Sci. Mob. Comput. **3**(7), 106–110 (2014)
15. Lämmel, R.: Googles MapReduce programming ModelRevisited. Sci. Comput. Program. **70**(1), 1–30 (2008)
16. ODriscoll, A., Daugelaite, J., Sleator, R.D.: Big data, hadoop and cloud computing in genomics. J. Biomed. Inform. **46**(5), 774–781 (2013)
17. Sagiroglu, S., Sinanc, D.: Big data: a review. In: International Conference on Collaboration Technologies and Systems (CTS), 2013, pp. 42–47. IEEE (2013)
18. Sangupamba, O.M., Prat, N., Comyn-Wattiau, I.: Business intelligence and big data in the cloud: opportunities for design-science researchers. In: Indulska, M., Purao, S. (eds.) ER Workshops 2014. LNCS, vol. 8823, pp. 75–84. Springer, Heidelberg (2014)
19. Strauch, C., Sites, U.L.S., Kriha, W.: NoSQL databases. Stuttgart Media University, Lecture Notes (2011)
20. Vaquero, L.M., Rodero-Merino, L., Caceres, J., Lindner, M.: A break in the clouds: towards a cloud definition. ACM SIGCOMM Comput. Commun. Rev. **39**(1), 50–55 (2008)
21. Xinhua, E., Han, J., Wang, Y., Liu, L.: Big data-as-a-service: definition and architecture. In: 15th IEEE International Conference on Communication Technology (ICCT), pp. 738–742. IEEE (2013)
22. Zheng, Z., Zhu, J., Lyu, M.R.: Service-generated big data and big data-as-a-service: an overview. In: IEEE International Congress on Big Data (BigData Congress), pp. 403–410. IEEE (2013)

Performance Comparison of Sensor Implemented in Smartphones with X-IMU

Juraj Machaj, Jan Racko, and Peter Brida[✉]

Department of Telecommunications and Multimedia, University of Zilina,
Univerzitna 1, 010 26 Zilina, Slovakia
{juraj.machaj,jan.racko,peter.brida}@fel.uniza.sk

Abstract. In this paper a comparison of inertial sensors in smartphones and X-IMU (Inertial Measurement Unit) is presented. The goal of the experiment is to compare the performance of inertial sensors implemented in smartphones with special IMU. The orientation of the devices will be compared. Measuring data from accelerometer and gyroscope provide orientation estimation in three dimensional space and for this purpose orientation in all three axes is needed. Accelerometer measures acceleration and gyroscope measures angular velocity. Orientation can be calculated by using one sensor, but both are affected by negative parameters which make estimation imprecise. Accelerometers measure all forces acting on it including gravitation. This fact can be used to estimate orientation, however, output data of accelerometer are quite noisy. Another possibility how to obtain orientation estimate is integration of gyroscopes data, but this estimation is insufficient due to bias. Combination of output data from both sensors, more precise orientation estimation can be obtained. Combination of sensors is called sensor fusion and is done by using Complementary filter based on Euler angles.

Keywords: Data fusion · Complementary filter · Orientation · Inertial measurement unit

1 Introduction

Nowadays smartphones become a part of daily usage. Because of implementation of relevant equipment, smartphones suits as appropriate devices in various fields. One of these fields is LBSs (Location Based Services) [1]. Currently, the most commonly GNSSs (Global Navigation Satellite Systems) are used for position estimation in LBSs. Although, GNSSs work correctly in outdoor environment, in indoor environment their performance is not sufficient. Problem with GNSSs positioning is that received radio signal from satellite has low power level which is caused by attenuation of radio signal by building walls. Another negative effect is that signal is propagated in complex multipath environment which leads to high position estimation errors. Due to these two drawbacks GNSSs cannot be used in the indoor environment, thus novel positioning systems are developed.

In last few years positioning systems based on gathering data from wireless networks to estimate position are being proposed. These systems commonly utilize different

© Springer International Publishing Switzerland 2016
N.T. Nguyen et al. (Eds.): ICCCI 2016, Part II, LNAI 9876, pp. 190–199, 2016.
DOI: 10.1007/978-3-319-45246-3_18

network technologies like ZigBee [2], WiFi [3–5] or Bluetooth [6]. Readings from magnetometer can also be used for mobile positioning, called magnetic fingerprinting [7, 8]. Mentioned systems can work separately but provide lack of localization accuracy. Combining systems with IMU can be achieved better accuracy.

Accelerometers and gyroscopes are basic components of IMUs. Data from both sensors can be separately used to estimate information about orientation. However, combining of their output data can provide significantly better estimation results. Combination can be performed by using different techniques of sensor fusion e.g. complementary or Kalman filtering. Kalman filter is more complex, thus needs higher computing power and can have longer computation time. Because of these facts complementary filter has been chosen in our work. Quality of sensors can significantly affect accuracy of position estimation. However, in recent years smartphones were equipped with more precise IMUs. In this paper, we analyze and compare the output data from inertial sensors of two smartphones and X-IMU. Data measured within given orientation provides us with information about differences between outputs from smartphone and IMUs sensors.

The rest of paper will be divided as followed: In Sect. 2 will be presented related work of positioning with IMU. In Sect. 3 sensors in smartphone and X-IMU used for measurement will be described. Section 4 explains principle of working of complementary filter. Discussion about scenarios and results will be in Sects. 5 and 6. Section 7 will conclude the paper and provide some information about future work.

2   Related Works

Smartphones equipped with appropriate inertial sensors have a wide range of use. In [9] is presented navigation system called LifeMap. This system combines inertial sensors in the smartphone with GPS and Wi-Fi positioning systems. Presented system provides solution for localization in both indoor and outdoor environments with room-level accuracy.

A new ZUPT (Zero Velocity Update) algorithm was presented in [10]. Algorithm is suited for the shoe-mounted IMU, cascade Kalman and particle filter were used. In the bottom layer Kalman filters is utilized to remove angle and position errors. Step length and heading angle changes are used by particle filer as the state variables, while heading angles coordinates and horizontal position are used as variables for PDR (Pedestrian Dead Reckoning) motion model. Presented algorithm achieved positioning error less than 0.5 % of traveled distance.

Similar to previous paper ZUPT is used in [11]. In the paper three different ZUPT methods namely acceleration moving variance detector, acceleration magnitude detector and angular rate energy detector were used. All presented methods were based on Kalman filter, and linear discrete Kalman filter was applied for ZUPT. Results show that presented detectors produce reliable data and linear discrete Kalman filter significantly reduces positioning error.

Another solution for indoor navigation using smartphone is shown in [12]. It is called SmartPDR and uses information from smartphones sensors for typical pedestrian reckoning approach without additional device. Proposed system was tested in several buildings with reasonable accuracy, keeping in mind low-cost noisy sensors.

Inertial measurement unit can be used not only in personal navigation but also for navigation of robots [13, 14], in aviation [15], and other usage [16, 17].

3 Sensors

This section deals with computation of orientation from data obtained by inertial sensors, such as accelerometer and gyroscope.

3.1 Gyroscope

Gyroscope is used to measure angular velocity in [rad/s]. The output of gyroscope is smooth but is negatively affected by bias. Bias is a negative effect which causes non-zero output when gyroscope is still. Integration of bias over time leads to increased error of estimated orientation. Bias can be compensated by a simple measurement when gyroscope is still and data are collected in a given period of time. Consequently correction constant is calculated as an average value of measured data. Correction constant is then subtracted from gyroscope's measured data.

To estimate the angle from the output data following formula can be used:

$$\theta = \int_0^t \omega(t)dt, \tag{1}$$

where θ represents angle of orientation in [] and ω is measured angular velocity [18].

3.2 Accelerometer

Accelerometer is a MEMS (Microelectromechanical system) sensor used for measuring all forces acting on it, gravity including. Measurement of gravity force is needed for estimation of orientation. On contrary, in case that movement acceleration is estimated gravity has to be subtracted.

For orientation in 3-D space roll Φ, pitch θ and yaw Ψ have to be estimated for all three X, Y, Z axes, respectively. In this paper we will focus on rotation around only one axis, which can be calculated as:

$$roll = \phi = \arctan\left(\frac{A_y}{A_z}\right) * 180 \Big/ \pi, \tag{2}$$

where A [m/s^2] is accelerometer's output data for Y and Z axes [18].

3.3 X-IMU

X-IMU is combination of IMU and AHRS (Attitude Heading Reference System). It consists of integrated on-board sensors: triple axis gyroscope, triple axis accelerometer in unit of gravitation, triple axis magnetometer in unit of Gauss and thermometer. Specifications of integrated sensors can be found in Table 1. Communication is enabled via USB or Bluetooth. X-IMU can be used as stand-alone data logger, in such case the data can be logged into SD card. The device does not require calibration before the measurement, because of factory calibration [19].

Table 1. Specification of X-IMU sensors

Sensor	ADC resolution	Selectable range
Gyroscope	16-bit	±2000°/s
Accelerometer	12-bit	±8 g
Magnetometer	12-bit	±8,1 G
Thermometer	16-bit	

4 Complementary Filter

The complementary filter can be described as a combination of high pass and low pass filters. It can be used to fuse output data from both accelerometer and gyroscope. Main advantage is that its use is quite simple and straightforward. Moreover, it has low computational complexity and therefore it has become very popular.

The complementary filter used in this work is based on Euler angle. The gyroscope is considered as main data source, since the data can be used to estimate the orientation of device. It is assumed that previous state can be used to predict the next state.

Angular change can be estimated by integration of data from gyroscope over time. Estimate of current orientation can be calculated as a sum of the previous orientation estimates and the angular change:

$$\theta_g = \theta_{g-1} + \Delta G, \tag{3}$$

where θ_g represents the prediction of the orientation estimate, θ_{g-1} represents previous state and ΔG is computed orientation change

$$\Delta G = \left(G_m * 180/\pi \right) dt, \tag{4}$$

where G_m represents measured output data from gyroscope and dt states for time difference.

Complementary filter can predict the next state based on high-frequency data from gyroscope, however, integration of small errors over time may cause errors in estimate of angle which are called drift. Addition of these errors cause that orientation estimate is useless. Low frequency data from accelerometer can be used to compensate the orientation drift.

It can be stated that orientation estimates from both sensors are used in order to calculate orientation of smartphone in the independent global coordinate frame. The first estimate represents the prediction of orientation, and the second is estimated from correction data. The final orientation is computed as weighted average of these estimates using:

$$\theta_e = k * \theta_g + (1 - k)\theta_m, \tag{5}$$

where θ_e is the final orientation estimate, θ_g is the angle estimated from gyroscope prediction (3), θ_m is a correction angle estimated from accelerometer data and k is weighting factor [20].

5 Experimental Setup

To obtain orientation of navigation device is a challenging task. In PDR heading angle can be calculated from gyroscope's Z axis data. Small tilt in X and Y axis lead to errors in heading estimation. One possible solution for heading angle calculation by using complicated motion model is presented in [21].

The main goal of the experiments is to evaluate the performance of inertial sensor implemented in different smartphones and external X-IMU. The performance of particular IMUs was evaluated by accuracy of their rotation in defined positions. Measurements were performed by rotation of devices around X axes. Center line of devices was aligned with central line of rotating table. This approach has been used, since there is no information of the sensors exact location. Outputs from both accelerometer and gyroscope were used to collect data. Data from magnetometer couldn't be used, due to use of measuring structure with metal parts. Output data from sensors were fused by means of complementary filter based on Euler angles.

Orientation measurements were performed in four different positions of devices. The starting position was set to 0°. After the measuring period the device (Smartphone or X-IMU) was rotated to next position. Measurements were repeated in positions with rotation of 30, 60 and 90°. At each position, data from the sensors were collected. Afterwards, measured data were processed with accuracy of seconds of angle. In the process angles of orientation were estimated.

Androsensors application was used for data collection. Measuring period was 10 s at each orientation. During the measurements, investigated devices were stable without any noisy movements around. The Newport M-471 Theta stage was used for different rotations. During the experiments the gimbal of the stage was fixed to the laboratory table, in the leveled position. Principal scheme of the setup is shown in the Fig. 1.

In the experiment, two smartphones and X-IMU were used. Smartphones were made by different producers: Samsung Galaxy S6 (SGS6) and Sony Xperia Z3 Compact (SX3ZC). These smartphones are equipped with different sensors, i.e. SGS6 is equipped with motion tracking sensor, MPU 6500 [14], made by InvneSense and SX3ZC is equipped with motion tracking sensors made by BOSH. For simplification, the measurements and analyze of a device orientation was performed only around X axis. We

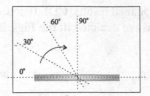

Fig. 1. Scheme of the experimental setup

assume that very similar results should be obtained in other axes, because of the fact that sensors have equal components implemented for measurements in each direction.

6 Experimental Results

In this part the results achieved in experiments are analyzed. The performance of the individual IMUs is evaluated by basic statistical characteristics and cumulative distribution function. Device orientation is estimated using complementary filter. The obtained estimates of orientation for all scenarios are shown in Fig. 2. The figure depicts mutual comparison of both smartphones with X-IMU. It can be seen that data were collected in four different positions (orientations) during 10 s intervals with sampling frequency 200 Hz. During the measurements the focus was on stable state, not on transitions between individual positions.

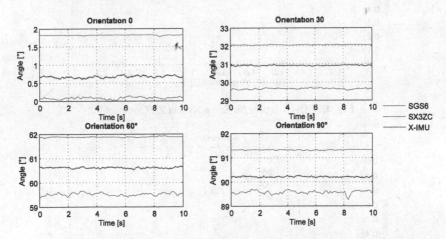

Fig. 2. Devices orientation in particular stages

From the Fig. 2, it can also be seen that each device estimated different position at each orientation. It can be seen, that for all cases angle estimated by X-IMU is between SX3ZC and SGS6. It is possible to state that SGS6 provided lower estimates compared to X-IMU. In contrast, estimates from sensor implemented in SX3ZC were always higher when compared to X-IMU.

In order to better analyze obtained results, CDF functions of achieved errors of estimates in all measured positions are depicted in the Figs. 3, 4 and 5 for SGS6, SX3ZC and X-IMU, respectively.

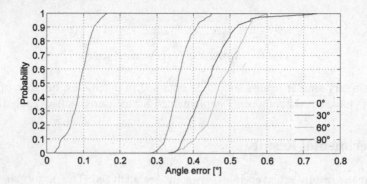

Fig. 3. CDF function for Samsung Galaxy S6

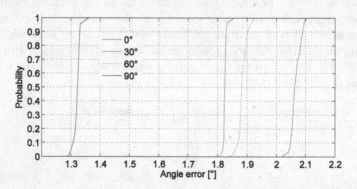

Fig. 4. CDF function for Sony Xperia Z3 Compact.

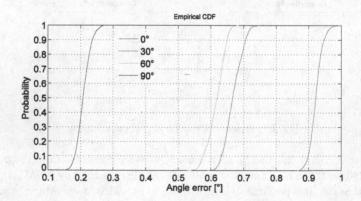

Fig. 5. CDF function for X-IMU

From the Fig. 3 it can be seen that SGS6 achieved the lowest errors in the starting position with orientation 0°. It can also be concluded that the localization error increased with the orientation. It is important to note that orientation estimates achieved with this device were extremely accurate, with errors lower than 0.5° in approximately 90 % of all estimates.

Orientation estimates achieved by SX3ZC were quite inaccurate, as can be seen from the Fig. 4. It can be seen that the lowest estimate error was achieved for orientation of 90°, however the error was still more than 1.3°. The error of estimates in the worst case, for orientation of 30°, was more than 2°. Such errors in orientation estimates may cause high localization errors in a significantly shorter time, compared to SGS6.

The results shown in the Fig. 5 were achieved by X-IMU. It can be seen that this IMU can provide angle estimates with accuracy up to 1°. The X-IMU achieved the best orientation estimates for orientation of 90°, where the error was approximately 0.2°. On the other hand, the worst estimates were provided in case that orientation was 30°. In such case the localization error was approximately 0.9°.

According to results shown in Fig. 3 to Fig. 5 it can be concluded that particular devices achieve different accuracy. At the beginning we assumed that the most accurate device will be X-IMU, because of the fact that it is the device developed directly for use in inertial navigation. However, more accurate results were obtained by SGS6 when compared to X-IMU. The mean rotation estimate error for SGS6 and for all stages is 0.3346°. For comparison the average error achieved by X-IMU is 0.604°. It can also be seen that results achieved by SX3ZC were shifted significantly compared to other devices.

For the better analysis statistical parameters of measured errors in whole investigated time period (40 s) in absolute values are presented in Table 2.

Table 2. Error of measured data

Error of orientation [°]					
Device	Min	Max	Median	Mean	Standard deviation
SGS6	0.02	0.7422	0.3828	0.3346	0.1597
SX3ZC	1.2888	2.1056	1.8483	1.7732	0.2767
X-IMU	0.1506	0.9935	0.6425	0.604	0.2572

From the values in the table above it is clear that SX3ZC smartphone achieved the lowest accuracy out of all investigated devices. It is also possible to conclude that SX3ZC achieved lowest consistency of orientation estimates, however, the standard deviation of estimates was almost the same as for X-IMU. It can also be seen that the best results provided by SGS6 significantly outperformed X-IMU by providing estimates with approximately 2 times lower errors.

7 Conclusion

In the paper comparison of outputs from inertial sensors in different smartphones were presented. Measurement was done on a compact surface with accuracy of seconds of degrees. All the measurements were performed in X axis since all axes are equipped with the same sensor. Hence, it can be assumed that achieved results will be similar for all three axes. The best results were achieved with Samsung Galaxy S6 with average error 0.3346° followed by X-IMU where the mean error was 0.604°. On the other hand, Sony Xperia Z3 Compact achieved the worst results with average error 1.7732°. The obtained results are a bit surprising. We expected the best results will be achieved by special device X-IMU. On the other hand, it confirmed our other assumption that inertial sensors implemented in high end smartphones are high quality and therefore can be utilized for tasks related to indoor positioning with acceptable accuracy. Relatively precise rotation estimation in X and Y axis can be also used for compensation of smartphone's tilt.

Acknowledgment. This work was partially supported by the Slovak VEGA grant agency, Project No. 1/0263/16, by EUREKA project no. E! 6752 – DETECTGAME: R&D for Integrated Artificial Intelligent System for Detecting the Wildlife Migration and by Centre of excellence for systems and services of intelligent transport, ITMS 26220120050 supported by the Research & Development Operational Programme funded by the ERDF.

Authors thank to assoc. prof. Daniel Kacik, PhD. University of Žilina, Faculty of Electrical Engineering, Department of Physics, Slovakia, for his support with the performed experiments.

References

1. Schiller, J., Voisard, A.: Location-Based Services. Morgan Kaufmann Publishers, San Francisco (2004). ISBN 1-55860-929-6
2. Hirakata, Y., Nakamura, A., Ohno, K., Itami, M.: Navigation system using ZigBee wireless sensor network for parking. In: 2012 12th International Conference on ITS Telecommunications, pp. 605–609 (2012)
3. Kasantikul, K., Xiu CH., Zang, D., Zang, M.: An enhanced technique for indoor navigation system based on WIFI-RSSI. In: 2015 Seventh International Conference on Ubiquitous and Future Networks, pp. 513–518 (2015)
4. Khanbashi, N.A., Alsindi, N., Al-Araji, S., Ali, N., Aweya, J.: Performance evaluation of CIR based location fingerprinting. In: 2012 IEEE 23rd International Symposium on Personal Indoor and Mobile Radio Communications, pp. 2466–2471 (2012)
5. Khademi, A.F., Zulkernine, M., Weldemariam, K.: An empirical evaluation of web-based fingerprinting. IEEE Softw. **23**(4), 46–52 (2015)
6. Subramanian, S.P., Sommer, J., Schmitt, S., Rosentiel, W.: SBIL: scalable indoor location and navigation service. In: Third International Conference on Wireless Communication and Sensor Networks, pp. 27–30 (2007). ISBN: 978-1-4244-1877-0
7. Shervin, S., Valaee, S.: GIPSy: geomagnetic indoor positioning system for smartphones. In: 2015 6th International Conference on Indoor Positioning and Indoor Navigation, pp. 1–7 (2015)
8. Chung, J., Donahoe, M., Schmandt, CH., Kim, I.: Indoor location sensing using geomagnetism. In: Proceeding MobSys 2011 Proceedings of the 9th International Conference on Mobile Systems, Applications and Services, pp. 141–154 (2011)
9. Ohon, Y., Cha, H.: LifeMap: a smartphone-based context provide for location-based services. IEEE Pervasivc Comput. **10**(2), 58–67 (2011)

10. Xiao-dong, Y., Ren, M.-r. Pu, W., Kai, P.: A new zero velocity update algorithm for the shoe-mounted personal navigation system based on IMU. In: 2015 34th Chinese Control Conference, pp. 5297–5302 (2015)
11. Li, X., Mao, Y., Xie, L., Chen, J., Song, C.: Applications of zero-velocity detector and Kalman filter in zero velocity update for inertial navigation system. In: 2014 IEEE Chinese Guidance, Navigation and Control Conference, pp. 1760–1763 (2014). ISBN: 978-1-4799-4700-3
12. Kang, W., Han, Y.: SmartPDR: smartphone-based pedestrian dead reckoning for indoor localization. IEEE Sens. J. **15**(5), 2906–2916 (2015)
13. Lee, T., Shin, J., Cho, D.: Positioning estimation for mobile robot using in-plane 3-axis IMU and active beacon. In: IEEE International Symposium on Industrial Electronics, pp. 1956–1961 (2009). ISBN: 978-1-4244-4347-5
14. North, E., Georgy, J., Tarbouchi, M., Iqbal, U.: Enhanced mobile robot outdoor localization using INS/GPS integration. In: International Conference on Computer Engineering & Systems, pp. 127–132 (2009). ISBN: 978-1-4244-5843-1
15. Hayward, R., Marchick, A., Powell, J.D.: Single baseline GPS based attitude heading reference system (AHRS) for aircraft applications. In: Proceedings of 1999 American Control Conference, vol. 5, pp. 3655–3659 (1999). ISBN: 0-7803-4990-3
16. Nemec, D., Hruboš, M., Pirník, R., Janota, A., Šimák, V.: Ergonomic remote control of the mobile platform by inertial measurement of the hand movement. In: ELEKTRO 2016, pp. 445–449 (2016). ISBN: 978-1-4673-8698-2
17. Mravec, T., Vestenický, P.: Increasing objects localization precision by determination of inertial sensor calibration constants using differential evolution algorithm. In: proceedings of ICCC 2014: 15th International Carpathian Control Conference, Velké Karlovice, Czech, pp. 362–366 (2014). ISBN 978-147993528-4
18. Groves, P.D.: Principles of GNSS, Inertial, and Multisensor Integrated Navigation Systems. Artech House, Boston (2008). ISBN 13: 978-1-58053-255-6
19. X-IMU user manual 5.2, x-io technologies (2013)
20. Nowicki, M., Wietrzykowski, J., Skrzypczynski, P.: Simply or flexibility? Complementary filter vs EKF for orientation estimation on mobile devices. In: 2015 IEEE 2nd International Conference on Cybernetics, pp. 166–171 (2015). ISBN: 978-1-4799-8320-9
21. Kang, W., Nam, S., Han, Y., Lee, S.: Improved heading estimation for smartphone-based indoor positioning systems. In: IEEE 23rd International Symposium on Personal, Indoor and Mobile Radio Communications, pp. 2249–2453 (2012). ISBN: 978-1-4673-2566-0

Design Solution of Centralized Monitoring System of Airport Facilities

Josef Horalek[✉] and Tomas Svoboda[✉]

Faculty of Electrical Engineering and Informatics, University of Pardubice,
Pardubice, Czech Republic
josef.horalek@upce.cz, tomas.svoboda5@student.upce.cz

Abstract. This paper introduces a specific proposal and the implementation of a centralized monitoring system covering resources for control and planning of flight traffic within Air Force air bases of Czech Republic. It presents various specific systems for managing and planning flight traffic and their interdependence. Based on this, the design and implementation of a centralized monitoring system covering separated systems, is introduced. In conclusion three surveillance solutions are introduced, which differ in degree of integrity and the involvement of individual systems.

Keywords: Centralized monitoring systems · Category monitoring systems · Principles monitoring systems

1   Introduction

Surveillance systems for ATC (air traffic control) represents the extensive problem [1]. Solution of this is not yet conceived comprehensively as a whole, but it is solved individually by each of the vendors of these systems. The result is a group of unequal and different visual conducted surveillance systems. The fact is that there is always an evaluation of errors of each separate system. Occurrence of possible problems on dependent systems are not monitored and therefore there is no indication of these problems. Decentralized monitoring is not ideal for technical supervision and operators of monitoring systems. They have to watch several screens of surveillance systems, but these systems often do not display all necessary information [5].

A separate chapter is systems that are not monitored at all, and reveals any faults announced by a third party, for example, the operator of the system [2].

All systems and infrastructure elements, which are presumed monitoring, use the computer network of the Ethernet type. The use of TCP/IP can be considered as relevant parameters, concerning the availability of the reference element, ICMP response with potential use of data obtained through the SNMP protocol [1, 3, 4].

For ATC's, the control of the functionality of the application, and eventually the control of the availability of the server itself, does not constitute the only thing that affects function of the systems. There are many other factors that affect reliability and they should not be overlooked. For example CPU temperature, hard drive temperature, temperature in the server cabinet, free space in the system partition, percentage of the

N.T. Nguyen et al. (Eds.): ICCCI 2016, Part II, LNAI 9876, pp. 200–208, 2016.
DOI: 10.1007/978-3-319-45246-3_19

CPU usage, amount of zombie processes, etc. [6, 7]. Modern views of monitoring systems, services and ICT systems as a comprehensive solution can be used [8, 9].

2   Present Systems and Their Monitoring

ATC merges several independent systems which are interconnected through a computer network of LAN type. Individual systems are interconnected directly, or through more switches, in the infrastructure of the individual subsystems to core switch Cisco 6500 series. Due to a better bandwidth utilization and broadcast domain separation the data network is separated into several VLANs. This comprehensive system includes several subsytems which are briefly presented hereafter.

DANESE System. DANESE system is a data oriented system that provides communication between air bases (AB) of the Army of the Czech Republic (ACR). A global view is shown in Fig. 1. Due to redundancy, the backbone link is designed as a double circuit, built on radio relay links using a leased data circuits. Individual elements of ATC are joined in this way. Thus AB, command and control of the army and supervision workplaces, including links with IATCC Prague which provides sharing of data. This data is necessary for the security of ATC in the airspace of the Czech Republic.

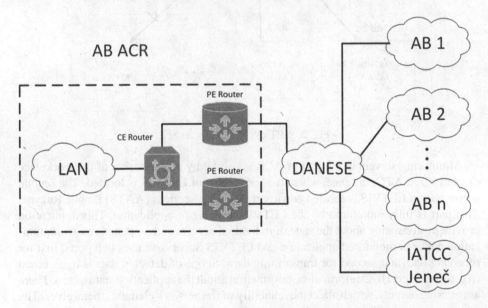

Fig. 1. DANESE ACR

LETVIS System. System LETVIS 99B-M4 is one of the key systems and it is designated to radar and procedural ATC. It also allows planning and coordination of air traffic control, airspace utilization planning at the time specified for administration requirements and coordination of airspace use in tactical phase. It creates automated integration links between military and civilian ATC. Serves as a source of surveillance data, flight plans and plans for utilization of airspace for slaved systems LETVIS MTWR and for other external configured systems. A global view is shown in Fig. 2. It is easy to deduce the total extensiveness of this application [10].

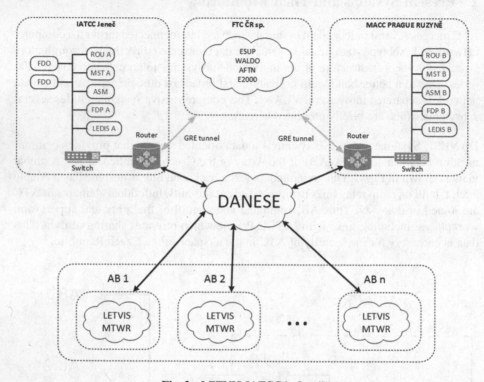

Fig. 2. LETVIS IATCC in Jeneč

Monitoring servers in the part of ACR is solved by a supervision of the workplace of centralized ATC, in Jeneč, where the server part of LETVIS is located. The second server part of LETVIS is located at Airport air traffic services (AATS) Prague Ruzyně. This part is fully monitored by the LETVIS Supervisor application. This application provides information about the state of individual servers and workstations. Communication between monitored application and LETVIS Supervisor uses two ports; first for receiving data and second for transmitting data. Scope of detected data is quite broad (e.g. CPU usage, HDD information, information about the application status, etc.). There are missing outputs, which affect the reliability of the network elements themselves. The missing outputs information on the temperatures of the components, and information of the established map for identifying moving targets. LETVIS system monitoring is not currently linked with other monitoring systems into one output, thus creating a separated traffic, which is, from the perspective of the overall concept, undesirable.

AMS System. AMS is a system for supervision of individual resources of long-range navigation system (LRNS) at the AB and it consists of several servers, workstation and power members. Its basic function is to indicate a status of lighting system, which is used to take off and land aircrafts, indicate the status of security systems (ESS), indicate the status of meteorological systems, indicate fire, supervise on the proper operation of radio beacons, provide basic information to ATIS, and indicate the status of system resources. The system has three key elements. The first is AMS Master - a control computer, that controls the activity of all the PCs in the network, secures communications with a DAP 128TC system, and provides communication with peripherals via serial lines. The second is DAP 128TC, which is a power element for controlling and monitoring elements involved in the system. The third element is the AMS workstation, which provides to the operator, information about the status of individual resources and the state of the AMS Master. The system is one of the critical elements of the ATC and any undetected failure is fatal. The problem is to detect failure of individual stations and the individual light stations without visual control. One proposed solution is to use surveillance with ICMP, which would have revealed this in a shorter time [11].

RCOM System. RCOM is a radio-communication system for communication between ground and aircraft, air-ground and ground-ground. Basic radio-communication system is divided into two parts – transmitting and receiving. These two sites are approximately 1000 m apart, and they are interconnected by a fiber-optic and radio relay backup link. The receiving site contains all of the receiving radiostations which includes 12 receivers (R&S4200 series), two scanning receivers (AR8600 series) and one analytics receiver (R&SEM100 series). All these components are connected into a LAN network via two Cisco C2960X switches. Monitoring of the state of backup power resources at the each site is very important, because they provide backup power in case of failure of the main power supply route. The application, MCU client, is used to display errors that indirectly threaten the radio communication, and it is necessary to be monitored. MCU indicates the state of controls of all radiostations and condition of backup power resources, which is critical in cases of power failure. On the other hand the failure of any backup or main route between the receiving and transmitting side is without indication.

AMS2 – AFTN System. It is a network designed for transmission of textual information, which originated as a worldwide network for exchange of reports of air traffic services. There exists one centre in each state which is responsible for international traffic. Workstations are the components of AFTN system. These are a clients that allow receiving and sending messages over the network. The application itself does not require special hardware and it can be executed on a personal computer with operating system Windows. No local monitoring tool is implemented, these are only end devices connected to the server. The operator can reveal unavailability of the server, when he is not able to send a message. Inability to send a message is not a critical operation and can be replaced by telephone coordination with the superior level.

3   Proposal of Centralized Monitoring

The proposal builds on the specification requirements. These include the possibility of independent supervision from multiple sites and thus the possibility to view the same information about the state of the components of two or more independent sites at one time. The platform on which it will be possible to view this data must be considered as independent. For this reason we selected a web platform that all above conditions are satisfied. For implementation of the monitoring system, PHP and MySQL databases have been used. Updating the displayed data is ensured by the automatic refresh of the page, set use HTML meta tag, after 10 s. Another requirement was to monitor the availability of components and temperature monitoring of relevant components. The recovery interval set to 30 s, with an indication of failure of both subsystems (ICMP, temperature) including data age. For centralized evaluation of monitored data, the best option is to use client-server architecture. The server must have IP access to each monitored component indiscriminately to which VLAN is connected. The server and the client side have been implemented in PYTHON.

3.1   Server Part

Proposal of a monitoring system operates with a proactive solution of application on the server site. It is not a typical client-server model, because in typical client-server model, the server is a passive element. But for the purpose of monitoring, it is also necessary to capture the elements that are unable to communicate directly with a server. Specifically, the detection of client availability is possible via ICMP and SNMP protocols. Status query is necessary to appropriately allocate network resources to avoid unplanned exploitation of the local network. This is largely built on the Gigabit Ethernet 1000BASE-T and this data traffic should be seamlessly handled. The application server is implemented as a service running in the background of the host computer, which is designated as a server. Application listen to a specified port and wait for an incoming call from a client application. These calls are served, processed and then the application sends back a response. After each request, the processing server returns the result of the operation. It is defined as a simple JSON object. Each answer contains a status code, thus it is an easily recognizable result of the operation. Following the successful outcome of the operation, the server returns a 200 status code, including short text information about the outcome to the client. To ensure greater efficiency in the processing of requests, a server application is subdivided into several threads after its launch and each thread handles a different type of requirement. Subsystems (ICMP and SNMP) are actively querying the status of monitored components based on predetermined rules. The rules are defined using predefined templates. Each template contains a list of identifiers that the SNMP protocol has to query. The server also checks whether information is received, at regular intervals, from a client applications. If data from the client exceeds the allowed tolerance, there is a reason to evaluate this data and initiate an indication of this status on the supervision screen.

3.2   Client Part

The client part provides an acquisition of monitoring data from the monitoring station and then sends it to evaluation to the server with TCP utilization. The client part (agent) can be executed on different platforms. It is therefore appropriate to define a set of parameters relevant to the determination of the monitored component state. Multiple versions run on multiple platforms, which means introducing unified communications (Application Programming Interface). Each client transmits information contains detected data and its identifying information.

3.3 Client-Server Communication

The agent periodically collects monitoring data and sends it to the server. The server receives the data, processes and evaluates them. Acceptance is subsequently confirmed. However, if the agent does not receive a predetermined percentage of confirmation to its messages, the information about exceeding the permitted number of unconfirmed reports will be inserted into the data packet fields. This information is not the reason for indication of component error.

3.4 Destination Availability Test

The basic test is to verify the availability of the physical component, with messages "Echo Request" and "Echo Reply". If a client responds to the message "Echo Request" with the message "Echo Reply", shall be considered as an available component for IP traffic, and at the same time it can be said that the network layer of the operating system works within standard parameters. In one tracking cycle five messages "Echo Request" are sent to the target. The answer is analyzed and the result is stored in the database. Finally, the need for monitoring ICMP take into account the phenomenon that is referred to, as a packet "flapping". This is a jump from one condition to another. In our case its about a periodic evaluation of the lost and received packets. If the problem with flapping is not resolved there could be frequent indications of a network problem caused only, by stray responses from individual components.

4 Proposal of Models for Implementation of Monitoring

This chapter introduces three models designed for monitoring. Their implementation depends on the technical and financial resources of the contracting authority.

4.1 Basic Model

The primary purpose of the model is to monitor the availability of network components, which are ensured using "Echo Request" messages and ICMP response. Querying is conducted periodically every twenty seconds. To increase the efficiency of the monitoring cycle, the server starts querying in several threads (the number of threads is definable) and the cycle is considerably faster. The default number of threads is set to five. A count is based on the number of monitored IP addresses. Each component is monitored at, at least one of its interfaces. If the component has multiple interfaces each of these interfaces can be monitored. After each monitoring cycle the result of availability, (including the time when the record to determine the condition occurred), is recorded. This information is then used to check the correct function of the ICMP subsystem. When the limit is exceeded, the difference between the current time and the last response time is larger. And, there is an indication of this condition. A parameter, which emerged as the least suitable for monitoring of the components, is monitoring the temperature inside the cabinet. Usually, CPU temperature and temperature of motherboard can be monitored. An agent runs as a scheduled job every ten minutes and it receives information about the temperature. The client portion detects temperatures by tools of an operating system. Collected data is sent to the server, which process and stores them. Information about the time of response is also stored, as well as gathered information about the availability. An obsolete data is visualized after exceeding the permitted limit. The course of temperature changes allows graphical representation of history records over a previous period of up to twelve hours. Records of temperatures and availability status is archived, and can be viewed in the table with a specified period of time, or exported to a file for any further statistical processing. These functions are implemented in the current version of the monitoring system. Due to increased demands for monitoring the status of components, it is necessary to solve a new version of the monitoring system.

4.2   Extended Model

The parameters monitored in the basic model of the monitoring system are insufficient and it is appropriate to extend the number of monitored parameters and methods. The main change is full integration, via a SNMP request including receiving and processing SNMP traps, and introducing of evaluated data obtained by a long-term monitoring. SNMP can monitor only components that supports SSL or components which can easily enable SNMP support. Generally these are active components (routers, switches, modems) workstations or servers. Most of the components which haven't installed SNMP support have an option to install it. Only the supplier of the systems has the possibility of additional installation. One or more templates are assigned to each component that supports SNMP. Requested information may be obtained for a given component by periodical querying and analyzing the responses. Data validity is observed in a similar way to monitoring availability. Exceeding the limit is probably due to the target unavailability or failure in the SNMP subsystem at a server. To receive SNMP traps, the server is configured similarly as in the case of querying. A template is assigned to the component that, unlike polling, does not serve as a list of addresses, but as an identifier according to which the server is able to distinguish the trap sender. Individual messages are stamped with the time of received. This value does not serve as a control parameter but it's intended for further statistical processing. A trap can not be expected in a specific interval. It is sent at random time depending on the change of state of the component, and therefore there is no point in evaluating intervals of received data. The server stores all data received during monitoring from that server, and can get a graphical view of the parameter progress in the chart. A time interval from which the resulting graph is generated is definable according to user input. Prepared intervals are: daily, weekly, monthly, yearly. There can be also combined multiple values in the graph, and progress of dependent parameters can be viewed in one graph. It is possible to combine up to six of these indicators.

4.3   Model with Minimum Limitations

The previous proposal offers big possibilities for obtaining information about the status of the whole monitored section, but does not offer an opportunities that would be beneficial in monitoring. They require necessary cooperation of third parties (suppliers, manufacturers of the system). Without that, the selected methodology can not be used. AMS - according to information supplied by company, SNMP implementation to AMS is possible and it is not a big intervention to the application. After starting SNMP support, outputs from sensors of all resources managed by the AMS would be available. This represents a next step towards a centralized monitoring system. LETVIS - installation support of SNMP on all stations is also associated with the implementation of information available on the SUP to MIB, and is also a step towards to centralized monitoring. Because most sites of the system are built on modern hardware, installations of SNMP support are not a fundamental problem. Adding other information to MIB already represents a necessary intervention of supplier. RCOM – monitoring the availability of data paths, is nit implemented, especially if the network architecture is designed for that including well-chosen graphical presentation. Introducing a hierarchy of dependencies components and algorithms for automated diagnosing problems closely related with the monitoring. Consequences of a problem for the reference component tried to capture the hierarchy of dependencies, and the part that is affected by the outage is highlighted. There are two affecting: direct (physical) and indirect (logical). Monitoring is trying, using a hierarchy of dependencies, to capture the consequences of a problem on the reference component so the part that is affected by the outage is highlighted. There are direct (physical) and indirect (logical) affecting. Physical affecting - a fault in the component immediately causes inaccessibility of the subordinate part. Usually, it's an active component (switch, router). Logical affecting – is not visible at first glance, the results will come usually within a few minutes or hours. This is usually the unavailability of services, free space borders, the incidence of physical errors on the hard disks etc. To solve these problems, algorithms, which offer possible causes (scenarios) and their subsequent remedies, are introduced. For each error a scenario can be created and that will help solve a similar problem in the future. Scenarios also allow automatic execution of remedial tasks (reports about the of the disk status, inaccessibility of services, absence of a specific data, etc.). A scenario in which it is possible to define executive actions is not available for every operator. This is an operation with direct implications for ATC. Therefore, this needs to be addressed, depending on the authentication and authorization, including the introduction of records of these activities. Statistical outputs in the form of tables or graphs can be accessed without authentication. For these outputs most of the parameters that the server gets, can be used. Everything is dependent on the configuration.

5   Conclusion

The aim of this article was to create a design and implementation of a centralized monitoring system covering resources for managing and planning air traffic within AB ACR.

Various systems for managing and planning air traffic together with an indication of their interdependence were introduced first. Based on theoretical background and especially the technical design of individual systems, centralized monitoring system design and implementation were created including three proposals which differ in the depth and the extent of monitoring. The proposed monitoring system was successfully tested in an environment of AB ACR and now it is fully integrated into operation.

Acknowledgment. This work and contribution is supported by the project of the student grant competition of the University of Pardubice, Faculty of Electrical Engineering and Informatics, Intelligent Smart Grid networks protection system, using software-defined networks, no. SGS_2016_016.

References

1. Koga, T., Lu, X., Mori, K.: Autonomous decentralized high-assurance surveillance system for air traffic control. In: 2014 IEEE 15th International Symposium on High-Assurance Systems Engineering (HASE), Miami Beach, FL, pp. 154–157 (2014). doi:10.1109/HASE. 2014.29
2. Đurišić, M.P., Tafa, Z., Dimić, G., Milutinović, V.: A survey of military applications of wireless sensor networks. In: 2012 Mediterranean Conference on Embedded Computing (MECO), Bar, pp. 196–199 (2012)
3. Lee, S., Levanti, K., Kim, H.S.: Network monitoring: present and future. Comput. Netw. **65**, 84–98 (2014). doi:10.1016/j.comnet.2014.03.007
4. Sinha, A., Sejwal, L., Kumar, N., Yadav, A.: Implementation of ICMP based network management system for heterogeneous networks. In: 2015 2nd International Conference on Computing for Sustainable Global Development (INDIACom), New Delhi, pp. 382–387 (2015)
5. Yin, J., Li, Y., Wang, Q., Ji, B., Wang, J.: SNMP-based network topology discovery algorithm and implementation. In: 2012 9th International Conference on Fuzzy Systems and Knowledge Discovery (FSKD), Sichuan, pp. 2241–2244 (2012). doi:10.1109/FSKD.2012.6233879
6. Martinez, M., Javier, F., Ardestani, E.K., Renau, J.: Characterizing processor thermal behavior. In: Proceedings of the Fifteenth Edition of ASPLOS on Architectural Support for Programming Languages and Operating Systems (ASPLOS 2010), p. 193. ACM Press, New York (2010)
7. Horalek, J., et al.: Proposal to centralize operational data outputs of airport facilities, pp. 346–354. doi:10.1007/978-3-319-24306-1_34
8. Komarek, A., Pavlik, J., Sobeslav, V.: Network visualization survey. In: Núñez, M., Nguyen, N.T., Camocho, D., Trawinski, B. (eds.) ICCCI 2015. LNCS, vol. 9330, pp. 275–284. Springer, Heidelberg (2015). doi:10.1007/978-3-319-24306-1_27
9. Komarek, A., Sobeslav, V., Pavlik, J.: Enterprise ICT transformation to agile environment. In: Núñez, M., Nguyen, N.T., Camacho, D., Trawinski, B. (eds.) ICCCI 2015. LNCS, vol. 9330, pp. 326–335. Springer, Heidelberg (2015). doi:10.1007/978-3-319-24306-1_32
10. LETVIS ASM/ACM Systems. ALES ATM, Praha: ICZ a.s., 199n. http://www.ales-atm.com/en/market-and-products/air-traffic-control-management-systems/letvis-asm-amc-systems/
11. AMS. About the system AMS. Transcon (2016). http://www.transcon.cz/index.php?option=com_content&task=view&id=66&Itemid=132

Room-Level Indoor Localization Based on Wi-Fi Signals

Filip Maly, Pavel Kriz$^{(\boxtimes)}$, and Martin Jedlicka

Department of Informatics and Quantitative Methods,
Faculty of Informatics and Management, University of Hradec Kralove,
Hradec Králové, Czech Republic
{Filip.Maly,Pavel.Kriz}@uhk.cz

Abstract. This paper deals with the indoor localization using Wi-Fi networks. It reacts to the increasing trend in smart devices containing a large number of sensors and modules. The work deals with the use of a Wi-Fi module and a camera. The Android application which uses these modules for the localization and acquisition of fingerprints of wireless networks was created. The application consists of a client and a server side. The approach suggested enables a mobile device to be localized in a building for which a database with fingerprints of wireless networks is created. The application suggested can be used for every building provided that the following conditions are fulfilled; QR codes or other unique visual tags are distributed, the map of the building is available and the fingerprints of wireless networks are captured.

Keywords: Room-level localization · Fingerprinting · QR codes · Wi-Fi positioning · Android · GPS

1 Introduction

Mobile devices, especially smartphones and tablets, play a more and more important role in human life and people cannot imagine their life without them. Thanks to their portability, sufficiently large display, performance and a number of internal built-in sensors, they become important helpers. In many scenarios, they already replaced dedicated navigation systems (devices).

Positioning and navigation using GPS (Global Positioning System) has become a common part of human life. Many people including drivers, cyclists or tourists cannot imagine traveling without having it. The problem arises inside buildings where GPS does not work.

A lot of people get lost in big buildings or shopping malls. Based on the statistical data, one spends more than 80 % of his/her time in buildings per day [1]. That is why the indoor localization is a hot topic among researchers and a demanded feature among users of mobile devices. The existing indoor localization techniques are based on several technologies; radio frequencies (RF) – for example Wireless Local Area Networks (WLANs, also known as Wi-Fi), cellular

© Springer International Publishing Switzerland 2016
N.T. Nguyen et al. (Eds.): ICCCI 2016, Part II, LNAI 9876, pp. 209–218, 2016.
DOI: 10.1007/978-3-319-45246-3_20

networks, Bluetooth, Radio Frequency Identification (RFID), Ultra-wideband beacons, lasers etc. These technologies can provide localization with the sufficient accuracy but an issue arises regarding creation and maintenance of the up-to-date database [2]. The approach based on Wi-Fi signal strengths [3] is the most promising and will be elaborated in this paper.

Almost every shop, household or office needs an internet connection and uses one or more Wi-Fi access points for wireless connection of mobile devices. Today, even cheap mobile phones have an built-in wireless Wi-Fi adapter that may be used for the localization purpose [4]. The Bluetooth technology also serves for localization and is very popular nowadays [5]. The Bluetooth technology is broadly supported by mobile phones, but the prevalence of Bluetooth transmitters (beacons) in a public space is still low today though it is increasing. Besides this, other technologies can be used for localization, but they are not widespread. For example, technologies based on the laser localization are far too expensive and mobile phones are not capable of working with them yet.

The indoor localization is not available for the daily use yet. In most cases, navigation in shopping malls, airports or companies is solved by touch LCD information panels. From this point of view it is beneficial to create an application which could be used in any building and which would make orientation easy for users.

The main objective of this work is to create the application which can determine the position of the user in a building. His/her position will be determined using the surrounding Wi-Fi networks. The application will be composed of two parts; the client side and the server side.

The client side of the application will be created for the Android mobile operating system. This platform has been chosen due to the highest share at the market of mobile operating systems [6]. The server side will be written in the Java programming language with Spring MVC extension and MySQL database.

Because the estimation of the position will be performed using Wi-Fi, there has to be a sufficient number of Wi-Fi access points (APs) in the building to provide as accurate measurements as possible. In case of low density of Wi-Fi APs, the combination with Bluetooth beacons can be used [5]. We will focus on the unique identifiers of the access point (MAC address) and the Received Signal Strength (RSS) of the networks. From the already obtained results it is known that the most accurate way to determine the position is with the aid of fingerprints of wireless networks [7].

The application will be designed in order to be used in any building provided that the map of the building is available and a radio map with fingerprints of Wi-Fi networks is created. This map will be saved in the database and has to be updated continuously to allow the accurate determination of the position of the user.

The rest of this paper is organized as follows. We formulate the problem in Sect. 2. Section 3 describes the existing techniques and applications. In Sect. 4, we propose a new solution. Several details regarding the implementation are shown in Sect. 5. We present the results of the evaluation and testing in Sect. 6. Section 7 concludes the paper.

2   Problem Formulation

This work deals with a design and implementation of the application for the Android mobile platform. The main objective of the application is to facilitate the localization of the user's mobile device inside a building in a real time. We will use indoor localization techniques based on fingerprints of wireless networks. The localization process (working with a huge fingerprint database) should be offloaded to the server in order not to burden the mobile device with a high memory, CPU and input/output load (which may result in fast battery drain). For a smaller database, one should consider energy budget for local processing compared with energy budget for communication with a server. Local processing may be more energy efficient in some cases. Then, after the localization, the user will be informed about his/her position by e.g. a marker in a building's map. The fingerprint database should be shared among users. We will also focus on the issue of updating the radio map of the building allowing to keep the high accuracy while the environment is changing. Crowd-sourcing approach to updating the database will be elaborated.

3   Related Work

A lot of developers deal with similar applications; among them there are big and well-known companies such as Google, Cisco and companies belonging to the In-Location Alliance but also smaller companies and individuals.

Thaljaoui [8] deals with the localization using the Bluetooth Low-energy (BLE) and iBeacon technology. This work describes the principle of the localization based on the Bluetooth technology and focuses on the transmitting power of individual transmitters (beacons). This technology is quite accurate but needs a large number of beacons.

Other commonly used methods for the localization inside buildings are based on the use of access points of Wi-Fi wireless networks. The localization technique is based on the strength of the Wi-Fi signals. Mahiddin [9] deals with the localization based on a so called Wi-Fi trilateration method. Fingerprinting is another method based on the Wi-Fi signal [10].

From the above stated techniques using the Wi-Fi signal, the method based on fingerprints of individual access points is more commonly used because it achieves a higher accuracy and is more robust in an environment with a lot of reflections with multipath propagation of a signal (typical indoor environment). Fingerprints of access points are acquired either in fixed reference (calibration) points or by walking through a given building in a predetermined way. This approach does not require the knowledge of the position of access points (transmitters). In contrast to this, we have to know the position of these points in the trilateration method to be able to localize the user. The Wi-Fi signal strength is attenuated after passing through the barrier or after reflection [11]. This is the reason why the trilateration method is less commonly used indoors; there is a large scatter when localizing the user.

Nowadays, there are several tens of mobile applications focused (not only) on the indoor localization. As an example, we can mention several applications currently available in the market.

Google Maps is a popular multi-platform map application which is freely available. Most people know it as an application which can navigate people when traveling and display certain places in a town. If the building plan exists, it is necessary to zoom the building to the maximum to display its plan. Then, the application is able to navigate indoors. But unfortunately, support for indoor localization is not ubiquitous. For example, there has not been any indoor map of a building in the Czech Republic in Google Maps. That is why the accuracy of this application could not have been tested. In the Google's documentation [12], there is a list of countries and objects which enable this service. In the paper [13] there is an example of the localization implemented using Google Maps.

NAO Campus is an application developed by a French company PoleStar. Indoor navigation was developed for MWC (Mobile World Congress) in Barcelona in 2014. It is developed for the Android and iOS platforms and uses the combination of GPS, Wi-Fi, Bluetooth LE and motion sensors of mobile devices for the localization. Based on reviews, the application is quite accurate but also unstable, slow and the maps are loaded reluctantly. The application could not have been tested due to support of MWC Barcelona's area only.

Wi-Fi Indoor Localization is a free Android application. It enables to load a map (image) of a floor in which the user wishes to be navigated. The application supports more floors. After adding the map it is necessary to create fingerprints of wireless networks. This is done in a way that the map is divided into four squares of the same size. Measurements have to be done in each square so that the application can navigate fluently. After obtaining the fingerprints, the application can be used. The application shows the user's position very inaccurately and furthermore, the fingerprints of wireless networks have to be created for each device or a CSV (Comma-Separated Values) file has to be exported and shared manually. There is no central database of the access points.

Indoor GPS is an application for the Android platform. It exists in two editions – premium and classic. Similarly to the previously described application, it enables to load the chosen map. In both versions it is possible to load only one floor. On top of that, the premium version provides a service which enables to export a map. After insertion of the map, the user is obliged to create the database of fingerprints. The user estimates his/her position on the map and then he/she creates the fingerprint (which can be inaccurate due to a bad estimation of the position done by the user). Again, there is no central database of access points for the application and new fingerprints of wireless networks have to be created for every device. When testing the application, the user's position was determined very inaccurately.

While there are existing solutions, none is easily applicable, reliable or widespread in order to be used on daily basis by ordinary users.

4   Proposed Solution

Our paper deals with the issues mentioned in the previous section. In this section we will describe the way how to create a new application for the indoor localization. We will use the approach based on Wi-Fi fingerprints and design the system that will be divided into the server and the client side.

The client side can be divided into two modules – one for obtaining the fingerprints (for the calibration phase) and one for the localization of the user.

The calibration-phase module, which will serve for the acquisition of the fingerprints of wireless networks, will use the QR (Quick Response) code scanner. The scanner will be implemented due to the fact that when creating fingerprints it is also necessary to find out the device's position precisely in order to know where the given fingerprint was obtained. From this point of view it is necessary to distribute QR codes throughout the building which will be associated with particular positions. In an optimal situation, one QR code should be located in each room. If it is a bigger room or a corridor, there should be more QR codes. The QR code should also be situated at each door so that the application can be extended in the future and serve as navigation. These codes near doors play an important role so that navigation does not lead the user through a wall but through the door. Kaushik [14] describes the principle of the operation of these codes in detail. After scanning the code, the code's position and the corresponding fingerprint of the wireless networks will be obtained. Besides the QR code, other visual tags unambiguously determining the position can be used – e.g. numeric designation of the door which can be read using the Optical Character Recognition (OCR) or pattern-matching in general.

The second module of the client-side application is designated for the localization itself. The position will be determined based on Wi-Fi networks found during the scanning. The procedure is the following. The Wi-Fi networks and their RSS will be found. Then, the request will be sent to the server which will process the data and send back the user's estimated location.

The fingerprints of wireless networks will be saved on the server. This means the fingerprints will not have to be stored in the client nor synchronized among several devices. In contrast to some existing solutions, the user will not have to measure the fingerprints on his/her own in order to be localized immediately after the application's installation. All users simply access the same data which are saved in one place.

Theoretically, the provider could perform only one measurement at each QR code but then, the localization would not be accurate enough. But the more measurements are done, the more accurately the position can be determined. That is why it is important to figure out how to obtain these data easily and effectively. One of the promising ways could be to combine QR codes with e.g. discount coupons or leaflets with discounted products in shopping malls so that customers themselves will fill and update the database willingly. Thanks to this crowd-sourcing approach, the database will be kept up-to-date and will contain several hundreds to thousands measurements (fingerprints).

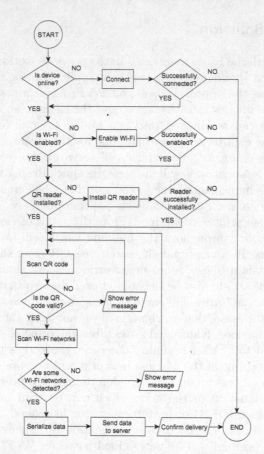

Fig. 1. Flow chart of the fingerprint acquisition

The suggested process of the fingerprint acquisition is depicted in Fig. 1. The suggested localization process is depicted in Fig. 2.

In this section, we have introduced the suggestion of the new solution which aims to choose the best way from the existing principles and procedures and to add knowledge and experience of the authors. Implementation of the solution outlined here will be described in the following section.

5   Implementation

The application is based on a client-server architecture. The application uses the internet connection for mutual communication and can be installed at the devices with the Android version 4.0. The compatibility with the older devices is important because they are still quite widely used. The server side is implemented in the Java programming language utilizing the Spring MVC web framework.

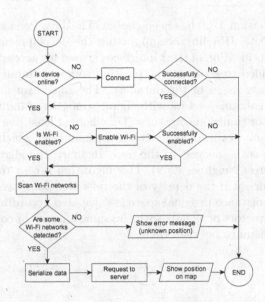

Fig. 2. Flow chart of the localization process

The client-server architecture has been chosen to make the application as simple as possible and to offload all complex calculations from the client to the server. The possibility to use one repository of fingerprints available to all users was another reason to choose this architecture. This means the client side does not have to contain the local database. The use of the database which is situated at the server side is advantageous for every user. Anyone will be able to create the fingerprints of networks and improve and update the shared database.

5.1 Client Side – The Mobile Application

The client side of the application enables to perform two key activities. It enables volunteers to scan the QR code and to find out neighboring Wi-Fi networks and consequently, to send these data to the server in the JSON format (see Fig. 1). The database of fingerprints is filled this way. The second purpose of the client side is to localize the user based on the fingerprint of surrounding Wi-Fi networks, if requested. This fingerprint will be sent to the server where its position will be estimated and sent back to the application (see Fig. 2). The application also finds out the manufacturer, the model of the device and the version of the Android operating system. These additional features will be included in the estimation process in the future making the localization more accurate.

5.2 Server Side

The server side of the application is implemented using the MVC (Model View Controller) design pattern. The Java programming language with the Spring

MVC framework version 4.2.0 has been chosen. The Java Persistence API (JPA) with the Spring Data JPA library supporting the development of data access objects by mere specification of their interfaces is used for access to the database.

The class intended to estimate the position compares input values with the values in fingerprints saved in the database. The Euclidean distance in signal space (SS), which contains vectors with signal strengths of individual transmitters, is calculated for similar fingerprints (i.e. where at least one MAC address is the same). According to this distance it will be found out which reference point from the database is the nearest to the user. In short, the algorithm is looking for a so called Nearest Neighbor (NN). This algorithm is easy to implement and provides good accuracy, if the density of the reference points is high [7].

The Euclidean distance in signal space is calculated according to the formula (1) where a, b are vectors of signals with the same amount of components and d is the calculated distance between them.

$$d = \sqrt{\sum_{i=1}^{n}(a_i - b_i)^2} \tag{1}$$

6   Evaluation and Testing

While testing the application, the question arose how to optimally distribute the QR codes. Two possibilities of their distribution were tested. In the first case, the QR codes were placed into the center of the room. In the second case, additional QR codes were placed near each door. The second case was supposed to correspond more to a school building or a shopping mall where it is required to lead the user towards the room itself and the orientation inside the terminal room is not necessary.

Testing of the final application was divided into several parts. The client side was tested first. First of all, detection of surrounding Wi-Fi networks and their signal strengths, type and manufacturer of the device and the version of the operating system were tested. In the next step, the server side was tested – particularly saving and updating of records in the database and its administration. End-to-end testing was the last step – i.e. verifying of the complex functions and accuracy of the localization of the device.

The localization using the QR codes situated in the centers of the rooms is shown in the first part of Table 1. Measurement was carried out in each room in a certain distance which is stated in the table. Three measurements were performed for each distance. If the algorithm estimated correctly that the user is located in the given room, the *correct* estimation is recorded; if not, the *incorrect* estimation is recorded. Measurement in the distance of 2 m could not have been carried out in the Room 2 due the small size of the room.

The second part of Table 1 shows the localization using the QR codes which are placed at the door of each of these rooms. Measurement was taken in the distance of 1 m from the given QR code only. Again, three measurements were performed for each spot and the values were recorded into the table in the same

Table 1. Estimation results of the "same place test" in the certain distance from the QR code (number of correct and incorrect cases)

	1m distance		2m distance	
	Correct	Incorrect	Correct	Incorrect
Room 1	3	0	3	0
Room 2	3	0	N/A	N/A
Room 3	3	0	2	1
Room 4	3	0	3	0
Room 5	3	0	3	0
Room 6	3	0	3	0
Room 7	3	0	3	0
Entrance to room 1	2	1		
Entrance to room 2	2	1		
Entrance to room 3	3	0		
Entrance to room 4	2	1		
Entrance to room 5	3	0		
Entrance to room 6	2	1		

way. Table 1 shows that incorrect estimations have been observed mainly at the entrances to the rooms. Variance of these values was not severe. In these incorrect cases the algorithm estimated that the user was located in one of the rooms which were connected by the door tagged by the scanned QR code.

Two different devices were used for acquisition of the data; HTC Desire X (Android 4.1.1) and ASUS Nexus 7 2013 (Android 6.0.1). When scanning Wi-Fi networks at the same places the differences caused by several factors may appear. Thus, the average value is calculated at the server from the measured fingerprints of wireless networks and the position is determined based on this value. This approach should decrease the probability of deviation and estimate the user's position as accurately as possible.

The results of the overall test have confirmed that the application is relatively accurate and could also be used in the real environment. The application was also highly responsive.

7   Conclusion

The fully functional application for the indoor localization at the level of individual rooms based on fingerprints of Wi-Fi networks is the result of this work. The solution can be used in any building where there is a sufficient number of Wi-Fi networks and QR codes are distributed or other already existing tags are used (e.g. unique tables next to the door tagging the individual rooms).

The advantage of the application, in contrast to the existing solutions available in the Google Play store, is the fact that even when the user enters the building for the first time he/she can immediately have his/her position estimated. Or, he/she can voluntarily participate in creation of fingerprints at individual places inside the building. The user can be motivated to collect fingerprints by incorporating gamification aspects (bonuses in the form of discounts etc.).

In the future, we will include the fingerprints of Bluetooth Low Energy beacons which are becoming increasingly widespread.

Acknowledgements. The authors of this paper would like to thank Tereza Krizova for proofreading. This work was supported by the SPEV project, financed from the Faculty of Informatics and Management, University of Hradec Kralove.

References

1. Li, H., Chen, X., Jing, G., Wang, Y., Cao, Y., Li, F., Zhang, X., Xiao, H.: An indoor continuous positioning algorithm on the move by fusing sensors and Wi-Fi on smartphones. Sensors **15**(12), 29850 (2015)
2. Zhang, P., Zhao, Q., Li, Y., Niu, X., Zhuang, Y., Liu, J.: Collaborative WiFi fingerprinting using sensor-based navigation on smartphones. Sensors **15**(7), 17534 (2015)
3. Yu, R., Wang, P., Zhao, Z.: The location fingerprinting and dead reckoning based hybrid indoor positioning algorithm. In: Sun, L., Ma, H., Fang, D., Niu, J., Wang, W., Pavlidis, M. (eds.) CWSN 2014. CCIS, vol. 501, pp. 605–614. Springer, Heidelberg (2015). doi:10.1007/978-3-662-46981-1_57
4. Liu, H.H., Lo, W.H., Tseng, C.C., Shin, H.Y.: A WiFi-based weighted screening method for indoor positioning systems. Wireless Pers. Commun. **79**(1), 611–627 (2014)
5. Kriz, P., Maly, F., Kozel, T.: Improving indoor localization using bluetooth low energy beacons. Mobile Information Systems (2016, in Press). http://www.hindawi.com/journals/misy/aip/2083094/
6. IDC: Smartphone OS market share, 2015 Q2 (2016). http://www.idc.com/prodserv/smartphone-os-market-share.jsp
7. Machaj, J., Brida, P.: Using of GSM and Wi-Fi signals for indoor positioning based on fingerprinting algorithms. AEEE **13**(3), 242–248 (2015)
8. Thaljaoui, A., Val, T., Nasri, N., Brulin, D.: BLE localization using RSSI measurements and iRingLA. In: 2015 IEEE International Conference on Industrial Technology (ICIT), pp. 2178–2183, March 2015
9. Mahiddin, N.A.: Indoor position detection using WiFi and trilateration technique. In: The International Conference on Informatics and Applications (ICIA 2012) (2012)
10. Chen, L., Li, B., Zhao, K., Rizos, C., Zheng, Z.: An improved algorithm to generate a Wi-Fi fingerprint database for indoor positioning. Sensors **13**(8), 11085 (2013)
11. Cook, A.: 5 phenomena that impact Wi-Fi signal (2015). http://www.mirazon.com/5-phenomena-that-impact-wi-fi-signal/
12. Google: Indoor maps availability. https://support.google.com/gmm/answer/1685827
13. Ramani, S.V., Tank, Y.N.: Indoor navigation on google maps and indoor localization using RSS fingerprinting. CoRR abs/1405.5669 (2014)
14. Kaushik, S.: Strength of quick response barcodes and design of secure data sharing system. Int. J. Adv. Comput. Sci. Appl. (IJACSA) **2**(11), 28 (2011)

IPv6 Firewall Functions Analysis

Josef Horalek[⊠] and Vladimir Sobeslav[⊠]

Faculty of Informatics and Management,
University of Hradec Kralove, Hradec Kralove, Czech Republic
{josef.horalek,vladimir.sobeslav}@uhk.cz

Abstract. Currently, the most security solutions are based on technologies realted to the old IPv4 protocol. Although the new protocol requires a different approach, the network security solution often does not correspond and the network protection may be affected. One of the results of the IPv6 implementation is the end of network address translation (NAT). Despite its disadvantages, NAT can act as a security element of IPv6 protocol. The goal of this paper is to analyse, present and compare firewall functions at the most used Windows and Linux distribution along with the detailed packet analysis.

Keywords: IPv6 · RFC · Firewall · Testing · Firewall6

1 Introduction

One of the IPv6 design intentions is to make a clean break with the need of the network address translation (NAT). Despite its drawbacks, NAT operates also as a security element [1] and its extraction creates a breach that needs to be filled. However, the transition to IPv6 will not occur immediately and all at once; the IPv4 and IPv6 protocols will coexist for some time.

Known problems emerging from the dualness of the environment are e.g. IPv4-mapped IPv6 addresses which leads to a security breach in the firewall [4]. This can be problematic especially for the home and small business/office networks without experienced administrators. RFC 6092 [5] provides a set of recommendations for simple security of IPv6 gates also via packet filters functions for device manufacturers particularly aimed at given users. IPv6 packet filtering in big "enterprise" networks originally designed as "IPv4-only" networks is covered by RFC 7123 [6]. The security scenarios related to the firewall issues that are necessary to be taken into consideration are stated as:

- IPv4 firewall might not be able to enforce same security policy for IPv6 operation.
- The firewall can support both IPv4 and IPv6, but it might not be correctly configured for IPv6 traffic control.

The aforementioned document RFC 4942 [3] is relatively extensive and from the purely IPv6 problems point of view it mentions also routing headers issues (defined in RFC 2460 [7]) and their potential abuse for firewall bypass or amplification attack. Such an attack is serious and RFC 5095 [8] recommends all the IPv6 nodes to deprecate the use of Type 0 Routing Headers and firewalls to filter this kind of header.

© Springer International Publishing Switzerland 2016
N.T. Nguyen et al. (Eds.): ICCCI 2016, Part II, LNAI 9876, pp. 219–228, 2016.
DOI: 10.1007/978-3-319-45246-3_21

The firewall can potentially be bypassed by means of Type 2 Routing Header. Forged packets in ICMPv6 error reports, traffic filtering with anycast addresses, IPSec, procedures for firewall processing of extended or unknown headers, firewall defence against misuse of Pad1 and PadN options or misuse of link-local addresses, are also all introduced in RFC 4942 including the fragmentation issue.

Security problems are especially packets with overlapping fragments. The detailed description of the issue is offered by RFC 5722 [9] suggesting an IPv6 specification modification in terms of overlapping fragment prohibition. IPv6 specification also allows the use of a Fragment Header in packets that are in fact not fragmented. As RFC 6946 [10] mentions, these so called atomic fragments often result from the host reaction on message ICMPv6 "Packet Too Big." An attacker can artificially fabricate these messages and then launch any fragmentation-based attacks against such traffic. It therefore suggests to process atomic fragments independently on other fragments. RFC 6980 [11] subsequently explicitly forbids use of fragmentation in Neighbor Discovery (ND) messages.

IPv6 brings new Internet Control Message Protocol (ICMP) implementation, so called ICMPv6, defined in RFC 4443 [12]. ICMPv6 is necessary part for correct IPv6 function, although when uncontrolled this protocol presents potential security risk. Similar situation is valid for ICMP in IPv4, although same filtering strategies can not be applied on its IPv6 equivalent. RFC 4890 [13] contains recommendations for ICMPv6 firewall filter, for it is necessary to seek balance between too aggressive and too benevolent filter policy.

Filtering rules also have to take into consideration the specific ICMPv6 message type. For instance ICMPv6 error messages have to be let through by the firewall, but if the firewall serves also as a router, the messages that might be used in configuring a router interface should not be transited through. RFC 4890 filtering recommendations furthermore differentiate goals of ICMPv6 traffic – the rules for traffic directed to the firewall interfaces and the rules for traffic transiting the firewall. ICMPv6 messages are categorized into classes:

- Messages that must not be dropped.
- Messages that should not be dropped.
- Messages that may be dropped.
- Messages that administrators may or may not want to drop depending on local policy.
- Messages that administrators should consider dropping.

Messages are within their classes filtered according to their type and scope of source and destination addresses. In comparison with the original ICMPv4 vs. Firewall relationship it is necessary to create more granular set of rules. Firewalls also need to be adapted for so called multihoming – Shim6 protocol defined in RFC 5533 [14]. It offers the host an option to use multiple IPv6 addresses at the same time. Thus firewall has to correctly keep the state in situation, when the host uses different IP address for the same connection. This problem is also discussed in RFC 6629 [15].

The firewall might not be able to process whole long chain and thus allows the attacker to bypass the filtering rules – RFC 7112 [16] suggests a solution in which the first packet fragment (i.e. offset equals to 0) has to contain complete Header Chain.

Field tests of fragmentation and routing header problems can be found in [17], where the authors were really able to bypass the firewall in this way. The importance of proper configuration of packet fragment filtering rules is shown [18] on the illustrative case where answers of DNSSEC (extension for DNS) to name inquiries can be fragmented and dropped by firewall due to their size. The Internet zone becomes unavailable for clients behind firewall configured like that due to the impossibility of domain name translation into IP address.

The problem of extension headers/fragmentation is also their practical utilization and implementation – as examined in the study [19], in real world environment extension headers are frequently dropped. IETF discusses the possible prohibition of fragmentation as such in IPv6. Lai et al. [20] researched the possibilities of host firewall and network firewall cooperation in order to secure IPv6 network. Since packets using the ESP header are encrypted and hence it is not possible to inspect them by firewalls (the content of the packet is known only to the source and destination, which owns the secret key), they propose deployment of dual firewall controlled by a single central server. It is interesting approach to the solving of the security issues in IPv6 by actualisation of the idea of so called distributed firewalls for IPv6. Firewall issue in IPv6 is considerably complex.

2 Problematic Areas of IPv6 Employment

This part represents main areas of the IPv6 employment and utilization issues and presents the impacts of specific IPv6 features on firewalls. These impacts are introduced here and tested in the following part.

2.1 IPv6 Address-Based Filtering

Stateless packet filtering based on source or destination address works in IPv6 on the same principle as in IPv4, but it brings higher complexity, for single network interface can have multiple IPv6 addresses. The problem also arises during the change of internet service provider (ISP) for in the first 64 bits of IPv6 address, in so called network prefix, a change occurs. Hosts could have at the same time new and old "deprecated" addresses for certain specified duration, during which are old addresses being discarded. The packet filter administrator has to properly react on such a situation by modification of filtering rules.

2.2 Multihoming

Multihoming stands for a state in which a network is connected to multiple internet service providers and it is desirable and typical especially in cases, where a high connection reliability is required. It is not a feature that would not be already incorporated in IPv4, however in IPv6 it has new queries. The suggested multihoming solution for IPv6 is Shim6 protocol defined in RFC 5533 [17]. It is able to reroute the communication to proper paths for multihoming utilization which on the other hand creates problematic relation with firewalls. A stateful firewall creates a state for each connection. Since Shim6 changes so called locators, which communicating hosts use for connection and which not adapted stateful filter uses for specific data stream identification, it is necessary to adapt the firewall in order for it to be able to respond to the change. The administrator of the stateless packet filter has to take into consideration new packet types generated by Shim6.

2.3   Internet Control Message Protocol Version 6

ICMPv6 is used by IPv6 nodes for reporting of errors originating from packet processing and for managing of other Internet1 layer functions, such as diagnostics (ICMPv6 ping). ICMPv6 is integral IPv6 part and the basic protocol (all the messages and behavior required by the technical specification) has to be fully implemented by every IPv6 nod. [13] ICMPv6 uses for its functioning messages which can be divided into two groups, namely error and information messages. The message type is defined by the value in the ICMPv6 message Header Type field. Field Type is 8 bits long; the highest bit for error messages is always set to value 0 (range of possible values 0–127) whereas for information messages it has always the value 1 (128–255 range).

2.4   Extended IPv6 Headers

IPv6 packet is made of two main parts – the header and payload data. The first 40 bytes of the packet belongs to the header. Extended headers can be inserted and chained between the header and the data. Every IPv6 firewall has to be able to process a chain of extended IPv6 headers, since their function might be undesirable from the security policy point of view of the given organization. The minimum is the ability to skip the extended headers and move to the higher layer header. The firewall should be able to set and apply rules due to mere presence of some of the extended headers in the packet. Firewall rules do not have to be applied on all the extended headers; it is possible to add or block packets even based on specific choices inside the headers.

3   IPv6 Firewall Functionality Efficiency Testing

In this part the results of the executed tests focused on efficiency and functionality of the firewalls over IPv6 will be presented.

3.1   Tested Operating Systems

All the operating systems (and firewall implementations they are distributed with) runned within the scope of testing on the hosts are summarized in the Table 1. All the tested operating systems were always fully updated before the testing. 64 bit operating systems variants were used for the testing.

Table 1. Tested operating systems

Operating system	Firewall	Kernel version
MS windows 7 SP1	Windows firewall	6.1.7001
MS windows 8.1	Windows firewall	6.3.9600
MS windows server 2008 R2	Windows firewall	6.1.7601
MS windows server 2012 R2	Windows firewall	6.3.9600
Debian "Wheezy"	Netfilter/iptables	3.2.0-4-amd64
Ubuntu 14.04 LTS	Netfilter/iptables	3.13.0-24-generic
Fedora 21 workstation	Netfilter/iptables - service iptables service replaced by firewall	3.17.4-301-fc21. x86_64
CentOS	Netfilter/iptables - service iptables service replaced by firewall	3.10.0-229.el7. x86_64
FreeBSD	IP firewall	10.0-RELEASE
OpenBSD	Packet filter	5.6
NetBSD	IP filter	6.1.5
Oracle solaris	IP filter	5.11

3.2   Hardware Used for the Testing

As a source in the generated packets testing figured physical personal computer with the following hardware configuration:

- Processor - AMD Phenom II X4 955 Deneb Quad-Core 3.2 GHz
- Memory - 8 GB RAM
- Operating system - Xubuntu 14.04 LTS

Virtual machines employed as a destination or a proxy were on the same PC virtualized by means of virtualization software Oracle VM VirtualBox version 4.3.24. Every virtual machine was assigned one virtual processor, 1024 MB RAM and turned on hardware virtualization support (AMD-V). Both on the source and on the packets destination there was always active software WireShark for network traffic detailed analysis during the testing.

3.3   Testing Scenario and Firewall6 Results

The tool of choice used for the IPv6 firewall testing is firewall6 from the The Hacker's Choice IPv6 Toolkit. It is the most sophisticated of the similar solutions. It was always launched on the source within the topology depicted in Fig. 1.

The firewall6 intention is to attempt to bypass the rules of access control (ACL) defined on the firewall. For purposes of this work firewall6 generated TCP traffic with destination port 22. The combination of protocol and destination port was blocked by every tested firewall. Firewall6 one by one puts to test scenarios summarized in the following list:

Fig. 1. Firewall6 testing topology

- FW1 A common SYN packet is sent to find out whether the port is blocked.
- FW2 Same as FW1, but with added data.
- FW3 Field Type on the 2nd layer in the ISO/OSI model is incorrectly set on the IPv4 value. This can lead to IPv6 rule bypassing via IPv6 network code in case the blocking decision is based only on layer 2.
- FW4 SYN packet has added 8 bytes long Hop-by-hop header, whereas its field Options is filled with zeroes. The packet should be processed as a packet without any extended header.
- FW5 The header for Destination Options is used in the same way as in FW4; again, it should not have any influence on the packet processing.
- FW6 The packet contains Hop-by-hop header with the option Router Alert (which indicates that it is expedient to examine packet closely since it can possess valuable information for the router).
- FW7 The packet contains 3 headers for Destination Address. We have mentioned that the header can occur maximally twice per packet, nevertheless the Options field is again filled with zeroes and hence these headers should be ignored. They should not have any impact on packet processing as in FW4 and FW5.
- FW8 The principle is the same as in FW7 and FW5 scenarios, only this time 130 headers for Destination Address is used. In such an amount they can cause difficulties to some devices.
- FW9 Atomic fragment is sent (i.e. packet with Fragmentation header, yet it is in fact not fragmented).
- FW10 Two atomic fragments with the same ID (Fragmentation headers in the packet have same 32bit identification number necessary for reassembling) are sent. Atomic fragments must not be processed in the same way as truly fragmented network traffic.
- FW11 2 atomic fragments with different IDs.
- FW12 3 atomic fragments with the same ID.
- FW13 3 atomic fragments with different IDs.
- FW14 130 atomic fragments with the same ID.
- FW15 130 atomic fragments with different IDs.
- FW16 260 atomic fragments with the same ID.
- FW17 260 atomic fragments with different IDs.
- FW18 The packet with 2 KB large Address Option is sent. Because of the size it has to be fragmented and can cause the firewall to collapse.
- FW19 The same principle as in FW18, only with addition of another Address Option.

- FW20 Another FW18 variant. The amount of Address Option headers is increased to 32.
- FW21 The packet with combination of two Address Option headers and 2 Fragmentation headers.
- FW22 The packet with combination of 4 Address Option headers and 3 Fragmentation headers.
- FW23 Only first two fragments out of three have the value of Next header field set at TCP.
- FW24 Next header of the first fragment is set at ICMPv6, of the second at TCP.
- FW25 The first two fragments are overlapping. Fragments are filled with Address Option headers.
- FW26 The packet is sent in three fragments. Next header of the first one is set at ICMPv6, the remaining two are set at TCP.
- FW27 The first of the three fragments has Next header set at TCP, the remaining two are set at ICMPv6.
- FW28 The last two fragments are overlapping.
- FW29 The ICMPv6 "ping" packet is fragmented and first fragment is overlapped by the second.
- FW30 The ICMPv6 "ping" packet is fragmented into three parts and the second fragment is overlapped by the third.
- FW31 In the fragment chain, after the second fragment the following one is sent with the same ID and completely overlaps the second fragment, while it is identical with the overlapped one except the altered Data Payload at the end of the fragment.
- FW32 Several packets with different source ports are sent.

After The Hacker's Choice IPv6 Toolkit installation the complete test set can be easily launched from the command line by `./firewall6 <Outgoing Interface> <Destination Address> <Destination Port>`.

The individual operating systems results and their firewalls tested by firewall6 tool are presented in the Table 2. As we already stated, all the traffic was generated as TCP traffic with destination port 22. This combination of the port and the protocol was always explicitly blocked. Successful packet/s blocking is indicated by Y, bypass of the firewall rules is NO. Since all the systems of the MS Windows family rendered same results, they are condensed into single column in the table.

The majority of the firewalls passed the tests with a clean slate. Detailed log analysis of the individual firewalls reveals that some types of fragmented traffic always passes through, however they are correctly not reassembled and system does not respond to them. This can be considered as a satisfactory result and these cases are evaluated as such in the Table 2. Firewall IP Filter had problems with some of the testing scenarios. The firewall is used by NetBSD and Oracle Solaris systems, whereas it has problems in different scenarios in each system. Over Oracle Solaris IP Filter does not manage correct atomic fragments processing, in NetBSD it does not manage the work with certain combinations of extended headers and both variants of overlapping ICMPv6 "ping." When sending packets with different source ports in both systems, some of the source ports were blocked while others were not. Moreover, the behaviour is inconsistent among both systems. Packets with source ports 21, 179, 443 and 8080

Table 2. The results for firewall 6

	MS Windows	Debian	Ubuntu 14.04	Fedora 21	CentOS	FreeBSD	OpenBSD	NetBSD	Solaris
FW1	Y	Y	Y	Y	Y	Y	Y	Y	Y
FW2	Y	Y	Y	Y	Y	Y	Y	Y	Y
FW3	Y	Y	Y	Y	Y	Y	Y	Y	Y
FW4	Y	Y	Y	Y	Y	Y	Y	Y	Y
FW5	Y	Y	Y	Y	Y	Y	Y	Y	Y
FW6	Y	Y	Y	Y	Y	Y	Y	Y	Y
FW7	Y	Y	Y	Y	Y	Y	Y	Y	Y
FW8	Y	Y	Y	Y	Y	Y	Y	Y	Y
FW9	Y	Y	Y	Y	Y	Y	Y	Y	Y
FW10	Y	Y	Y	Y	Y	Y	Y	Y	NO
FW11	Y	Y	Y	Y	Y	Y	Y	Y	NO
FW12	Y	Y	Y	Y	Y	Y	Y	Y	NO
FW13	Y	Y	Y	Y	Y	Y	Y	Y	NO
FW14	Y	Y	Y	Y	Y	Y	Y	Y	NO
FW15	Y	Y	Y	Y	Y	Y	Y	Y	NO
FW16	Y	Y	Y	Y	Y	Y	Y	Y	Y
FW17	Y	Y	Y	Y	Y	Y	Y	Y	Y
FW18	Y	Y	Y	Y	Y	Y	Y	NO	Y
FW19	Y	Y	Y	Y	Y	Y	Y	NO	Y
FW20	Y	Y	Y	Y	Y	Y	Y	NO	Y
FW21	Y	Y	Y	Y	Y	Y	Y	NO	Y
FW22	Y	Y	Y	Y	Y	Y	Y	NO	Y
FW23	Y	Y	Y	Y	Y	Y	Y	Y	Y
FW24	Y	Y	Y	Y	Y	Y	Y	Y	Y
FW25	Y	Y	Y	Y	Y	Y	Y	Y	Y
FW26	Y	Y	Y	Y	Y	Y	Y	Y	Y
FW27	Y	Y	Y	Y	Y	Y	Y	Y	Y
FW28	Y	Y	Y	Y	Y	Y	Y	Y	Y
FW29	Y	Y	Y	Y	Y	Y	Y	NO	Y
FW30	Y	Y	Y	Y	Y	Y	Y	NO	Y
FW31	Y	Y	Y	Y	Y	Y	Y	Y	Y
FW32	Y	Y	Y	Y	Y	Y	Y	NO	NO

were blocked over NetBSD, where in case of Oracle Solaris it was ports 53, 80, 111, 123, 179, 443 a 8080. This can not be considered as a good result, we can imagine e.g. misuse for so called "OS fingerprinting," meaning remote identification of the operating system running on the given host, which represents certain security risk because attacker can acquire more in-depth information about the potentional target of the attack. Then he can more accurately select means of the attack, for example by use of known security holes of the given operating system, even though one of the firewall responsibilities is to defend unauthorized information drain.

4 Conclusion

The aim of the article was to map the firewall issue within the IPv6 environment. Initially the current state of the issue was outlined building mainly on related RFC documents. The selected IPv6 specification areas were introduced alongside the necessary definitions of terms. IPv6 firewalls were then practically tested in virtualized laboratory environment. Several host firewalls standardly distributed with selected operating systems were tested. These firewalls will probably achieve even more significant importance considering the return of the truly end-to-end connectivity idea. Higher responsibility share in terms of security will be moved from centralized network perimeter defence to individual hosts. Every firewall was subjected to a series of tests via selected software tools. The tests were aimed at known problematic IPv6 specification areas. The behavior of individual firewalls was recorded and discussed.

Taking everything into account it can be concluded that firewalls tested in the aforementioned tests by means of firewall6 from The Hacker's Choice IPv6 Toolkit behaved inconsistently both between each other and in the relationship to relevant RFC. In order to achieve more correct behaviour closer to RFC it is necessary to create high-quality access control ruleset.

Acknowledgment. The support of Czech Science Foundation GACR 15-11724S DEPIES is gratefully acknowledged.

References

1. RFC 2993 - architectural implications of NAT. http://tools.ietf.org/html/rfc2993
2. RFC 4864 - local network protection for IPv6. http://tools.ietf.org/html/rfc4864
3. RFC 4942 - IPv6 transition/co-existence security considerations. http://tools.ietf.org/html/rfc4942
4. RFC 7059 - a comparison of IPv6-over-IPv4 tunnel mechanisms. http://tools.ietf.org/html/rfc7059
5. RFC 6092 - recommended simple security capabilities in customer premises equipment (CPE) for providing residential IPv6 internet service. http://tools.ietf.org/html/rfc6092
6. RFC 7123 - security implications of IPv6 on IPv4 networks. http://tools.ietf.org/html/rfc7123
7. RFC 2460 - internet protocol, version 6 (IPv6) specification. http://tools.ietf.org/html/rfc2460
8. RFC 5095 - deprecation of type 0 routing headers in IPv6. http://tools.ietf.org/html/rfc5095
9. RFC 5722 - handling of overlapping IPv6 fragments. http://tools.ietf.org/html/rfc5722
10. RFC 6946 - processing of IPv6 "atomic" fragments. http://tools.ietf.org/html/rfc6946
11. RFC 6980 - security implications of IPv6 fragmentation with IPv6 neighbor discovery. http://tools.ietf.org/html/rfc6980
12. RFC 4443 - internet control message protocol (ICMPv6) for the internet protocol version 6 (IPv6) specification. http://tools.ietf.org/html/rfc4443
13. RFC 4890 - recommendations for filtering ICMPv6 messages in firewalls. http://tools.ietf.org/html/rfc4890

14. RFC 5533 - Shim6: level 3 multihoming shim protocol for IPv6. http://tools.ietf.org/html/rfc5533
15. RFC 6629 - considerations on the application of the level 3 multihoming shim protocol for IPv6 (Shim6). http://tools.ietf.org/html/rfc6629
16. RFC 7112 - implications of oversized IPv6 header chains. http://tools.ietf.org/html/rfc7112
17. Kim, J., Cho, H., Mun, G., Seo, J., Noh, B., Kim, Y.: Experiments and countermeasures of security vulnerabilities on next generation network. In: Future Generation Communication and Networking (FGCN 2007) (2007)
18. Van Den Broek, G., van Rijswijk-Deij, R., Sperotto, A., Pras, A.: DNSSEC meets real world: dealing with unreachability caused by fragmentation. IEEE Commun. Mag. **52**, 154–160 (2014)
19. Gont, F., Linkova, L.: IPv6 extension headers in the real world v2.0. (2016)
20. Lai, Y., Jiang, G., Li, J., Yang, Z.: Design and implementation of distributed firewall system for IPv6. In: 2009 International Conference on Communication Software and Networks (2009)

Information Technology
in Biomedicine

Novel Edge Detection Scheme in the Trinion Space for Use in Medical Images with Multiple Components

Dawit Assefa[1,2] and Ondrej Krejcar[2(✉)]

[1] Center of Biomedical Engineering, Division of Biomedical Computing, Addis Ababa Institute of Technology, Addis Ababa University, Addis Ababa, Ethiopia
`dawit.assefa@aau.edu.et`
[2] Center for Basic and Applied Research, Faculty of Informatics and Management, University of Hradec Kralove, Hradec Kralove, Czech Republic
`ondrej@krejcar.org`

Abstract. Very recently we proposed a promising scheme for tissue classification of multi-parametric magnetic resonance images (MP-MRI) of the brain based on signal analysis in higher dimensional vector spaces. The method treats MP-MR images as colors represented holistically in three (trinion) or four (quaternion) algebraic spaces. Compared to the well known quaternions, the recently proposed three component trinions are more efficient in representation of images with three channels and the respective Fourier transforms allow visualization of their wavenumber spectra as a whole. The current study discusses an edge detection scheme based on statistical metrics derived from locally computed trinion Fourier transforms for use in robust edge detection of MP-MR images and other color medical images. Performance of the proposed scheme is compared against a quaternion formulation and with another vectorial approach. Application of the method is shown in edge detection of various color test images and scenes with different degrees of difficulty. Discussion and preliminary results on the application of the proposed scheme on MP-MR images of brain scans of patients treated for glioblastoma multiforme (GBM) have also been included.

Keywords: Color image processing · Trinion · Quaternion

1 Introduction

The inter-correlation information that is embedded among the monochromatic components of multi channel images is often considered more informative than just the individual serial images. Image processing approaches that make use of such inter-correlation information are proved to be efficient in many applications. One such interesting area of research in image processing is color edge detection. A successful edge detection is often considered as a good step towards effective image segmentation, registration, identification and many other applications.

© Springer International Publishing Switzerland 2016
N.T. Nguyen et al. (Eds.): ICCCI 2016, Part II, LNAI 9876, pp. 231–241, 2016.
DOI: 10.1007/978-3-319-45246-3_22

232   D. Assefa and O. Krejcar

Various monochromatic as well as vectorial approaches have been suggested in
the literature for use in edge detection [1–5]. Some of these include the Laplacian operator and vector gradients. A good review and detailed comparison of
different edge detection schemes can be found in [6]. Generally, in various applications, the vectorial approaches are shown to be more efficient in edge detection
of color images compared to most serial approaches. This is so as the vectorial
approaches account for the inter correlation between color channels that make
up the color image.

Integral transforms such as the 2D Gabor transform and wavelets are also
known to be one of the pillars in edge detection. Generally, for use in color
image processing, in which case edge detection is a major part, it is believed
that approaches that take into account the inter-correlation information that is
embedded among color components are more effective than serial approaches. In
this regard, a core issue is then finding an efficient color representation scheme.
A real or complex space representation that is used in the Gabor transform,
wavelets and others may not be adequate. A work around this issue, however,
has been realized recently through the use of higher dimensional algebras. One
such a holistic approach that showed great promises in color image processing
makes use of the well known quaternions [7] in four space and the recently
proposed trinions [8] in three space.

Based on Fourier transforms defined in trinion and quaternion spaces, a previous work by one of the authors (DA) showed a promising way of tissue identification and classification of multi-parametric magnetic resonance images (MP-MRI)
of the brain, applied to patients treated for the most common and aggressive
of the gliomas known by the name glioblastoma multiforme (GBM) [9,10]. Our
ongoing investigation aims to have a good automatic tool that efficiently and
uniquely characterizes/identifies different tissue structures, specifically to the
brain, mainly targeting tumors and surrounding structures, based on MP-MR
image processing. The previous attempt was mainly a signal analysis based approach that reveals useful signatures for glioma tumors which may have a potential to assist automatic tumor demarcation (segmentation). Inspired by recent
applications of higher dimensional Fourier transforms in color image analysis,
the scheme treats MP-MR images as multi band colors represented in a holistic
manner in higher dimensional algebraic spaces. The theme of the current study
is very similar but follows a different track and specifically makes use of edge
detection. A new color edge detection scheme for MP-MR images is proposed
in this paper making use of statistical features derived from a spatially localized analysis of colors in the trinion space. The rest of the paper is organized
into the following sections. Sections 2 and 3 present review of quaternions and
trinions and the respective Fourier transforms. The novel algorithm for use in
edge detection of images with multiple components is presented in Sect. 4 while
results with illustrative examples and discussions are included in Sect. 5. The
final section presents concluding remarks.

2   Quaternions vs Trinions in Color Image Processing

Use of quaternions offers the advantage to represent multi-channel color images vectorially and also permit the analysis of such images holistically by treating every multi-band color pixel as a single entity. Particularly the introduction of quaternion Fourier transforms has helped to realize a way for computing the Fourier transform of color images as one quantity [11–15]. This approach has been shown to be useful in many color image processing areas including color vector filtering [16,17], color image cross and auto-correlations [18], the design of quaternion principal component analysis [19], and the quaternion matrix singular value decomposition [20]. Their application in local color spectral analysis has also been shown including use of quaternionic Gabor filters [21], the quaternion wavelet transform [22,23], and the quaternion S transform [24]. Color edge detection also has been another area where holistic color analysis techniques such as the quaternion formulation proved to be useful. Different quaternionic schemes have been proposed previously in this regard including use of local quaternion phase [25], and quaternion moments [26], to mention few.

Different forms of the quaternion Fourier transform (QFT) exist which resulted mainly due to the non-commutative multiplication in the quaternion space. A QFT as defined in [27] is given by:

$$QFT1(u,v) = \int_{-\infty}^{\infty} \int_{-\infty}^{\infty} e^{-j2\pi ux} h(x,y) e^{-k2\pi vy} \mathrm{d}x \mathrm{d}y \qquad (1)$$

where $h(x,y)$ is given color image and u and v are the wavenumbers (spatial frequencies) in the x and y directions respectively. The ortho-normal bi-vectors $\{i,j,k\}$ satisfy: $i^2 = j^2 = k^2 = -1$, and $ij = -ji = k$, $jk = -kj = i$, $ki = -ik = j$. Another type of the QFT suggested in [13] is given by:

$$QFT2(u,v) = \int_{-\infty}^{\infty} \int_{-\infty}^{\infty} e^{-\mu 2\pi(ux+vy)} h(x,y) \mathrm{d}x \mathrm{d}y \qquad (2)$$

where μ is an arbitrary unit (norm equals unity) pure (zero real component) quaternion, $\mu^2 = -1$. A value of $\mu = (i+j+k)/\sqrt{3}$ is used in the literature and this choice coincides with the gray line on the RGB space with all three components equal. Both QFT types are invertible and the formulae for the inverses could be found by simply reversing the signs ($-$ to $+$) of the exponents in the respective exponential kernel terms.

There exist many modalities to capture color images such as digital cameras, scanners and other color photography techniques, which nowadays could be found almost everywhere in modern life. However, most protocols operate only with three color channels/bands generating three component colors, which could not be uniquely represented in four space. For example, an RGB color $h(x,y)$ is commonly mapped in the quaternion space as a pure quaternion as $h(x,y) = R(x,y)i + G(x,y)j + B(x,y)k$, setting the real part equal to zero. Even though it might increase the degrees of freedom, the extra fourth dimension makes quaternions redundant in representation of such colors. Hence, for a more efficient representation, trinions have been introduced recently and trinion based integral transforms have been proposed as a substitute to the QFT [8].

A trinion is defined with one real and two vector components as $t = a + ib + jc$ where i, j are operators satisfying: $i^2 = j$, $ij = ji = -1$, and $j^2 = -i$. It can easily be checked that trinions with the three base elements $\{1, i, j\}$ form an abelian (commutative) group where $1 = j^3$ is the multiplicative identity element. Distinct from quaternions, trinions with the above structure form a commutative ring. Trinions are associative, commutative, and distributive with respect to addition and multiplication. It is also easy to show that all other field axioms are satisfied by trinions except invertibility. Hence trinions are not fields. In comparison, the skew field of quaternions $\mathbb{Q}$ is a four dimensional, non-commutative field over the field of real numbers $\mathbb{R}$. Sum and product of trinions is defined analogous to the complex/quaternion case. Trinion exponentiation, conjugate and inverse (when it exists) have all quite complicated expressions. For example, the conjugate $\bar{t}$ of $t = a + ib + jc$ is given by:

$$\bar{t} = \frac{(a^2 + b^2 + c^2)(a^2 + bc - i(c^2 + ab) - j(ac - b^2))}{(a^3 - b^3 + c^3 + 3abc)} \tag{3}$$

Corresponding to their number of degrees of freedom, three attributes have been defined for trinions: amplitude, eigen axis and eigen angle. Accordingly, any trinion $t = a + ib + jc$ can be written in the form $t = |t|(\cos(\phi) + \mu \sin(\phi))$ where $|t| = \sqrt{a^2 + b^2 + c^2}$, $\mu = V(t)/|V(t)|$, and $\phi = \arctan(|V(t)|/S(t))$, $0 \le \phi < \pi$ are the amplitude, eigen axis, and eigen angle (phase), respectively, and S and V are the real and vector parts of t. t is unit when $|t| = 1$, and it is pure trinion when $a = 0$. More interesting properties of trinions can be found in [8].

3   Trinion Fourier Transforms

Two possible working definitions for the trinion Fourier transform (TFT) are suggested in [8]. We chose to present only the discrete versions in this study and readers can consult a previous study in [8] to learn more details. Accordingly, for $M \times N$ input color image $h(x, y)$, the discrete version of the type I TFT and its inverse are given by:

$$TFT1(u, v) = \frac{1}{MN} \sum_{x=0}^{M-1} \sum_{y=0}^{N-1} h(x, y) \left(\cos(2\pi(\frac{ux}{M} + \frac{vy}{N})) - \mu_1 \sin(2\pi(\frac{ux}{M} + \frac{vy}{N})) \right) \tag{4}$$

$$h(x, y) = \sum_{u=0}^{M-1} \sum_{v=0}^{N-1} TFT1(u, v) \left(\cos(2\pi(\frac{ux}{M} + \frac{vy}{N})) + \mu_2 \sin(2\pi(\frac{ux}{M} + \frac{vy}{N})) \right) \tag{5}$$

where μ_1 is a unit, pure trinion, and μ_2 is a trinion such that $\mu_1 \mu_2 = -1$. Note that μ_1 and μ_2 are chosen arbitrarily and the expression $\mu_1 \mu_2 = -1$ generally results in a system of non-linear algebraic equations (with respect to the trinion coefficients) and so μ_1 and μ_2 must be chosen such that this system has a real solution. For a RGB image, we customarily use $\mu_1 = (i - j)/\sqrt{2}$ and $\mu_2 = (-1 - i + j)/\sqrt{2}$. It can be seen that TFT1 resembles the euler form of QFT2. Otherwise the two are distinct. For example the conjugate relationship that exists between the kernels of the forward and inverse transforms in the case of the QFT doesn't hold in the TFT space. There are also other properties which are known to hold in the complex and quaternion spaces but not with the trinions. However, as shown in previous studies, such unique properties of the trinions do not pose much problems in using the TFTs for efficient color image analysis.

The discrete TFT of type II and its inverse are similarly defined as follows:

$$TFT2(u, v) = \frac{1}{MN} \sum_{x=0}^{M-1} \sum_{y=0}^{N-1} h(x, y)(\cos(2\pi\frac{ux}{M}) - \mu_1\sin(2\pi\frac{ux}{M}))(\cos(2\pi\frac{vy}{N}) - \mu_2\sin(2\pi\frac{vy}{N})) \qquad (6)$$

$$h(x, y) = \sum_{u=0}^{M-1} \sum_{v=0}^{N-1} TFT2(u, v)(\cos(2\pi\frac{ux}{M}) + \mu_3\sin(2\pi\frac{ux}{M}))(\cos(2\pi\frac{vy}{N}) + \mu_4\sin(2\pi\frac{vy}{N})) \qquad (7)$$

where μ_1, μ_2 are unit, pure trinions, and μ_3, μ_4 are arbitrary trinions satisfying $\mu_1\mu_3 = -1 = \mu_2\mu_4$. In the current work we assumed $\mu_1 = \mu_4 = i$ and $\mu_2 = \mu_3 = j$. As shown in [8], in terms of symmetry analysis, TFT2 is a desired choice than TFT1. However, it has also been shown previously that many operations of interest in image processing including convolutions and correlations are easier using TFT1 than TFT2, and hence TFT1 has been adopted in the current study. Given a color image $h(x, y)$, say an RGB with components $R(x, y)$, $G(x, y)$ and $B(x, y)$, by mapping $R(x, y)$ to the real component of a trinion and $G(x, y)$ and $B(x, y)$ to the two imaginary components, $h(x, y)$ can be written: $h(x, y) = R(x, y) + iG(x, y) + jB(x, y)$.

As demonstrated in [8], one main advantage of use of trinions over the quaternions is the ability to see color spatial frequency spectra as a whole rather than just its amplitude or phase. This is so as the TFT of a given three band image itself has three components and hence both the input image as well as its TFT can be plotted as color images. The case is different in the quaternion space as the QFT of a given three channel color image, often represented as a pure quaternion, has four components with generally non zero real part. In most color applications ranging from a low level (non-linear) filtering to the higher level ones such as segmentation, the extra fourth dimension in the quaternion space is only redundant and use of the trinions should alleviate such redundancy. In that regard, an efficient color edge detection scheme is proposed in the next section making use of features extracted from a spatially localized application of the TFT.

4   Trinion Edge Detection

The first step in our edge detection scheme utilizes a color space transformation. Each component of a RGB image is first normalized to its maximum and then the normalized RGB image is transformed to the Hue (H), Saturation (S) and Luminosity (L) color space. The reason for RGB to HSL conversion was based on

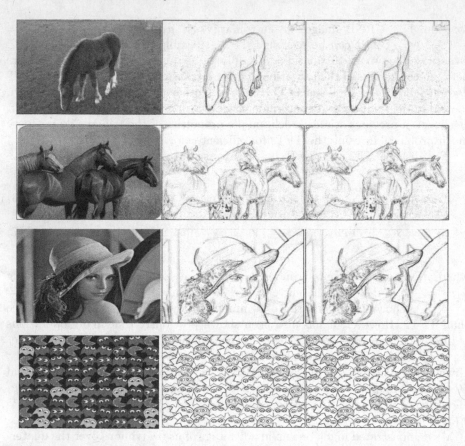

Fig. 1. Original color image (left), trinion edge map (middle), and quaternion edge map (right).

recommendations in previously published works suggesting the superiority of the HSL space in many useful image processing applications [28,29]. Following this step, the resulting color image in the HSL space is mapped to a trinion. The issue of the order of the mapping of a given three component color image to a trinion has been raised in the previous study [8]. The study already showed the order should not affect many image processing applications including texture analysis, pattern recognition and other spectral analysis applications. This is so as the conversion doesn't actually alter the color wavenumber spectra. The proposed edge detection scheme in the current study relies on robust features (texture, pattern) generated through holistic analysis of color pixles, and hence the issue of order of the mapping is not relevant, i.e. the order of the mapping from the HSL to the trinion space shouldn't affect our color edge detection. Then a spatially localized analysis was carried out on a pixel-by-pixel basis by computing the TFTs over a translating localizing window of size 3×3. On the resulting 3×3 trinion valued matrix, a second order feature was computed and assigned to the central pixel value. For the purpose of edge detection, the following feature F was computed:

$$F = Metric\Big(\frac{log(1 + P(x,y))}{log(1 + \max_{x,y}\{P(x,y)\})}\Big) \tag{8}$$

where $P(x,y)$ is the square amplitude ('power') of the TFT, and $Metric$ is a second order statistical feature (variance, standard deviation, etc.). The above step is repeated across all pixels that are included in the color image under consideration and the resulting matrix is plotted as an edge map. Note that all discrete TFT calculations were performed using a repeated use of the 2D Fast Fourier Transform (FFT).

The primary intent of the current study is to show the application and feasibility of the above edge detection scheme on MP-MR image sets. The MP-MR images in this study were taken from a cohort of 29 patients treated for glioblastoma multiforme(GBM), the most aggressive of the gliomas, and is the same data set used in previous studies [9, 10]. Contrast-enhanced T1W gradient-echo, T2-Fluid-Attenuated Inversion Recovery (FLAIR) spin-echo as well as ADC MR images of the patients acquired at baseline as well as within and after radiation therapy were available for analysis. The MR trios came with distinct image slice thickness and the number of slices were also different. Hence, for each patient, the normalized mutual information algorithm was used on all serial MR volumes and were registered to the orientation of the baseline T1W image volume and trilinear interpolation was used to re-sample them to a common resolution. The edge detection analysis on the MP-MR images was then carried out by combining the T1W, T2-FLAIR and the ADC images (parameters) as three channel 'RGB' color images. However, two main issues have to be considered when applying the proposed algorithm in edge detection of these data sets. Primarily, our scheme assumes that the different MR parameters behave well so that our color assumption and representation is acceptable. For example, not all glioma tumors show elevated ADCs, which can be problematic when trying to detect tumor edges. Hence such cases should be excluded from our analysis. This issue of course is not present in actual color images. Another issue is possible registration inaccuracy, which in most cases is unavoidable, accompanied by the fact that we have also applied interpolation as a pre processing step so that all MR image parameters assume a common resolution. These could potentially pose a challenge in implementing our edge detection scheme.

5   Results and Discussion

We start with some standard test images to evaluate the performance of our edge detection scheme. Note that only qualitative analysis is presented in this work while a more quantitative evaluation awaits further validation steps. The first two natural images in Fig. 1 contain horses with a grass background while there is an increased degree of difficulty on the second compared to the first

Fig. 2. Original color image (left), edge map using the max gradient method (second), trinion edge map (third), and quaternion edge map (right).

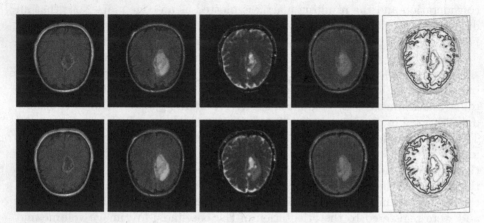

Fig. 3. Columns 1 to 3 are original T1, T2-FLAIR and ADC images respectively containing contrast enhanced glioma tumors; column 4 contains the respective combined color images and column 5 contains the corresponding edge maps generated in the trinion space.

(top) as the former contains more noise. In both cases, the edge maps generated using the proposed scheme picked the edges effectively. For comparison the edge map generated in the quaternion space (using $QFT2$) is also presented. We see similar edge map results generated in the two distinct spaces while the amount of computation involved and the memory requirement in the trinion space is at least an order less than the corresponding operation in the quaternion space. The next two images contain the standard lena image and another color image with visible patterns. The corresponding edge maps identified the edges quite well. Note that, in all the results presented in this paper, the statistical feature computed ($Metric$ in Eq. (8)) is standard deviation. Results found using other statistical features like second order variance were generally similar.

One efficient color edge detection scheme proposed in the literature is the max gradient method. Like the trinions, quaternions and other vectorial approaches, it treats colors as vectors. The edge detection results we found using the max gradient method were satisfactory for most of the test images we considered. Here we want to show with demonstrating examples that the trinion/quaternion formulation is much more powerful than standard vectorial approaches. The example shown in

Fig. 2 show a case where the max gradient method fails to detect edges appropriately. In this example what was mostly desired is to detect the bird sitting in the middle of the grass field. Both the trinion and the quaternion schemes effectively detected the bird and other visible edges while it was hardly with the max gradient method. Note that the max gradient implementation was based on a recent version of the algorithm by J. Henriques which is freely available on the Matlab Central website. The original color scene in this example and many other natural color scenes are freely available on the McGill calibrated color image database at http://tabby.vision.mcgill.ca/html/browsedownload.html. All tests carried out showed that the trinion/quaternion scheme is by far superior to other vectorial approaches including the max gradient operator indicating that an effective color edge detection requires not just a vectorial scheme but a more systematic manipulation of the inter correlation information that is embedded among the monochromes that make up the color image.

Figure 3 presents typical edge detection results we found using the proposed algorithm applied on representative MP-MRI slices. Only the trinion edge maps are presented as the results in the quaternion space were very similar. The brain tissue edges were very well detected with some difficulties around skull boundaries which may have resulted from some image registration inaccuracies. The images also present identified contrast enhanced glioma tumors. The algorithm performs well in detecting the tumor edges and surrounding edema. In most cases that we considered, the scheme robustly showed well defined edges for tumors and the surrounding edemas. However, only representative preliminary results are shown in this paper while further investigation of the method is still underway including clinical validations. For example, two relaxation (T1 and T2) and one functional (ADC) MR parameters are used in this study for the MP-MRI edge detection. Looking at just one set at a time, i.e. three relaxation or three functional parameters, could be an interesting case to try. Resolving these and similar other issues is awaiting further investigation.

6   Conclusions

An automatic color edge detection scheme that makes use of statistical features derived from locally computed trinion Fourier transforms (TFT) is proposed. The scheme is holistic in the sense that it takes into account the inter-correlation information that is embedded within color components. The various tests that we performed both on synthetic and natural color images as well as MP-MR images of the brain clearly showed that the proposed trinion/quaternion formulation offered promising results proving that a systematic use of the inter-correlation information that is embedded within the monochromes (separate color bands) is useful for efficient and effective edge detection of color images. The study is partly a continuation of our ongoing effort to find robust techniques that are able to uniquely characterize different tissues in MP-MRIs and also study their clinical significance. Though the edge detection results found were very similar for both the trinion and quaternion based schemes, the trinion formulation is shown to be more efficient. It is also demonstrated with examples that the proposed trinion/quaternion formulation is more powerful and efficient than other vectorial approaches like the max gradient operator. However, there are still rooms to improve the performance of our edge detection scheme particularly when considering more complex color images. After computing the windowed Fourier transforms, ways to extract more powerful features than the once used in this study (second order variance, standard deviation, etc.) can be beneficial. This for example could mean searching for statistical features which are less sensitive to certain subtle changes within the color image and also noise. Moreover, other than the trinion local amplitude/power, research on inclusion of the other two attributes, the eigen axis and eigen angle (phase), is a work in progress. In principle, other than MRIs, the proposed scheme can be applied to images acquired through other modalities such as CT, PET, and also other color medical images; this without implying that we have proposed an edge detection scheme that can work for any given color image.

Acknowledgements. This work was supported by the project "Smart Solutions for Ubiquitous Computing Environments" FIM (ID: UHK-FIM-SP-2016-2102), University of Hradec Kralove, Czech Republic.

References

1. Sonka, M., Hlavac, V., Boyle, R.: Image Processing Analysis and Machine Vision, 2nd edn., Thomson Asia Pte Ltd and PPTPH, Berlin (2001)
2. Canny, J.: A computational approach to edge detection. IEEE Trans. Pattern Anal. Mach. Intell. **8**(6), 679–698 (1986)
3. Trahanias, P.E., Venetsanopoulos, A.N.: Color edge detection using vector order statistics. IEEE Trans. Image Proc. **2**(2), 259–264 (1993)
4. Tang, H., Wu, E.X., Ma, Q.Y., Gallagher, D., Perera, G.M., Zhuang, T.: MRI brain image segmentation by multi-resolution edge detection and region selection. Comput. Med. Imaging Graph. **24**, 349–357 (2000)
5. Lukac, R., Smolka, B., Martin, K., Plataniotis, K.N., Venetsanopoulo, A.N.: Vector filtering for color imaging: opening a world of possibilities. IEEE Sig. Proc. Mag. **22**(1), 74–86 (2005)
6. Zhu, S.-Y., Venetsanopoulos, A.N., Plataniotis, K.N.: Comprehensive analysis of edge detection in color image processing. Opt. Eng. **38**(4), 612–625 (1999)
7. Hamilton, W.R.: Lectures on Quaternions. Hodges and Smith, Dublin (1853). http://historical.library.cornell.edu/
8. Assefa, D., Mansinha, L., Tiampo, K.F., Rasmussen, H., Abdella, K.: The trinion Fourier transform of color images. Sig. Proc. **91**, 1887–1900 (2011)
9. Assefa, D., Keller, H., Jaffray, D.A.: Signal analysis of multi-parametric MR images in higher order Fourier Spaces. Int. J. Comput. Biosci. **4**(1) (2013)
10. Assefa, D., Keller, H., Jaffray, D.A.: Multi-parametric MR image processing using higher dimensional vector algebra. ACTA Press, Proced. IASTED, ISPHT, pp. 24–31, May 2011
11. Sangwine, S.J., Ell, T.A.: Hypercomplex Fourier transforms of color images. In: IEEE International Conference on Image Processing (ICIP 2001), Thessaloniki, Greece, vol. 1, pp. 137–140, October 2001

12. Sangwine, S.J.: The problem of defining the Fourier transform of a color image. In: IEEE Proceedings of International Conference on Image Processing (ICIP 1998), Chicago, IL, USA, vol. 1, pp. 171–175, October 1998

13. Ell, T.A., Sangwine, S.J.: Hypercomplex Fourier transforms of color images. IEEE Trans. Image Proc. **16**(1), 22–35 (2007)

14. Chernov, V.M.: Some FFT-like algorithms for RGB-spectra calculation. Mach. Graph. Vis. Int. J. **11**(2/3), 139–151 (2002)

15. Labunets, V.: Clifford Algebras as unified language for image processing and pattern recognition. In: Byrnes, J., Ostheimer, G. (eds.) NATO Sciences Series II Mathematics, Physics & Chemistry, vol. 136. Kluwer (2003)

16. Sangwine, S.J., Ell, T.A., Gatsheni, B.N.: Color-dependent linear vector image filtering. In: Proceedings of the EUSIPCO 2004, 12th European Signal Processing Conference, Vienna, Austria, pp. 585–588, 6–10 September 2004

17. Denis, P., Carre, P., Fernandez-Maloigne, C.: Spatial and spectral quaternionic approaches for colour images. Comput. Vis. Image Underst. **107**(1–2), 74–87 (2007)

18. Sangwine, S.J., Ell, T.A.: Hypercomplex auto and cross-correlation of color images. In: IEEE Proceedings of International Conference on Image Processing (ICIP 1999), Kobe, Japan, vol. 4, pp. 319–322, October 1999

19. Le Bihan, N., Sangwine, S.J.: Quaternion principal component analysis of color images. In: IEEE International Conference on Image Processing, vol. 1, pp. 809–812, September 2003

20. Pei, S.C., Chang, J.H., Ding, J.J.: Quaternion matrix singular value decomposition and itsapplications for color image processing. In: IEEE International Conference on Image Processing, vol. 1, pp. 805–808, September 2003

21. Bülow, T., Sommer, G.: Quaternionic Gabor filters for local structure classification. In: Proceedings of the 14th Annual Conference on Pattern Recognition, vol. 11, pp. 808–810 (1998)

22. Bayro-Corrochano, E.: The theory and use of the quaternion wavelet transform. J. Math. Imaging Vis. **24**(1), 19–35 (2006)

23. Chan, W.L., Choi, H., Baraniuk, R.G.: Coherent multiscale image processing using dual-tree quaternion wavelets. IEEE Trans. Image Proc. **17**(7), 1069–1082 (2008)

24. Assefa, D., Mansinha, L., Tiampo, K.F., Rasmussen, H., Abdella, K.: Local quaternion fourier transform and color image texture analysis. Sig. Proc. **90**(6), 1825–1835 (2010)

25. Pei, S.-C., Hsiao, Y.-Z.: Colour image edge detection using quaternion quantized localized phase. In: EURASIP, Aalborg, Denmark, 23–27 August 2010

26. Pei, S.-C., Cheng, C.-M.: Color image processing by using binary quaternion-moment-preserving thresholding technique. IEEE Trans. Image Proc. **8**(5), 614–628 (1999)

27. Ell, T.A.: Quaternion-fourier transforms for analysis of two-dimensional linear time-invariant partial differential systems. In: IEEE Proceedings of the 32nd Conference on Decision and Control, San Antonio, TX, USA, pp. 1830–1841, December 1993

28. Bora, D.J., Gupta, A.K., Khan, F.A.: Comparing the performance of L*A*B* and HSV color spaces with respect to color image segmentation. Int. J. Emerg. Technol. Adv. Eng. **5**(2) (2015)

29. Manian, V., Vásquez, R.: Approaches to color and texture based image classification. J. Int. Soc. Opt. Eng. (SPIE) **41**(7), 1480–1490 (2002)

Simulations of Light Propagation and Thermal Response in Biological Tissues Accelerated by Graphics Processing Unit

Jakub Mesicek[1], Jan Zdarsky[1], Rafael Dolezal[1,2], Ondrej Krejcar[1(✉)], and Kamil Kuca[1,2]

[1] Faculty of Informatics and Management, University of Hradec Kralove,
Rokitanskeho 62, 50003 Hradec Kralove, Czech Republic
{jakub.mesicek, jan.zdarsky, rafael.dolezal,
ondrej.krejcar, kamil.kuca}@uhk.cz
[2] Biomedical Research Center,
Sokolska 581, 50005 Hradec Kralove, Czech Republic

Abstract. In this paper we report on a prototype program for laser-tissue interaction simulation accelerated by graphics processing unit (GPU). We developed a Monte Carlo (MC) model for photon migration in arbitrary shaped turbid media which simulates the light flux inside biological tissues to solve the thermal source term in Pennes' bioheat transfer equation (PBTE). Since both problems are highly parallelizable, we have transformed the underlying mathematical formalism into an OpenCL language code to reduce the computational time-costs. Comparing to sequential implementation, speedup of 210 was achieved in our simulation with GPU. Acceleration benefits are demonstrated separately for MC and PBTE and also for single simulation with both models. The simulation results were obtained in real-time allowing the effective usage in laser interstitial thermal therapy for thermal damage evaluation.

Keywords: Monte Carlo · Pennes' equation · GPU · Opencl · Biological tissues

1 Introduction

The laser-tissue interaction and the possible biomedical utilization have been studied since the invention of the laser. In medicine, specifically focused laser beam has allowed surgeons to achieve necessary precision to destroy tumors, polyps, and kidney stones, to seal nerve endings, blood vessels or to repair retinal detachment while avoiding damage of the healthy cells in the surrounding tissue. Actually, various types of photothermal therapy approaches have been introduced, for example, into surgical cancer treatments, pain relieving, inflammation reduction, tissue healing, or dermatological and cosmetic applications [1]. However, physical parameters like the laser wavelength and radiant energy have to be properly tuned up in order to achieve the best clinical results.

For better understanding and prediction of the physical effects that occur within the laser photons' propagation in the tissue, numerical analysis has been shown to be a

© Springer International Publishing Switzerland 2016
N.T. Nguyen et al. (Eds.): ICCCI 2016, Part II, LNAI 9876, pp. 242–251, 2016.
DOI: 10.1007/978-3-319-45246-3_23

valuable tool [2, 3]. Welch & van Gemert proposed 4-step model that describes the whole interaction [4].

In the first step, light flux inside the tissue is calculated from the radiative transfer equation (RTE). RTE can be solved numerically by diffusion approximation [5] or stochastically by Monte Carlo calculation method [6]. In the next step, the rate of heat generation is evaluated from the fluence and the absorption coefficient of the tissue. Third step is to estimate heat distribution in the irradiated region. The calculation is usually performed by solving Pennes' bioheat transfer equation (PBTE) [7]. In last step, the thermal damage caused to the tissue is evaluated. Processes that lead to the damage are characterized by the damage integral, based on the first order Arrhenius equation [8].

Visible and near-infrared radiation propagation in biological tissues is strongly driven by the scattering events [9]. After few hundred micrometers the light becomes diffusive and distributed in all directions. The fluence is usually calculated by the diffuse approximation (DA) or the Monte Carlo (MC) method. DA is a method based on the RTE with few simplifying assumptions reducing the accuracy [10]. On the other hand the MC method is a gold standard with high accuracy and easy-to-implement code. Due to its stochastic nature, the accuracy is redeemed with high computational costs. The calculated fluence is used as a source term in the Pennes' bioheat transfer equation (PBTE). PBTE is the standard heat equation with the metabolic heat term and the blood perfusion term [11]. Like other partial differential equations, PBTE can be solved by various numerical methods e.g. the finite difference [12], the finite element method [13] or the finite volume method [14]. The equation can be used to determine temperature, size and location of tumors due to thermal properties change [14].

Unfortunately, even substantially simplified simulations of the light interactions with complex biological tissues demand for enormous computational resources. Nowadays, the simulations can take advantages of the parallel computation to reduce the calculation time. Graphics processing units (GPU) containing a lot of computational cores have become available at acceptable price and as such they represent promising devices for acceleration of the biophysical simulations [15]. Soon after the release of the Compute Unified Device Architecture (CUDA) tools, the acceleration up to $1000\times$ of the MC method was reported in comparison with non-parallel CPU code [16].

In this paper we report on a computational model for light propagation and thermal response of the tissues coupled simulations using OpenCL. Section 2 describes the algorithms used for modeling light propagation and heat transfer in tissues. In Sect. 3 the simulation parameters and the measurement setup are described. Section 4 compares and discusses the results of the accelerated and non-accelerated code. Conclusions are made in the last section.

2 Problem Definition

Although the MC method has been found to be the most accurate for solving extensive coupled systems with many degrees of freedom, it is a considerably time consuming technique which renders this type of random sampling simulations practically infeasible for ordinary desktops. GPUs are one of the possible ways to accelerate the code executing at low expenses. Realtime photodynamic analysis can help physicians to design for instance laser interstitial thermal therapy for a particular patient with a tumor or even to determine the optimal parameters during the surgical procedure. Since the induced heat production has to be carefully regulated to prevent the laser ray from insufficient thermal impact on tumors or damaging the healthy tissue, a simulation program can be integrated directly within the laser modulating unit to automatically control the laser effect on the treated organs. Recently, such realtime feedback has been implemented with MPI acceleration [17]. GPU accelerated codes were also presented separately for both problems, light propagation and thermal distribution, respectively [16, 18, 19]. We connect both problems into a single simulations giving new easy-to-implement possibilities (time overlapping, laser intensity modulation etc.) accelerated by GPU.

3   Methods

In this section the algorithms for light propagation and modeling temperature distribution are described.

3.1   The Monte Carlo Model

The MC algorithm for light propagation in biological tissues has been one of the most popular ways to determine the light flux in the tissues since 1995. In that year the MC model for steady-state light transport in multi-layered tissues (MCML) was released allowing to model propagation in media with cylindrical symmetry [6]. The acceleration of the MCML code was done soon after the release of the CUDA tools [16].

The algorithm is based on the generation of pseudorandom numbers and tracing huge number of the photon packets (photons). Photons follow few basic rules and are characterized by their position $\vec{r} = (x, y, z)$, moving direction $\vec{\mu} = (\mu_x, \mu_y, \mu_z)$ and so-called weight w. First the random number ξ in range $(0,1)$ is used for the calculation of the step length s

$$s = \frac{-\log \xi}{\mu_t} \tag{1}$$

where μ_t is the sum of the scattering coefficient μ_s and the absorption coefficient μ_a. Since the space is divided into voxels in the simulation, there are two possibilities that could happen. If the step length is longer than the distance to the nearest border of the voxel the photon is moved to the border. The nearest border is determined from the shortest time-of-flight (TOF) from the current photon position to the borders. When the photon hits the border remaining step is recalculated and lose weight according to Lambert-Beer's law [20]

$$w = w_0 \exp(-\mu_a L) \qquad (2)$$

where L is the traveled distance. If the step length is shorter than the nearest border, the position is updated according to

$$\vec{r} = \vec{r_0} + \vec{\mu} \cdot s$$

and other random numbers are used for calculating new direction due to scattering. For that purpose the deflection angle is calculated using Henyey-Greenstein phase function [6] and the azimuthal angle, which is uniformly distributed over 2π, is determined. The equations for the new direction vector can be found in the referenced literature [6].

On the interface of two media with refractive index, mismatch reflection/refraction occurs. To predict the event, a random number is compared with the reflectance calculated from Fresnel's formulas [6]. If the random number is greater, then the photon is refracted. Otherwise reflection occurs.

Time-resolved simulation is performed by recording photon's TOF. During the propagation, lost energy is recorded from all the photons into the matrix. When the TOF exceed the chosen time gate another matrix is taken. The propagation continues until the photon's weight drops below certain value. In that case, the photon is terminated.

3.2   The Pennes' Bioheat Transfer Equation

PBTE is the most common model used for the temperature distribution calculation. The equation in one of its form can be written as [14]:

$$\rho c_p \frac{\partial T}{\partial t} = \nabla(k\nabla T) + \omega_b \rho_b c_{pb}(T_b - T) + Q_m + \mu_a \Phi \qquad (3)$$

where ρ (ρ_b) is the tissue (blood) density, c_p (c_{pb}) is the specific tissue (blood) heat capacity, k is the thermal conductivity, T is the temperature, T_b is the arterial blood temperature, ω_b is the blood perfusion rate, Q_m is the volumetric metabolic heat generation rate and Φ is the photon fluence. The equation has to be supplemented with the boundary condition $-k\frac{\partial T}{\partial n} = h(T - T_{sur})$, where h is the heat transfer coefficient, T_{sur} is the temperature of the surrounding media (air, tissue). In our study we simplified Eq. (3) to the condition that the temperature at the boundaries is equal to the body temperature, i.e. 36.5 °C. The initial condition at time 0 is similar, i.e. $T(\vec{r}, 0) = 36.5..$

The steady-state solution of Eq. (3) can be found if the left side is set to zero. Then the equation is solved iteratively with finite difference method (n denotes n-th iteration)

$$
\begin{aligned}
T_{i,j,k}^{n+1} = \frac{1}{6} \Big(& \Big(T_{i+1,j,k}^n + T_{i-1,j,k}^n + T_{i,j+1,k}^n + T_{i,j-1,k}^n + T_{i,j,k+1}^n + T_{i,j,k-1}^n \Big) + \\
& \frac{\omega_b \rho_b c_{pb}}{k} \Big(T_b - T_{i,j,k}^n \Big) + \frac{Q_m}{k} + \frac{\mu_a}{k} \Phi_{i,j,k}^n \Big)
\end{aligned} \qquad (4)
$$

where n denotes n-th time step, i,j,k denotes i-th, j-th, k-th voxel in x,y,z directions respectively and other coefficients without an index remain constant. This calculation runs until the condition $\varepsilon = \frac{\left| T_{i,j,k}^{n+1} - T_{i,j,k}^{n} \right|}{T_{i,j,k}^{n+1}} < \text{const.}$ is satisfied.

4   Testing Setup

The main goal of this paper is to present benefits of the parallel code that is used in laser-tissue interaction simulations. This section contains descriptions of the parameters used in the model.

4.1   Validation with MCML

First a validation of the obtained results has to be performed. For that purpose the tissue optics community tests various codes against the first and the most popular one, MCML [6]. We compared the results of the absorption in three layered media of our software with MCML code using parameters presented in Table 1.

Table 1. A validation test. Parameters in the test: μ_a – absorption coefficient, μ_s – scattering coefficient, g – anisotropy parameter, d – thickness of the layer, n – refractive index

	μ_a [mm^{-1}]	μ_s [mm^{-1}]	g [-]	d [mm]	n [-]
Layer 1	1	100	0.9	0.1	1.37
Layer 2	1	10	0.0	0.1	1.37
Layer 3	2	10	0.7	0.2	1.37

4.2   Steady-State Simulations

For lasers running in the quasi-continuous wave regime, the thermal equilibrium is established after few miliseconds. In this case, steady-state model can be used. In the MC model the time resolving is turned off and the steady-state fluence corresponds to the sum of the fluence in particular times. The steady-state temperature distribution is calculated from Eq. (4) with the condition $\varepsilon < 0.05$. Steady-state simulation was performed with the same optical parameters as in Table 1. The thermal coefficients are written in Table 2.

Table 2. Thermal coefficients used in simulation: k – thermal conductivity, ρ – tissue density, c – specific heat capacity, ω – perfusion of blood rate

	k [W m^{-1}K^{-1}]	ρ [kg m^{-3}]	c [J kg^{-1}K^{-1}]	ω [ml s^{-1}ml^{-1}]
Layer 1	0.3	1200	2000	0.0
Layer 2	0.5	1050	3600	0.5
Layer 3	0.2	900	2350	0.012
Blood [22]	0.52	1060	3617	

4.3   Time-Resolved Simulations

When the pulse duration or the time resolution of a detector is less than tens of nanoseconds, time-resolved modeling can be a useful tool providing new possibilities e.g. in medical imaging. In our simulation we monitored the time-resolved MC model. The parameters are the same as in the previous tests, i.e. Table 1. PBTE wasn't tested because of the similarity with the steady-state solution so the same results would be obtained.

4.4   Speedup

The comparison of the simulation duration was done between all cores of AMD FX-8320 and GPU AMD Radeon R9 290X. We individually tested the steady-state MC model, the time-resolved MC model and the steady-state PBTE model. The simulations run in the space made of $150 \times 150 \times 150$ voxels. In MC simulations 1,000,000 photons were used. MWC64X was employed as a random number generator for OpenCL [21].

5   Results and Discussion

In this section the results of the simulations with parameters described above are presented. The main effort is focused on the comparison of the time consumption of the accelerated and non-accelerated code.

5.1   Validation Test

The validation test was performed according to Sect. 4.1. Absorbed energy in the center of the beam along propagation axis is compared in Fig. 1. As can be seen in Fig. 1 the results for many-times experimentally validated MCML software and our software (PPTT) are almost the same. So our obtained results have been assumed to be valid.

5.2   Steady-State Simulations

First steady-state MC simulation duration was measured and compared in Fig. 2. The MC method allows straightforward parallelization. Each photon packet can be dedicated to one computational core due to its independent propagation. Theoretically speedup (ratio between execution time of the sequential and parallel algorithm) rises linearly with the increasing number of used cores. Since the absorbed energy is often scored into the same address in memory we also included AtomicAdd operation that synchronize threads during writing to avoid errors in results. We achieved 208 speedup in comparison with single core CPU code without the usage of atomic operations. Even

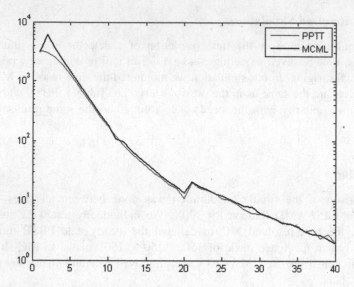

Fig. 1. A validation test with MCML; x-axis displays traveled distance in mm, y-axis represents light fluence [W/mm^3].

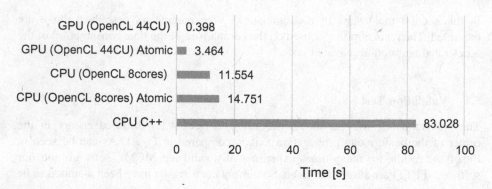

Fig. 2. Calculation time of steady-state MC simulation comparison.

when all 8 CPU cores were employed the calculation time was 29 times slower than GPU. Using atomic operation the speedup was 4.2. A previous work by Alerstam et al. achieved 1080 speedup but only 1 layer was involved and only medium with cylindrical symmetry was considered [16]. Fang et al. achieved 300 times acceleration using their own MC model with similar photon tracing algorithm as our software [18]. However both results were achieved on CUDA capable GPUs so our OpenCL software can be launched on much wider range of devices. Also their programs lacked of heat calculation capabilities.

Similarly steady-state PBTE is solved using GPU and CPU. One computational core is dedicated for temperature calculation for every voxel. Since forward Euler

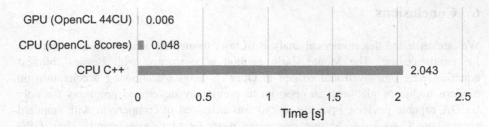

Fig. 3. Calculation time of steady-state PBTE simulation comparison.

method is employed every voxel is calculated in one iteration independent of other voxels. The obtained calculation times are displayed in Fig. 3. The achieved speedup was 340. Generally temperature calculation is much less time consuming than MC light propagation.

The whole simulation (MC + PBTE) lasts in the slowest case 85 s on single-core CPU and 0.4 s on the GPU in the fastest case which corresponds to a speedup of around 210.

5.3   Time-Resolved MC Simulations

For time-resolved simulations similar results as for steady-state simulations were obtained. Calculation time for MC simulations are displayed in Fig. 4. In time-resolved MC model 222 speedup was achieved. These results are similar to steady-state solution indicating that scoring time-of-flight is a negligible operation in terms of time consumption. The GPU in time-resolved simulations is 30 times faster than the 8-core CPU. Including the atomic operations the speedup 3.2 was achieved. It is important to notice that atomic operations significantly slow GPU computation (8-9 times) in comparison with CPU slowdown. The difference occurs due to much higher number of threads that have to be synchronized simultaneously.

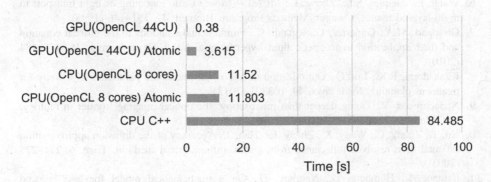

Fig. 4. Calculation time of time-resolved MC simulation comparison.

6 Conclusions

We demonstrated that numerical analysis of laser-tissue interactions are quite suitable for parallelization. The Monte Carlo method accompanied with Pennes' bioheat equation were employed and written in OpenCL language allowing acceleration on modern multicore platforms in opposite to previously developed programs for only CUDA capable devices. Speedup to 210 was achieved in comparison with standard single core C ++ code. Atomic operations were found to significantly slow GPU computation but for CPU the slowdown was negligible. Whole simulation can be calculated in a fraction of seconds making a step towards real-time simulations usable in clinical praxis. This approach offers a possibility to design patient-specific laser interstitial thermal therapy much more effectively than current practices in the clinics, e.g. in cancer therapy.

Acknowledgment. This work and the contribution were supported by project "Smart Solutions for Ubiquitous Computing Environments" FIM, University of Hradec Kralove, Czech Republic (under ID: UHK-FIM-SP-2016-2102). The work was also supported by project 16-13967S.

References

1. Farivar, S., Malekshahabi, T., Shiari, R.: Biological effects of low level laser therapy. J. Lasers Med. Sci. **5**, 58–62 (2014)
2. Afrin, N., Zhou, J., Zhang, Y., Tzou, D.Y., Chen, J.K.: Numerical simulation of thermal damage to living biological tissues induced by laser irradiation based on a generalized dual phase lag model. Numer. Heat Transf. Part A Appl. **61**, 483–501 (2012)
3. Eggener, S., Salomon, G., Scardino, P.T., De la Rosette, J., Polascik, T.J., Brewster, S.: Focal therapy for prostate cancer: possibilities and limitations. Eur. Urol. **58**, 57–64 (2010)
4. Welch, A.J., Van Gemert, M.J.C.: Optical-thermal response of laser-irradiated tissue (2011)
5. Tarvainen, T., Vauhkonen, M., Kolehmainen, V., Kaipio, J.P.: Finite element model for the coupled radiative transfer equation and diffusion approximation. Int. J. Numer. Methods Eng. **65**, 383–405 (2006)
6. Wang, L., Jacques, S.L., Zheng, L.: MCML—Monte Carlo modeling of light transport in multi-layered tissues. Comput. Methods Programs Biomed. **47**, 131–146 (1995)
7. Giordano, M.A., Gutierrez, G., Rinaldi, C.: Fundamental solutions to the bioheat equation and their application to magnetic fluid hyperthermia. Int. J. Hyperthermia. **26**, 475–484 (2010)
8. Kannadorai, R.K., Liu, Q.: Optimization in interstitial plasmonic photothermal therapy for treatment planning. Med. Phys. **40**, 103301 (2013)
9. Ntziachristos, V.: Going deeper than microscopy: the optical imaging frontier in biology. Nat. Methods **7**, 603–614 (2010)
10. Xu, T., Zhang, C., Wang, X., Zhang, L., Tian, J.: Accuracy of the diffusion approximation for total time resolved reflectance from a semi-infinite turbid medium. Time. **6**, 271–275 (2003)
11. Fasano, A., Hömberg, D., Naumov, D.: On a mathematical model for laser-induced thermotherapy. Appl. Math. Model. **34**, 3831–3840 (2010)

12. Jiang, S.C., Zhang, X.X.: Effects of dynamic changes of tissue properties during laser-induced interstitial thermotherapy (LITT). Lasers Med. Sci. **19**, 197–202 (2005)
13. Cvetković, M., Poljak, D., Peratta, A.: FETD computation of the temperature distribution induced into a human eye by a pulsed laser. Prog. Electromagn. Res. **120**, 403–421 (2011)
14. Das, K., Singh, R., Mishra, S.C.: Numerical analysis for determination of the presence of a tumor and estimation of its size and location in a tissue. J. Therm. Biol **38**, 32–40 (2013)
15. Eklund, A., Dufort, P., Forsberg, D., LaConte, S.M.: Medical image processing on the GPU - Past, present and future. Med. Image Anal. **17**, 1073–1094 (2013)
16. Alerstam, E., Svensson, T., Andersson-Engels, S.: Parallel computing with graphics processing units for high-speed Monte Carlo simulation of photon migration. J. Biomed. Opt. **13**, 060504 (2008)
17. Fuentes, D., Oden, J.T., Diller, K.R., Hazle, J.D., Elliott, A., Shetty, A., Stafford, R.J.: Computational modeling and real-time control of patient-specific laser treatment of cancer. Ann. Biomed. Eng. **37**, 763–782 (2009)
18. Fang, Q., Boas, D.A.: Monte Carlo simulation of photon migration in 3D turbid media accelerated by graphics processing units. Opt. Express **17**, 20178–20190 (2009)
19. Reis, R.F., dos Loureiro, F.S., Lobosco, M.: 3D numerical simulations on GPUs of hyperthermia with nanoparticles by a nonlinear bioheat model. J. Comput. Appl. Math. **295**, 35–47 (2016)
20. Watté, R., Aernouts, B., Van Beers, R., Herremans, E., Ho, Q.T., Verboven, P., Nicolaï, B., Saeys, W.: Modeling the propagation of light in realistic tissue structures with MMC-fpf: a meshed Monte Carlo method with free phase function. Opt. Express **23**, 17467 (2015)
21. Thomas, D.B.: The MWC64X Random Number Generator. http://cas.ee.ic.ac.uk/people/dt10/research/rngs-gpu-mwc64x.html#overview
22. IT'IS Foundation: Database of Tissue Properties. http://www.itis.ethz.ch/virtual-population/tissue-properties/overview/

Comparison of RUST and C# as a Tool for Creation of a Large Agent-Based Simulation for Population Prediction of Patients with Alzheimer's Disease in EU

Richard Cimler[2(✉)], Ondřej Doležal[1], and Pavel Pscheidl[1]

[1] Faculty of Informatics and Management, University of Hradec Králové,
Rokitanského 62, 500 03 Hradec Králové, Czech Republic
[2] Center for Basic and Applied Research (CZAV), University of Hradec Králové,
Hradec Králové, Czech Republic
richard.cimler@uhk.cz
http://www.uhk.cz

Abstract. Introduction: During a creation of an agent based simulation is important to choose appropriate tools suitable for simulated topic. Large scale agent based simulations with a wast number of agents are very demanding on computational resources.

Aim: Aim is to compare two approaches (1) Object oriented written in C# and (2) low level in Rust for creation of agent-based simulation and point out pros and cons of these tools.

Methods: Simulation with over 500 millions agents is created in RUST and C#. Performance results from repeated simulation runs are compared. Runtime, memory consumption, multithreading, random number generation and agent synchronization is discussed.

Results: Significant savings of computational resources were observed in Rust simulation. It's advanced options of memory management and concurrent programming options resulted in fractional execution time and memory consumption compared to C#. However, implementation in C# took less time to create and can be easily understood by domain experts.

Keywords: Agent-based · Simulation · C# · Rust · Programming language comparison · Alzheimer's disease

1 Introduction

Choosing appropriate language and tools for creating a simulation can save significant amount of time during creation process as well as computational resources necessary for the simulation run. Programing languages and their libraries offer different levels of syntactical complexity allowing different level of control over exact instructions processed by computer, called level of abstraction. With higher levels of abstraction, resulting code become shorter and more

© Springer International Publishing Switzerland 2016
N.T. Nguyen et al. (Eds.): ICCCI 2016, Part II, LNAI 9876, pp. 252–261, 2016.
DOI: 10.1007/978-3-319-45246-3_24

problem-oriented, however user's ability to influence exact simulation program behavior is reduced.

In large-scale simulations with many agents, the resulting program, or more specifically the machine code, may not perform in an optimal way in terms of simulation speed and memory usage if non-optimal modeling language, tool, algorithm or model architecture is used [7,8]

In simulation of Alzheimer's disease spread prediction in the European Union population (500 millions), two programming languages: (1) C# .NET Framework 4.5 [18] and (2) Mozilla's Rust [2], with different complexity and level of control were used. More computationally optimal model was written in Rust. Rust allows precise control of memory management [1] and overall simulation flow, resulting in complex, less problem-oriented syntax with many optimization-specific parts present in code. Parallelism without additional performance cost in a form locks and other thread synchronization can be achieved in Rust [9], further reducing execution time. Although C# programming language was used as a more problem-oriented alternative, enabling it's user to create formal description of the problem without the necessity of declaring low-level operations directly in code.

2   Model Background

Aim of the simulation is to predict number of patients with Alzheimer's disease (AD) in European Union(EU) until year 2083. Model was created based on data from Eurostat [11] and characteristics of AD [10,13,15,16]. Results of model which has been created using system dynamics modeling approach can be found at [24]. Gained results can be used for determining economic burden of AD in ongoing years [17]. In current research agent-based modeling approach has been used. Each citizen of EU is modeled as an agent and its age and AD occurrence is monitored. Population of EU is predicted between 507 and 525 millions in years 2013–2083. Basic simulation characteristics are in the Table 1.

Table 1. Model description

Characteristics	Values
Initial population:	based on Eurostat
Number of agents:	507–525 millions
Simulated years:	70 (2013–2083)
Observed parameters:	Age, AD occurrence

2.1   RUST

Rust is a new language started by Mozilla Research with an open development process. It is intended to be a drop replacement for low level C-like languages.

Rust offers compile-time memory and thread safety with no sacrifices to control and speed [2].

2.2 C#

C# (C sharp) is a simple, modern, type-safe, object-oriented programming language developed by Microsoft within its .NET framework technology. both C and C++ were used as a source material for C# thus knowledge of these two languages might be helpful for basic understanding. There is strong base of developers using C#. Also intensive development, constant support and documentation are strong advantages of this language [6,23,25]

C# has been chosen as a typical objected-oriented language that provides its user very convenient balance between easily build model based on real system and control over usage of computer resources. Agents in model are represented by objects of standard class that offers necessary variables regarding each agent, that has to be stored.

3 Evaluated Model Areas

In the following part different aspects of simulations created in RUST and C# are compared. Final results can be found at conclusion in a Table 3.

3.1 Agent Invocation and Synchroniztion

In real world, every agent performs it's behavior in parallel, having exactly the same amount of time to perform their behavior. In a computer simulation, guarantee of parallelism for each agent is impossible. Agent's behavior is represented as a routine in the simulation program, sharing limited computing resources with other agents and most likely other programs. [22] This results in inability of all agents to operate at once. Valid simulation requires agents to have equal opportunity to perform their defined behavior at simulation runtime.

Different agent simulations require different level of agent synchronization. Presented simulation with Alzheimer's disease prediction has agents with behavior that doesn't directly affect environment or other agents. Therefore, simple approach of sequential agent execution in each step was used, which later proved to be beneficial in agent invocation optimization. Sequential execution presumes agent's behavior and success in environment is invariant to agent invocation order. Some agent simulation programs like AnyLogic provide random agent execution mechanism in every simulation [14]. More complex methods exist, e.g. rating time-consumption of atomic tasks performed by agents [12].

In Rust, the ability to virtually split the population into n chunks and invoke agents in those chunks sequentially in separate thread results in big performance gain due to lack of locks or any thread synchronization in general. Simulation uses CPU only. Due to approximately half a billion of agents present in this simulation, further improvement by using GPU computation is possible [5]. Similar results can be achieved in C#, however compiler checks related to thread safety make Rust better choice for highly parallelized agent-based simulations.

Rust. In presented AD simulation, there is no need to order agent invocation. Agents are stored as one logically continuous array (operating system handles physical memory allocation) on the heap. Pointer to the beginning of such array is copied and held in each function's stackframe. Each year, array of agents is splitted into n logical chunks of even size (if possible), where n is the number of threads used by the simulation. Each thread than iterates all agents from the chunked linked to it. This behavior is achieved with an pointer to the very first agent of the chunk, which is in fact a pointer to nth agent in original array. No copying is performed.

Each thread has it's own random number generator, as stated in previous chapters. All threads read from common statistical data gathered at the end of every year, represented by separate block of memory. Threads access statistics in read-only regime, therefore no locking mechanism is necessary and no thread-safety overhead is introduced.

Each year, dead agents are removed from the population. This is the only action performed sequentially by one thread. Agents marked as dead are removed from memory and whole array is shrinked into a continuous logical block of memory.

C#. All agents are stored in standard List collection, thus invocation would also be performed with standard C# collection item invocation syntax. Due to demand of storing more than half a billion of agents, manual invocation of every single one would be inefficient at least. Therefore the Parallel.ForEach loop [19] was used. Not only that this method taking care of list items invocation, it is also providing multithreaded execution to the code within.

Comparison. One logically continuous array of agents can be achieved both in Rust and in C#. In C# however, implementing population as an array of structures takes away simplifications given by build-in system for parallel approach presented in Sect. 3.4 and forces the programmer to abandon object-oriented code. In Rust, optimal solution can be achieved with thread safety guaranteed by the compiler and less code.

3.2   Agent Memory Footprint

Both C# and Rust programs perform memory allocation differently. In AD simulation, environment memory footprint is minimal and predictable compared to total amount of memory allocated by agents. Each agent in AD simulation has following parameters defined Age, AD occurrence, Is alive, Total size of objects. Agent's memory requirements are not the same for C# and Rust. See Table 2.

Memory allocation and deallocation strategies differ. Both languages offer multiple ways to achieve optimal storage. However, simulation in C# was created with emphasis on program simplicity. Rust provides environment more suitable for absolute control over memory allocation in terms of size. Agents were identified as the main source of memory consumption. Memory requirements M of simulations with large numbers of agents can be accurately predicted.

Table 2. Datatype representation in different models

Property	C#	Rust
Age	Int32 (32 bits)	byte (8 bits)
Has AD	Boolean (8 bits)	bool (4 bits)
Is alive	Boolean (8 bits)	bool (4 bits)
Total size of object	24 bytes	3 bytes incl. padding

Agent Memory Requirement. The total memory footprint M can be calculated as follows.

$$M = n \cdot (P_a + P_i + A)$$

P_a represents pointer size in bits for each agent. Pointer size depends on target platform and it's typical size is 32 or 64 bits. If no pointers are used to address agents, then the value is zero.

$$P_i = \sum_{1}^{k} p_s$$

P_i represents total size of pointers used to address agent's properties in bits, where k is number of properties of an agent addressed using a pointer and p_s is target platform-dependent pointer size.

$$A = \left(\sum_{1}^{k} a_k\right)$$

A is total size of agent's properties in bits, where k is number of agent's properties and a_k is k'th property size in bits.

Memory Prediction Limits. Memory prediction doesn't consider environment and presumes agents are the main source of memory allocation. In AD simulation, environment is represented by previous year's statistics gathered at the end of each year. It's size is constant and is not affected by agent quantity.

Agent memory footprint is measured with average number of 520,000,000 agents. Table summarizing memory consumption of both simulations is presented in Table 3.

Rust Memory Consumption.

$$M = 520000000 \cdot (0 + 0 + 24)$$

$$P_a = 0$$

Pointer size in bits for each agent, where p_s is the target platform pointer size (typically 32 or 64 bits) and p_e represents whether separate pointer is used to address each agent.

$$P_i = \sum_1^2 4$$

Rust's bool datatype allocates 8 bits of memory (8 bit pointer).

$$A = 8 + 1 + 1$$

Age is represented by an unsinged integer, one byte large, called $u8$. Two boolean states allocate one byte each.

$$M \approx 1560MB$$

Total simulation runtime memory overhead varies from 1630 MB to 1650 MB, depending on used operating system and number of threads. This corresponds with expected memory allocation $M = 1560MB$ plus simulation overhead.

C# Memory Consumption.

$$M = 520000000 \cdot (64 + 0 + 192)$$

$$P_a = 64$$

In C# class is a reference type of variable. That means a variable itself contains only reference to real location of object. Therefore not only data size of object but also size of reference to object must be considered especially with this much agents used in simulation. Size of reference depends on used platform. Reference size for x64 platform is 64 bits.

$$P_i = 0$$

Unlike class the Int32 and Boolean are values types. Therefore no pointer or reference is needed and variable contains targeted value itself.

$$A = 32 + 8 + 8 + 144$$

Age is represented by an 32 bit integer. Two boolean variables, 8 bits each. Additional 144 bits is consumed by object structure itself.

$$M \approx 16640MB$$

Comparison. In Rust, creation of one logically continuous block of memory containing agents is easy to achieve. Additional of the memory savings come from Rust's zero cost abstractions [3]. These are hard to achieve in C#. One option to reduce C#'s memory footprint is to use arrays of structures, this however reduces programmer's comfort. The resulting code of C# is more domain oriented with less optimization-specific details.

3.3   Random Number Generation

Different simulations need different random number generators. Often, congruent generators are used due to the need of result reproducibility. In AD simulation, agents are invoked in parallel. This implies the need for a thread-safe random number generator or usage of multiple generators, one for each thread.

C#. There is class Random [20] that servers as a standard random number generator for majority of purposes. Unfortunately combination of multithread approach and just one instance of Random class causing very unpleasant behavior such as returning 0 value each time. The reason is that method Next that supposed to return randomly generated value is not exactly thread save and the ThreadLocal must be used in order to revive correct sequence of random values. Although this solution has significant effect on simulation performance. Next solution could be to create own Random class object for each thread separately, but this solution also seem inefficient for bigger amount of threads.

As acceptable solution the RNGCryptoServiceProvider class [21] that inherits Random was chosen. Finally, this type is thread safe and can be used without any additional locks adjustments. On the other hand resource consumption of RNGCryptoServiceProvider is still very high.

Rust. Multiple pseudo-random number generators approach is used in Rust. Each thread responsible for executing chunk of n agents takes ownership of single generator with approximately uniform distribution.

Different thread synchronization and management are known to be utilized for different problems. Thread joining, different kinds of locks and their use cases related to agent modeling. This approach was chosen to avoid need for locking and thus increase performance at the cost of storing state of n random number generators in memory, where n is number of threads created by the simulation. Rust provides no random number generator as a part of standard library. Used library in a form of an external crate provides [4] random number generator seeded by OS-dependent source of randomness. Every 32 KiB of generated random data, generator is seeded again automatically.

Generator used in Rust was tested with Kolmogorov-Smirnov distribution test against uniform distribution. Generator passed at 5 % level of significance.

3.4   Multithreading

Rust. Each year, rust creates n threads to iterate chunks of logically continuous memory with agents. Gathering statistic for each year is not performed in a parallel way, as well as giving birth to new agents. Giving birth to new agents is a simple calculation with advantage in parallelism. Rust simulates following agent behavior in parallel: (1) Probability of starting Alzheimer's disease (2) Death probability.

Both steps need read-only access to previous year's statistics. This implies that every thread in both steps needs read-only access to these statistics. This is simply solved by each thread having a copy of a pointer to the statistics structure.

Before death probability calculation may start, probability of Alzheimer's disease outbreak calculation must be successfully finished for all agents. First, k threads are created, iterating chunks of approximately $\frac{n}{k}$ agents, where n represents total number of agents each year. During this process, all agent's behavior is invoked. Before function's stack frame is deleted, all threads are joined and destroyed. This process is then repeated for death probability calculation.

Agents are distributed among simply by diving number of agents alive n by number of current threads t, always rounding up to the nearest integer. This results in one thread possibly operating with lower number of agents. Performance effect of equalizing number of agents invoked in each thread was not examined.

C#. Creation of multiple threads was supported by .NET platform from its beginning. Framework used for this application is version 4.5, therefore naturally also multithreading was highly improved since first version. Currently developer has several possibilities how to provide several threads to his application depending on what level of control over these threads he need.

In our case, we chose one of the latest approaches that is also one of the easiest ones, sacrificing developer's control. Parallel. ForEach loop works similar to standard ForEach which goes through every single object in collection or array. Advantage of Parallel. For Each if that without any additional settings, it use multiple threads for going through collection, thus it is very useful if the collection has a large number of items or action code within loop is demanding. Despite the fact that developer does not have to manage threads run by himself, there is still issue with non exclusive access to objects by multiple threads, therefore the "lock" politics must be applied in that case to avoid additional complications.

Comparison. Agents in Rust language are iterated by n threads, splitting the agent population into n chunks. Chunks do not overlap. Each threads sequentially iterates over it's chunk of agents. This ensures no locking mechanism is needed, resulting in much faster execution. In C#, locking mechanisms and reference counting is necessary, slowing down agent iteration process.

In Rust, no JoinGuard mechanism ensuring all spawned threads must be finished before function's is implemented yet. This implies usage of raw pointers and blocks of code marked as unsafe, operating with raw pointers. This way, threads are harder to work with.

Table 3. Average simulation performance results

Property	C#	Rust
Time of run	15 804 sec	1028 sec
Memory consumption	19 722 MB	1 650 MB
Average mem. cons. of agents	12 493 MB	1 560 MB
Max num. of agents	524 787 161	524 818 205

4   Conclusion

Non-trivial agent-based simulation models require a programming language involved in the process of model creation. Often, the behavior of agents and simulation environment is difficult to express using WYSIWYG (what you see is what you get) editors.

Immense impact on performance and resource usage was observed when efficient algorithms were utilized, however model creation complexity and time were increased. With Rust, efficient algorithms provided performance increase and lower resource usage, however the time needed for simulation creation was increased and code became less problem-oriented.

Major cause for Rust's shortened execution time is control over memory management and the way threads work with agents in memory. Each thread iterates over a fix-sized chunk of agents. Set of agents iterated by each threads are unique, therefore no synchronization is needed when threads are reading from and writing to the large, logically continuous array of agents.

C# as a higher level language, abstracting memory management and providing easy to use concurrency utilities resulted in significantly shorter code that is easier to read, understand, repair and is more domain oriented.

In Rust however, more low-level problems have to be addressed directly in code, making the final domain oriented code interfere with thread synchronizations, reference checking and other low-level mechanisms. Time needed to create the simulation is hardly quantifiable due to differences among programmer skills, but the absolute time required to build the simulation in Rust was considerably higher. Rust's compiler's approach to concurrency, enforcing compile time thread safety with optional unsafe blocks provides better assurance for the programmer that the final simulation works correctly. Although C# is a good and easy choice when the relatively high resources consumption is not an issue.

Acknowledgment. The support of the grant project GAČR #14-02424S and the specific research project SPEV at FIM UHK is gratefully acknowledged.

References

1. Programming Rust. Oreilly & Associates Inc (2015)
2. RUST (2016). https://www.mozilla.org/en-US/research/

3. RUST ownership (2016). https://doc.rust-lang.org/book/ownership.html
4. RUST rng library (2016). https://github.com/rust-lang-nursery/rand
5. Aaby, B.G., Perumalla, K.S., Seal, S.K.: Efficient simulation of agent-based models on multi-gpu and multi-core clusters. In: Proceedings of the 3rd International ICST Conference on Simulation Tools and Techniques, p. 29 (2010)
6. Albahari, J., Albahari, B.: C# 5.0 in a Nutshell: The Definitive Reference. In a nutshell, O'Reilly Media, Incorporated (2012)
7. Bakar, N.A., Selamat, A.: Agent-based model checking verification framework. In: 2012 IEEE Conference on Open Systems (ICOS), pp. 1–4. IEEE (2012)
8. Bakar, N.A., Selamat, A.: Runtime verification of multi-agent systems interaction quality. In: Selamat, A., Nguyen, N.T., Haron, H. (eds.) ACIIDS 2013, Part I. LNCS, vol. 7802, pp. 435–444. Springer, Heidelberg (2013)
9. Balbaert, I.: Rust Essentials. Packt Publishing Ltd, Birmingham (2015)
10. Brookmeyer, R., Corrada, M.M., Curriero, F.C., Kawas, C.: Survival following a diagnosis of alzheimer disease. Arch. Neurol. **59**(11), 1764–1767 (2002)
11. EUROSTAT: (2015). http://ec.europa.eu/eurostat/statistics-explained/index.php/People_in_the_EU_E28093_population_projections
12. Fougères, A.J.: Modelling and simulation of complex systems: an approach based on multi-level agents. arXiv preprint (2012). arXiv:1201.3880
13. Ganguli, M., Dodge, H.H., Shen, C., Pandav, R.S., DeKosky, S.T.: Alzheimer disease and mortality: a 15-year epidemiological study. Arch. Neurol. **62**(5), 779–784 (2005)
14. Grigoryev, I.: AnyLogic 6 in three days: a quick course in simulation modeling. AnyLogic North America (2012)
15. Klimova, B., Maresova, P., Valis, M., Hort, J., Kuca, K.: Alzheimer's disease and language impairments: social intervention and medical treatment. Clin. Interv. Aging **10**, 1401–1407 (2015). http://europepmc.org/articles/PMC4555976
16. Maresova, P., Mohelska, H., Dolejs, J., Kuca, K.: Socio-economic aspects of alzheimer's disease. Curr. Alzheimer Res. **12**(9), 903–911 (2015)
17. Mohelska, H., Maresova, P., Kuča, K.: Economic and managerial aspects of alzheimer's disease in the czech republic. Procedia Soc. Behav. Sci. **109**, 674–678 (2014)
18. MSDN: (2016). https://msdn.microsoft.com/en-us/library/w0x726c2%28v=vs.110%29.aspx
19. MSDN: (2016). https://msdn.microsoft.com/en-us/library/system.threading.tasks.parallel.foreach%28v=vs.110%29.aspx
20. MSDN: (2016). https://msdn.microsoft.com/en-us/library/system.random.aspx
21. MSDN: (2016). https://msdn.microsoft.com/en-us/library/system.security.cryptography.rngcryptoserviceprovider.aspx
22. Silberschatz, A., Galvin, P.B., Gagne, G., Silberschatz, A.: Operating System Concepts, vol. 4. Wesley, Reading (1998)
23. Stellman, A., Greene, J.: Head First C#, 1st edn. O'Reilly, Sebastopol (2007)
24. Tomášková, H., Kühnová, J., Cimler, R., Doležal, O.: Prediction of population with alzheimers disease in eu using system dynamics model. Neuropsychiatric Dis. Treat. **12**, 1589 (2016)
25. Troelsen, A.: Pro C# 5.0 and the.NET 4.5 Framework. Apress, Berkely (2012)

Dorsal Hand Recognition Through Adaptive YCbCr Imaging Technique

Orcan Alpar and Ondrej Krejcar[✉]

Faculty of Informatics and Management,
Center for Basic and Applied Research, University of Hradec Kralove,
Rokitanskeho 62, 500 03 Hradec Kralove, Czech Republic
orcanalpar@hotmail.com, ondrej@krejcar.org

Abstract. Dorsal hand recognition is a trending topic in biometrics and human computer interactive systems. The characteristic and unique shape of the dorsal side of users' hands could be identified and discriminated for continuous authentication or could be tracked for second security option as a keyboard passwords. Therefore we propose a novel recognition system that deals with users' hands on the keyboard using adaptive YCbCr color space. The images are extracted from a video recorded by a camera mounted on the monitor and the Cb and the Cr color intervals of the dorsal hands are identified and stored. In contrast with the common algorithms that deal with the static interval, we propose an adaptive system which initially identifies the Cb and Cr values of the users' hands and subsequently recognize the dorsal hands throughout the frames of the video.

Keywords: Dorsal hand · Identification · Ycbcr color space · Image processing · Video processing

1 Introduction

Hand recognition is a mutual topic of biometrics and human-computer interaction though recognition and identification of the palm region is mostly researched instead of dorsal. The main reason is the uniqueness of the palm prints consisting of principal and secondary lines that can be extracted and discriminated, in biometric authentication and forensics. Additionally, the palms are which enables the usage of the palms in human-computer interaction. Moreover, hand gesture recognition algorithms start with the identification of the hands, yet generally deal with the palms as well [1–6].

However, if we consider continuous authentication of the users, realized by extraction of unique characteristics of hands or password entering style validation, the dorsal hand recognition is a promising methodology. Through the identification, it is possible to validate the biometric features of the users which could be either physical (hand shape, finger shape, hand geometry, finger ratios) or behavioral (key-pressing styles, finger choices, general keyboard usage characteristics). The basic feature of the palm or dorsal hand is the skin color [7] and hand geometry characteristics [8] that can be extracted for identification. Outside of the visible spectrum, it is still possible to recognize hands by near-infrared lenses and multispectral imaging techniques [9–11].

© Springer International Publishing Switzerland 2016
N.T. Nguyen et al. (Eds.): ICCCI 2016, Part II, LNAI 9876, pp. 262–270, 2016.
DOI: 10.1007/978-3-319-45246-3_25

Most skin color searching algorithms lay their ground on computing color intervals in a suitable color space such as red-green-blue (RGB), hue-saturation-lightness (HSL), or luminance-blue difference Chroma-red difference Chroma (YCbCr) as used in the current study. YCbCr color space actually is a manipulated RGB that is gathered by encoding red green and blue data on corresponding layers of images and yet it is simpler to search for the required color component in the layers of Cb and Cr, regardless of Y. However, the skin colors may vary so the confidence intervals should be determined as wide as possible. Even in this case, some misidentification may happen, on the contrary if the interval is too narrow, it sometimes is impossible to recognize the hands.

Therefore we propose an adaptive algorithm that firstly calculates the Cb and Cr values of the user's hand and subsequently assigns confidence intervals for each. To recognize the skin color, assuming that it is homogenous, the unoccupied keyboard is captured in the very first frame of the RGB video process and one more snapshot is taken when the hands of the user are on the keyboard. The common differentiation methods won't have successful results considering the structure of the keyboards so the matrices of each RGB layer is compared to find the region of interest.

Afterwards, since the ROI revealed by RGB layer comparison is noisy, the hands on the corresponding grayscale image are eroded until achieving only the skin tissue. Mapping the grayscale pixel coordinates to Cb and Cr layers will give us the Chroma values of the user's hand. The values subsequently stored to be the reference point for the rest of video processing. The final algorithm searches the interval, designated by the Cb and Cr values, on each frame of the video for recognizing the hand.

Different researchers have previously proposed mechanisms on how to use the YCbCr color space for various image processing applications: Qiu-yu et al. [12] dealt with YCbCr imaging as well and used this imaging technique with k-means clustering for segmentation purposes. Kaur and Kranthi [13] proposed a comparison between two color spaces YCbCr and CIE-Lab for skin color segmentation to be used in face and hand recognition. Chitra and Balakrishnan [14] made a similar comparative study but with HSCbCr and YCbCr color spaces for skin color identification. Shen and Wu [15] presented a system that can segment lips using YCbCr color spaces.

Considering the new developments on authentication methodology and novel methods, we previously presented four various articles [16–19]. These proposed methods could be combined with the dorsal hand recognition to have an extra security. For instance in [17], we presented a novel methodology for keystroke recognition using inter-key durations. The authentication system proposed in that paper could be enhanced by adaptive dorsal hand recognition for a new fusion layer in biometric authentication systems.

This study proposes a novel adaptive YCbCr imaging technique which only focuses on dorsal hand recognition of users' hands on computer keyboard. Mainly we narrow down the interval of Cb and Cr values of the dorsal hands by identifying the skin color and subsequently our system manages to recognize dorsal hands based on these values stored.

2 Methodology

Three major subsystems are introduced in this paper namely: identification, Cb/Cr extraction, and continuous recognition as shown in the schematic presented in Fig. 1 below.

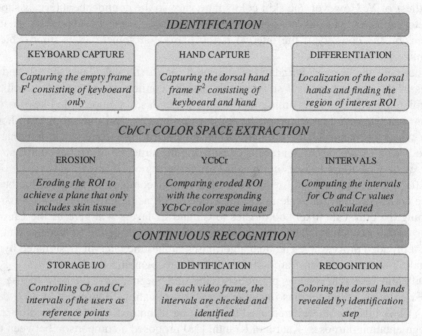

Fig. 1. Workflow of adaptive YCbCr color space recognition methodology of dorsal hands of computer keyboard users. (Color figure online)

The continuous recognition system starts with 640 × 480 resolution in RGB format video recording by A4tech PK-333E LED Lighting Web Cam, mounted on Dell Inspiron 15R laptop. In the first frame of the video F^1, the camera captures the keyboard without the hands. Within a desired duration, in this project it is 5 s, the camera takes one more photo F^2 including the dorsal hand or hands on the keyboard. Any pixel on these frames could be expressed as follows:

$$p^1_{i,j,k} \in F^1_{i,j,k}(i = [1 : w], j = [1 : h], k = [1 : 3]) \tag{1}$$

$$p^2_{i,j,k} \in F^2_{i,j,k}(i = [1 : w], j = [1 : h], k = [1 : 3]) \tag{2}$$

where all of the parameters are integers, w is the width, h is the height of the image; i represents the row number, j the column number and k is the layer of color channel. Figure 2 presents two sample images.

Fig. 2. Unoccupied Keyboard on the left, Hands on the keyboard on the right

However, $F_{i,j,k}^1$ and $F_{i,j,k}^2$ themselves don't enable the extraction of Cb and Cr values, even if we manipulate the color layers or grayscale images. Therefore we define a binary layer $B_{i,j}$ as:

$$b_{i,j} \in B_{i,j}(i = [1:w], j = [1:h]), b_{i,j} \in [0,1] \tag{3}$$

where $w = 640$ and $h = 480$ for this research. So that if the change in color layers is greater than a threshold (for this project $T = 50$) the corresponding pixel in the binary layer turns to 1, otherwise remains zero

$$b_{i,j} \in B_{i,j} \begin{cases} 1, & \left| p_{i,j,k}^1 - p_{i,j,k}^2 \right| > T \\ 0, & otherwise \end{cases} \tag{4}$$

The threshold $T = 50$ is determined after pre-analysis of the image processing in Fig. 2. However it is still not possible to achieve a perfect identification of ROI, and therefore, in the binary layer, a kind of erosion is applied for noise removal. Erosion or dilation is made by eroding the corresponding neighborhood pixels for all pixels on the binary image produced by subtraction. For each pixel on the image, all 4 left and right neighbors are compared with the current pixel, the lowest which means the black one is chosen and replaced and new eroded image $E_{i,j}$ including the skin tissues is gathered as follows:

$$\dot{b}_{i,j} \in E_{i,j} = min(I(i-1,j), I(i,j), I(i+1,j)) \tag{5}$$

$$\dot{b}_{i,j} \in E_{i,j} = min(I(i,j-1), I(i,j), I(i,j+1)) \tag{6}$$

The main reason for the sub-algorithm mentioned above is noise removal due to pixel differences between the reference and final image. The iteration number could be adjusted yet for a resolution of 640×480, we figured out that 50 as the epoch number is enough for noise removal and reveal only the skin tissue. The example images of creating the new binary layer and erosion is shown in Fig. 3.

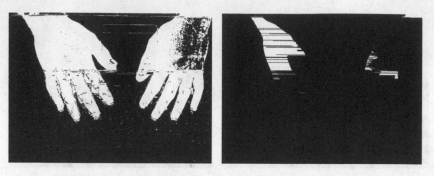

Fig. 3. Binary layer, before erosion (left), after erosion (right)

Although Fig. 3 (right) looks like an image, it actually is a binary matrix consisting zeros and ones, nd therefore if this matrix is multiplied by Cb and Cr layers of the original image $F^2_{i,j,k}$ we get the Cb and Cr values of ROI. The RGB images are turned into YCbCr format as follows:

$$p^2_{i,j,k} \in F^2_{i,j,k}(i = [1:w], j = [1:h], k = [Y, Cb, Cr]) \tag{7}$$

Afterwards, the corresponding pixels are accumulated to achieve the average of Cb and Cr values of the users' hands.

$$H_{Cb} = \sum_{i=1}^{640}\sum_{j=1}^{480} p^2_{i,j,Cb}.\dot{b}_{i,j} / \sum_{i=1}^{640}\sum_{j=1}^{480} \dot{b}_{i,j} \tag{8}$$

$$H_{Cr} = \sum_{i=1}^{640}\sum_{j=1}^{480} p^2_{i,j,Cb}.\dot{b}_{i,j} / \sum_{i=1}^{640}\sum_{j=1}^{480} \dot{b}_{i,j} \tag{9}$$

The average of the $H_{Cb} = 106.3763$ and $H_{Cr} = 150.3123$ are calculated for the images. For $H_{Cb} \pm 5$, - $H_{Cr} \pm 5$ and $H_{Cb} \pm 10$, - $H_{Cr} \pm 10$ confidence intervals are tested for these images and results are presented in Fig. 5. The confidence intervals are chosen just to give idea about the results of the methodology (Fig. 4).

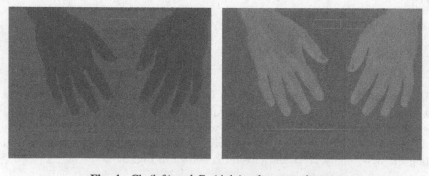

Fig. 4. Cb (left) and Cr (right) color space images,

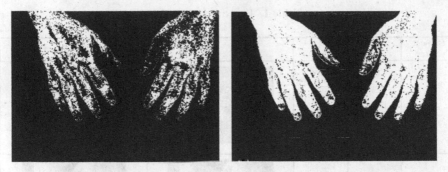

Fig. 5. $H_{Cb} \pm 5$ - $H_{Cr} \pm 5$ (left) and $H_{Cb} \pm 10$ - $H_{Cr} \pm 10$ (right)

Based on the algorithms presented above, the experiments are made with different hand colors to calculate the Cb and Cr values for identifying the dorsal hands of the subjects.

3 Experiments

The experiments are conducted with 6 subjects in the University of Hradec Kralove. In each experiment, the process started over from the beginning of unoccupied keyboard. Afterwards the subjects are told to put their hands on the keyboard as if they are writing something. The second snapshot is taken in the fifth second of the video to identify the Cb and Cr codes of the subject's hands by our algorithm. The results for $H_{Cb} \pm 5$ - $H_{Cr} \pm 5$ and $H_{Cb} \pm 10$ - $H_{Cr} \pm 10$ intervals are shown in Table 1.

It seems that the Cb and Cr values are significantly different within the subject group which is the main feature that could be extracted. This hypothetical approach is based on tailor-made adaptive dorsal hand recognition therefore the methodology uses the Cb and Cr values of the specific user under the current illumination conditions. Given this fact, the intervals are adaptive and determined by the average values which makes the difference significant enough since there are no equal values. Further processing is just alike that after the proposed second the camera captures the dorsal hands again to compare with the original image.

4 Evaluation of the Results

The results of the experiment are greatly encouraging that our algorithm perfectly calculates the Cb and Cr values of the dorsal hands for adaptive recognition along the video. The Chroma codes of the subjects' hands are correctly estimated and the confidence intervals around the average Cb and Cr values fit the requirements while identifying the hands.

Furthermore, the shapes are so apparent, mostly in $H_{Cb} \pm 10$ - $H_{Cr} \pm 10$ interval that enables the segmentation easier, when necessary. On the other hand, the algorithm of $H_{Cb} \pm 5$ - $H_{Cr} \pm 5$ interval doesn't recognize every pixels yet the shadows and

Table 1. Experimental results

Subject Name	Cb	Cr	$H_{Cb} \pm 5$ - $H_{Cr} \pm 5$	$H_{Cb} \pm 10$ - $H_{Cr} \pm 10$
Anna	116.8542	150.9333		
Selin	111.5068	147.6682		
Aydin	116.3470	153.9180		
Ilhan	112.4673	156.9682		
Mustafa	109.7743	154.4370		
Marta	98.5477	161.1963		

highlights are more evident. Therefore it could be easily stated that, narrower intervals reveal more characteristic features of the hands while wider intervals identify more pixels and entire shapes of the hands better.

Although the $H_{Cb} \pm 10$ - $H_{Cr} \pm 10$ could be considered as a very wide interval, our experiments show that there could be more gap between some Cb and Cr values of the subject's hands. In other words, within a range of computed Cb and Cr values, hands of some users wouldn't be identified since their values are out of the range. Illumination change is not affecting the process at all unless it happens in the same video recording session.

After recognition, the eroding algorithm we used in (5) and (6) could be reversed for filling the black pixels on the hands. Therefore the black pixels will turn into white if they are surrounded by white pixels. This kind of erosion is crucial when the whole hand image will be used in the related systems.

5   Conclusions and Discussions

The adaptive approach in this paper resulted in promising identification methodology of the dorsal hands yet the parameters or the methodology could be changed based on the requirements. The results of this research could be used in identification of users through hand geometry or the skin colors. Moreover it is possible to implement a new type of keystroke authentication by identifying finger-key matches along an alphanumeric password.

As the future research possibilities, the imaginary lines could be sketched to estimate the fingers' length and width of the hand for biometric purposes. Using wider intervals, the hands are identified better therefore biometric systems consisting of hand geometry could utilize the outputs of this research so easily.

Regarding the hand shapes and habits of using the keyboard, it is very clear to see the differences among the subject group. The ratios are useful to be computed to train a classifier that controls the users' hands for authentication. On the other hand, with adaptive dorsal hand recognition, habitual biometrics could be processed.

In medical point of view, dorsal hand recognition could be utilized for avoiding nerve entrapments in the wrists. It is also possible to combine the outcomes of this research with other types of user identification to have a multimodal biometric authentication system. Besides, various features could be extracted from the video processing mentioned in the article, like the behavioral or habitual traits of the users.

Acknowledgment.   This work and the contribution were supported by project "Smart Solutions for Ubiquitous Computing Environments" FIM, University of Hradec Kralove, Czech Republic (under ID: UHK-FIM-SP-2016-2102).

References

1. Frolova, D., Stern, H., Berman, S.: Most probable longest common subsequence for recognition of gesture character input. Cybern. IEEE Trans. **43**(3), 871–880 (2013)
2. Ghotkar, A.S., Kharate, G.K.: Vision based real time hand gesture recognition techniques for human computer interaction. Int. J. Comput. Appl. **70**(16), 1–6 (2013)
3. Weber, H., Jung, C.R., Gelb, D.: Hand and object segmentation from RGB-D images for interaction with planar surfaces. In: 2015 IEEE International Conference on Image Processing (ICIP), pp. 2984–2988. IEEE (2015)

4. Feng, K.P., Wan, K., Luo, N.: Natural gesture recognition based on motion detection and skin color. Appl. Mech. Mater. **321**, 974–979 (2013)
5. Plouffe, G., Cretu, A.M., Payeur, P.: Natural human-computer interaction using static and dynamic hand gestures. In: 2015 IEEE International Symposium on Haptic, Audio and Visual Environments and Games (HAVE), pp. 1–6. IEEE (2015)
6. Tu, Y.J., Kao, C.C., Lin, H.Y., Chang, C.C.: Face and gesture based human computer interaction. Int. J. Sig. Process. image Process. Pattern Recogn. **8**(9), 219–228 (2015)
7. Jeong, J., Jang, Y.: Max–min hand cropping method for robust hand region extraction in the image-based hand gesture recognition. Soft. Comput. **19**(4), 815–818 (2015)
8. Ahmad, I., Jan, Z., Shah, I.A., Ahmad, J.: Hand recognition using palm and hand geometry features. Sci. Int. **27**(2), 1177–1181 (2015)
9. Zhang, D., Guo, Z., Gong, Y.: Dorsal hand recognition. In: Multispectral Biometrics, Springer International Publishing, pp. 165–186 (2016)
10. Zhang, D., Guo, Z., Gong, Y.: Comparison of Palm and Dorsal Hand Recognition. Multispectral Biometrics. Springer International Publishing, Heidelberg (2016)
11. Zhang, D., Guo, Z., Gong, Y.: Multiple Band Selection of Multispectral Dorsal Hand. Multispectral Biometrics. Springer International Publishing, Heidelberg (2016)
12. Qiu-yu, Z., Jun-chi, L., Mo-yi, Z., Hong-xiang, D., Lu, L.: Hand gesture segmentation method based on YCbCr color space and k-means clustering. Int. J. Signal Process. Image Process. Pattern Recogn. **8**(5), 105–116 (2015)
13. Kaur, A., Kranthi, B.V.: Comparison between YCbCr color space and CIELab color space for skin color segmentation. IJAIS **3**(4), 30–33 (2012)
14. Chitra, S., Balakrishnan, G.: Comparative study for two color spaces HSCbCr and YCbCr in skin color detection. Appl. Math. Sci. **6**(85), 4229–4238 (2012)
15. Shen, X.G., Wu, W.: An algorithm of lips secondary positioning and feature extraction based on YCbCr color space. In: International Conference on Advances in Mechanical Engineering and Industrial Informatics. pp. 1472–1478. Atlantis Press (2015)
16. Alpar, O.: Intelligent biometric pattern password authentication systems for touchscreens. Expert Syst. Appl. **42**(17), 6286–6294 (2015)
17. Alpar, O.: Keystroke recognition in user authentication using ANN based RGB histogram technique. Eng. Appl. Artif. Intell. **32**, 213–217 (2014)
18. Alpar, O., Krejcar, O.: Biometric swiping on touchscreens. In: Saeed, K., Homenda, W. (eds.) Canadian AI 2013. LNCS, vol. 9339, pp. 193–203. Springer, Heidelberg (2015)
19. Alpar, O., Krejcar, O.: Pattern password authentication based on touching location. In: Jackowski, K., et al. (eds.) IDEAL 2015. LNCS, vol. 9375, pp. 395–403. Springer, Heidelberg (2015). doi:10.1007/978-3-319-24834-9_46

Designing QSAR Models for Promising TLR4 Agonists Isolated from *Euodia Asteridula* by Artificial Neural Networks Enhanced by Optimal Brain Surgeon

Rafael Dolezal[1,2(✉)], Jan Trejbal[1], Jakub Mesicek[1,2], Agata Milanov[1], Veronika Racakova[1], and Jiri Krenek[1]

[1] Center for Basic and Applied Research, Faculty of Informatics and Management, University of Hradec Kralove, Rokitanskeho 62, 50003 Hradec Kralove, Czech Republic
{rafael.dolezal,jan.trejbal,jakub.mesicek}@uhk.cz
[2] Biomedical Research Center, University Hospital Hradec Kralove, Sokolska 581, 500 05 Hradec Kralove, Czech Republic

Abstract. Undoubtedly, computer aided drug design (CADD) has gained important position in medicinal chemistry thanks to balancing random approaches to discovery of new drugs by prioritizing rational insight into the development process. From many CADD methods, quantitative structure activity relationships (QSAR), which are able to exploit chemical and biological information hidden in chemical structures through utilization of numerous machine learning and artificial intelligence methods, are expected to provide the necessary assistance in mechanistic interpretation and prediction of biological activities. In the present work, 56 derivatives of a natural adjuvant euodenine A, which occurs in *Euodia asteridula*, were selected for a QSAR study with the use of artificial neural networks (ANN). Since building of robust QSAR models is still a challenging research area, several methods had to be utilized to achieve a robust solution. Among various backpropagation based algorithms, much effort has been devoted to research of an optimal brain surgeon (OBS) method, which attempts to prune unimportant ANN elements according to the second derivation of the output signal error with respect to the weights. Herein, the performance of OBS in QSAR analyses is discussed and compared with other ANN learning methods.

Keywords: QSAR · Euodenine A · ANN · Pruning · OBS · Molecular descriptors

1 Introduction

Eodia asteridula (Rutaceae, synonym: *Melicope denhamii*) is an understorey tree or shrub growing up to 25 meters in mixed dipterocarp and swamp forests especially in Borneo, the Philippines, Papua New Guinea. Mainly, its wood is used as building or decorative material for construction of local houses, although leaves of the tree have been well known for its beneficial effects against numerous skin diseases and bacterial

© Springer International Publishing Switzerland 2016
N.T. Nguyen et al. (Eds.): ICCCI 2016, Part II, LNAI 9876, pp. 271–281, 2016.
DOI: 10.1007/978-3-319-45246-3_26

infections for ages [1]. Recently, it has been discovered that *E. asteridula* leaf oil exhibits significant activity against Gram positive and negative bacteria, and shows also noteworthy anticancer *in vitro* activity towards Dalton's lymphoma ascites (DLA) cells *via* nuclear damage and apoptotic cascade induction. Importantly, Neve et al. published in 2014 a high throughput screening (HTS) of 750 pure natural products which has led to identification of euodenine A as a potent and selective small-molecule agonist of human toll-like receptor 4 (TLR4) [2].

TLR4, as the most studied TLR, is a trans-membrane receptor present in human immune cells that is able to recognize microbe-associated molecular patterns (MAMPs) attached to bacteria cell surface or damage-associated molecular patterns (DAMPs) produced within inflammation or tissue injuring processes and to start efficient immune response. Through activation of signaling cascades triggering cytokines and chemokines production, TLR4 agonists may stimulate pathologically insufficient immune response and help to tackle such diseases as asthma. Accordingly, it was experimentally proved that the immune response in the acute phase of asthma is ruled by a strong activation of Th2 lymphocytes which can be counterbalanced by an adequate response of TLR4. At present, euodenine A derivatives are important pre-clinical candidates for TLR4 agonist due their high biological potency and selectivity which outperform current adjuvant drugs such as Tak-242, (+)-naloxone or dimethyl 2-(2-nitrobenzylidine)malonate.

Herein, we will focus on development of quantitative structure activity relationship (QSAR) model for known euodenine A derivatives as TLR4 agonists. The principles of QSAR were established in 60's of 20th century by Hansch and we can simplify them as an assumption of a continuous and neat mathematical relationship between quantitative molecular descriptors and biological activity [3, 4]. Once the function, i.e. QSAR model, is established and properly validated, novel drug candidates belonging to the same structure class as the training compound set can be proposed. Currently, many supervised and unsupervised machine learning methods such as multiple linear regression (MLR), partial least square regression (PLS), principal component analysis (PCA), support vector machines (SVM), random forests (RF), k-nearest neighbor, etc. have been employed in QSAR analyses utilizing plethora of various *in silico* molecular descriptors as the data input.

In the present paper, we have developed a QSAR model for 39 + 17 euodenine A derivatives by artificial neural networks (ANN). 4883 molecular descriptors as independent variables have been generated by molecular modeling in HyperChem and descriptor calculations in Dragon. The objective of the work is to design a pruning algorithm for building a robust ANN QSAR model. As a source of inspiration we have recycled the concept of optimal brain surgeon (OBS) pruning technique developed originally by Hassibi and Stork and improved by Magnus Norgaard [5]. The efficiency of the pruning algorithm is compared to other 10, more or less, classical backpropagation based algorithms for ANN training.

2 Variable Selection and Pruning in ANN

ANN represent a powerful tool for approximating unknown and generally nonlinear QSAR models from excessive and highly inter-correlated input variables. Even though many linear QSAR or quantitative structure-property relationships (QSPR) models have been described in the literature, biological processes are often very complex being composed of various consequent and parallel physico-chemical interactions which can be linearized only for a limited range of cases. Commonly, the issue of ANN QSAR designing is solved empirically either by letting ANN grow from a minimal size to the least model complexity that provides acceptable robustness or, conversely, by starting with a large network which is gradually simplified by eliminating useless elements. Recently, numerous algorithms for removing uninformative, noisy, irrelevant, or redundant input variables have been described and recommended for utilization in statistical and data mining methods. The taxonomy of input variable selection algorithms can be divided into three main classes [6]:

Wrapper algorithms – the model optimization algorithms select the whole input variable set, or a subset of it, evaluate the performance and continue with different set of variables until the best generalization is yielded. The machine learning method used is handled as a black box, only the output is decisive. Examples are: forward selection, backward elimination, nested subsets, systematic exhaustive search, heuristic genetic search, single variable ranking, generalized regression neural network input determination algorithm (GRIDA), etc. *Embedded algorithms* – a model involving the whole set or a combination of input variables is evaluated by a machine learning method and the redundant or irrelevant variables are indicated through analysis of the resulting model parameterization. For example, variables multiplied with low weights or near-zero regression coefficients have only minor effect on the model outputs and, thus, they can be eliminated, while variables associated with significant parameters can be prioritized. Embedded algorithms consider the effect of each individual variable on the model performance. They are represented by: recursive feature elimination, stepwise selection, genetic and evolution algorithms, optimal brain damage (OBD), Lasso, etc. *Filter algorithms* – input variables are investigated with additional calculations which attempt to measure their predictive relevance. Currently, bivariate or trivariate statistics are frequently employed to rank the input variables. The first main subclass of filter algorithms is based on the calculation of linear correlations with the target variable(s), the second one is derived from information theory measures such as mutual information. After the irrelevant input variables are filtered out, the model is built up from the remaining data. Filter algorithms are exemplified by: Pearson and Spearman maximum correlation ranking, forward partial correlation selection, minimum entropy ranking, mutual information ranking and feature selection, partial and joint mutual information, input variable selection by independent component analysis (ICA), etc.

All the mentioned algorithms have their merits and drawbacks, although they may be combined to improve the final model quality. In the present work we particularly deal with pruning algorithms for ANN which consider the effect of each involved input variable, and, therefore they belong to the class of embedded algorithms. In simple terms, ANN can be introduced as a complex mathematical function $\hat{y}$ which depends on

x_i, activation function f, weights w, biases b, and provides a mean square error (MSE) function E (Eq. 1). Pruning in development of ANN can be characterized as determination and elimination of such network units which do not contribute significantly to the solution. Lately, several pruning approaches such as local sensitivity analysis, local variance sensitivity analysis or cross validated pruning algorithms have been developed to reduce the ANN size while to maintain satisfactory accuracy and performance of the model [7].

$$\hat{y}_j = f\left(a_j\right), a_j = \sum_j w_{ij}x_i + b_j, x_i = f(a_i), E = \frac{(\hat{y}-y)^T(\hat{y}-y)}{n}. \tag{1}$$

A simple pruning technique is based on deletion (i.e. zeroing) of those neural interconnections which have lower weights comparing to an arbitrary defined threshold. However, Hassibi and Stork pointed out that elimination of low weights is a problematic approach which does not necessary lead to decreasing the overall network error due to the effect of surviving higher weights which may enhance the influence of noisy input variables and increase wrong output bias [8].

The optimal brain damage (OBD) pruning algorithm developed by Le Cun et al. relies on calculation of second order derivation of the error function, which radically challenges the idea that "magnitude equals saliency" [9]. OBD approximates the error network function E by a truncated Taylor series and analytically predicts the effect of the weight vector perturbation δW (Eq. 2):

$$\delta E = \sum_i g_i \delta w_i + \frac{1}{2} \sum_i h_{ii} \delta w_i^2 + \frac{1}{2} \sum_{i \neq j} h_{ij} \delta w_i \delta w_j + O\left(\|\delta W\|^3\right), \tag{2}$$

where δW has δw_i components, the gradient G of E with respect to W has g_i components and the Hessian H of E with respect to W has h_{ij} components (Eq. 3):

$$g_i = \frac{\partial E}{\partial w_i}, h_{ij} = \frac{\partial^2 E}{\partial w_i \partial w_j}. \tag{3}$$

The aim of OBD is to find those parameters whose deletion would bring minimal increment of the error function E. Due to enormous computational demands imposed by calculation of the H matrix, simplifying approximations like neglecting cross terms perturbations $\delta w_i \delta w_j$, assumption that E is nearly quadratic with the parameters trained to a minimum of E had to be introduced. Thus, Eq. 2 can be reduced to the second term which can be solved by evaluating only the diagonal elements h_{ii} of H (Eq. 4):

$$\delta E = \delta w^T g + \frac{1}{2} \delta w^T H \delta w + \ldots \rightarrow \ = \frac{1}{2} \sum_i h_{ii} \delta w_i^2. \tag{4}$$

It was proved by experiments that OBD can substantially improve the speed of ANN and increase predictive accuracy on the test set. Unfortunately, the diagonal approximation of H in OBD has been criticized as it enables elimination of wrong weights.

A slightly more complex version of OBD, named optimum brain surgeon (OBS), focuses on a recursive calculation of the inverse Hessian H^{-1} to take into account the off-diagonal elements. Basically, elimination of the ith weight w_i is expressed as a constrain $e_i^T(\delta w + w) = 0$ that is introduced into the optimization of E (Eq. 5):

$$\min_{1 \le i \le I} \left\{ \min_{\delta w} \left(\frac{1}{2} \delta w^T \frac{\partial^2 E}{\partial w^2} \delta w \right) | e_i^T(\delta w + w) = 0 \right\}, \tag{5}$$

where e_i is a unit vector with the ith element equal to 1 and other components being 0. OBS also supposes the first gradient term in Eq. 2 to vanish in a local minimum of E and ignores third and higher terms of the Taylor expansion, but through a Lagrangian transformation L of the reduced E function (Eq. 4) it introduces a general term of "weight saliency" based upon weighting the weights by H^{-1} elements (Eq. 6) [10]:

$$L = \frac{1}{2} \delta w^T H \delta w + \left(e_i^T(\delta w + w) \right), \tag{6}$$

where λ is an undetermined Lagrange multiplier. The weighting of weights is a key concept of OBS which allows not only to establish a saliency criterion L_i for w_i but also to adjust the other weights w to reach the least increase of E. According to Hassibi and Stork, the iterative weigh adjustment and L_i saliency of w_i are given as follows (Eq. 7):

$$\delta w = -\frac{w_i}{\left[H^{-1}\right]_{ii}} H^{-1} \cdot e_i, L_i = \frac{1}{2} \frac{w_i}{\left[H^{-1}\right]_{ii}}. \tag{7}$$

Therefore, OBS does not assume the Hessian matrix nor its inverse to be diagonal but it correctly evaluates all off-diagonal elements. Due to weight recalculations, OBS processes need not theoretically be followed by a training phase. It was proved that OBS is able to eliminate more than 50 % of weights while stably decreasing the error of ANN. Comparing to OBD, OBS is more general providing more robust network complexity reduction but it itself is highly complicated and quite demanding for computer time and memory.

3 Proposed Solution for QSAR Analyses by ANN

For the QSAR analyses, the chemical structures of 56 euodenine A derivatives were taken from the literature and processed in computational chemistry programs to obtain molecular descriptors for building ANN model [2]. As a supervisory signal, concentrations causing 50 % activation (EC_{50}) of human TLR4 receptor expressed by HEK-293 cells were used. All the studied compounds were synthesized and biologically evaluated according to a standardized laboratory protocol, which is one of the necessary conditions for establishing unbiased QSAR analyses [2]. Since EC_{50} were quantitatively determined only for 39 euodenine A derivatives, 27 compounds were randomly chosen for the training set in each training iteration, similarly 6 compounds for the validation set, 6 compounds for the test set, and the remaining 17 compounds with unknown biological activities towards TLR4 were taken as an external set for *in silico* prediction.

The objective of the present work is to design a pruning algorithm based on OBD and OBS, and to develop QSAR models for euodenine A derivatives. The OBD/OBS based ANN QSAR analyses are compared with other ANN training algorithms based on backpropagation. Reliability of the ANN QSAR models is estimated by prediction of EC_{50} for the validation and test set compounds.

3.1   Data Input Preparation

Chemical structures of 56 euodenine A derivatives were manually drawn in HyperChem 8.0 program and then geometrically pre-optimized in vacuum by semi-empirical quantum chemistry PM3 method with the use of Polak-Ribiére conjugate gradient technique. In order to obtain as accurate 3D chemical structures as possible, the resulting chemical models were resubmitted to an advanced geometrical optimization by a DFT/m062x/6-311 ++G(d,p) method with implicit simulation of aquatic environment by a Self Consistent Reaction Field (SCRF) model at 323.15 K in Gaussian 09 program. The demanding quantum chemistry calculations on the density functional theory level were performed distributively over 4 physically different computer machines employing 4CPU/10 GB RAM per node. The resulting chemical structures, corresponding to a local energy minimum with no imaginary frequencies, as well as the calculated Mulliken partial atomic charges were converted into a suitable file format and imported into Dragon 6 program for calculation of molecular descriptors.

In Dragon 6, 4885 molecular descriptors were calculated for each optimized structure of the euodenine A derivatives. The calculated molecular descriptors can be classified into 29 different blocks such as constitutional indices, topological indices, charge descriptors, drug-like indices, etc. Nonetheless, 1945 molecular descriptors (e.g. number of triple bonds) assumed zero/constant values and, thus, they were removed from the input variable X matrix by a script developed in Matlab 8.3. The experimentally determined EC_{50} values were taken from the literature and stored as a supervisory signal variable (y) [2]. For QSAR analyses, 2940 input independent variables in real number format were utilized.

3.2   ANN Development and Pruning Algorithms

For the QSAR analyses, we decided to utilize the assets of Matlab 8.3 (2014a) program suite with the support of Parallel Computing and Neural Toolboxes. The calculations were carried on a machine with 16 logical CPU cores (2 × Intel Sandy Bridge E5-2470, 2.3 GHz, 96 GB RAM) and one NVIDIA Kepler K20 card (2496 shaders, 5 GB VRAM) to achieve a higher performance. In the Matlab editor, several M-script were developed to import the data file, to remove constant descriptors, to extract the compounds without measured biological activities, to initialize ANN, to train ANN with 10 different training algorithms: (Broyden–Fletcher–Goldfarb–Shanno quasi-Newton backpropagation – BFG, conjugate gradient backpropagation with Powell-Beale restarts – CGB, conjugate gradient backpropagation with Fletcher-Reeves updates – CGF, conjugate gradient backpropagation with Polak-Ribiére updates – CGP, gradient descent with adaptive learning rate backpropagation – GDA, Gradient descent with momentum backpropagation – GDM, gradient descent with momentum and adaptive learning rate backpropagation – GDX, resilient propagation – RP, scaled conjugate gradient backpropagation – SCG, one-step secant backpropagation – OSS) with ten random resampling, and to evaluate the performance stability of re-trained ANN QSAR models. The following simplified code summarizes the undertaken technical and mathematical operations:

```
import_data (X, Y);
clean (X, Y, [const, NaN]);
process_input (mapminmax);
net = feedforward (user_defined_topology(2 or 3 layers));
algorithms = {BFG,CGB,CGF,CGP,GDA,GDM,GDX,RP,SCG,OSS};
for i=1:size(algorithms, 2)
    for j=1:10
        net.trainFcn = algorithms{i}
        net_configured = configure (net, X', Y');
        net_configured = init(net_configured, GPU);
        [net_trained,tr] = train(net_configured, X', Y', useGPU);
        Y_pred = (net_trained (X', useGPU))';
        correlation (Y_pred, Y);
        evaluate_performance (tr);
    end
end
```

By means of the above script, 10 training algorithms were examined for optimization of a three-layer feedforward perceptron network with 2940 neurons in the input layer (i.e. number of descriptors in X matrix), 59 neurons in the hidden layer, and with one linear neuron (i.e. pureline transfer function) in the output layers. For the input and hidden layers, hyperbolic tangent sigmoid as the transfer function was chosen. To control the training process, the input dataset containing 39 cases with known EC_{50} values was randomly split in every loop initialization into training, validation and test set in ratio 70-15-15. To achieve an insight into the stability of the found solutions, each training process was repeated 10 times and the resulting performance spectrum was statistically analyzed. Regarding the training process, MSE was taken as the objective function of the performance. The training stopped after 6 consecutive increases of validation error, selecting the iteration with the lowest validation error. Other parameters such as gradient threshold, learning rate, etc. were used in the default mode.

Similar conditions were used for a next experiment where the hidden layer was removed, that is a two-layer feedforward perceptron network having only one input layer with hyperbolic tangent sigmoid transfer function and one output neuron with pureline transfer function was examined. The overall setting, training methods and other parameters were kept the same.

In order to evaluate the effect of OBS pruning algorithm, a methodology developed by Magnus Norgaard was adopted in the present work (http://www.iau.dtu.dk/∼pmn/). Marking OBS as the most important strategy in regularization of ANN, Magnus Norgaard suggested a subtle improvement on handling neurons whose input connections are very weakly weighted while having significantly weighted connections toward the output layer, or *vice versa*. The approach therefore extends the concept of weight saliences to neuron saliences giving a rule for removing the entire neuron. The saliency L_k of neuron is a product of all weights w_k leading to and from it (Eq. 8):

$$L_k = \frac{1}{2} \prod_k \frac{w_k}{\left[\boldsymbol{H}^{-1}\right]_{kk}}. \tag{8}$$

Nonetheless, obtaining neuron saliences becomes very demanding if the inverse Hessian matrix is to be calculated for many thousands weights like in our case. Therefore, we transformed the original code suggested by Norgaard for calculation on GPU using Parallel Computing Toolbox in Matlab. A simplified code for implementation of Norgaard OBS approach was developed as outlined below:

```
import_data (X, Y);
clean (X, Y, [const, NaN]);
[Train, Test]=split_set (X, Y);
process_input (mapminmax);
net = define_layers (tansig, pureline, 6, 1);
[W1, W2]=backprop(net,Train,maxiter,1000,eta,0.01);
initialize_parameters (Train, Test);
for i=1:max_iterations
  net_retrain (Train, LMB);
  calculate the derivative of y with respect to W1, W2;
  calculate the Hessian inverse matrix with GPU;
  eliminate weights and neurons with low saliences (P);
  update the inverse Hessian;
  if evaluate_net (Train, Test)=!1;
     break;
  end
end
```

To make OBS approach feasible on available hardware, only a two-layer feedforward network consisting of one input layer (6 neurons) with hyperbolic tangent sigmoid transfer function and one output neuron with pureline transfer function was designed. Within the OBS, the ANN was retrained by Levenberg-Marquardt backpropagation (LMB), which is a very time-consuming Jacobian based technique. Within the while loop, it was set to remove 5 % (i.e. P parameter) of low saliency elements in maximally 200 iterations. The pruning efficiency of OBS on ANN QSAR model was properly analyzed and compared with result of previously mentioned experiments.

4 Results and Discussion

ANN QSAR analyses of agonistic activities of 39 euodenines for TLR4 with a 2/3-layer network revealed significant differences with respect to the training algorithm used. Although each training showed considerable instability after resampling the input sets, the optimal prediction in terms of high positive Pearson R was achieved through CGF, CGB, and CGP, all of them being conjugate gradient backpropagation variants.

From Fig. 1. it is evident that 2-layer ANN QSAR models provided lower mean R's and, therefore, 2-layer models can be regarded as a less efficient topology for the studied data, though the conjugate training algorithms still retained relative best performance. Accordingly, the records on MSE values in the validation set after 10 training repetition revealed that CGF and CGB, particularly along with a 3 layer ANN topology, outperformed the other training algorithms (TA), and several times very drastically (i.e. GDM, RP, GDX, GDA) (see Table 1). According to the lowest MSE in the validation sets, CGF algorithm provided the best ANN QSAR model if 3-layer ANN was used (MSE = 0.53 ± 0.45). On the other hand, OBS algorithm revealed, trough iterative pruning the total 17 653 weights, that the global minima of MSE in the validation set was reached in the **149-th iteration** with 5 938 weights remaining in the ANN model.

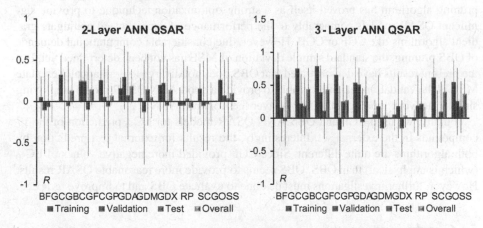

Fig. 1. 3- and 2-layer ANN QSAR models trained by 10 different algorithms, characterized by mean Pearson R correlation coefficients and standard sample deviation bars after 10 repetitions.

Table 1. Mean MSE's and their sample standard deviations s after 10 training repetitions

Performance in validation set – 2 – layer ANN QSAR										
TA	BFG	CGB	CGF	CGP	GDA	GDM	GDX	RP	SCG	OSS
$\overline{MSE}$	337.5	**26.4**	60.9	22.6	1.2e + 3	6.3e + 32	1.5e + 3	1.8e + 6	20.9	89.6
$s(\overline{MSE})$	309.4	**26.1**	76.8	29.5	2.0e + 3	6.8e + 32	1.9e + 3	1.4e + 6	23.9	81.9
Performance in validation set – 3 – layer ANN QSAR										
$\overline{MSE}$	4.53	0.68	**0.53**	0.72	185.51	2.26e + 11	25.58	1.07e + 3	1.13	2.50
$s(\overline{MSE})$	5.07	0.56	**0.45**	0.97	554.47	3.93e + 11	21.50	2.13e + 3	2.86	3.00

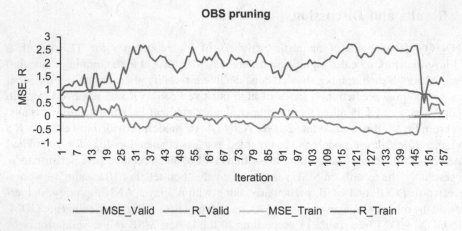

Fig. 2. Performance of OBS pruning in building a 2-layer ANN QSAR model for euodenine A derivatives. According to MSE_Valid, the best model resulted in the 149-th iteration.

Thus, 11 715 weights with low saliencies were eliminated by OBS providing the best ANN model with MSE = **0.5236** in the validation test (Fig. 2). Accordingly, the OBS pruning algorithm has proved itself as a strong optimization technique to provide significant QSAR models comparably to the performance of sophisticated conjugate gradient algorithms like CGF or CGB. However, due considerable computational demands of OBS pruning, the standard sample deviation of MSE has not been determined yet. But the present results have demonstrated that OBS does not diverge to dramatically escalate MSE in the validation set, and it is able to provide a good ANN QSAR model employing simpler network topology, that is 2 – layer feedforward ANN.

Finally, we used CGF and OBS ANN QSAR models for EC_{50} prediction of the 17 compounds in the external set. Interestingly, the results for external test prediction by both algorithms are quite different. Since CGF provided more negative values of EC_{50} (which is unphysical) than OBS, OBS seems to provide more reasonable QSAR results. However, further investigations must be done to evaluate OBS and to improve its speed.

5   Conclusions

OBS represents an efficient regularization method for improving pre-trained ANN. As a backward type elimination approach based on mathematical analysis of the error function, OBS starts up with the most complex ANN structure which is systematically reduced to reach the global minimum of MSE. Application of OBS to QSAR analysis of euodenine A derivatives proved that OBS is a well suited optimization technique for establishing robust QSAR models with predictive potency. A comparable performance to OBS was exhibited mainly by GCF algorithm. Currently, the OBS ANN QSAR model obtained in the present study is used for evaluation of a virtual compound library to propose novel TLR4 agonist lead structures. Regarding further perspectives, hybridization of OBS with conjugate gradient methods as well as powerful parallelization may be very beneficial to drug research and discovery by ANN based QSAR analyses.

Acknowledgements. This work was supported by project "Smart Solutions for Ubiquitous Computing Environments" FIM UHK, Czech Republic (under ID: UHK-FIM-SP-2016-2102). The work was also supported by: Czech Science Foundation (GA15-11776S), the IT4Innovations Centre of Excellence project (CZ.1.05/1.1.00/02.0070), and the MŠMT project (LM2011033).

References

1. George, S., Nair, S.A., Venkataraman, R., Baby, S.: Chemical composition, antibacterial and anticancer activities of volatile oil of Melicope denhamii leaves. Nat. Prod. Res, **29**, 1959–1962 (2015)
2. Neve, J.E., Wijesekera, H.P., Duffy, S., Jenkins, I.D., Ripper, J.A., Teague, S.J., Campitelli, M., Garavelas, A., Nikolakopoulos, G., Le, P.V., de, A.L.P., Pham, N.B., Shelton, P., Fraser, N., Carroll, A.R., Avery, V.M., McCrae, C., McCrae, C.: Euodenine A: a small-molecule agonist of human TLR4. J. Med. Chem. **57**, 1252–1275 (2014)
3. Waisser, K., Dolezal, R., Palat, K., Cizmarik, J., Kaustova, J.: QSAR study of antimycobacterial activity of quaternary ammonium salts of piperidinylethyl esters of alkoxysubstituted phenylcarbamic acids. Folia Microbiol. **51**, 21–24 (2006)
4. Nesmerak, K., Dolezal, R., Hudska, V., Bartl, J., Sticha, M., Waisser, K.: Quantitative structure-electrochemistry relationship of 1-Phenyl-5-benzyl-sulfanyltetrazoles and their electrooxidation as a metabolic model. Electroanalysis **22**, 2117–2122 (2010)
5. Maggi, G., Giaccone, G., Donadio, M., Ciuffreda, L., Dalesio, O., Leria, G., Trifiletti, G., Casadio, C., Palestro, G., Mancuso, M., et al.: Thymomas. A review of 169 cases, with particular reference to results of surgical treatment. Cancer **58**, 765–776 (1986)
6. May, R., Dandy, G., Maier, H.: Review of input variable selection methods for artificial neural networks. In: Kenji, S. (ed.) ANN-MABA, InTech (2011). doi:10.5772/16004
7. Sabo, D., Yu, X.-H.: A new pruning algorithm for neural network dimension analysis. 2008 IEEE, pp. 3313–3318, (2008). doi:10.1109/IJCNN.2008.4634268
8. Hassibi, B., Stork, D.G.: Second order derivatives for network pruning: Optimal brain surgeon. In: Cowan, J.D. et al. (ed.) NIPS 1993, pp. 164–171, Morgan Kaufmann (1993)
9. LeCun, Y., Denker, J.S., Solla, S.A., Howard, R.E., Jackel, L.D.: Optimal brain damage. In: Touretzky, D. (ed.) NIPS 1989, 2, pp. 598–605, Morgan Kaufman (1990)
10. Hassibi, B., Stork, D.G., Wolff, G.J.: Optimal brain surgeon and general network pruning. IEEE ICNN **1**, 293–299 (1993). doi:10.1109/ICNN.1993.298572

Exploration of Autoimmune Diseases Using Multi-agent Systems

Richard Cimler[1,2], Martina Husáková[1(✉)], and Martina Koláčková[3]

[1] Faculty of Informatics and Management, University of Hradec Králové,
Hradec Králové, Czech Republic
{richard.cimler,martina.husakova.2}@uhk.cz
[2] Center for Basic and Applied Research (CZAV), University of Hradec Králové,
Hradec Králové, Czech Republic
[3] Faculty of Medicine, Department of Clinical Immunology and Allergy, Charles University,
Hradec Králové, Czech Republic
kolackovam@lfhk.cuni.cz

Abstract. Autoimmune disease is a group of pathological events identified by abnormal reactions of the immune system against self-structures of the organism. Pathogenesis of autoimmune diseases is multi-factorial. Genetics, infections, and environmental factors can support the progress of the autoimmunity. We investigated this mechanism using in vivo or in vitro approaches. Main aim of this manuscript is to explore the autoimmunity with in-silico approach - multi-agent systems. The preliminary research finds out which results can be acquired using factual data applied for building the multi-agent-based model. Preliminary computational model integrates one of the common aspects of autoimmune diseases - abnormal behaviour of B-cells during their organogenesis.

Keywords: Autoimmunity · B-cell organogenesis · Multi-agent system

1 Introduction

The biological immune system (BIS) is an adaptive system which plays a substantial role during homeostasis. The maintenance of homeostasis is very important for sustaining life of humans. BIS disposes of wide spectrum of "weapons" against self and foreign objects (antigens) that initiate the activity of the BIS and can cause serious injuries. Specific types of immune white blood cells (lymphocytes) are called B-cells. Production of antibodies directed against antigens is the characteristic feature of these cells. B-cells also remember encounters with antigens, thereby being able to react faster and eliminate antigens effectively during the next encounter. Direct elimination of pathogens is realised by T-cells which are also able to regulate functions of other immune cells, especially B-cells. Investigation of the BIS is difficult due to the complexity and ambient nature of the BIS. Complexity of the BIS is caused by the amount of players which sustain homeostasis of the body and interactions between them. BIS is highly interconnected with other biological subsystems, such as the endocrine, neural and the reproduction system. Computational immunology (CI) is a subarea of computational

© Springer International Publishing Switzerland 2016
N.T. Nguyen et al. (Eds.): ICCCI 2016, Part II, LNAI 9876, pp. 282–291, 2016.
DOI: 10.1007/978-3-319-45246-3_27

biology exploring amazing world and complexity of the BIS using in-silico-based approaches. It interconnects biology, medicine, chemistry, immunology, modelling, simulations, mathematics, physics, statistics, and computer science (artificial or computational intelligence) to create efficient strategies for influencing behaviour of the BIS. The paper presents the preliminary research that applies multi-agent systems (MAS) for investigation of the autoimmune diseases. The investigation is mainly focused on modelling development of B-cells and their behaviour which is deviated during the majority of autoimmune diseases. The paper is structured as follows. Section 2 introduces applications of multi-agent systems in computational immunology. The Sect. 3 presents the most important facts about the autoimmune diseases. Section 4 demonstrates systematic framework for building of the multi-agent model trying to represent and simulate particular aspects of the autoimmune diseases. Future directions are mentioned in the Sect. 5 and Sect. 6 concludes the paper.

2 Multi-agent Systems in Computational Immunology

Multi-agent systems (MAS) are stochastic systems consisting of autonomous entities called agents able to solve particular problem and interact with each other to achieve required goals. They are mainly used for modelling interactions that occur in the cellular and tissue level of organisation of a biological system in the view of the CI. This fact is not so surprising, because the nature of the MAS predestine them for modelling aggregate, cooperative or coordinate behaviour, where the emergence phenomena occur. We can find really huge amount of MAS-based applications used in the CI. We focus only on the applications of MAS that are the most relevant to our research interest, i.e. autoimmune diseases and development of cells.

Experimental autoimmune encephalomyelitis (EAE) is complex autoimmune disease occurring in mice. The disease plays the role of a murine model for the multiple sclerosis in humans. EAE is caused by the decomposition of the myelin covering the axons of neural cells. UML-based domain model of the EAE is presented in [1]. Exploration of the EAE is extended in the [2, 3] where the EAE is modelled with the multi-agent-based approach. The Agent Modelling Language (AML) is the UML-based conceptual language used for modelling immune cells and interactions between them which occur in the lymph node [23]. Overview and comparison of various conceptual languages and computational approaches used in the CI are presented in [24]. Simulator ARTIMMUS (Artificial Murine Multiple Sclerosis Simulation) is built for doing experiments with the in-silico model of the EAE [4, 5]. Agent-based approach is also applied for modelling autoimmune diseases where the immune response on the viral infection is investigated [6]. Model represents T-cells, B-cells, dendritic cells, tissue cells, macrophages and infectious agents (viruses) together with interactions between them in the level of the innate immunity. The agent-based model simulating autoimmune response in Type 1 diabetes is presented in [7]. Interactions between B-cells and macrophages, in pancreas of a mouse and motility of cells are explored in this preliminary study.

Hematopoiesis is a process of production and maintenance of all types of blood cells which are differentiated into various blood cell types of myeloid or lymphoid lineage.

This complex process is also modelled by MAS in [8]. The research mentioned in [9] advocates the usefulness of cellular automata and MAS for modelling development of blood cells where the UML is applied for visualization of the most important "players" of hematopoiesis. Normal and pathological hematopoiesis (leukemia) is simulated and presented in [10]. Epitheliome project is focused on modelling normal formation of epithelium and investigation of possible anomalies in behaviour of cells leading to pathologies during epithelium formation [11, 12]. The agent-based approach is used and integrates stem cells, transit amplifying cells, (post)mitotic cells and dead cells in the model. Cell cycle is modelled in the initial studies on the basis of data received in-vitro.

3   Autoimmune Diseases

There is a vast range of autoimmune diseases of different origin among the population. It seems that their incidence has increased, but it is difficult to assess the increase because they might have been formerly underdiagnosed. The overall prevalence in Western countries ranges from 4.5 % to 9.4 %. Various autoimmune disorders have not only different origins but also prevalence, gender predominance, and age-of-onset [14–16]. Hayter and Coók identified 81 types of the disorders with more than half being rare [16]. These researchers also characterised the most common structures at which autoimmunity is aimed: they include repeated and coiled-coiled structures. In some autoimmune diseases the immunodominant structures have not been determined. Rheumatoid arthritis, systemic lupus erythematosus, Sjögren's syndrome, celiac disease, Crohn's disease, diabetes mellitus 1, and multiple sclerosis are among the most prevalent autoimmune diseases, although sometimes the classification could be difficult. It is an interesting point that there may not be a discrete border between some of the symptoms as well as autoimmune mechanisms behind them. As an example, diabetes mellitus 1 is associated with autoimmune thyroiditis and celiac disease.

As the BIS protects us against invaders, it has to recognise foreign and self-structures. Deviations of the recognition could lead to detrimental outcomes, either in sense of insufficient protection or harmful attack on the body. Although there are some characteristic features of foreign structures, the border between them and self-structures might be rather blurry for more reasons; often there are similarities between antigens or tactics that an invader uses to disable differentiation between self and non-self. Broken tolerance of self-antigens causes autoimmunity, thus the body of autoimmune patients is damaged by the reaction which can have diverse backgrounds. Some deviations in the function of their immune system and the mechanisms that are behind these deviations are well-known, other mechanisms or even deviations are the mystery. Prevailing mechanisms that damage the body can be either cellular, mediated by T-cells, or humoral which are based on production of autoantibodies. Although innate immunity and T-cell-mediated immunity were found disturbed in diseases such as systemic lupus erythematosus, ankylosing spondylitis, and rheumatoid arthritis, it is now presumed that the main mechanism which destroys tissues and organs of these patients involves autoreactive B-cells.

4   Development of Multi-agent Model of Autoimmune Disease

Investigation of complex biological systems requires systematic approach. CoSMoS (Complex Systems Modelling and Simulation) is a process and infrastructure used for the exploration of complex systems using modelling and simulation. The framework has been proposed within the CoSMoS project with the aim to support analysis, design, development and validation of models dealing with complex systems [13]. CoSMoS process is applied for building of the MAS-based model that investigates the autoimmunity.

4.1   Discovery Phase – Research Context

As the immune system protects us against invaders, it has to recognize foreign and self-structures. Deviations of the recognition could lead to detrimental outcomes, either in the sense of insufficient protection or harmful attack on the body. Although there are some characteristic features of foreign structures, the border between them and self-structures might be rather blurred for more reasons; often there are similarities between antigens or tactics that an invader uses to disable differentiation between self and non-self. Broken tolerance of self-antigens causes autoimmunity, thus the body of autoimmune patients is damaged by the reaction which can have diverse backgrounds. Some deviations in the function of their immune system and the mechanisms that are behind these deviations are well-known, other mechanisms or even deviations are a mystery. Prevailing mechanisms that damage the body can be either cellular, mediated by T-cells, or humoral which are based on production of autoantibodies. Although innate immunity and T-cell-mediated immunity were found disturbed in diseases such as systemic lupus erythematosus, ankylosing spondylitis, and rheumatoid arthritis, it is now presumed that the main mechanism which destroys tissues and organs of these patients involves autoreactive B-cells. Therefore, we focused our preliminary research on the modelling of particular stages of B-cell development.

This research is highly dependent on factual data and expert knowledge. The data is extracted from scientific literature. Unfortunately, most data arise from mouse models of autoimmunity and they are extrapolated to the human immune system. Based on this data, our primary computational model helps to generate possible outcomes of B-cell behaviour using MAS-based approach.

4.2   Discovery Phase – Domain Model

The preliminary model is focused on the investigation of particular phases of B-cells development, see Fig. 1. B-cells mature in the bone marrow where they undergo several phases during which a B-cell receptor (BCR) is formed on the surface of all B-cells. It is suggested that B-cells in the bone marrow are taught not to recognise self-structures. Therefore only B-cells which recognise foreign structures are released into peripheral blood; other B-cells are eliminated within the bone marrow. Yet, some B-cells maintain a weak degree of autoreactivity because their BCR recognises self-structures. Partially, the reason lies in the nature of released B-cells: B-cells that migrate from the bone marrow to the spleen have either a phenotype of naïve B-cells or immature B-cells [17]. Both are considered to be transitional (T) B-cells, but immature B-cells have not experienced the selection process in the bone marrow. It has been reported that the high percentage of transitional B-cells are autoreactive, although it is difficult to test the functional autoreactivity of these cells due to the enormous range of possible antigens. Moreover, these autoreactive cells are thought to be mostly anergic which means that they are set not to display their autoreactivity. Further development of transitional B-cells continues in the spleen [18, 19]. B-cells undergo other stages at the end of which they either become memory B-cells or plasmablast/plasma cells. These cells form highly specific clones because their BCR (memory B-cells) or Ig ('antibody'; plasmablast/plasma cells) is capable of binding the antigen with high affinity. They can be found in the peripheral blood and secondary lymphoid organs, such as individual lymph nodes, tonsils, and gut-associated lymphoid tissue. However, many B-cells die during the development in the spleen and other B-cells never leave this compartment but stay here as marginal zone B-cells [20]. Considering autoreactivity, it is currently unclear which cells could be the most dangerous ones; it is expected that memory B-cells and plasmablast/plasma cells play the major role.

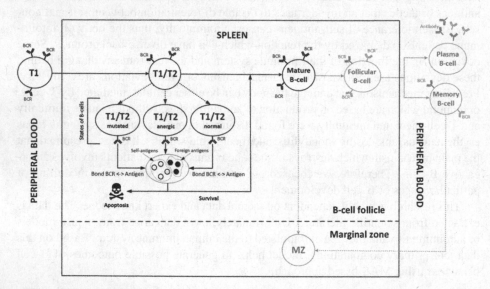

Fig. 1. Hypothetical states of B-cells development in the view of the research interest

The B-cell binds to the antigen using the BCR (B-cell receptor) that is placed on the B-cell membrane. Level of signalling and the B-cell state determine the fate of the B-cell. According to signalling, the T1/T2 B-cell becomes either apoptotic or survives and transits into other stages; see Fig. 1. The survival depends on whether the T1/T2 B-cell (mutated, anergic, or normal) recognises (self/foreign) antigens. Either strong or weak signalling or no signalling are the results of the interaction between B-cells and antigens in the preliminary domain model. Table 1 summarises hypothetical states of B-cells after the recognition of the antigenic peptides.

Table 1. Hypothetical states of transitional B-cells after the interaction with antigens

Type of antigen	BCR-signalling	Developmental states following transitional B-cells		
		T1/T2 mutated	*T1/T2 anergic*	*T1/T2 normal*
Foreign	Strong	Autoreactive cell	Death	Death
	Weak	Autoreactive cell	Anergic cell	Normal cell
	No interaction	Death	Death	Death
Self	Strong	Autoreactive cell	Death	Death
	Weak	Autoreactive cell	Anergic cell	Death
	No interaction	Death	Death	Death

4.3 Development Phase – Platform Model

Program AnyLogic has been selected as a tool for modelling the environment. Models in AnyLogic can be created using system dynamics, agent-based or discrete event tools. In this particular research, agent-based approach has been chosen. Each B-cell is modelled as an agent whose life cycle can be observed from its creation till death. Life cycle of each cell is created by state chart diagram where various conditions have to be fulfilled in order to transit between states. The model has been created based on the logic, see Fig. 1 which is described in the part 4 of this paper.

Time step of the model is 1 hr. Cells are created at a certain rate according to the parameter "Generation rate". In relation to this parameter, a specific number of new cells enters the system each hour with a given probability to be one of three types: mutated, anergic, or normal. As a cell enters the system (a state "T1") and stays in the spleen, there is a probability of an interaction between the cell and either self or foreign antigen (0–100 %). These probabilities and efficiency of the BCR signalling (0–100 %) are also the input parameters of the model. Possible results of the interaction with the antigen can be found in Table 1. If there is no interaction between the cell and any given antigen within 24-48 h, the cell dies. Following the interaction with the antigen, the cell transits into the other stages. After 24–48 h the cell is transformed into either memory B-cells (94 %) or plasma cells (6 %). 70 % of these cells move to the marginal zone and rest goes to the peripheral blood. We can observe number of cells in different parts of the system in experiments.

4.4 Model Calibration

Calibration of the model parameters is the first step before conducting experiments. Table 2 shows results of medical research [17] which explores number of different populations of B-cells in 1 μl of blood. The aim of calibration is to find such a combination of input parameters that we obtain corresponding results in our model environment.

Parameter variation experiment is used for testing different setups of the model. Range of the concentration of self and foreign antigen is set from 0 to 1 with a step of 0.1 (1 step = 1 h). Range for the cell generation rate is set from 1 to 10 with a step of 1.

Table 2. Count of cells in 1 μl of the peripheral blood

Cell type	Count
Immature	8.8
Naive	101
Memory	52
Plasma	3
SUM	164.8

Efficiency of the BCR signalling is set between 0 and 1 with a step of 0.1. Generation rate of new cells is set to 3, as based on previous tests.

The result of the simulation is the death rate of cells, as well as the number of naïve (and immature) B-cells, plasma cells, and memory B-cells in the blood and their autoreactivity. In the literature, percentage of autoreactive naïve (and immature) B-cells that has been measured varies. Based on the assumptions that central tolerance in autoimmune diseases is broken-down and maximum of B-cells which escaped from the bone marrow without being checked for autoreactivity is 10 %, number of abnormal B-cells (termed "mutated" for the purpose of this experiment) is set to 10 % of the pool of B-cells that leave the bone marrow. Other B-cells that make up this pool are anergic B-cells (45 %) and normal B-cells (45 %), [17, 21, 22]. Each model run simulates 10 000 h of bio-system activity.

Around 8000 different setups has been tested. Result are filtered according to the values in Table 3. Number of immature and naïve B-cells differs ± 5 from results observed by Perez-Andrez [17]. Resulting death rate and number of memory and plasma cells in the peripheral blood is limited by upper and lower boundaries.

Table 3. Boundaries of variables

Mem. up	59	Plas. up	5	Death up	97%
Mem. down	45	Plas. down	1	Death down	70%
Immature + Naive	109,8 ± 5				

There are only 52 setups which fulfils all these conditions. Various system characteristics and dependencies can be observed. For example, BCR signalling and number of autoreactive cells are correlated, as seen from Pearson's product-moment correlation, see Fig. 2. Confidence interval is 0.79-0.93.

Pearson's product-moment correlation
t = 12.649, df = 50, p-value < 2.2e-16
Alternative hypothesis: true correlation is not equal to 0
95 % confidence interval: 0.7875755; 0.9253468
Sample estimates: cor 0.8728702

Similar experiments can be used for testing various scenarios and simulating different setups of the system. Proliferation of autoreactive cells may lead to an

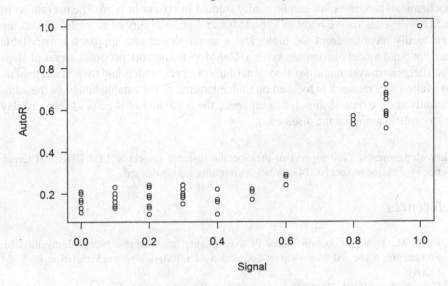

Fig. 2. Scatterplot - Signal vs. autoreactivity

autoimmune disease. As proliferation of autoreactive B-cells may lead to an autoimmune disease, experiments can be focused on parameter variations in which is the higher probability that the number of autoreactive cells grows.

5 Discussion

Proposed multi-agent model is the preliminary research that is focused on exploration of autoimmune diseases. The model is simplified. It observes only one type of lymphoid cells - B-cells and their transitions into the particular states during their development. Investigation of B-cells in the view of autoimmunity is crucial, because it is not known why B-cells behave abnormally, i.e. as self-reactive cells in most of autoimmune diseases. Transitions of B-cells into the next phases are influenced by various stimuli. The main attention is paid to interactions of B-cells with (self/foreign) antigenic peptides which can influence survival of the immune cells. The MAS-based model is based on factual data received from the literature. The future research is going to substantially extend the MAS-based model in a sense that it might integrate next stages of B-cells development along with the interactions of B-cells and dendritic cells.

6 Conclusion

The paper investigates the autoimmune diseases with the application of the multi-agent systems. Multi-agent systems bring added value into the exploration of complex systems, especially biological systems, because MAS can visualise particular biological (biochemical) processes that can be hardly studied in vivo or in vitro. The usefulness of the MAS depends on the body of knowledge and accessibility of factual data that are often hardly received from the biological system. Systematic approach is inevitable when we build model of complex system. CoSMoS framework proposes series of steps which helps with systematic development of multi-agent models and their management. Our preliminary research is focused on understanding of the establishment of the auto-immunity and the type of stimuli that influence the behaviour of B-cells which may play the key role in autoimmune diseases.

Acknowledgements. The support of the Specific research project at FIM UHK and Czech Science Foundation project Nr. 14-02424S is gratefully acknowledged.

References

1. Read, M., et al.: A domain model of experimental autoimmune encephalomyelitis. In: Proceedings of the 2nd Workshop on Complex Systems Modelling and Simulation, pp. 9–44 (2009)
2. Greaves, R.B., Read, M., Timmis, J., Andrews, P.S., Kumar, V.: Extending an established simulation: exploration of the possible effects using a case study in experimental autoimmune encephalomyelitis. In: Lones, M.A., Smith, S.L., Teichmann, S., Naef, F., Walker, J.A., Trefzer, M.A. (eds.) IPCAT 2012. LNCS, vol. 7223, pp. 150–161. Springer, Heidelberg (2012)
3. Williams, R.A., et al.: In silico investigation into dendritic cell regulation of CD8Treg mediated killing of Th1 cells in murine experimental autoimmune encephalomyelitis. BMC Bioinform. **14**(9), 1918–1929 (2013)
4. Read, M., et al.: Determining Disease intervention strategies using spatially resolved simulations. PLoS ONE 10(5) (2013). http://journals.plos.org/plosone/article?id=10.1371/journal.pone.0080506
5. Read, M: Statistical and modelling techniques to investigate imunology through agent-based simulation. Ph.D. thesis University of York, Computer Science Department (2011). http://etheseswhiterose.ac.uk/2174
6. Possi, M. A., et al.: An in-silico immune system model for investigating human autoimmune diseases. In: Proceedings of the 37th Conferencia Latino-Americana de Informatica (CLEI) 37 (2011). www.elsevier.nl/locate/entcs
7. Martínez, I.V., et al.: Definition of an agent-based model of the autoimmune response in Type 1 diabetes. In: Proceedings of the 7th Iberian Conference on Information Systems and Technologies (CISTI), IEEE, pp. 1–4 (2012)
8. Montagna, S., et al.: Modelling hematopoietic stem cell behaviour: an approach based on multi-agent systems. In: Proceedings of the 2nd Conference Foundations of Systems Biology in Engineering (FOSBE), Stutgart, Germany, pp. 243–248 (2007)
9. D´Inverno, M., et al.: Agent-Based Modeling of Stem Cells. Chapter 13: Multi-Agent Systems: Simulation and Applications. CRC Press, Taylor and Francis Group (eds. Uhrmacher, A. M., Weyns, D.), pp. 389–418 (2009)
10. Bessonov, N., et al.: Multi-Agent Systems and Blood Cell Formation. Chapter 18: Multi-Agent Systems – Modeling, Interactions, Simulations and Case Studies. InTech, (eds. Alkhateeb, F., et al.), pp. 395–424 (2011). DOI:10.5772/1936

11. Walker, D.C., et al.: The epitheliome: agent-based modelling of the social behavior of cells. Biosystems **76**, 89–100 (2004)
12. Smallwood, R., Holcombe, M.: The Epitheliome project: multiscale agent-based modeling of epithelial cells. In: Proceedings of the 2006 IEEE International Symposium on Biomedical Imaging: From Nano to Macro, pp. 816–819 (2006)
13. Andrews, P.S., et al.: CoSMoS process, models, and metamodels. In: Proceedings of the 2011 Workshop on Complex Systems Modelling and Simulation, pp. 1–13. France, Luniver Press (2011)
14. Moroni, L., Bianchi, I., Lleo, A.: Geoepidemilogy, gender and autoimmune disease. Autoimmun. Rev. **11**(6–7), 386–392 (2012)
15. Cooper, G.S., Bynum, M.L., Somers, E.C.: Recent insights in the epidemiology of autoimmune diseases: improved prevalence estimates and understanding of clustering of diseases. J. Autoimmun. **33**(3–4), 197–207 (2009)
16. Hayter, S.M., Cook, M.C.: Updated assessment of the prevalence, spectrum and case definition of autoimmune disease. Autoimmun. Rev. **11**(10), 754–765 (2012)
17. Perez-Andres, M., et al.: Human peripheral blood B-cell compartments: a crossroad in B-cell traffic. Cytometry B Clin. Cytom **78**(Suppl. 1), 47–60 (2010)
18. Loder, F., et al.: B cell development in the spleen takes place in discrete steps and is determined by the quality of B cell receptor-derived signals. J. Exp. Med. **190**(1), 78–89 (1999)
19. Petro, J.B., et al.: Transitional type 1 and 2 B lymphocyte subsets are differentially responsive to antigen receptor signalling. J. Biol. Chem. **277**(50), 48009–480019 (2002)
20. Niiro, H., Clark, E.A.: Regulation of B-cell fate by antigen-receptor signals. Nat. Rev. Immunol. **2**(12), 945–956 (2002)
21. Shlomchik, M.J.: Sites and stages of autoreactive B cell activation and regulation. Immunity **28**(1), 18–28 (2008)
22. Merrell, K.T., et al.: Identification of anergic B cells within a wild-type repertoire. Immunity **25**(6), 953–962 (2006)
23. Husáková, M.: The usage of the agent modeling language for modeling complexity of the immune system. In: Proceedings of the 7th Asian Conference on Intelligent Information and Database Systems (ACIIDS). Springer-Verlag LNCS, pp. 323–332 (2015)
24. Husáková, M.: Combating infectious diseases with computational immunology. In: Núñez, M., et al. (eds.) ICCCI 2015. LNCS, vol. 9330, pp. 398–407. Springer, Heidelberg (2015). doi:10.1007/978-3-319-24306-1_39

Decision Support Biomedical Application Based on Consistent Optimization of Preference Matrices

Richard Cimler[1,2(✉)], Martin Gavalec[1], Karel Mls[1], and Daniela Ponce[1]

[1] Faculty of Informatics and Management,
University of Hradec Králové, Hradec Králové, Czech Republic
richard.cimler@uhk.cz
[2] Center for Basic and Applied Research (CZAV),
University of Hradec Králové, Hradec Králové, Czech Republic
http://www.uhk.cz

Abstract. In designing a medical application, big emphasis must be given on correct understanding of experts' opinions. This article deals with the decision making support based on Analytic Hierarchy Process (AHP). Inconsistency measures of preference matrices and the problem of consistent optimization are studied. A new method for computing the optimal consistent approximation of a given preference matrix is described. Furthermore, algorithms for finding a consensus of several experts are discussed. The proposed methods are presented and explained on numerical examples.

Keywords: Analytic Hierarchy Process · Inconsistency · Orthogonal consistent optimization

1 Introduction

Smart technologies are nowadays an inherent part of human life. In the last decades, the increase in computational power and miniaturization enabled to create various devices simplifying human life. Health care is one of branches where modern devices are widely used. There are various technologies for collecting data [3,8,27], data processing and, based on the results, evaluating the health status of a person. Complex algorithms are used to process the data and built-in devices make decisions about actions. The devices may only be used as an information source and as a support for decision making, or they can also perform autonomous actions and operate without human interaction.

During creation of every application, there must be experts who share their knowledge with the programmers of the application. The aim of every expert system is to create an autonomous application which substitutes the expert's decisions, and is able to make decisions based on expert's knowledge and experience, even without his/her personal presence.

© Springer International Publishing Switzerland 2016
N.T. Nguyen et al. (Eds.): ICCCI 2016, Part II, LNAI 9876, pp. 292–302, 2016.
DOI: 10.1007/978-3-319-45246-3_28

In designing a medical application, big emphasis must be given on correct understanding and programming of experts opinions. Nobody is perfect, and even an expert on a particular domain might be sometimes not 100 % sure. The decision can be influenced by different perturbations. Creation of medical applications requires that experts from different domains discuss the problem and try to find consensus during the creation of the rules for an expert system. Finding the right way to the optimal solution might be sometimes very difficult. There might by different views on the same thing by various experts. This article deals with the decision making based on Analytic Hierarchy Process (AHP). Matrices used in this process are created by many experts. Algorithms for finding a consensus of several experts are introduced.

The article is divided into the following sections. The Analytic Hierarchy Process is introduced in the next section. Inconsistency of preference matrices and ways to measure it are discussed in Sect. 3. Our proposed method for finding the optimal consistent approximation of a given preference matrix is described in Sect. 4. The problem of aggregation of multiple experts' opinions is discussed in Sect. 5. Proposed methods are presented on implementation examples in Sect. 6. The last section of the article summarizes the results.

2 Analytic Hierarchy Process

In the AHP approach to modeling decision situations, priority scales are derived formally after pair-wise subjective judgments are made. According to Saaty [26], there are at least three reasons to make comparisons instead of direct measurements:

1. the lack of instrument or scale to make a measurement,
2. our belief that the outcome of comparisons using our judgments would be better than using some general scale of measurement,
3. a way to measure something at present - happiness, popularity or aesthetic appeal.

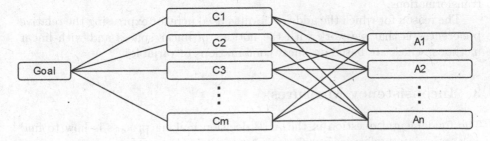

Fig. 1. AHP model with criteria C_i and alternatives A_j

Moreover, thinking about values in economics is mostly based on utility theory, implicitly subsuming benefits, opportunities, costs and risks. The construction and evaluation procedure of the AHP model can be summarized as follows [24], see Fig. 1:

- Model the problem as a hierarchy containing the decision goal, the alternatives for reaching it, and the criteria for the alternatives evaluation.
- Establish priorities among the elements of the hierarchy by making a series of judgments based on pairwise comparisons of the elements.
- Synthesize these judgments to yield a set of overall priorities for the hierarchy.
- Check the consistency of the judgments.
- Reach the final decision based on the results of this process.

Given alternatives $A_1, A_2, .., A_n$ will be considered till the end of the paper. N will denote the set $\{1, 2, \ldots, n\}$ and $\mathcal{R}$ the set of all real numbers.

For every pair A_i, A_j, the real number a_{ij} is interpreted as an evaluation of the relative preference of A_i with respect to A_j, in the additive sense. The matrix $A = (a_{ij})$, $i, j \in N$ is called *additive preference matrix* (for short: preference matrix) of the alternatives $A_1, A_2, .., A_n$. The basic properties of preference matrices are defined as follows

A is *antisymmetric* if $a_{ij} = -a_{ji}$ for every $i, j \in N$,
A is *consistent* if $a_{ij} + a_{jk} = a_{ik}$ for every $i, j, k \in N$.

Clearly, if A is consistent, then A is antisymmetric, but the converse implication is not true. E.g., $A = \begin{pmatrix} 0 & 1 & 1 \\ -1 & 0 & 1 \\ -1 & -1 & 0 \end{pmatrix}$ is antisymmetric, but it is not consistent, because $a_{12} + a_{23} = 1 + 1 = 2 \neq a_{13}$.

Another frequently used form for the relative preferences of alternatives are *multiplicative preference matrices*. If $M \in \mathcal{R}^+(n, n)$ is a multiplicative preference matrix, then every entry m_{ij} with $i, j \in N$ is considered as an multiplicative evaluation of the relative preference. The multiplicative preference matrices have analogous properties as the additive preference matrices. In fact, they can be equivalently transferred to each other by the logarithmic and exponential transformations.

The reason for which the additive form is used here for expressing the relative preferences is that we work with the notions of linear spaces and with linear methods, which are based on linear combinations of variables.

3   Inconsistency Measures

The fundamental question in the AHP decision making process is how to find an appropriate preference matrix for a set of alternatives. The preferences given by human experts are often inconsistent and do not reflect the deep relations between the processed notions, see [6, 16, 20–22].

One way of solving the inconsistency problem for a preference matrix is to define the *consistency index* of A and the *consistency ratio* of A

$$\text{CI}(A) = \frac{\lambda_{\max} - n}{n - 1}, \quad \text{CR}(A) = \frac{\text{CI}(A)}{\text{ARI}(n)}, \tag{1}$$

where $\lambda_{\max}$ is the principal eigenvalue of A, n is its size and $\text{ARI}(A)$ is the average consistency index of randomly generated reciprocal matrices of size n. Then the preference matrix A is considered to be acceptable if CR does not exceeds the empirical value 0.1, see [23, 25]. Many further inconsistency measures were defined like: simple normalized column sum, preference weighted geometric mean, least square deviation and/or least absolute error deviation sum (see [2], for a discussion on these and other inconsistency measures). Based on a numerical study of several inconsistency measures, the authors in [4] point out that there is no agreement in the research community on which measure of inconsistency should be used, and moreover, little attention is paid to similarities and differences of various measures.

When the preference matrix is not consistent, the decision maker faces a dilemma: either to reject the preference matrix and replace it by a new preference matrix or to modify the preference matrix in such a way that it becomes consistent. This decision can be supported by the inconsistency measure of the matrix, that is, when the inconsistency measure of the preference matrix is acceptably low for the decision maker, then the consistent approximation of the preference matrix can be reasonable. Put this way, the measure of inconsistency of a matrix A can be interpreted as a measure of acceptability of the approximating matrix X. The smaller the measure of inconsistency of A is, the more acceptable the consistent approximation X to matrix A is for the decision maker, because it is the nearest consistent matrix to A from all consistent matrices.

The search for consistent approximation of an inconsistent preference matrix has been investigated in some previous works, see e.g. [10, 13, 17]. Inconsistency measure may be based on distance of preference matrices (see e.g. [15]), but not necessarily. For example, error-free correctness methods [7] give 0 for consistent matrices and positive values otherwise, using some formula with the inconsistent matrix as input value. In their work [5], Cao et al. propose an algorithm that gradually modifies the given inconsistent matrix till the inconsistency measure of the matrix drops below an acceptable value.

Another approach to the inconsistency problem is to take the values in the expert's preference matrix A as a starting point for computing a good consistent approximation of A. Such computations have been suggested e.g. in [12, 18, 19]. Various distance metrics are used and compared in [18]. The additive form of expressing the relative preferences is applied. The additive form is more convenient for the optimization purposes than the multiplicative form, as the notions and methods from linear algebra can directly be applied.

A new method for measuring the inconsistency of preference matrices is described in this paper. The method is based on the orthogonal projection of a given (possibly inconsistent) preference matrix to the linear subspace of all consistent matrices.

In biomedical applications, usually more than one expert participates in creating the preference matrix. Every expert creates his/her personal version of preferences and the individual judgments are then merged to make the common preference matrix.

In practical applications, the preference matrix is created by a human expert in the given field. While the antisymmetricity is easy to verify, the consistency of a preference is not directly seen from the data. As a consequence, the preference matrices given by experts are often inconsistent.

The following approximation problem is investigated in this section: given a (possibly inconsistent) preference matrix A, find a consistent matrix X which will be as close to A as possible. Matrix X is then called *the optimal consistent approximation* of A.

Clearly, every approximation depends on the distance measure that is used in the optimization. A general family of distances l^p with $1 \leq p \leq \infty$ is known in the literature. For chosen value p, the distance of vectors $x, y \in \mathcal{R}(n)$ is

$$l^p(x,y) = \left(\sum_{i \in N} |x_i - y_i|^p \right)^{1/p} \tag{2}$$

For $p = 1$, $p = 2$, or $p = \infty$, (2) gives the so-called Manhattan distance $l^1(x,y) = \sum_{i \in N} |x_i - y_i|$, Euclidean distance $l^2(x,y) = \sqrt{\sum_{i \in N} |x_i - y_i|^2}$, or Chebyshev distance $l^\infty(x,y) = \max_{i \in N} |x_i - y_i|$.

The optimal consistent approximation according to the above three distance functions have been compared in [18]. In this contribution, the Euclidean distance will only be considered, and denoted simply by l.

4   Orthogonal Consistent Optimization

In the AHP theory, the consistent preference matrices are closely related with preference vectors showing the importance for every of alternatives. In the additive notation, vector $w \in \mathcal{R}(n)$ is called a *balanced preference vector* (for short: a balanced vector) if $\sum_{i \in N} w_i = 0$. When alternatives $\mathcal{A}_1, \mathcal{A}_2, \ldots, \mathcal{A}_n$ are considered, then w_i is interpreted as the preference degree of $\mathcal{A}_i$ for every $i \in N$. The differences of preferences are the entries of the corresponding matrix of relative preferences $A(w)$ with $a_{ij}(w) = w_i - w_j$ for $i, j \in N$. We say that $A(w)$ *is induced by* w. The relation between preference matrices and balanced vectors has been described in [19].

Theorem 1. *[19] Let $A \in \mathcal{R}(n,n)$ be a preference matrix.*

(i) *If $w \in \mathcal{R}(n)$, then the induced preference matrix $A(w)$ is consistent.*
(ii) *If $w, w' \in \mathcal{R}(n)$ and $A(w) = A(w')$, then $w' = w + \delta$ for some $\delta \in \mathcal{R}$.*

(iii) *If A is consistent, then there is a unique balanced vector w such that $A = A(w)$.*

We are interested in solving the following optimization problem:

OCA (optimal consistent approximation)

given $A \in \mathcal{R}(n, n)$, minimize

$$z = l(A, X) \longrightarrow \min \tag{3}$$

such that $X \in \mathcal{C}$.

Suppose n alternatives $\mathcal{A}_1, \mathcal{A}_2, \ldots, \mathcal{A}_n$ are considered. Denote by $\mathcal{B}$ the set of all balanced vectors $w \in \mathcal{R}(n)$, and by $\mathcal{C}$ the set of all consistent matrices $A \in \mathcal{R}(n, n)$. $\mathcal{C}$ is sometimes called the *consistency* subspace. According to Theorem 1(iii), any $w \in \mathcal{B}$ induces a uniquely determined consistent matrix $A(w)$. Moreover, any linear combination $\sum_{i \in K} c_i w_i$ with $c_i \in \mathcal{R}$ and $w_i \in \mathcal{B}$, for every $i \in K = \{1, 2, \ldots, k\}$ induces the linear combination

$$A\left(\sum_{i \in K} c_i w_i\right) = \sum_{i \in K} c_i A(w_i) \tag{4}$$

of consistent matrices in $\mathcal{C}$ with the same coefficients. Using Theorem 1 (iii) and (4), we get the following result.

Theorem 2. *Suppose n alternatives are considered. Then*

(i) $\mathcal{B}$ *is a linear subspace of $\mathcal{R}(n)$,*
(ii) $\mathcal{C}$ *is a linear subspace of $\mathcal{R}(n, n)$,*
(iii) $(w_1, w_2, \ldots, w_k)$ *is a base in $\mathcal{B}$ if and only if*
    $(A(w_1), A(w_2), \ldots, A(w_k))$ *is a base in $\mathcal{C}$,*
(iv) $\dim(\mathcal{B}) = \dim(\mathcal{C}) = k = n - 1$.

In view of Theorem 2, the solution to the optimization problem OCA can be found as the orthogonal projection $X = P(A)$ of the input matrix A into $\mathcal{C}$.

Assume $B_1, B_2, \ldots, B_k$ is a base in $\mathcal{C} \subset \mathcal{R}(n, n)$. Then the solution can be expressed as linear combination $X = \sum_{j \in K} c_j B_j$ with $c_j \in \mathcal{R}$. The coefficients must satisfy the orthogonality conditions

$$B_i \cdot \left(A - \sum_{j \in K} c_j B_j\right) = 0 \quad \text{for every } i \in K \tag{5}$$

where the dots in (5) indicate the sumproduct operation in $\mathcal{R}(n, n)$. By easy modification we get

$$\sum_{j \in K} c_j (B_i \cdot B_j) = B_i \cdot A \quad \text{for every } i \in K \tag{6}$$

If $c_1, c_2, \ldots, c_k$ satisfy conditions (6), then $Y = A - X$ is orthogonal to every B_i from the base of $\mathcal{C}$. That is, Y is orthogonal to subspace $\mathcal{C}$ and X is the orthogonal projection of A to $\mathcal{C}$. Therefore, X is determined uniquely, and does not depend on the base $B_1, B_2, \ldots, B_k$. On the other hand, for a fixed base, the coefficients $c_1, c_2, \ldots, c_k$ are determined uniquely, and (6) has exactly one solution.

The above considerations are formulated in the following theorem.

Theorem 3. *Assume $A \in \mathcal{R}(n, n)$ is a preference matrix. Then*

(i) *the optimization problem OCA with input A has unique solution X,*
(ii) *X is the orthogonal projection of A to the consistency subspace $\mathcal{C}$,*
(iii) *if $B_1, B_2, \ldots, B_k$ is a base in $\mathcal{C}$, then X can be computed as a linear combination with coefficients satisfying (6).*

The orthogonal consistent optimization method described in Theorem 3 is called **OCO**.

Remark 1. It is clear that if the input matrix A is consistent, then $X = A$.

Remark 2. Along with other optimization problems, OCA has been studied in [18]. Solution has been found by the least squares method, under the assumption that A is antisymmetric.

5   Methods of Aggregation

There are various approaches how to reach consensus from several (usually inconsistent) experts judgments in group decision making. Dong in [11] proved that both consistency and consensus may not be perfect in general and so there is a need of some measure or degree for them. The consistency measure is then used to quantify the difference among individual decision makers and to help in finding an acceptable solution of the consensus model. Considering AHP as a platform for decision making support, opinions of individual experts can be merged in several ways.

Aggregation of individual opinions can be done at the initial stage of the decision making process by searching consensus on the given judgments (aggregation of individual judgments AIJ) or at the last stage by individually induced preferences with importance weights of the group members (aggregation of individual priorities - AIP) [11]. Between these stages, other approaches, e.g. voting, can be applied to the decomposed intermediate judgments aggregation [14].

We use the Synergy Aggregation Method (SAM) that takes preference matrices (i.e. individual judgments of experts involved), approximates them by the nearest consistent matrices if needed and produces an aggregated consistent preference matrix (i.e. collective judgment). We furthermore suppose that each expert excels in a particular part of the problem and therefore for every expert we assign weights to pairwise alternative comparisons formulated by that expert. These weights are then used in the calculation of the weighted arithmetic mean of preference matrices. Because of the specific features of the problem, we can suppose that experts work in synergy and they agree on the rank of alternatives.

As a consequence, the preference matrices of all experts can be all transformed into matrices such that they contain non-negative elements only in the upper triangle matrices. In the case of discordance, other aggregation methods are applied, see [1].

6 Implementation Examples

The same procedures as in [9] are used in the examples below describing a health-monitoring application. We assume that experts, in our case doctors, have prepared a preference matrix A. The submitted matrix may be neither consistent nor antisymmetric.

Example 1 In the first example there is only one expert who created an inconsistent and non-antisymmetric matrix A. For implementation into the application we need to find a consistent and antisymmetric matrix which is as close as possible to A. The method OCO described in Theorem 3, is used for computing such a matrix. The alternatives in this example are the levels of alert: Normal, Warning level 1, Warning level 2, and Alarm. The entries of the matrix have been suggested by the expert to express which level of alert is the most appropriate in the given situation (such as: blood pressure = increased, heart rate = slightly increased, body temperature = high).

$$A = \begin{pmatrix} 0 & -2 & -3 & 1 \\ 2 & 0 & -1 & 2 \\ 3 & 1 & 0 & 2 \\ 3 & 2 & -2 & 0 \end{pmatrix}$$

The computed matrix X is consistent and antisymmetric. The Euclidean distance from the original matrix A is 3.

$$X = \begin{pmatrix} 0 & -2.25 & -3.0 & -0.75 \\ 2.25 & 0 & -0.75 & 1.5 \\ 3.0 & 0.75 & 0 & 2.25 \\ 0.75 & -1.5 & -2.25 & 0 \end{pmatrix}$$

Example 2 In the second example we assume that several experts are creating preference matrices related to the same problem. There might be different opinions on the same problem thus preference matrices can be different. Using the methods OCO and SAM described in Sects. 4 and 5, we can get one consistent antisymetric matrix which will be the closest one to all matrices created by the individual experts. According to the experts' competences there are different weights q of experts' suggestions. In this example: $q^{(1)} = 4$, $q^{(2)} = 3$, $q^{(3)} = 1$.

$$A^{(1)} = \begin{pmatrix} 0 & -2 & -3 & 1 \\ 2 & 0 & -1 & 2 \\ 3 & 1 & 0 & 2 \\ 3 & -2 & -2 & 0 \end{pmatrix} \quad A^{(2)} = \begin{pmatrix} 0 & -2 & -3 & -2 \\ 2 & 0 & 1 & 4 \\ 3 & -1 & 0 & 4 \\ 2 & -4 & -3 & 0 \end{pmatrix} \quad A^{(3)} = \begin{pmatrix} 0 & -2 & -4 & -3 \\ 3 & 0 & -3 & -2 \\ 4 & 3 & 0 & 4 \\ 2 & 3 & -4 & 0 \end{pmatrix}$$

In the first phase of the solution we compute the approximations by consistent and antisymmetric matrices using the methods described in Sect. 4.

$$X^{(1)} = \begin{pmatrix} 0 & -2.25 & -3.0 & -0.75 \\ 2.25 & 0 & -0.75 & 1.5 \\ 3.0 & 0.75 & 0 & 2.25 \\ 0.75 & -1.5 & -2.25 & 0 \end{pmatrix} \quad X^{(2)} = \begin{pmatrix} 0 & -3.5 & -3.125 & -0.375 \\ 3.5 & 0 & 0.375 & 3.125 \\ 3.125 & -0.375 & 0 & 2.75 \\ 0.375 & -3.125 & -2.75 & 0 \end{pmatrix}$$

$$X^{(3)} = \begin{pmatrix} 0 & -1.5 & -5.0 & -2.5 \\ 1.5 & 0 & -3.5 & -1.0 \\ 5.0 & 3.5 & 0 & 2.5 \\ 2.5 & 1.0 & -2.5 & 0 \end{pmatrix}$$

By the aggregation method SAM, we then get from several consistent and antisymmetric matrices one merged matrix X, which can be used in the biomedical application as a criterion matrix. The weighted average of experts' individual preference matrices based on the experts' particular competences has been used in our case.

$$X^{(\text{final})} \approx \begin{pmatrix} 0 & -2.63 & -3.30 & -0.83 \\ 2.63 & 0 & -0.67 & 1.8 \\ 3.3 & 0.67 & 0 & 2.47 \\ 0.83 & -1.80 & -2.47 & 0 \end{pmatrix}$$

7   Conclusions

The decision making in a medical application requires big emphasis on experts' opinions and experience. Solving the inconsistency problem for preference matrices is therefore of vital importance.

The euclidean distance of a given preference matrix to the linear subspace of all consistent matrices has been suggested in the paper as the inconsistency measure for preference matrices. Thus, the nearest consistent matrix can be computed as the orthogonal projection of a given preference matrix submitted by an expert.

For a group of experts, the individual consistent approximations are merged into one final result, taking in view the experts' competence levels. The proposed methods have been illustrated by numerical examples.

Acknowledgment. The support of the grant project GAČR #14-02424S and the specific research project SPEV at FIM UHK is gratefully acknowledged.

References

1. Abel, E., Mikhailov, L., Keane, J.: Group aggregation of pairwise comparisons using multi-objective optimization. Inf. Sci. **322**, 257–275 (2015)
2. Barzilai, J.: Consistency measures for pairwise comparison matrices. J. Multi-Crit Decis. Anal. **7**(3), 123–132 (1998)

3. Blazek, P., Krenek, J., Kuca, K., Krejcar, O., Jun, D.: The biomedical data collecting system. In: Radioelektronika (RADIOELEKTRONIKA), 2015 25th International Conference, pp. 419–422. IEEE (2015)
4. Brunelli, M., Canal, L., Fedrizzi, M.: Inconsistency indices for pairwise comparison matrices: a numerical study. Ann. Oper. Res. **211**(1), 493–509 (2013)
5. Cao, D., Leung, L.C., Law, J.: Modifying inconsistent comparison matrix in ahp: a heuristic approach. Decis. Support Syst. **44**(4), 944–953 (2008)
6. Carlsson, C., Fullér, R.: Fuzzy multiple criteria decision making: Recent developments. Fuzzy Set Syst. **78**(2), 139–153 (1996)
7. Choo, E.U., Wedley, W.C.: A common framework for deriving preference values from pairwise comparison matrices. Comput. Oper. Res. **31**(6), 893–908 (2004)
8. Cimler, R., Matyska, J., Balík, L., Horalek, J., Sobeslav, V.: Security issues of mobile application using cloud computing. In: Abraham, A., Krömer, P., Snasel, V. (eds.) Afro-European Conf. for Ind. Advancement. AISC, vol. 334, pp. 347–358. Springer, Heidelberg (2015)
9. Cimler, R., Mls, K., Gavalec, M.: Decision support smartphone application based on interval AHP method. In: Nguyen, N.T., et al. (eds.) ICCCI 2015. LNCS, vol. 9330, pp. 306–315. Springer, Heidelberg (2015). doi:10.1007/978-3-319-24306-1_30
10. Dahl, G.: A method for approximating symmetrically reciprocal matrices by transitive matrices. Linear Algebra Appl. **403**, 207–215 (2005)
11. Dong, Y., Zhang, G., Hong, W.C., Xu, Y.: Consensus models for ahp group decision making under row geometric mean prioritization method. Decis. Support Syst. **49**(3), 281–289 (2010)
12. Dopazo, E., González-Pachón, J.: Consistency-driven approximation of a pairwise comparison matrix. Kybernetika **39**(5), 561–568 (2003)
13. Elsner, L., Van Den Driessche, P.: Max-algebra and pairwise comparison matrices. Linear Algebra Appl. **385**, 47–62 (2004)
14. Entani, T.: Interval ahp for group of decision makers. In: IFSA/EUSFLAT Conference, pp. 155–160 (2009)
15. Fedrizzi, M.: Distance-based characterization of inconsistency in pairwise comparisons. In: Greco, S., Bouchon-Meunier, B., Coletti, G., Fedrizzi, M., Matarazzo, B., Yager, R.R. (eds.) Advances in Computational Intelligence. CCIS, vol. 300, pp. 30–36. Springer, Heidelberg (2012)
16. Gavalec, M., Ramík, J., Zimmermann, K.: Decision Making and Optimization: Special Matrices and Their Applications in Economics and Management, vol. 677. Springer, Heidelberg (2014)
17. Gavalec, M., Tomášková, H., Cimler, R.: Computer support in building-up a consistent preference matrix. In: Sulaiman, H.A., Othman, M.A., Iskandar Othman, M.F., Rahim, Y.A., Pee, N.C. (eds.) Advanced Computer and Communication Engineering Technology. LNEE, vol. 362, pp. 947–956. Springer, Heidelberg (2016)
18. Gavalec, M., Tomášková, H.: Comparison of consistent approximations for a matrix of pair preferences. In: Czech-Japan Seminar, pp. 45–57 (2015)
19. Gavalec, M., Tomášková, H.: Optimal consistent approximation of a preference matrix in decision making. Int. J. Math. Oper. Res. (2016) (to appear)
20. Leung, L.C., Cao, D.: On consistency and ranking of alternatives in fuzzy ahp. Eur. J. Oper. Res. **124**(1), 102–113 (2000)
21. Ramík, J., Korviny, P.: Inconsistency of pair-wise comparison matrix with fuzzy elements based on geometric mean. Fuzzy Set Syst. **161**(11), 1604–1613 (2010)
22. Ramík, J., Perzina, R.: A method for solving fuzzy multicriteria decision problems with dependent criteria. Fuzzy Optim. Decis. Ma **9**(2), 123–141 (2010)

23. Saaty, T.L.: The Analytic Hierarchy Process: Planning, Priority Setting, Resources Allocation. McGraw, New York (1980)
24. Saaty, T.L.: Decision Making for Leaders: The Analytic Hierarchy Process for Decisions in a Complex World. RWS Publications, Pittsburgh (1990)
25. Saaty, T.L.: Decision-making with the ahp: why is the principal eigenvector necessary. Eur. J. Oper. Res. **145**(1), 85–91 (2003)
26. Saaty, T.L.: Theory and applications of the analytic network process: decision making with benefits, opportunities, costs, and risks. RWS Publications, Pittsburgh (2005)
27. Sobeslav, V., Maresova, P., Krejcar, O., Franca, T.C., Kuca, K.: Use of cloud computing in biomedicine. J. Biomol. Struct. Dyn. **32**, 1–25 (2015). (just-accepted)

Implementation of Artificial Neural Network on Graphics Processing Unit for Classification Problems

Syahid Anuar[1], Roselina Sallehuddin[1], and Ali Selamat[1,2,3(✉)]

[1] Faculty of Computing, Universiti Teknologi Malaysia,
81310 Skudai, Johor, Malaysia
syah2105@yahoo.com, roselina@utm.my
[2] UTM-IRDA Digital Media Centre, MaGIC-X (Media and Games Innovation
Centre of Excellence), Universiti Teknologi Malaysia, 81310 Skudai, Johor, Malaysia
aselamat@utm.my
[3] Faculty of Informatics and Management, Center for Basic and Applied Research,
University of Hradec Kralove, Hradec Kralove, Czech Republic

Abstract. The artificial neural network (NN) is widely use in pattern recognition related area such as classification. After all this time, the computational process of NN is done using central processing unit (CPU). In recent years, the introduction of graphics processing unit (GPU) has opened another way to perform calculations with the advantage to speed up the calculation. In this paper, the computational process of multilayer perceptron neural network be tested on GPU using classification datasets. The performance of NN model with different number of input, hidden and output neurons are explored and compared based on the computational between GPU and CPU. The experimental result shows that the computational on GPU is much faster than CPU.

1 Introduction

Artificial neural network is widely used in pattern recognition based on the concept of the workings of the human brain [10]. The basic computational unit of NN is the artificial neuron consists of three main components such as input synapses, hidden units and output value of the neurons [1]. The most basic equation for a single neuron is given by Eq. 1.

$$f(x) = W \bullet X + B \tag{1}$$

where $f(x)$ consists of multiplication of weight, w and input value, x with added bias, b value. The basic Eq. 1 is still used as a part of a larger network model. Multi layer perceptron (MLP) is the simplest neural network with single input layer, several hidden layer and an output layer. It is usually trained by back propagation (BP) algorithm. One of the most important advantages of training NN using BP algorithm is the high degree of accuracy when used to generalize

© Springer International Publishing Switzerland 2016
N.T. Nguyen et al. (Eds.): ICCCI 2016, Part II, LNAI 9876, pp. 303–310, 2016.
DOI: 10.1007/978-3-319-45246-3_29

over a set of unseen examples of the problem domain [13]. However, NN modelling is not yet widely accepted in industry due to it slow learning process [2]. The introduction of GPU has attracted attention of researchers for performing calculation from CPU to GPU. The calculation on GPU is said could give dramatic speedups over CPU for certain algorithms [11]. Instead, implementation of NN on GPU has shown some improvements in term of speed for text detection and handwritten digits recognizing systems [3,10].

Therefore, the objective of this study is to investigate the performance of NN on GPU in terms of processing time. To achieve this goal, the comparisons between GPU and CPU using two different classification dataset scales namely large scale and small scale are carried out. After the Introduction section, this paper continues with Sect. 2 which discusses the implementation of NN on GPU. The next section describes the experiment setup together with the definition of the datasets used throughout this study. The findings obtained from the experiments are presented in Sect. 4. Finally, Sect. 5 concludes the work done and proposes future works of this paper.

2   Implementation of Artificial Neural Network on Graphics Processing Unit

The concept of NN is based on how human brain and functions. The NN's architecture consists of many types such as multilayer perceptron, self organizing map and radial basis function network. However, the basic structure of NN consists of input layer, hidden layer and output layer with various number of nodes in each layer. Assume W is the weight, X is the input and B is the bias term. Based on Eq. 1, the elements of W, X and B can be represented in matrices form with row i and column j as Eq. 2.

$$f(x) = \begin{bmatrix} w_{11} & \cdots & w_{1,j} \\ \vdots & \ddots & \vdots \\ w_{i,1} & \cdots & w_{ij} \end{bmatrix} \bullet \begin{bmatrix} x_{11} & \cdots & x_{1,j} \\ \vdots & \ddots & \vdots \\ x_{i,1} & \cdots & x_{ij} \end{bmatrix} + \begin{bmatrix} b_{11} & \cdots & b_{1,j} \\ \vdots & \ddots & \vdots \\ b_{i,1} & \cdots & b_{ij} \end{bmatrix} \tag{2}$$

The NN model usually use back propagation algorithm as a learning mechanism. The matrices operations as in Eq. 2 is usually done on CPU. A simple calculation using small matrix does not really effect the performance in term of processing time for CPU and GPU [7]. However, as the data complexity increase in terms of volume and velocity, the structure of NN also become more complex with the increasing number of neurons and hidden layers. Therefore, the matrices operations becoming more complicated thus compromise the performance on CPU which requires more processing time to accomplish the training process [10]. Therefore, a solution is needed to overcome this limitation. The graphic hardware was initially used for rendering the graphic within the last few decades [10]. However, the advanced development of graphical hardware has contributed to the creation of graphics processing unit for general purpose computation, called GPU computing [5]. Figure 1 illustrates the architecture of Intel Haswell CPU and Nvidia GTX 980 GPU.

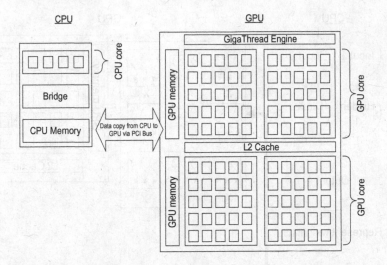

Fig. 1. The architecture of CPU and GPU

The GPU supports parallel process which can speed up the computational time [4]. Hence, using this advantage of GPU, the problem in handling complex matrices operations of NN on CPU can be done. Figure 1 demonstrates how GPU solved the problems in CPU. Based on Fig. 1, the matrices on CPU will be transferred to GPU. The calculation is done on GPU and the result will be transferred back to CPU. The CUDA programming will be used to allow the execution of calculation on GPU [9]. Figure 2 illustrates the representation of NN on GPU.

Based on Fig. 2, the elements of the matrix are mapped on the GPU core. We choose CUDA to fully exploit the available state-of-the-art GPU Maxwell architecture. The concept is based on kernels (represented as the curly arrows), which functions are executed in parallel by CUDA threads. All of these threads are grouped together into blocks and will be distributed on the GPU core to be executed independently. By default, the CPU architecture will process the element using a single core processor or to one of the multi-core processor, and it differs from the GPU which performs the parallel process on many core [6].

3   Experiment Setup

The purpose of this experiment is to simulate the difference processing implementation between CPU and GPU. The performance of GPU and CPU are analysed using two different scales of classification datasets: large-scale and small-scale. The classification datasets are gathered from UCI Machine Learning website [8]. As been suggested by [12], conjugate gradient algorithm will be used as the transfer function between neurons because it more suitable for parallel processing and can provide large speed-up. For a fair comparisons, both CPU and GPU will use conjugate gradient algorithm.

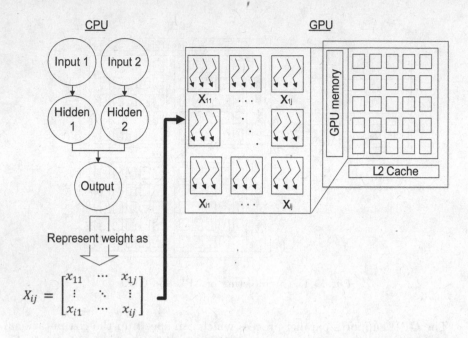

Fig. 2. Representation of artificial neural network on GPU

To utilize the parallelism of GPU, a lots of input and weight vectors will be used. Based on Table 1, the datasets consist of difference size and dimension. The small-scale datasets consist of the number of inputs less than 1000 whilst the large scale-datasets consist of the number of inputs more than 1000. Each dataset is divided into training and testing with the ratio of 80:20 respectively. The experiment is conducted to compare the performance of artificial neural network running on CPU (CPUNN) and artificial neural network running on GPU (GPUNN). The Nvidia Gefore GTX970 is been chosen as the GPU and Intel Core i7-4770 for CPU. The comparison results are discussed in the next section.

Table 1. Datasets description [8] where NOI is referring to number of instances.

Dataset	NOI	Number of inputs	Number of targets	Description
PEMS	440	138672	7	The occupancy rate of different car lanes of the San Francisco bay area freeways across time
Dorothea	1950	100000	2	Drug discovery dataset
Arcene	900	10000	2	Distinguish cancer versus normal patterns from mass-spectrometric data
Gisette	13500	5000	2	Handwritten digit recognition problem
Thyroid	7200	21	3	Thyroid function dataset
Breast	699	9	2	Breast cancer dataset
Glass	214	9	2	Glass chemical dataset
Iris	150	4	3	Iris flower dataset

4 Result and Discussion

For both CPU and GPU processing, the experiments are done for 500 epochs. The experiments with various number of hidden neurons to diversify the complexity of NN are carried out. The important point to be highlighted for comparative performance between CPU and GPU is the processing time. The results of processing time are given in form of line graph as in Fig. 3. For the graph, the x-axis indicates the time in seconds(s) and the y-axis indicates the number of hidden neurons.

Based on Fig. 3, the results indicate that the process of GPUNN is faster than CPUNN for large-scale datasets namely PEMS, Dorothea, Arcene and Gisette. For small scale datasets, two different results are obtained. First, GPUNN processing time is faster than CPUNN for Thyroid dataset which has many instances. However, the usage of GPU is less significant for small scale dataset like Breast, Glass and Iris where the CPUNN is much faster than GPUNN. This result indicates that it is enough to use CPU instead of GPU and the performance of CPU is much better for small-scale datasets that have small number of instances.

Based on the GPU architecture as in Fig. 1, many processing cores assist the computational process because the elements of matrices can easily be mapped

Table 2. Comparison of testing accuracies between CPUNN and GPUNN on classification datasets

Datasets	Methods	Number of hidden neurons				
		100	200	300	400	500
PEMS	CPUNN	**96.63**	**95.13**	92.88	96.25	93.63
	GPUNN	91.39	90.26	**94.76**	**97.75**	**97.75**
Dorothea	CPUNN	**90.88**	**90.50**	90.50	**91.13**	90.38
	GPUNN	90.75	**90.50**	**90.75**	90.00	**92.88**
Arcene	CPUNN	93.00	**90.00**	94.00	89.00	87.00
	GPUNN	**95.00**	88.00	**96.00**	**94.00**	**89.00**
Gisette	CPUNN	**99.33**	98.27	98.62	97.83	**99.02**
	GPUNN	98.90	**99.00**	**98.78**	**98.55**	98.80
Thyroid	CPUNN	**92.58**	**97.56**	**97.96**	92.58	**96.99**
	GPUNN	**92.58**	97.47	95.36	**97.53**	92.58
Breast	CPUNN	97.71	**98.28**	**98.71**	98.57	**98.86**
	GPUNN	**97.85**	98.14	97.00	**99.28**	98.00
Glass	CPUNN	94.86	95.33	**93.46**	91.59	97.20
	GPUNN	**95.79**	**97.20**	**93.46**	**98.60**	**99.07**
Iris	CPUNN	**97.33**	98.67	**98.67**	**98.67**	**97.33**
	GPUNN	94.67	**99.33**	97.33	96.00	**97.33**

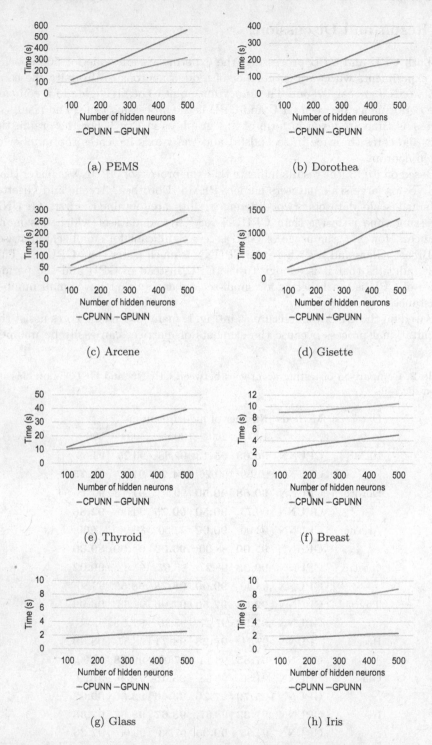

Fig. 3. Comparison results of processing time between CPUNN and GPUNN on classification datasets

into the cores. The usage of GPU is really effective for large-scale datasets because all the values can be mapped across the many number of cores. Hence, the calculation can be executed in parallel. The parallelism of GPU has speed up the process and reduced the computational time. However, for small-scale datasets, the use of CPU to perform the calculation is more than enough since there is no need to consider the transferring time between GPU and CPU. In fact, the processing time of GPU on small-scale datasets are quite high because of the transferring time encountered between CPU and GPU. The small-scale datasets which can be processed on CPU alone will give much lowest processing time because the calculation is done directly on CPU which mean that there is no need to consider any transferring time. In order to validate the classification performance, Table 2 shows the testing accuracy for CPUNN and GPUNN.

Based on Table 2, the accuracies of GPUNN and CPUNN on test datasets are quite similar. Occasionally, GPUNN produced higher accuracy than CPUNN and vice versa. Performance wise, in term of accuracy, both methods can achieve highest accuracy for all kinds of datasets. However, in term of processing time, GPUNN is faster than CPUNN for large-scale datasets.

5   Conclusion

The purpose of this paper is to investigate the performance of CPU and GPU in training NN Therefore, the difference between processing of NN on CPU and GPU in terms of architecture and implementation are discussed. Two different kind of classification datasets namely large-scale and small scale are chosen for testing the capability of both CPU and GPU. In order to assist the reader to understand the process involved, the implementation of NN on GPU compared to CPU is also given and explained. The results indicate that the performance of GPUNN on large-scale datasets is much better than CPUNN where the processing time is drastically reduced. For small-scale datasets, it is enough to use CPU instead of GPU because the processing time of CPUNN is much better. However, if the small dataset consists of large number of instances, using GPU is recommended. The purpose of processing using GPU is to reduce the processing time of CPU when handling the complex data. The experiment results indicate that the processing time of NN has been reduced by using GPU. When the data become more complex, the NN needs more hidden unit and it will increase the complexity of NN's structure and force CPU to takes more time to train NN. However, in GPU, the more hidden layers involved the more effective performance produced. Or in other words, the parallelism concept in GPU is suitable for training complex data and it can improve the processing time of CPU. The extensive experiments of the implementation of NN on GPU includes experiments on various domains with various scale of datasets is proposed as the future works.

Acknowledgment. This study is supported by the Fundamental Research Grant Scheme (FRGS vots: 4F738 & 4F550) that sponsored by Ministry of Higher Education (MOHE). Authors would like to thank Research Management Centre (RMC)

Universiti Teknologi Malaysia, for the research activities and Soft Computing Research Group (SCRG) for the support and motivation in making this study a success.

References

1. Almási, A.D., Woźniak, S., Cristea, V., Leblebici, Y., Engbersen, T.: Review of advances in neural networks: neural design technology stack. Neurocomputing **174**, 31–41 (2016)
2. Boger, Z., Guterman, H.: Knowledge extraction from artificial neural network models. In: IEEE International Conference on Systems, Man, and Cybernetics, Computational Cybernetics and Simulation, vol. 4, pp. 3030–3035. IEEE (1997)
3. Brito, R., Fong, S., Cho, K., Song, W., Wong, R., Mohammed, S., Fiaidhi, J.: GPU-enabled back-propagation artificial neural network for digit recognition in parallel. J. Supercomput., 1–19 (2016). doi:10.1007/s11227-016-1633-y
4. Brodtkorb, A.R., Hagen, T.R., Sætra, M.L.: Graphics processing unit (GPU) programming strategies and trends in GPU computing. J. Parallel Distrib. Comput. **73**(1), 4–13 (2013)
5. Bustos, B., Deussen, O., Hiller, S., Keim, D.: A graphics hardware accelerated algorithm for nearest neighbor search. In: Alexandrov, V.N., van Albada, G.D., Sloot, P.M.A., Dongarra, J. (eds.) ICCS 2006. LNCS, vol. 3994, pp. 196–199. Springer, Heidelberg (2006)
6. DeléVacq, A., Delisle, P., Gravel, M., Krajecki, M.: Parallel ant colony optimization on graphics processing units. J. Parallel Distrib. Comput. **73**(1), 52–61 (2013)
7. Lee, V.W., Kim, C., Chhugani, J., Deisher, M., Kim, D., Nguyen, A.D., Satish, N., Smelyanskiy, M., Chennupaty, S., Hammarlund, P., et al.: Debunking the 100x GPU vs. CPU myth: an evaluation of throughput computing on CPU and GPU. In: ACM SIGARCH Computer Architecture News, vol. 38, pp. 451–460. ACM (2010)
8. Lichman, M.: UCI machine learning repository (2013). http://archive.ics.uci.edu/ml
9. Luebke, D.: CUDA: scalable parallel programming for high-performance scientific computing. In: 5th IEEE International Symposium on Biomedical Imaging: From Nano to Macro, ISBI 2008, pp. 836–838. IEEE (2008)
10. Oh, K.S., Jung, K.: GPU implementation of neural networks. Pattern Recogn. **37**(6), 1311–1314 (2004)
11. Owens, J.D., Houston, M., Luebke, D., Green, S., Stone, J.E., Phillips, J.C.: GPU computing. Proc. IEEE **96**(5), 879–899 (2008)
12. Tarsa, S.J., Lin, T.H., Kung, H.: Performance gains in conjugate gradient computation with linearly connected GPU multiprocessors. In: USENIX HotPar 12 (2012)
13. Thrun, S.B.: Extracting symbolic knowledge from artificial neural networks. Revised Version of Technical research report TR-IAI-93-5, Institut für Informatik III-Universität Bonn (1994)

Impact of Smart and Intelligent Technology on Education

Ontology-Based Education Support System for Solving Emergency Incidents

Martina Husáková[✉]

Faculty of Informatics and Management, University of Hradec Králové,
Hradec Králové, Czech Republic
martina.husakova.2@uhk.cz

Abstract. Crisis situations influencing human lives require fast and faultless solutions. Decision making during emergency incidents is burdened with high uncertainty, complexity and risks. These situations can be effectively solved on the basis of complex knowledge that is received during education, training and experience. Non-professional rescuers often have a problem to decide which steps in which order have to be done in case of injured persons. Education of non-professionals is crucial and inevitable because of saving human lives. The main aim of the paper is to present the OWL ontology-based education support prototype using SWRL rules suggesting suitable solutions for particular emergency incident.

Keywords: Emergency incident · Education · OWL ontology · SWRL

1 Introduction

Emergency situations require fast and faultless solutions. Rescuers deal with various events occurring suddenly and unpredictably during saving human lives. They often have incomplete information about the actual situation. The sub-optimal solutions are the outputs of their activities in most cases. Efficient decision-making, often under time limits and stressful situations, is inevitable for ensuring quality of life. Emergency events can be effectively solved with the intensive theoretical education and training for receiving valuable experience. Fundamental knowledge of strategies for sustaining basic life functions is necessary also for non-professional rescuers. Ontologies are knowledge-based schemata playing an important role in modelling context and complex knowledge. A lot of applications prove that they can be applied as knowledge bases for solving emergency incidents requiring complex knowledge. The paper continues in the research that has been introduced in [14] where the framework for development of the ontology-based decision support system for solving emergency incidents is presented. This conceptual framework is practically used for building of the OWL ontology-based prototype supporting education in the first aid application domain. The paper is structured as follows. The Sect. 2 introduces the ontologies in the context of education and emergency. The Sect. 3 deeply investigates development of the ontology-based education support system for solving emergency situations. Future directions are mentioned in the Sect. 4 and the Sect. 5 concludes the paper.

© Springer International Publishing Switzerland 2016
N.T. Nguyen et al. (Eds.): ICCCI 2016, Part II, LNAI 9876, pp. 313–322, 2016.
DOI: 10.1007/978-3-319-45246-3_30

2   Ontologies for Education of Solving Emergency Incidents

Ontologies have the origin in philosophy (metaphysic) where they are interested in human being and answering the elementary questions related to living and non-living things existing in our world. At present, ontologies are more frequently mentioned in the context of ontological engineering as formal and explicit specification of shared conceptualisation [3].

A lot of case studies and software solutions prove that ontologies are useful for educational purposes. The ontology is able to visualise knowledge in graph-based format that can be helpful for a learner if his (her) learning style is based on visual memory. Various educational ontologies (focused on physics, maths or marksmanship) are mentioned e.g. in [4]. The OWL ontology is developed for education of java programming language and presented in [5]. Ontologies can also be used for structuring content of educational courses [6, 7] or for building e-learning web portals [8]. Ontologies are not often cited in the context of emergency education, but they are frequently applied in emergency management where the ontology-based applications can also be used for educational purposes. Emergency management is a multi-disciplinary area applying science, decision-making, planning and technology for reducing possible emergency events and minimisation of losses. The paper [9] focuses on a design of general schema for decision support in response operations during biological (biochemical) incidents. OWL ontology-based conceptual framework is proposed for improving shared situation awareness among teams of rescuers in case of emergency incidents in [10]. Mass evacuation during tsunami event is the case study for framework demonstration. Han, Y. and Xu, W. propose the framework-based decision support system used for solving crisis situations. Framework combines four types of ontologies for provision of the final solution [11]. Practical usage of the framework is demonstrated with the Tianjin Port Explosion case study. The paper [12] proposes ontology-based solution for resolving semantic heterogeneity of data in case of emergency responses. The authors of the paper explain how to build the ontology for the needs of the emergency management.

3   Development of Ontology-Based Prototype

Systematic approach is inevitable for building ontologies integrating complex knowledge. The approach is based on the six phases for building of the ontology-based education support prototype. These phases are the following: knowledge acquisition, concept mapping, knowledge base development including building of an ontology, instance data implementation together with integration of rules, and testing of the software solution [14]. These phases are deeply explained in the following sub-sections.

3.1   Knowledge Acquisition

Knowledge acquisition phase is focused on the selection of the most important facts and knowledge of the application domain. It has to be obvious which part of the reality is going to be represented by the ontology. Provision of the first aid consists of many interrelated activities. Non-professional rescuers have a problem to decide about the order of rescue operations. C-ABCDE algorithm offers concrete steps ordered according to the priority with the aim to increase the chance for survival [1]. Concepts of this algorithm are modelled with the OWL ontology for education of the first aid and support of decision-making during emergency incidents. The algorithm consists of six phases in the extended version. The first phase (C – Catastrophic Haemorrhage Control) is focused on the control of the haemorrhage. Stopping of bleeding has the highest priority because loss of blood can cause a shock or a death. If the blood stream is under control, it is necessary to check the state of the air passages (A – Airway). If the air passages are free, but injured person cannot breath sufficiently, the artificial respiration is applied (B – Breathing). The fourth phase is focused on monitoring of blood circulation. Blood circulation is evaluated on the combination of consciousness, breath and hearthbeat. If these characteristics are not present, the indirect hearth massage is applied (C – Circulation). The fifth phase (D – Disability) is aimed at the checking the level of consciousness. It is obvious that this characteristic is often monitored during each of the previous phase. The final phase called Environment (E) is the secondary examination where the global examination is applied. The SAMPLE history system is used for this purpose where symptoms, allergies, medication, past medical history, last oral intakes or events leading up to the illness or injury are investigated.

3.2   Concept Mapping

Concept mapping is a technique for non-formal knowledge modelling and knowledge visualisation with the usage of graph-based structures – concept maps [2]. Concept maps are applied for the analysis of the application domain - provision of the first aid. Four concept maps are developed. Each one represents one central topic and related concepts of the C-ABCDE algorithm, i.e. *Heamorrhage, Breathing, Blood circulation* and *Consciousness*. Concept maps cover the first five phases of the C-ABCDE algorithm (C-ABCD). Interconnections between these four concept maps are realised only in

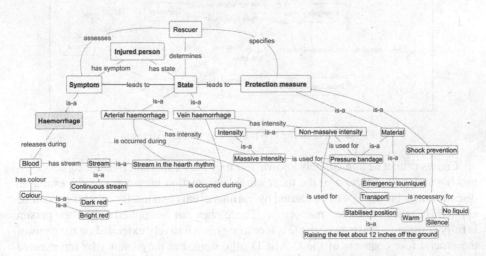

Fig. 1.   Concept map representing knowledge about haemorrhage

human mind. One global concept map integrating these four partial maps is not easily readable. VUE editor is used for concept mapping. The example of the concept map modelling *Haemorrhage* concept is depicted in the Fig. 1 (see grey rectangle). The general core of the all concept maps is identified during concept mapping (see bold underlined concept *Injured person*, *Symptom*, *State* and *Protection measure* in the Fig. 1). These concepts are used as a basis for the OWL ontology development in a modified form.

3.3   Ontology Development

OWL ontology development is based on concept maps where concepts correspond to OWL classes (or individuals) and relations correspond to object properties. Core of the concept maps is the core of the OWL ontology with minor modifications, see the Fig. 2. The state (the OWL class *State*) of the injured person (the OWL class *Emergency case*) is judged according to the symptoms (the OWL class *Symptom*) (e.g. colour of blood or skin, stream of blood, intensity of reactions on the stimuli, etc.). The state is evaluated and the suitable protection measures (the OWL class *Protection measure*) are applied. Core of the OWL ontology is extended with the OWL class representing a cause of symptoms or states (the OWL class *Cause*), see the Fig. 2.

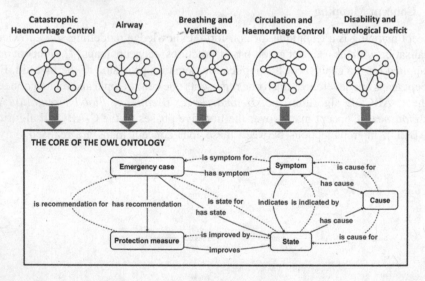

Fig. 2.   Core of the OWL ontology and its extension

Causes cannot be immediately known, but they play important role during decision-making about the selection of the most suitable protection measures. As the example, discontinuous breathing can be caused by various events and not for all of them can be applied Heimlich or Gordon manoeuvre. These ones can be applied only if the person is fully conscious. The core of the OWL ontology is iteratively extended on the basis of the crucial four concepts of the C-ABCD algorithm, but they cannot be represented without the relations with each other. This tendency appeared in the preliminary stages of the OWL ontology modelling.

Protégé modelling tool is the most cited environment used in ontological engineering. Protégé 4.3.0 (build 304) is used in case of the conceptualisation of the C-ABCD algorithm. Protégé 4.x offers the OWL 2 for knowledge modelling (in comparison to Protégé 3.x) that is based on the real applications, use cases and experience of users. OWL 2 extends the OWL 1 in the view of the more efficient processing in reasoners and provides new "semantics constructs", for more details see [13]. Protégé 4.3.0 is stable and provides sufficient possibilities for visualisation knowledge structures (OWLViz, OntoViz), supports inference with the open source reasoner HermiT and FACT++. Structure of the normalised OWL ontology visualised with the OntoGraf plugin is depicted in the Fig. 3.

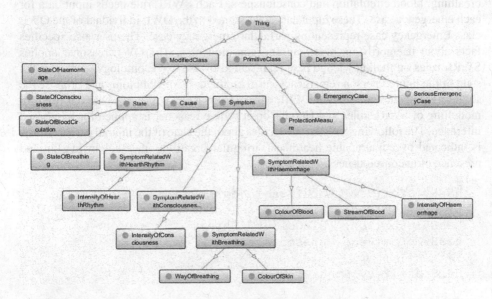

Fig. 3. OWL ontology of the first aid visualised with the OntoGraf plugin (Protégé 4.3.0)

Structure of the OWL ontology is extended with the OWL individuals representing particular symptoms, states, causes and protection measures. The statistics of the ontology is the following:

- number of OWL classes: 32,
- number of OWL object properties: 12 (including inverse object properties),
- number of OWL individuals: 41.

3.4 Extension of Ontology with SWRL Rules

OWL ontology of the first aid integrates the most important concepts and relations of the C-ABCD algorithm in a well arranged way, but this is not sufficient in the view of the intended aim. The OWL ontology should recommend the suitable combination of protection measures for solving emergency events, but this cannot be done only by the ontology itself. Rules are able to use and combine "ontological knowledge" for making implicit knowledge visible and offer the solution(s) for particular situation. SWRL (Semantic Web Rule Language) is a rules-based language based on the Datalog, the RuleML and the OWL DL (Lite) languages. SWRL rules of the two types are proposed for the OWL ontology of the first aid:

- SWRL rules facilitating to decide about the state of the emergency incident,
- SWRL rules facilitating to select suitable protection measures.

Systematic approach is not only applied for the OWL ontology development, but it is also used for design of SWRL rules. SWRL rules are proposed with respect to the four concepts of the C-ABCD algorithm, i.e. SWRL rules examine haemorrhage, breathing, blood circulation and consciousness. Each SWRL rule needs input data for each emergency case. These input data are assigned to the OWL individual of the OWL class Emergency case representing particular emergency case. The user also specifies facts about haemorrhage, breathing and consciousness. The OWL reasoner applies SWRL rules on the knowledge base represented by the OWL ontology, evaluates the state of the injured person (emergency case) on the basis of symptoms and propose the solution (protection measures). SWRL editor (View Rules) of Protégé 4.3.0 is used for modelling of SWRL rules and HermiT open source reasoner is applied for rule-based inference. The following SWRL rule decides about the state of the injured person. Stasis is indicated by impalmpable hearthbeat, irregular breathing, pale colour of skin and presence of unconsciousness.

```
hasSymptom(?ec, ImpalmpableHearthBeat),
hasSymptom(?ec, IrregularBreathing),
hasSymptom(?ec, PaleColourOfSkin),
hasSymptom(?ec, Unresponsive)
->
hasState(?ec, Stasis)
```

The second example demonstrates SWRL rule using the information about the state of the injured person and proposes the protection measures. Calling the emergency has to be the first step during the first aid provision. The next steps should be done after that, e.g. usage of the AED (automated external defibrillator) or application of hearth massage.

```
hasState(?ec, Stasis),
hasSymptom(?ec, IrregularBreathing)
->
hasRecommendation(?ec, Emergency),
hasRecommendation(?ec, AutomatedExternalDefibrillator),
hasRecommendation(?ec, HearthMassage)
```

3.5   Testing of Prototype

The OWL ontology-based prototype uses Protégé 4.3.0 as a "testbed" for design of the first version of the knowledge-based system facilitating decision-making during emergency incidents. The actual version of the OWL ontology is extended by 19 SWRL rules. The functionality and correctness of the inference process is tested with the HermiT – open source OWL reasoner based on the "hypertableau" calculus.

The following two scenarios demonstrate correctness of the inference by the HermiT. The first scenario solves the emergency case (Case01-NonSeriousHaemorrhage) where the five symptoms are investigated: dark red colour of blood, non-massive intensity of haemorrhage, alertness (according to the AVPU scale for evaluation of the consciousness level), normal breathing and continuous stream of blood. Definable class *Serious-EmergencyCase* helps to identify which emergency case is serious on the basis on the following definition:

```
EmergencyCase and
    hasState value DisorderOfBreathing or
    hasState value Stasis or
    hasState value Unconsciousness or
    hasSymptom value MassiveIntensityOfHaemorrhage
```

The output of the inference process by HermiT reasoner is depicted in the Fig. 4. HermiT reasoner correctly infers that vein haemorrhage is obvious according to the symptoms. Consciousness and normal breathing together with vein haemorrhage lead to the application of recovery position and usage of the pressure bandage, see the Fig. 4.

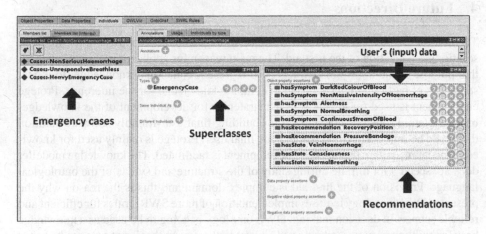

Fig. 4.   Test scenario 01: emergency case with non-serious haemorrhage

The second scenario solves the emergency case (Case02-UnresponsiveBreathless) where the two serious symptoms are investigated: injured person without the ability to response on stimuli and breathless. The output of the inference process by HermiT reasoner is depicted in the Fig. 5.

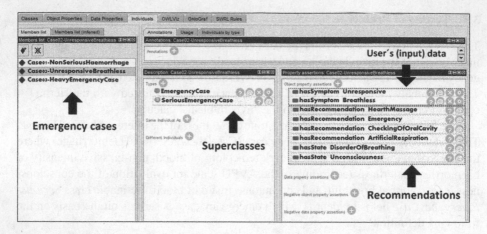

Fig. 5. Test scenario 02: emergency case with unresponsivity and breathless

HermiT reasoner correctly infers that this emergency case is serious according to the definition of the OWL class SeriousEmergencyCase. Disorder during breathing and unconsciousness is evident because of the breathless symptom and unresponsivity (according to the AVPU scale for evaluation of the consciousness level). Calling of the emergency is self-evident. If there is some problem during breathing, checking of the oral cavity has to be done. Artificial respiration and hearth massage have to be applied together until the emergency will arrive.

4   Future Directions

The main aim of the research is to investigate strategies for development of the software solution for education of the first aid. The main reason for application of ontologies for this purpose is the ability of ontologies to represent contextual knowledge, to model semantics of particular concepts and to combine knowledge for the inference. Protégé environment is used for testing selected strategies for development of the knowledge-based system. Protégé is not suitable for building final user-friendly software solution that could be immediately available to the final user. Protégé is mainly used for knowl-edge modelling where the ontology development is facilitated. The knowledge modeller does not spend a lot of time during study of the structure and syntax of the ontological language. Provision of the first aid is complex domain and this is the reason why the presented OWL ontology requires implementation of more SWRL rules for efficient and reliable support in decision making. OWL ontology is going to be used as a knowledge base for building of more sophisticated educational software that is not going to be based on the usage of Protégé environment. There are various semantic frameworks that facil-itate building of the semantic web-based applications. We can mention e.g. OWL API, Jena, OWLReady, CubicWeb or Sesame. Two semantic web-based frameworks are taken into account on the basis of the realised investigation of possible solutions. OWL API is mainly used for building and managing OWL ontologies. Jena is the RDF-based framework, but the usage of the OWL syntax is also possible. These two solutions are going to be investigated more deeply for development of the semantic web-based application used for the educational purposes covering provision of the first aid.

5   Conclusion

The paper investigates strategies for development of the education support system where knowledge related with the provision of the first aid is modelled in Protégé environment. SWRL rules are applied for recommendation of the suitable combination of protection measures according to symptoms and states of the particular emergency case. The OWL ontology with SWRL rules is useful approach for intended educational application, but the more sophisticated software solution has to be developed because Protégé environment is not suitable for building the final semantics-based system. Two semantic web-based frameworks are taken into account for building more sophisticated educational tool – OWL API and Jena. Necessity to build this system is evident, because everybody should be able to know what to do if someone encounters with sudden health incident.

Acknowledgements.   The support of the Specific research project at FIM UHK is gratefully acknowledged.

References

1. Thim, T., et al.: Initial assessment and treatment with the Airway, Breathing, Circulation, Disability, Exposure (ABCDE) approach. Int. J. Gen. Med. **5**, 117–121 (2012)
2. Novak, J.: Concept maps and Vee diagrams: two metacognitive tools to facilitate meaningful learning. Instr. Sci. **19**, 29–52 (1990)
3. Borst, W. N.: Construction of engineering ontologies for knowledge sharing and reuse. PhD thesis, Universiteit Twente (1997). http://doc.utwente.nl/17864/
4. Baker, E.L.: Ontology-Based Educational Design: Seeing is Believing: Resource Paper No. 13. The Regents of the University of California. (2012). https://www.cse.ucla.edu/products/resource/cresst_resource13.pdf
5. John, S.: Development of an educational ontologz for java programming (JLEO) with a hybrid methodology derived from conventional software engineering process models. Int. J. Inform. Educ. Technol. **4**(4), 308–312 (2014)
6. Boyce, S., Pahl, C.: Developing domain ontologies for course content. Educ. Technol. Soc. **10**(3), 275–288 (2007)
7. Ke, Z.: Engineering education knowledge management based on Topic Maps. World Trans. Eng. Technol. Educ. **11**(4), 376–381 (2013)
8. Olševičová, K.: Topic maps e-learning portal development. Electron. J. e-Learning **4**(1), 59–66 (2006)
9. Cech, P., et al.: Ontological models and expert systems in decision support of emergency situations. Mil. Med. Sci. Lett. **80**, 21–27 (2011)
10. Javed, Y., Norris, T., Johnston, D.: Ontology-based inference to enhance team situation awareness in emergency management. In: Proceedings of the 8th International ISCRAM Conference, Lisbon, Portugal, p. 9 (2011)

11. Han, Y., Xu, W.: An ontology-oriented decision support system for emergency management based on information fusion. In: Proceedings of the 8th ACM SIGSPATIAL International Workshop on Location-Based Social Networks, Bellevue, USA, p. 8. (2015)
12. Fan, Z., Zlatanova, S.: Exploring ontologies for semantic interoperability of data in emergency response. Appl. Geomatics 3(2), 109–122 (2011)
13. W3C OWL Working Group: OWL2 Web Ontology Language Document Overview (Second Edition), W3C Recommendation (2012). https://www.w3.org/TR/owl2-overview/
14. Husáková, M.: Systematic development of ontology-based decision support system for solving emergency incidents. In: Proceedings of the 27th International Business Information Management Association Conference (IBIMA), Milan, Italy, 152–156 (2016)

Innovations in Software Engineering Subjects

Petra Poulova[✉] and Ivana Simonova

Faculty of Informatics and Management, University of Hradec Králové,
Rokitanského 62, 500 03 Hradec Králové, Czech Republic
{Petra.Poulova,Ivana.Simonova}@uhk.cz

Abstract. The topic of software engineering has been emphasized in the curriculum of the Faculty of Informatics and Management, University of Hradec Kralove, thus reflecting the call for new knowledge and skills required from the graduates. The quality of highly professional training programme was ensured by the cooperation with IT companies so as to provide students with theoretical and practical professional skills and prepare them for new positions recently created on the labour market.

Keywords: Software engineering · Programing · Modeling · e-Learning · LMS · Efficiency · Success rate · Visit rate · Tracking

1 Introduction

The success on the labour market is one of the assessment criteria which reflect the quality of the educational institution. In case of the Faculty of Informatics and Management, University of Hradec Kralove (FIM), the integration of graduates on the national and international labour markets means to equip them with the latest theoretical and applied knowledge in the software engineering area. Therefore, study programmes are tailored to the requirements of specialists from the practical environment to teach students in accord with needs of the labour market.

Reflecting the proposals of practitioners, the curricula of study programmes are innovated to have a modular structure and appropriate professional competences were developed with graduates to be easily employable. The proposal of a new design is assessed by the HIT Cluster (Hradec IT Cluster), i.e. by the society of important IT companies in the region, and by the University Study Programme Board. The main objective is to create a transparent set of multidisciplinary courses, seminars and online practical exercises which can provide students the opportunity to gain both theoretical knowledge and practical skills, and to develop the key competences.

Subjects in each study programme are structured into five groups – data engineering, computer networks, software engineering, enterprise informatics and knowledge technologies.

2 Innovations in the Group of Software Engineering Subjects

As listed above, students of the Computer Science study programmes at FIM are expected to have rather wide knowledge all five groups of subjects, including the field

N.T. Nguyen et al. (Eds.): ICCCI 2016, Part II, LNAI 9876, pp. 323–332, 2016.
DOI: 10.1007/978-3-319-45246-3_31

of software engineering. Most software development companies expect the graduates to have both good theoretical knowledge and practical experience in this field, which is the reason why changes have been made in the curricula and new knowledge required from the FIM graduates.

The software engineering courses were embraced in the curricula of bachelor and master study programmes of Information Management and Applied Informatics more than 25 years ago.

Following subjects have been included in the group of software engineering:

- Algorithms and Data Structures,
- Programming I, II, III,
- Introduction to Object Modeling,
- System Programming,
- Advanced Programming,
- Computer Graphics I, II,
- Mobile Technologies
- Optimization of Web Applications,
- Theoretical Informatics.

This structure reflected the proposal designed in co-operation with companies AG COM, ORTEX, GIST, FG Forrest and DERS.

2.1 Company Requirements on Graduates

The expert group consisting of 11 members of the Hradec IT Cluster set new competences required from FIM graduates. The experts assessed the original state and designed a new concept of subjects; the instruction of subjects was supported by online courses in the Learning Management System (LMS) Blackboard. All the study materials were created and uploaded there. Each online course was equipped with didactic instructions, study materials, tests and communication tools [1–4].

The expert team was headed by the vice-dean for study affair. They defined 20 roles of IT specialists (e.g. Analyst, Business Analyst, De-signer, Flash Developer, Internet Marketer PPC/SEO, Programmer, IT Manager, IT Sales, IT Administrator etc.); for each role the description of the working position was defined and consequent requirements on the position described. The expert team detected following competences to be required from graduates:

- problem analysis, i.e. to detect the width and depth of the problem, consider it in consequences, to apply the abstraction, or decomposition;
- system approach, i.e. to have a general view of all aspects of the problem, to be able to identify them and respect their mutual consequences and within the whole project;
- orientation towards output, i.e. to prefer the output to the process, to reach the objective/s and apply the active approach;
- effective communication, i.e. to express thoughts explicitly and timely, to listen and comprehend, to interact in an appropriate way;

- co-operation and adaptability, i.e. to work in team, to bear responsibility, accept changes and others' opinions, cope with stress, negative states, accept other's critical evaluation, present own criticism in an appropriate manner, to have active approach to lifelong/further education.

Further on, each graduate should have:

- general overview in the IT field, trends and latest ICT development;
- good knowledge of office software;
- good knowledge of technical (professional) English, particularly IT texts;
- good knowledge of Czech language;
- willingness to learn;
- flexibility;
- ability to work in team, to work autonomously;
- analytic thinking;
- creativeness, invention, innovation;
- presentation skills and good knowledge of presentation tools;
- communication skills;
- sharing and providing own knowledge;
- responsibility towards own work, consistency and an active approach;
- ability to analyze problems and generalize requirements.

Particularly in the field of software development experts emphasize:

1. analytic and system thinking;
2. knowledge of the UML notation for reading and modeling (e.g. in Enterprise Architect);
3. knowledge of OOP and Java language (knowledge of other languages is appreciated – Scala, Groovy, Ruby, Python, JavaFX);
4. Design Patterns (on the GOF level), MVC, AOP;
5. Java/J2EE-based technologies – Servlets, JSP, JDBC, Log4j, ORM (Hibernate/iBatis),
6. web frameworks (Spring MVC/Stripes/JSF/others), Spring Framework etc.;
7. SQL, relational database – MySQL, PostgreSQL, Oracle, or others;
8. appropriate knowledge of web technologies – HTML, CSS, JavaScript,ASP.NET;
9. development techniques and tools (e.g. issue tracker, Ant/Maven 2, integration server, source repository CVS/SVN/GIT);
10. knowledge of C++, platform Linux (Solaris, AIX are appreciated);
11. high level of knowledge of OOP principles; experience in application, we and database servers is appreciated; knowledge of MS.NET platform and C# language;
12. knowledge of architecture and problems relating to web applications;
13. knowledge of principles of secure application development;
14. knowledge of mathematics, algorithmization and the applications;
15. knowledge of technical English;
16. creative thinking, graphic feeling, ability to create appropriate design for required functionality

17. focus on appropriate, functional but innovative design, accent on user friendliness and intuitiveness of the design;
18. basic knowledge of typography principles;
19. and last but not least, knowledge of methodology of testing applications – the design and preparation of testing scenarios and testing environments.

2.2 Innovative Contents in Single Subjects

As listed below, following learning objectives and innovative design and contents were included.

The main objective of the '*Algorithms and Data Structure*' subject is to develop logical and algorithmic thinking and autonomously design the algorithms. The semestral project was added to the syllabus, when students choose the topic/problem by their own and introduce the algorithmic solution.

Reflecting the outputs within the HIT Cluster discussions, the learning content of subjects of '*Programming I, II, III*' was differentiated for students of Applied Informatics (AI) study programme and Information Management (IM). Whereas these subject for AI students will provide them deeper knowledge in Computer Science, Information techno-logy ACM, algorithmization and data structures, the IM students will undergo basic courses of programming only (Algorithmization and Introduction to programming).

As resulted from the HIT Cluster discussions, the knowledge of C language was added to the subject of '*System Programming*'. Therefore, excluding introductory lessons, practical activities are running in the operation system GNU/Linux in C language.

The subject of '*Advanced Programming*' was supported by virtual desktop. Reflecting the specific requirements of this subject the so called personalized desktops were prepared for students, i.e. they have more source and data space available.

Within practical lessons of '*Computer Graphics*' the object approach was emphasized, as well as another topical paradigms of the software design (inversion of control – dependency injection). It means with students the emphasis was paid on the development of analytic and system thinking, application of mathematics knowledge and deepening skills to design, implement, test and present solution of a particular problem. Within the autonomous semestral project both the listed know-ledge and skills are required, as well as to write a journal article on the topic.

The innovations within the subject of '*Optimization of Web Applications*' reflected fast development of this field, particularly according to recommendations of ACM curricula for software engineering and information technologies.

Large innovations were applied within '*Mobile Technologies*' which reflected the latest development in the field. Topics of selected operation systems were omitted (Symbian, Palm OS), and other parts were strengthened, e.g. OS Android, mobile OS Microsoft (Windows Phone 7), updated parts on RIM Blackberry etc. Moreover, the exploitation of new cloud services was implemented in the subject.

The subject of '*Theoretical Informatics*' covers the field of automats and formal languages and complexity theory. Reflecting the labour market requirements, the emphasis is paid on analytic thinking and autonomous problem solution. Therefore, the semestral project was added to this subject, where students focus on a practical problem.

2.3 e-Learning Support to the Subjects

The whole programme is supported by e-learning lessons. All the study materials are created and uploaded in the FIM LMS BlackBoard [5–7].

Each module in the e-learning course is equipped with its own guiding instructions, study materials and modified tests with comprehensive sets of questions. Study materials were developed not only as standard presentations; they contain audio-visual materials, animations and instructional video-recordings. Students' attendance in the courses and lectures is monitored in order to evaluate not only the results of the individual tests but also to detect the fields of students' interest.

Since smart phones were launched, the mobile market has rapidly changed. As a result, the field of education is concerned with delivery of knowledge through smart devices (Smart Education). Smart Education mechanism can be seen as an integrated educational environment in which cooperative, interactive, participative, sharing, and intelligent learning are available through new forms of teaching–learning content, environment, and ICT.

2.4 Expert Evaluation of Subjects

The evaluation of e-learning software engineering subjects was provided by the HIT Cluster, particularly companies ORTEX, GIST, GMC and DERS.

Experts had both the complex information available on innovated subjects and access to e-subjects within the LMS BlackBoard. For the evaluation the questionnaire consisting of 25 statements was applied. Their dis/agreement was expressed on the four-level Likert scale (strongly agree – agree – disagree – strongly disagree), and they could add any comments in the form of open answer. Totally 17 evaluation reports were provided for the group of software engineering subjects. The results can be summarized as follows:

- Learning objectives were clearly defined through measurable outputs. They reflect the labour market requirements on graduates' competences (knowledge and skills) in the field of computer networks; they are available to students on several places within the e-subjects.
- The learning content reflects the pre-defined learning objectives; it is structured into appropriate modules and presented in logical order. Multimedia elements are properly implemented into single parts and chapters. Students are expected deeper understanding of the learning content which is explained via practical examples/models and best practices.
- Each e-subject contains the didactic instruction on how to study the subject effectively.
- Clear assessment criteria are available to students from sample semestral projects and each testing relates to the pre-defined learning objective.

- Various methods of testing knowledge are applied (questions of different types, projects, discussions etc.) and students have self-evaluation tools available providing immediate feedback on their knowledge.

Experts also provided oral evaluation. Selected samples are presented below:

'This group of subjects aims at "mastering the profession" to some extent, i.e. the content should be more practically and less theoretically focused, as it really is. Nevertheless, the amount of subjects and their content covers all common disciplines (algorithms, programming, graphics, web), which are necessary for practice.'

ORTEX

'Innovations reflect our requirements. Even before, the subjects focused on education for practice. Through current innovations the usability of developed knowledge and skills was increased. The learning contents of single subjects cover the methods and procedures of software production to larger extent and are much closer to activities of our company; which enables graduates to faster adapt to conditions in the IT company. The highly positive for us is the fact that requirements have been differentiated for AI and IM students; this will help better prepare the graduates in both study programmes. Subjects are logically placed in single years of study.'

GIST

3   Success Rate in the Subject Versus Visit Rate to the Course

The research question was whether there exists any correlation between students' frequency of the visits to the course and their success in the course (i.e. meeting the requirements and passing the course) In other words, are students more successful, if they more frequently access the course? Does the learner's low frequency of visits to the course correlate to poor study results, i.e. to failing the course?

All courses were taught for 13 weeks, and as mentioned above, all subjects were supported by online courses in LMS Blackboard which pro-vides tracking services, so it was easy to collect data on students' performance in the courses.

The research process started in the online course *Programming I* (PRO I) where the methodological guide containing didactic instructions was available to students, study materials, assignments settings, tests, etc. Totally 186 students enrolled in this subject; 116 of them successfully passed the course (Fig. 1).

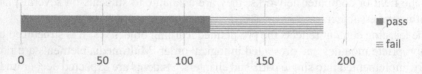

Fig. 1.   Students' assessment after the course of Programming I (n)

The correlation of getting the credit and the visit rate to the PRO I course is displayed in Fig. 2. The visit rate oscillated from zero to 77; the mean visit frequency was 13.93; median 9. These data show that (in average) students accessed the course once per week, which demonstrates not very intensive use.

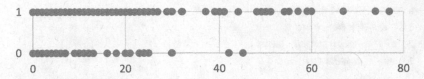

Fig. 2. Correlation between the success rate and visit rate to the course of Programming I (1 = pass the course, 0 = fail the course)

The received results prove the certain correlation was detected: whereas students who successfully passed the course, accessed the e-courses 19× in average, the unsuccessful students who failed, reached significantly lower frequency – their mean visit rate was 5.7×. Similarly to this, the identical correlation was detected in medians (Table 1, line 4).

Table 1. Correlation between the success rate and visit rate to the course of Programming I

	Passed	Failed
Mean	19.0	5.7
Min	0	0
Max	77	45
Median	16	0

The research process continued in the online course *Computer Graphics*. Totally 106 students enrolled in this subject; 61 of them successfully passed the course. Complete evaluation reflecting the ECTS is displayed in Fig. 3.

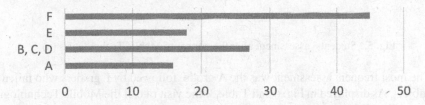

Fig. 3. Students' assessment after the course of Computer graphics (n)

Most students failed (F) the subject. Approximately one quarter of them got B, C or D grade, a few ones were excellent (A), or satisfactory (E). The frequency of the visit rate to the course was different. As displayed in Fig. 4 and Table 2, the visit rate to the course oscillated from zero to 753; the mean visit frequency was 132; median 123. These data show that during the semester period of 13 weeks students accessed the course ten times per week, which also demonstrates highly intensive use.

These results did not prove direct correlation; whereas the mean visit rate of students having A (Excellent) was 172, the unsuccessful students who failed (F), reached three times lower frequency (58×). Higher correlation was detected in medians (five times, see Table 2, line 4).

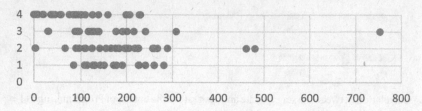

Fig. 4. Correlation between the success rate and visit rate to the course of Computer Graphics (1 = A, 2 = B, C, D, 3 = E, 4 = F)

Table 2. Correlation between the success rate and visit rate to the course of Computer Graphics

	A	B, C, D	E	F
Mean	172.4	19.6	185.9	57,9
Min	84	3	29	0
Max	281	480	753	230
Median	161.5	196	137	38

The final phase of the research was conducted in the online course *Mobile Technologies.* Totally 26 students enrolled in this subject; 18 of them successfully passed the course. Complete evaluation is displayed in Fig. 5.

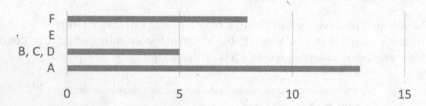

Fig. 5. Students' assessment after the course of Mobile Technologies (n)

The most frequent assessment was the A grade, followed by F graders who failed in the subject. As displayed in Fig. 6 and Table 3, the visit rate of the Mobile Technologies course oscillated from zero to 40; the mean visit frequency was 14.3; median 11.5. These data show that (in average) students accessed the course, which was also taught for 13 weeks, once per week.

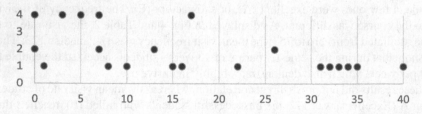

Fig. 6. Correlation between the success rate and visit rate to the course of Mobile Technologies (1 = A, 2 = B, C, D, 3 = E, 4 = F)

The received results prove the certain correlation was detected: whereas students having A (Excellent) accessed the e-courses 22× in median, the unsuccessful students who failed (F), reached frequency was 15 times lower (1.5×). Similarly to this, the identical correlation was detected in mean values (Table 3).

Table 3. Correlation between the success rate and visit rate to the course of Mobile Technologies

	A	B, C, D	E	F
Mean	22.5	9.2	–	4.0
Min	1	0	–	0
Max	40	26	–	17
Median	22	7	–	1.5

4   Summary of Results and Conclusions

New requirements for graduates' knowledge and skills are appearing, as the Faculty of Informatics and Management, University of Hradec Kralove (FIM) started the co-operation with local IT companies within the HIT Cluster. The IT companies see a great potential in curricula which reflect requirements of graduates' potential employers. However, the assessment of curricula is a long time and demanding process so the FIM strongly appreciates the efforts the HIT Cluster companies devoted to this process. Reflecting the expert recommendations, large changes were made in the group of software engineering subjects, i.e. in Algorithms and Data Structures; Programming I, II, III; Introduction to Object Modeling; System Programming; Advanced Programming; Computer Graphics I, II; Mobile Technologies; Optimization of Web Applications and Theoretical Informatics.

To illustrate the educational process, three examples of subjects are presented – Programming I, Computer Graphics and Mobile Technologies – where the visit frequency to the e-subjects supporting the face-to-face instruction is monitored. In subjects of Programming I (Pro I) and Mobile Technologies (MT) the mean visit rates do not differ substantially (Pro I 13.93; median 9; MT 14.3; median 11.5), whereas in Computer Graphics (CG) the mean visit rate reached 132 (median 123). The total number of students enrolled in single subjects differ (Pro I 186; CG 106; MT 26). However, the amounts of successful students are rather identical (Pro I 62.4 %; CG 57.6 %; MT 69.2 %). As all subjects were taught in the form of blended learning (face-to-face lessons plus e-learning support in LMS BlackBoard), there was no theoretical reason of such a high visit rate in CG. CG is very difficult subject demanding on both the theoretical knowledge and practical skills. Therefore, numerous interactive animations demonstrating single principles are included in the learning content of the e-subject, and each practical task is structured into several practical steps for easier comprehension. It is not comfortable for students to download the study materials and work with off-line version; they access each study material online for several times. That is the reason why the visit rate to this course is so high.

After all changes in learning contents of the subjects had been made, students' feedback was collected; it was positive - students having more experience from practice, in agreement with the HIT Cluster experts, expressed high appreciation.

Acknowledgement. This research was financially supported by the SPEV 2016. In addition, the authors thank Zdenek Mlcoch for support within the research.

References

1. Poulova, P, Simonova, I.: Flexible e-learning: online courses tailored to student's needs. In: Capay, M., Mesarosova, M., Palmarova, V. (eds.) DIVAI 2012: 9th International Scientific Conference on Distance Learning in Applied Informatics: Conference Proceedings, pp. 251–260. UKF, Nitra (2012)
2. Tomaskova, H., Nemcova, Z., Simkova, M.: Usage of virtual communication in university environment. Procedia Soc. Behav. Sci. **28**, 360–364 (2011)
3. Zentel, P., Bett, K., Meiter, D.M., Rinn, U., Wedekind, J.: A changing process at German universities—innovation through information and communication technologies? Electron. J. e-Learn. **2**(1), 237–246 (2004)
4. Kotzian, J., Konecny, J., Krejcar, O.: User perspective adaptation enhancement using autonomous mobile devices. In: Nguyen, N.T., Kim, C.-G., Janiak, A. (eds.) ACIIDS 2011, Part II. LNCS, vol. 6592, pp. 462–471. Springer, Heidelberg (2011)
5. Bradford, P., Porciello, M., Balkon, N., Backus, D.: The blackboard learning system. J. Technol. Syst. **35**, 301–314 (2007)
6. Stepanek, J., Simkova, M.: Design and implementation of simple interactive e-learning system. Procedia Soc. Behav. Sci. **83**, 413–416 (2013)
7. Poulova, P., Simonova, I., Manenova, M.: Which one, or another? Comparative analysis of selected LMS. Procedia – Soc. Behav. Sci. **186**, 1302–1308 (2015)

Collective Unconscious Interaction Patterns in Classrooms

Roberto Araya[✉] and Josefina Hernández

Centro de Investigación Avanzada en Educación, Universidad de Chile, Santiago, Chile
roberto.araya.schulz@gmail.com,
josefina.hernandez@ciae.uchile.cl

Abstract. Students' unconscious interactions can be estimated from their gaze patterns. They signal who they are paying visual attention to, which is not necessarily conscious as they could express in surveys, but it reveals the students' unconscious preferences and decisions. A student's gaze reveals who captures their attention among a class full of students. Using two months of video recordings taken from a fourth grade class, where, every day, a sample of 3 students wore a mini video camera mounted on eyeglasses, we analyzed students' gaze tendencies throughout the sessions. We found that low GPA students' gaze to the teacher decreases much more than high GPA students after 40 min. On the other hand, popular students, high GPA students, attractive boys, and girls without upper body strength receive much more gazes from peers systematically throughout the sessions. However there are groups with some combinations of these characteristics that unexpectedly receive more gazes. On the other hand, in some cases there is a clear pattern on gender and popularity of the peers that do more of the gazes to the previous groups than the rest.

Keywords: Students' unconscious preferences · Interaction patterns · Visual attention · Classroom practices · Video analysis · Educational process mining · Collective intelligence

1 Introduction

Student's unconscious interactions are a powerful tool to improve teaching and learning practices. Their point of view can give very important cues about what is happening in the classroom. Analyzing their gaze patterns while they attend to their classroom activities is even more powerful since there is no delay, and therefore there is much less forgetting. For example, the Experience Sampling Method [8] uses a beeper or a smartphone to periodically ask each student to complete a mini survey on their current subjective experience and to state with whom they are interacting. This is performed several times a day and continues for several weeks.

However, "What a person says is not necessarily an accurate representation of what he/she thinks" [14]. Watching students' behavior instead of listening to them can be even more truthful. Since most of the students' choices are not consciously decided, most of the time, they are not aware of their own decisions. For example, Gazzaniga [7] proposes the existence of a left hemisphere module that seeks explanations for internal and external events, but several of them are far from what really

N.T. Nguyen et al. (Eds.): ICCCI 2016, Part II, LNAI 9876, pp. 333–342, 2016.
DOI: 10.1007/978-3-319-45246-3_32

happens and Dennett [6] proposes a Multiple Draft Editor that interprets part of distributed information and produces a narrative stream.

Surveys are powerful, but watching the students during classroom sessions can give more information about their true interaction patterns. For several years video recordings of classroom activities have been analyzed [2, 9]. However there is yet another powerful form of obtaining information: we can watch the class as they see it, from their point of view. This first person perspective can be obtained with cameras mounted on students' eyeglasses. With this kind of information it is much simpler to compute who the students are visually attending to, and therefore get a better estimation of their unconscious preferences.

There are some challenging data processing tasks to be performed to obtain reasonable estimations of the students' focus of their visual attention. For example, the amount of information that is rapidly generated or that students' attention is constantly changing between competing sources. The good news is that using digital technologies we can get a reasonable approximation to their focus of visual attention. First person videos captured from cameras mounted on eyeglasses can record students' "unique" points of view of classroom activities, providing precise information to understand what takes place in the classroom and what they do.

1.1 What Features of the Focus of Attention Reflect Student's Preferences in Classrooms?

Understanding students' collective behavior, decisions and interaction patterns is very important to improve classroom practices and education quality. According to Cuban [5] during the last decades "The 'what' of teaching has indeed changed, but when it comes to the how- the pedagogy- few major changes have occurred". Cuban says there is a fundamental attribution error policy makers make. They focus on teacher's characteristics and not in the situation. The context is important to understand what really happens in the classroom, what the students do, their interactions, their preferences, and how they depend on the strategies used by the teacher. Therefore we need to analyze the class's dynamics.

The main challenge of a teacher is to emotionally connect with the students [10]. The teacher has to consider not only the individual psychological motivational and cognitive mechanisms of the students but also the constant and intense social interactions present in the classroom.

Students' attention is subject to a constant competition between evolutionary meaningful and evolutionary neutral features of people and objects [8, 9]. The evolutionary meaningful features are biologically primary ones like fighting ability and sexual attraction. In species with aggressive social interactions, natural selection selects cognitive mechanisms for assessing physical formidability (fighting ability or resource-holding potential) [15]. In the case of the human evolutionary history, upper-body strength has been a major component of fighting ability [12]. Also, in the view of culture–gene coevolutionary theorists, [4] humans possess several learning biases which robustly enhance the fitness of cultural learners. In this work we propose to measure popularity as a measure of sociometric status—the respect and admiration one has in face-to-face groups. According to Anderson et al. [1], in adults sociometric status is very important and for example has a stronger effect on subjective well-being than does socioeconomic status.

The evolutionary neutral features or biologically secondary features like peers' GPA or the subject matter being taught are product of recent cultural developments. Thus the students' social interactions in the classroom must be under the same evolutionary forces. In this work we will consider the effect of three evolutionary meaningful features: upper body strength, physical attractiveness and popularity.

In this work we continue the study [3] of the effect of these features on their unconscious preferences revealed on their focus of attention in their social interaction patterns in classrooms. Particularly we analyze the dynamic flow of a session, divided in 9 segments of 10 min each.

2 Methodology

In this paper we studied the gaze patterns of students throughout their regular school days. Particularly, we were interested in analyzing (1) if students pay the same amount of attention on their teacher throughout an entire 90 min class period, and (2) if there are students that attract more attention of their peers and are therefore more observed, regarding five specific characteristics: gender, popularity, attractiveness, upper body strength and grade point average (GPA). For this, we worked with one fourth grade class, taught by one male teacher from Santa Rita School from Santiago, Chile, where, from September 26th until November 27th of 2012, a sample of three students were asked to wear a mini video camera mounted on eyeglass frames for the entire school day. The parents of the students gave consent to wear these eyeglasses, as well as agreeing to allow any information that was obtained to be disseminated both in professional conferences and in journal articles. The data has been anonymized.

By the end of the experiment, more than 150 h of video recording were obtained. Considering that the videos had a recording quality of thirty frames per second (30 fps), for each video, every second a frame was sampled and processed in order to detect the presence of faces. In this project we define subject 1 as looking at subject 2 if the video frame recorded by subject 1 reveals the face of subject 2 at a given time interval. With this definition, a subject can be considered to be looking at several subjects at any given moment. A total of 24,148 faces were detected and each face was saved as an image file, along with the information of who observed that student at that precise moment. Each facial image was then processed in order to identify the subject.

This fourth grade class had 36 students, 21 boys and 15 girls, and the average age of the students was 10.5 years. The GPA of each student was handed to us by the school for each subject, and in order to know each student's popularity, the class was asked to respond a survey (individually) where they were to grade the popularity of each classmate on a scale from 1 (least popular) to 7 (most popular). Additionally, three independent teachers from the school rated each student's upper-body strength and physical attractiveness in a scale from 1(lowest) to 7 (highest). For these characteristics, we averaged all the individual grades each evaluator gave each student, and obtained a final grade for each student respecting popularity, attractiveness and upper body strength. Furthermore, we analyzed the obtained data in order to look for correlations between the five characteristics in study, but found no correlations between each pair of variables, meaning that the five characteristics are independent one from another. Therefore, we categorized each student as: boy or girl, high GPA or low GPA, popular or non-popular, attractive or non-attractive and finally, strong or weak. For this, we considered the students as above or below the average class grade in each category.

Given that there are many more male students than female students, and that the amount of time that each student wore the eyeglasses also varied from one user to another, we normalized the data in order to estimate the intrinsic tendency to look at a certain group of students. This is defined as the proportion an observer looks at the group in the ideal case that each group comprises the same number of subjects. Finally, to compute the tendency of a group of observers we average the tendencies of the individual observers.

Also, each class period lasted 90 min. Therefore, we separated the periods in 10 min intervals, in order to get a better understanding of what happens during each class.

3   Results

We obtained information about 36 students' gazes during 22 entire school days. We looked for all the gazes on the teacher during his lectures, and found that in the five matters in study, students drop their visual attention on the teacher from the 4th to the 5th interval (minute 40 to minute 50 of his lectures),. When joining all subjects, this drop is statistically significant with a p-value of 0.01, and if we separate the observers in high GPA and low GPA, we found that this drop is only statistically significant in low GPA students (p-value 0.04), meaning that both high and low GPA students loose concentration, but high GPA students don't drop visual attention on the teacher as low GPA students do from this precise moment of the class, as can be seen in Fig. 1.

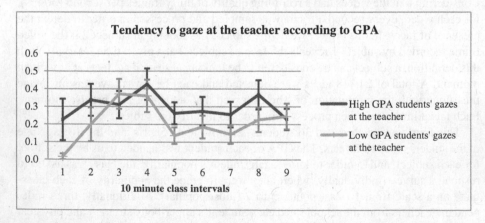

Fig. 1. Tendency to gaze at the teacher during 90 min classes according to the observers' GPA, with the 95 % confidence intervals

Most studies of classroom activities gather information on student attention to the teacher and the student's learning. However, it is reasonable to suspect that there is an intense social interaction dynamic even in classes where the teacher is lecturing and the students are passively listening or taking notes. The students' gazes are a very important form of their unconscious interaction patterns. This type of information is signaling their true interests, motivations, and focus of attention.

Regarding which students attract most the attention of their peers by being more observed, for each pair of characteristics in study, we considered each student as a crossing of these, resulting in four "types" of students for each observation. For example, considering gender and GPA, a student could be a boy with high GPA, a boy with low GPA, a girl with high GPA or a girl with low GPA, and the tendencies to look at these four students must add 1 in each interval (moment).

When analyzing gender and popularity, popular boys and girls are more observed than non-popular boys and girls in all intervals, and this difference is statistically significant from minute 40 to 80 of the session. Also, non-popular boys are more observed than non-popular girls in the 9 class intervals. Regarding gender and attractiveness, we found that attractive boy students are much more observed than non-attractive boy students, and this difference is statistically significant in 8 out of 9 intervals, whereas girls are not more observed if attractive or not. We also discovered that attractive boys are much more observed than attractive girls, and this difference is statistically significant in 5 intervals (consecutively from minute 21 to 60, and also the last 10 min of the class), as is presented in Fig. 2. When crossing gender and upper body strength, we found that strong boys are much more observed than weak boys, which is statistically significant in 6 intervals, and weak girls are more observed than strong girls, which is also statistically significant in 5 intervals, as can be observed in Fig. 3. Finally, regarding gender and grades, we found that boys are more observed than girls at all moments of a class indistinctly if they are high or low GPA students, and also that high GPA students

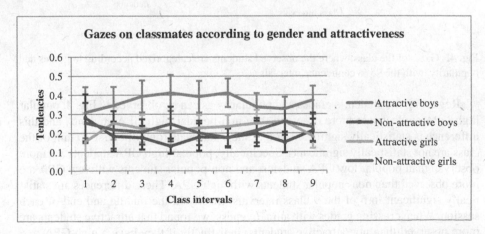

Fig. 2. Gazes of the class, where the observed students are categorized according to gender and attractiveness, with the 95 % confidence intervals

are more observed than low GPA students at all moments of a class, indistinctly if they are boys or girls, but neither of these differences are statistically significant.

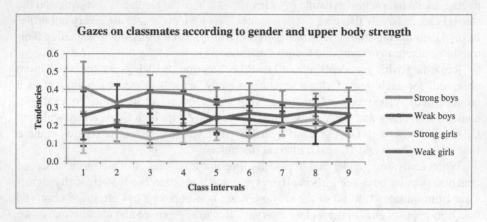

Fig. 3. Gazes of the class, where the observed students are categorized according to gender and upper body strength, with the 95 % confidence intervals

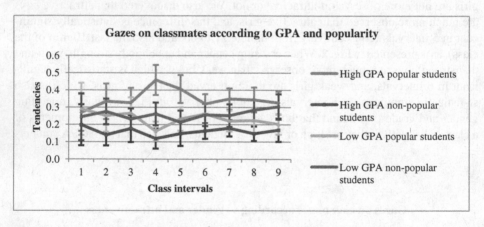

Fig. 4. Gazes of the class, where the observed students are categorized according to grades and popularity, with the 95 % confidence intervals

Regarding the student's grades and popularity as can be observed in Fig. 4, popular high GPA students are more observed than non-popular high GPA students, and this difference is statistically significant in 8 of the 9 intervals (only the first 10 min of the class are not statistically significant). Specifically, popular high GPA students are more observed than popular low GPA students, and non-popular students with low GPA are more observed than non- popular students with high GPA. These differences are statistically significant in 5 of the 9 class intervals, mostly in the middle and end of each session. When crossing grades with attractiveness, we found that attractive students are more observed than non-attractive students, indistinctly if they have a high GPA or a low GPA. This difference is statistically significant in 7 of 9 intervals in low GPA students (only the first and last 10 min of the class are not statistically significant), whereas it is only significant in 3 intervals in high GPA students, in the middle of the session. Also, non-attractive high GPA students are much more observed than low GPA non-attractive students throughout the entire 90 min classes.

When looking at popularity and attractiveness, presented in Fig. 5, we found that popular attractive students are more observed than non-popular attractive students in the 9 class intervals, and it is statistically significant in 5 (consecutively from minute 21 to 70 of the sessions), and popular attractive students are also more observed than popular non-attractive students, and this is statistically significant in 8 of the 9 intervals (only the first 10 min are not significant). Regarding popularity and strength, popular strong students are much more observed than popular weak students and non-popular strong students, and both differences are statistically significant in 8 intervals (only the second and first 10 min respectively of the class are not significant). Also, non-popular weak students are more observed than non-popular strong students and popular weak students, which is statistically significant in 6 and 5 intervals respectively, mainly during the second half of the sessions.

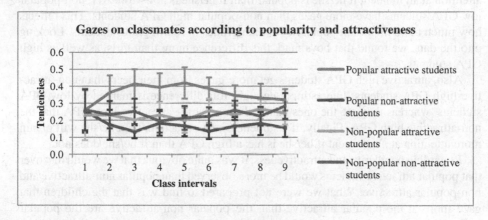

Fig. 5. Gazes of the class, where the observed students are categorized according to popularity and attractiveness, with the 95 % confidence intervals

Finally, when analyzing attractiveness and strength, we found that strong attractive students are more observed than strong non-attractive students, and also, that non-attractive strong students are less observed than non-attractive weak students, and both differences are statistically significant in 8 intervals (only the first 10 min are not significant).

4 Discussion

Prieto et al. [13] highlights the great value to study how the classroom process unfolds, not only at the temporal scale of whole sessions but in more fined-grained episodes. This type of information is critical to improve teaching, since it provides teachers with means of controlling the flow of activities and how to adapt them.

For the first 40 min of each class period, students' attention on the teacher tends to increase every 10 min, reaching its highest level after 40 min past the start of the lecture in almost all subjects. But for the next 10 min after this peak, students lose their concentration drastically, especially low GPA students. This finding is a clear indication that class lectures cannot last more than 45 min, or should always contain a break at this moment if are considered as 90 min periods in order for the teacher to maintain his/her students' attention at the highest.

Fourth grade boys and girls are very different. Boys are still children, whereas girls are beginning their adolescence stage. Throughout this study, boys are much more observed than girls, and especially attractive and strong boys. When looking through the data to find out who the observers are in these cases, we discovered that girls are the ones that mark this difference. Boys look at boys indistinctly if they are attractive and strong or not, but girls gaze much more at attractive strong boys than non-attractive weak boys. On the other hand, weak girls are much more gazed at than strong girls, but the observers in this case are both boys and girls.

Our findings reveal that if a student has a high GPA, than he/she will obtain more attention at all moment if he/she is popular than if he/she is not. Moreover, non-popular low GPA students have more gazes than non-popular high GPA students. This reflects how much more important popularity is than grades in fourth grade students. Looking into the data, we found that boys mark this difference more than girls, as well as high GPA students.

Also, attractive high GPA students get more gazes from their peers than non-attractive high GPA students. Interestingly enough, this difference is marked by low GPA students, whereas they are the ones that look more at the attractive high GPA that the non-attractive high GPA. Finally, if a student is non-attractive, than he/she will obtain more attention at all moment if he/she is has a high GPA than if he/she does not.

Regarding popularity and attractiveness, it was quite obvious that we would discover that popular attractive students would be more observed than popular non-attractive and non-popular attractive. What we were not prepared to find was that the children that gaze more at the popular attractive than the popular non-attractive are the popular students, because the non-popular do not distinguish between attractive and non-attractive when gazing at the popular. Also, the students that gaze more at the popular attractive than the non-popular attractive are the attractive students, because the non-attractive do not distinguish between popular and non-popular peers when gazing at the attractive.

Something similar was revealed with popular strong students. The children that gaze more at the popular strong than the popular weak are the popular students and the girls. Neither the non-popular nor the boys distinguish between strong and weak when looking at the popular peers. Also, the students that gaze more at the popular strong than the non-popular strong peers are the attractive students, because the non-attractive don't distinguish between popular and non-popular when gazing at the strong students of their class.

Finally, the strong attractive students are more observed than the weak attractive students, and this difference is marked mainly by the girls and not so much by boys, which is consistent with our first hypothesis about the difference between fourth grade boys and girls. In our previous paper [3] we stated that the tendency for male students to look at attractive female classmates was 0.537, higher than chance (0.5) with p-value less than 0.001. When double checking these results during this analysis, we realized that this finding wasn't correct, whereas boys do not look significantly more at attractive girls than non-attractive girls.

Most of the time awareness comes after unconscious decisions, and this can have at least a quarter of a second of delay [11] in very simple decisions. In the case of more complex cognition, the delay can be much longer. Siegler et al. [16] for example, proposes a model for conscious discoveries of strategies that occurs much later (even days or weeks) after the unconscious discovery of the strategies. Therefore, throughout this project we have come to discover that student's gazes can reveal part of their unconscious collective preferences and interaction patterns.

Thanks to these results, teachers now have a better understanding of their students' attention throughout a class period. Even more, teachers can now focus their attention on students that are more gazed by their peers for some activities where they might need the entire class's attention, and also should consider paying special attention to those students that are not gazed as much by their classmates, in order for them to feel more welcomed in their class.

From the unconscious interactions in classrooms emerge important patterns that affect the collective intelligence of the class and the effectiveness of the teacher. It is critical to detect and identify these hidden patterns in order to be able to improve classroom management. The teacher has to connect with the students, attract their attention, coordinate their actions, and make sure they follow his/her plan for the class. The unconscious gaze patterns reveal a whole dynamic mostly hidden to the teacher and not captured by surveys. In this paper we have identified some of them that show the dynamic of students' gazes to the teacher and classmates.

Acknowledgements. We are thankful to all the Santa Rita School staff; in particular, the enthusiasm and collaboration of the fourth grade teacher that was the subject of this paper. We also thank Paulina Sepúlveda and Luis Fredes for the development of the software; to Avelio Sepúlveda, Johan van der Molen, and Amitai Linker for preliminary statistical analysis; to Marylen Araya and Manuela Guerrero for manual classification of faces obtained from the videos; and to Ragnar Behncke for his participation in the design of the measurement strategy and data gathering process. We also thank the Basal Funds for Centers of Excellence Project BF 0003 from CONICYT.

References

1. Anderson, C., Kraus, M., Galinsky, A., Keltner, D.: The local-ladder effect: social status and subjective well-being. Psychol. Sci. **23**, 764–771 (2012)
2. Araya, R., Dartnell, P.: Video study of mathematics teaching in Chile. In: Proceedings 11th International Conference on Mathematics Education Conference, Monterrey, Mexico (2008)
3. Araya, R., Behncke, R., Linker, A., van der Molen, J.: Mining social behavior in the classroom. In: Núñez, M., Nguyen, N.T., Camacho, D., Trawinski, B. (eds.) ICCCI 2015. LNCS, vol. 9330, pp. 451–460. Springer, Heidelberg (2015). doi:10.1007/978-3-319-24306-1_44

4. Chudek, M., Heller, S., Birch, S., Henrich, J.: Prestige-biased cultural learning: bystander's differential attention to potential models influences children's learning. Evol. Hum. Behav. **33**, 46–56 (2012)
5. Cuban, L.: Inside the Black Box of Classroom Practice. Change without Reform in American Education. Harvard Education Press, Cambridge (2013)
6. Dennett, D.: Consciousness Explained. Back Bay Books, Boston (1991)
7. Gazzaniga, M.: The Ethical Brain. Dana Press, New York (2005)
8. Hektner, J., Schmidt, J., Csikszentmihalyi, M.: Experience Sampling Method: Measuring the Quality of Everyday Life. Sage Publications, Thousand Oaks (2007)
9. Hiebert, J., Gallimore, R., Garnier, H., Givvin, K.B., Hollingsworth, H., Jacobs, J.: Teaching mathematics in seven countries: results from the TIMMS 1999 video study. U.S. Department of Education, National Centre for Educational Statistics (2003)
10. Labaree, D.: Someone has to Fail. Harvard University Press, Cambridge (2010)
11. Libet, B.: Mind Time: The Temporal Factor in Consciousness. Harvard University Press, Cambridge (2004)
12. Petersen, M., Sznycer, D., Sell, A., Cosmides, L., Tooby, J.: The ancestral logic of politics: upper-body strength regulates men's assertion of self-interest over economic redistribution. Psychol. Sci. **24**(7), 1098–1103 (2013)
13. Prieto, L.P., Sharma, K., Dillenbourg, P.: Studying teacher orchestration load in technology-enhanced classrooms. In: Conole, G., Klobucar, T., Rensing, C., Konert, J., Lavoué, E. (eds.) EC-TEL 2015. LNCS, vol. 9307, pp. 268–281. Springer, Heidelberg (2015). doi: 10.1007/978-3-319-24258-3_20
14. Schiller, B., Gianotti, L., Baumgartner, T., Nash, K., Koening, T., Knoch, D.: Clocking the social mind by identifying mental processes in the IAT with electrical neuroimaging. In: Proceedings of the National Academy of Sciences of the United States of America, 22 February 2016
15. Sell, A., Tooby, J., Cosmides, L., Sznycer, D., von Rueden, C., Gurven, M.: Human adaptations for the visual assessment of strength and fighting ability from the body and face. Proc. R. Soc. London Ser. B **276**, 575–584 (2009b)
16. Siegler, R.S., Araya, R.: A computational model of conscious and unconscious strategy discovery. In: Kail, R.V. (ed.) Advances in Child Development and Behavior, vol. 33, pp. 1–42. Elsevier, Oxford (2005)

Smart Mobile Devices in the Higher Education: Comparative Study of FIM UHK in 2013/14 to 2015/16

Petra Poulova[✉] and Ivana Simonova

Faculty of Informatics and Management, University of Hradec Kralove,
Rokitanskeho 62, 50003 Hradec Kralove, Czech Republic
{petra.poulova,ivana.simonova}@uhk.cz

Abstract. This paper introduces results of two surveys focused on the use of smart mobile devices within the higher education. Data were collected by the method of questionnaire in the research sample of 205 students of the Faculty of Informatics and Management, University of Hradec Kralove, Czech Republic, matriculated in IT and Management study programmes in 2013/14 and 2015/16 academic years.

Keywords: Higher education, e-learning · m-Learning, mobile devices · Smart learning, survey

1 Introduction

Latest technological devices such as smart phones or tablets has provided strong impact on the whole society. The young generation cannot imagine their living without them anymore. All changes, which mobile devices bring to all spheres of human activities, are also reflected in education.

Mobile learning can be defined as learning exploiting means of wireless technology devices that can be pocketed and utilized wherever the learner's device is able to receive unbroken transmission signals [1].

In addition, Traxler lists the core characteristics that define mobile learning and differ it from e-learning [2].

These include (in alphabetic order): bite size, connectivity, context awareness, informality, interactivity, light weight, personalized device, privacy, portability, situation and spontaneity,

Furthermore, it is important to note that mobile device is a device which a user can carry with all the time; and thus he can search for information anytime and anywhere as been autonomous form electric supply [3] or [4]. Smart learning is not only education supported by mobile phones, but in fact, by all types of mobile devices. Below the authors of this article present the most suitable mobile devices for smart learning that were also monitored in the surveys: notebook, netbook, smart phone, tablet, multimedia player, e-book reader, portable playing console and others.

, A few years ago, in the Czech Republic, traditional non-portable (immobile) didactic means were widely explored in education. One of the reasons was that mobile

© Springer International Publishing Switzerland 2016
N.T. Nguyen et al. (Eds.): ICCCI 2016, Part II, LNAI 9876, pp. 343–353, 2016.
DOI: 10.1007/978-3-319-45246-3_33

devices were not largely available as in other technically developed countries. This state changed substantially within last few years, when mobile devices and applications have been applied on all levels of education, quickly moving from small-scaled to important didactic means.

The authors of this article see smart learning as a natural technological and didactic expansion of eLearning, which is gaining a new added value by this.

Therefore, the main objective of this article is to monitor the state-of-art in the use of smart mobile devices, and consequently reflect it in the concept of mobile learning (m-learning) at the Faculty of Informatics and Management, University of Hradec Kralove, Czech Republic. Partial research questions were defined as follows. How did the situation change from 2013/14 to 2015/16 academic years under these criteria:

1. Students' ownership of mobile devices;
2. purposes students exploit the mobile devices:
   (a) private communication compared to education related communication;
   (b) private purposes (other than communication);
   (c) entertainment;
3. mobile devices for higher education compared to further education;
   (a) sources of information exploited within the higher education compared to sources of information exploited within the further education;
4. social networks students exploit for the higher education (three most frequently used networks were monitored)
   (a) other social networks students exploit for the higher education

2 Theoretical Background

The paper deals with a current topic in education relating to the use of latest technologies in higher education. Its main objective is to contribute to the m-learning didactics, i.e. how to implement mobile devices in the process of instruction reflecting their advantages and limits. In the research running in the sample group of students, first, the ownership and use of mobile devices was detected, both for private and education purposes; second, the collected data were compared to other related researches.

Nowadays learning, particularly the university learning, is supported with modern information and communication technologies. Moreover, at present eLearning is being taken over by the so-called mobile learning (m-learning), which is possible thanks to the rapid growth of mobile devices such as notebooks, smartphones or tablets. In comparison with eLearning, smart learning provides further opportunities for more effective learning in the sense of its wireless connections, mobility and portability, full ubiquity or instant information sharing. The aim of this article is to explore whether university students at the Faculty of Informatics and Management in Hradec Kralove are well-equipped for this new smart learning and whether they use mobile technologies for their studies or not.

Higher Education institutions exploit mobile services to improve students' performance and the quality of education by supporting various approaches to wireless devices implementation in the e-learning process (e.g. [5, 6]).

Nevertheless, at present thanks to the rapid development of wireless technologies eLearning moves to mobile learning [7].

In fact, Park, Nam & Cha [8] see mobile learning (m-learning) as a new and independent part of eLearning where the education contents are handled solely by mobile technological devices. Table 1 below then presents paradigm shifts between eLearning and m-learning.

Table 1. Paradigm shifts between eLearning and m-learning

eLearning	m-learning
Wired	Wireless
Static	Mobile
Semi-ubiquitous	Fully ubiquitous
Personalized	Situation-based (solving real-life tasks)
Providing fast feedback	Providing instant feedback
Delayed information sharing	Instant information sharing
	(Authors' own source)

Since 2012/13 academic year virtual desktops running on VMware tools have been used at the Faculty of Informatics and Management, University of Hradec Kralove (FIM UHK), the leader in the process of the ICT implementation in higher (technical and engineering) education in the Czech Republic, to support the process of instruction in selected subjects. Currently, all IT laboratories have had their "twins" in virtual desktops, which enabled students to have anywhere/anytime access to specialized software. From the FIM's point of view smart learning is defined as such an approach which uses portable information and communication technologies within the process of instruction, and it does not matter what device is used - smartphone, notebook, netbook, tablet, PDA, e-reader, mp3 player, or game console etc. The question is whether/to what extent these devices are available to students and what purposes they use them for. We expected the monitoring results would show us the mobile devices were widely owned and used by students. It means students have both the mobile devices and m-competences, and therefore the mobile devices can be used for education purposes.

The process of ICT implementation into education started in 1997 at FIM; and it became widely spread after 2000, when the LMS WebCT (since 2006 called Blackboard) started to be used. In this academic year, more than 250 online courses support single subjects and are available to students, either for the part-time and distance forms of education, or to assist the face-to-face teaching/learning process. Moreover, all online courses are also available on mobile devices. It means the blended learning model is applied which combines three approaches: (1) the face-to-face instruction, (2) work in online courses and (3) individualized approach to them through mobile devices. This solution satisfies learners' time/place preferences and bridges formal and informal learning [5].

3   Methodology

3.1   Research Sample

Both surveys were held at the University of Hradec Kralove, Faculty of Informatics and Management (FIM). In the institution students can enroll at three-year bachelor study programmes in Applied Informatics (AI3), Financial Management (FM3), Tourism Management (MCR3), Sports Management (SM3), or follow-up two-year master study programmes in Applied Informatics (AI2) and Information Management (IM2), or doctoral study programmes in Knowledge Management (IZM) and Applied Informatics (AI). Totally 205 FIM students (male 60 %; female 40 %) participated in 2013/14 academic year and 270 students (male 46 %; female 54 %) in 2015/16. Reflecting the study programmes and respondents' age the research sample was structured as follows (Figs. 1 and 2):

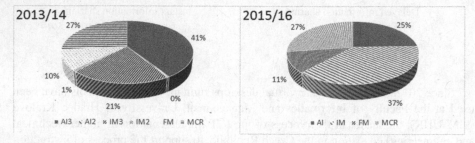

Fig. 1. Research sample: study programmes structure

The figures show both research samples are rather similar, having 62 % of IT respondents (AI, IM), 27 % of Tourism and Management (MCR) and 10 % or 11 % of Financial Management (FM) students.

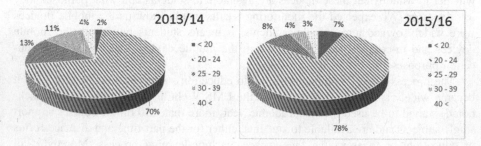

Fig. 2. Research sample: age structure

From the view of age, 70 % and 78 % of respondents were within the 20-24 year olds, having 4 % and 3 % of those above 40 years and similar amounts of other age groups.

3.2 Methods and Tools

Data were collected by the online questionnaire within the LMS Blackboard of Google Doc. The questionnaire contained 12 items focusing on two main fields of interest – the ownership, use and preferences of (1) mobile and other devices and (2) social networks.

Respondents provided answers of the multiple-choice type; four choices could have been made in items 1 and 2, all choices could be marked in items 3–8, one choice was in items 9 – 12. The NCSS2007 statistic software was used for processing the collected data; the method of frequency analysis was exploited and the results were analyzed.

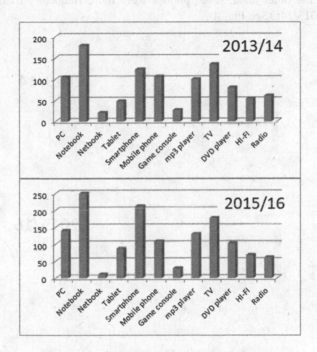

Fig. 3. Students' ownership of mobile devices

4   Results

The survey results are structured according to the criteria of partial research questions. If two separate figures are displayed, the former one relates to 2013/14 academic year, the latter figure to 2015/16.

4.1 Students' Ownership of Mobile Devices

Intentionally, the respondents' ownership is not expressed in percent but in absolute amounts. As the total amounts of respondents differed (205: 270) in the academic years but both figures have rather identical shape, we can conclude that there is no difference in students possession of mobile devices - most students own notebooks and smart phone. (See Fig. 3).

4.2 Purposes Students Exploit the Mobile Devices: Private Communication, or Education/Work-Related Communication

Under this criterion, the decrease was detected in the use of notebooks for education and work. On the other hand, smart phones were more frequently used in 2015/16 compared to 2013/14. (See Fig. 4).

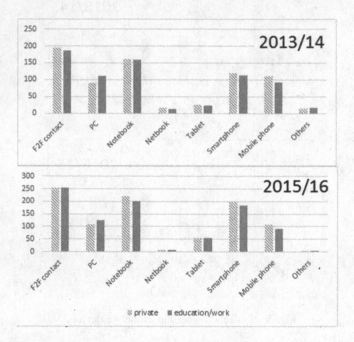

Fig. 4. Frequency comparison: mobile devices for private, or education/work-related communication

4.3 Mobile Devices for Higher Education Compared to Further Education

Identical changes can be seen under this criterion, i.e. the increase of smart phones and tablets use for both the university and further education, as well as the decrease in the use of notebooks. (See Fig. 5).

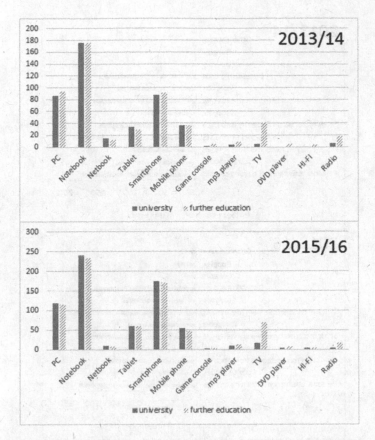

Fig. 5. Frequency comparison: mobile devices for university and/or further education

4.4 Sources of Information Within the Higher Education Compared to Further Education

What is really surprising under this criterion is that in 2015/16 some respondents declare they do not use e-subjects in LMS. Through these online courses the process of instruction in each subject at FIM is managed, the study of materials in courses is necessary for understanding the field (and passing credits and exams). So this result does not correspond to the teachers' requirements and general students' practice. Therefore, more detailed insight should be made in the future in this area (See Fig. 6).

4.5 Social Networks Students Exploit for the Higher Education

Three most frequently used networks were detected as frequently exploited, and therefore monitored. (see Fig. 7).

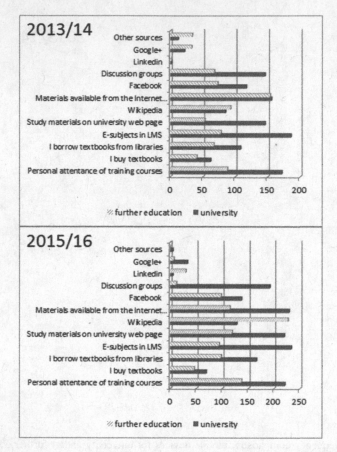

Fig. 6. Frequency comparison: sources of information for university and/or further education

Both figures show very similar results; slight decrease was detected in 'never' use of LinkedIn; however, the 'never' use increased with Google+.

From other social networks students exploit for the higher education only Twitter (19 %) and Skype (8 %) were detected. The frequency of exploitation other than the three above mentioned social networks was close to zero in 2013/14. In 2015/16 the data were nearly identical with Twitter (17 %) and Skype (12 %; the increase was detected with Whats Up only (12 %). Hardly any f other social networks were used by the respondents.

(Exceptionally: fewer than twice times per month; Seldom: irregularly, approximately 4 times per month; Not frequently: 2-3 times per week; Rather frequently: 4-6 times per week).

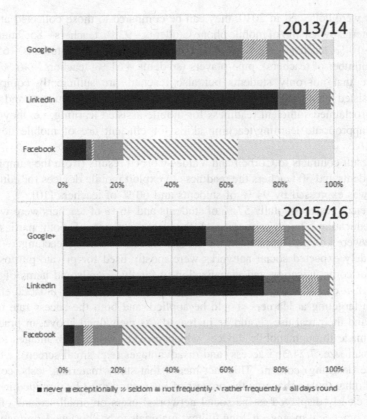

Fig. 7. The exploitation of three frequently used social networks

5 Discussions, Recommendations and Conclusions

As presented above, the data show students equipment in the area of mobile devices is sufficient; therefore these can be implemented in the teaching/learning process. Students rather frequently use various types of mobile devices for both the private and educational purposes. This result means the practical level of students' literacy is high and mobile devices can be implemented in the higher education to provide students the access to social networks. However, this field has not been adequately worked out from the didactic point of view in the Czech Republic.

There are hardly any researches; up to now only one valuable result of research on mobile devices implementation into higher education has been published – in 2011 by Lorenz [9] who analyzed the concept of mobile education under the conditions of developing university environment, particularly focusing on the process of learning enhanced by library services. He answered the research question dealing with students and teachers learning/teaching skills to use the potential of mobile-assisted learning and social networking for higher education. Particularly he focused on the criterion whether they are willing and able to bear financial expenses to cover relating services and what their attitudes to the mobile-assisted teaching and learning are. Been aware of the fact that his results were collected in 2010, they can be compared to those collected at FIM in several criteria: ownership of mobile phone (students – 92 %, teachers – 85; smartphone: students – 10 %, teachers – 27 % (in 2010); notebooks or laptops (students – 65 % and the same number of teachers); mp3 players (students – 61 %, teachers – 46 %). These data prove that not only students but also teachers are sufficiently equipped for mobile-assisted learning implementation. However, only 65 % of students and 42 % of teachers proclaimed sufficient readiness for mobile-assisted learning, i.e. they thought they had appropriate learning/teaching skills for efficient use of mobile devices in education.

This result contracts to Corbeil and Valdes-Corbeil results (from the sample group of 107 students and 30 teachers the readiness to exploit mobile devices for educational purposes was expressed by 94 % of students and 60 % of teachers [10].

In Lorenz's research totally 57 % of students and 46 % of teachers were willing to pay for education-related services and the same amount of both parties would appreciate/were going to implement mobile devices into learning/teaching.

As widely expected social networks were mostly used for private purposes, particularly for communication and visualization of family, friends and items of interest. So as this fact could be exploited for educational purposes, strong didactic effort and motivation targeting at learners should be applied, and both the access rate to social networks and their real use should be increased. As mentioned above, in practice the access is made through mobile devices which provide various advantages (e.g. low weight, small size, 7/24/365 access) and disadvantages (e.g. small screen), when presenting the (learning) content. This fact means that study materials, tests, communication and other tools enhancing the process of instruction must be provided in formats which are clearly displayed to the social networks' users on small screens of mobile devices. In practice it means e.g. long fulltext materials to be shortened, animations and video-sequences should meet requirements of good technical and technological quality and size on the screen, simple presentations with bulleted texts are preferred, as well as tests in multiple-choice, true/false, yes/no formats etc. Thus the mobile-assisted learning didactic principles are reflected within social networking. Thus they can be considered an inseparable part of this phenomenon.

Currently, in the globalized world, differences in technical and technological development are quickly fading in the field of mobile devices and Czech students and teachers are sufficiently equipped with them. However, the scientifically-verified methodology (didactics) on how to implement mobile devices into education, particularly how to start and apply mobile-assisted learning into the process of instruction, is still missing.

In a world that is increasingly reliant on connectivity and access to information, mobile devices are not a passing fad. As mobile technologies continue to grow in power and functionality, their use as educational tools is likely to expand and, with it, their centrality to formal education [11]. In addition, mobile devices are revolutionary because they transcend the boundaries of the structural status of classrooms and lecture halls and their associated modes of communication – they do not have to be confined to one particular place and time in order to be effective [12, 13].

Furthermore, smart learning is fully student-center in comparison with traditional classes and in some respect also with e-learning (cf. [8] or [9]).

Further research should thus examine implication of smart learning for the design of teaching and learning [14, 15].

Acknowledgment. The paper is supported by the Excelence 2016 project.

References

1. Attewell, J., Smith, C.: Mobile learning anytime everywhere. Learning and Skills Development Agency, London (2005)
2. Traxler, J.: Defining mobile learning. Retrieved September 9, 2014. http://www.academia. edu/2810810/Defining_mobile_learning
3. Liang, L., Liu, T., Wang, H., Chang, B., Deng, Y., Yang, J., Chou, C., Ko, H., Yang, S., Chan, T.: A few design perspectives on one-on-one digital classroom environment. J. Comput. Assist. Learn. **21**(3), 181–189 (2005)
4. Benson, V.: Unlocking the potential of wireless learning. Learn. Teach. High. Educ. **2**, 42–56 (2008)
5. Benson, V., Morgan, S.: Student experience and ubiquitous learning in higher education: impact of wireless and cloud applications. Creative Educ. **4**(8A), 1–5 (2013)
6. De Laat, M., Lally, V., Lipponen, L., Simons, R.J.: Online teaching in networked learning communities: A multi-method approach to studying the role of the teacher. Instr. Sci. **35**, 257–286 (2007)
7. Keegan, D.: The future of learning: From elearning to mlearning. ERIC. 2002. Retrieved April 1, 2015. http://files.eric.ed.gov/fulltext/ED472435.pdf
8. Park, S.Y., Nam, M.W., Cha, S.B.: University students' behavioral intention to use mobile learning: Evaluating the technology acceptance model. Br. J. Educ. Technol. **43**(4), 592–605 (2012)
9. Lorenz, M., de nechala škola díru, K: m-learning aneb Vzdělání pro záškoláky [Where school let a hole yawn: m-learning or education for truants]. http://pro.inflow.cz/kde-nechala-skola-diru-m-learning-aneb-vzdelani-pro-zaskolaky
10. Corbeil, J.R., Valdes-Corbeil, M.E.: Are you ready for mobile learning? Educause Q. **30**(2), 51–58 (2007)
11. UNESCO. Draft policy Guidelines for Mobile Learning. Retrieved September 9, 2014. http://www.naace.co.uk/cpd/tablets
12. El-Hussein, M.O.M., Cronje, J.C.: Defining mobile learning in the higher education landscape. Educ. Technol. Soc. **13**(3), 12–21 (2010)
13. Uden, L.: Activity theory for designing mobile learning. J. Mob. Learn. Organ. **1**(1), 81–102 (2007)
14. Guralnich, D.: The importance of the learner's environmental context in the design of m-learning product. Int. J. Interact. Mobile Technol. **2**(1), 36–39 (2008)
15. Sharples, M., Taylor, J., Vavoula, G.: Towards a theory of mobile learning, 2005. Retrieved November 2, 2014. http://www.mlearn.org/mlearn2005/CD/papers/Sharples-%20Theory% 20of%20Mobile.pdf

Motivation System on Prediction Market

Mikuláš Gangur[✉]

Faculty of Economics, University of West Bohemia,
Hradební 22, 350 02 Cheb, Czech Republic
gangur@kem.zcu.cz

Abstract. Prediction Market serves as an alternative tool mainly applied to gather the information widespread among the numerous experts. This tool can be used as a supplementary teaching aid in the financial engineering courses. The outcomes of selected markets give also the useful continuous feedback to the teachers. The contribution is the focus on motivational and incentive system. The prediction market inflation is introduced as motivation tool. The participants' activity is analyzed by the influence of inflation engagement. Two groups of market participants are compared with respect to participants' activity and inflation administration in the experiment. The comparison is maintained on the same market in the same conditions for all participants. The implemented system of signals allows to apply inflation only to the selected group and in the selected periods during the experiments. Finally, the increased number of the active shares on the counts of the participants' activity is considered.

Keywords: Prediction market · Motivation tools · Incentive system · Information aggregation · Financial engineering · Status quo bias · Endowment effect

1 Introduction

Prediction markets (PM) are speculative markets created for the purpose of making predictions. Experts use them as alternative tools for the information data gathering. Prediction markets simulate the activity of stock exchanges at which such instruments are traded and are used to forecast the probability of some particular event (e.g. Donald Trump will win the presidential elections), or those, which are related to a value of an estimated parameter (e.g. the percentage of votes won in the parliamentary election by a given political party). The value of an instruments is determined by the extent of confidence of the sellers and buyers in a given event or a value of a parameter. The current market price can be interpreted as an estimate (forecast) of the probability of an event or an estimated value of parameters. People who buy low and sell high are rewarded for improving the market prediction, while those who buy high and sell low are penalized for degrading the market prediction.

The prediction market described in this work is a supplement to the financial courses at the Faculty of Economics in Pilsen. Experimental PM FreeMarket has

© Springer International Publishing Switzerland 2016
N.T. Nguyen et al. (Eds.): ICCCI 2016, Part II, LNAI 9876, pp. 354–363, 2016.
DOI: 10.1007/978-3-319-45246-3_34

become one of the new approaches in education not only in this type of courses, but it can be put alongside other forms of education, such as e-learning, m-learning etc. (see [1,2]).

In education process the prediction market can serve in two areas.

- Teaching aids - the students familiarize themselves with the basic principles of financial markets. They have to analyze share prices and make decisions whether to sell or to buy shares. Less obvious operations in the market trade, e.g. short trade, can be clearly explained to the students. The failure in the experimental market is a better personal experience than hours of reading.
- Tool for information aggregation - the principal goal of prediction markets is to gather implicit information about the course management, training management, eventually exams management in any course. The gathered information provide valuable feedback for the teachers. The instructor can set up a new markets with different evaluation of lectures or practical lessons. Students should buy the shares corresponding to their evaluation of course activity. Prices of particular shares inform teachers about student evaluation.

The latter functionality is very useful in an interim teaching evaluation, i.e. the course of lectures, seminars, the final tests and assessments. Unlike conventional disposable evaluation of teaching quality, which often takes place after the classes, used technology of predictive market enables not only evaluation of various educational activities during the semester, but the shares price of offered predictions (teaching evaluations) allows the teachers to change the course of these activities.

Quality of predictive market, especially the accuracy of forecasts and credibility of the ongoing teaching evaluations and the resistance to price manipulation on the part of participants depends on the activity of market participants, i.e. their willingness to trade actively. For these reasons, the activity of traders and their motivation are the key factors the administration of predictive market is focused on.

In prediction market implementation we follow the basic structure of market. The setting of this structure designates the quality and fruitfulness of market. According to [3] the construction of PM can be divided up to three areas: *Choice of Forecasting Goal, Incentives for participation and information revelation* and *Financial market design*. This contribution focuses on incentives and motivation system.

We implemented several tools for the support of the activity participants. These tools motivate the market participants to increase their trade activity. One of the tools called PM inflation is introduced. The goal of the PM inflation implementation is to increase the trading volume, trading frequency, and thus the liquidity of the market. In this way, PM Inflation increases the accuracy of embedded predictions, which evaluates the progress of education while simultaneously increases the system resistance to the price and as well as of match predictions.

The main objective of this paper is to research the impact of proposed and implemented tool on the behavior of market participants and its influence on market activity. The experiments investigate whether inflation in the form of a cash penalty of inactive participants promotes more active participants in trading on the market and thus increases the accuracy of teaching evaluations.

The first section provides a theoretical background and literature review including the discussion of the existing motivation tools. The importance of motivation and incentive system is shown. The following section introduces the proposal and the application of the PM inflation as a possible motivation tool. The influence of PM inflation application is illustrated on the comparison of market participants with and without PM inflation on the same market. The results of these experiments are presented and finally, the possible influence of active shares count is discussed.

2   Theoretical Background and Literature Review

PMs were applied for the first time in the form of the political stock exchange to forecast the results of the 1988 U.S. presidential election launched as Iowa Electronic Market (IEM) [4]. The principles of prediction markets are also described in an article of R. Hanson [5]. The most famous example of PM is Iowa Electronic Markets [6,7]. Since 1988 this market has forecast the American presidential election results more accurately than traditional polls in 75 % of cases [8]. Pennock [9] describes the principle of the dynamic pari-mutuel markets in his contributions, Hanson [10,11] presents the idea of combinatorial information markets. The summarizing monograph "Information markets: A new way of making decisions" by Hahn and Tetlock (Eds.) [12] provides an excellent guide to the issues of prediction markets.

There are several studies dealing with the use of PMs in the area of education. For example Ellis and Sami [13] in political science courses, or Damnjanovic et al. [14] in project management courses. Buckley, Garvey, and McGrath [15] use the PM principle in the social science and economics courses to develop the orientation and decision making processes. Passmore, Cebeci, and Baker [16] use PMs to solve problems of innovations in the technology of education.

2.1   The Existing Motivation Tools

In the history of the prediction markets the problems of incentives system have been dealt with in various ways. R. Hanson [10,11] proposed and described an automatic market maker (AMM) which enables the buyer (or the seller) to trade without the necessity of the simultaneous online presence of the market participants.

The Dynamic Pari-mutuel Market Maker (DPM) is solution introduced by Pennock [9]. The principle of DPM links the advantages of the Continuous Double Auction (CDA) and the principle of the pari-mutuel market. This way it eliminates the disadvantage of the low liquidity CDA and also the problem of the pari-mutuel market which does not enable to respond to the new information by the price change.

The problem of the low level traders activity according to their absence is also dealt with by the introduction of records and keeping the book of orders. This measure also enables asynchronous trading without the necessity of the permanent presence on the market.

S. Luckner [17] analyzed the impact of different monetary incentives on the prediction accuracy. The results show that payment schemes related to performance do not necessarily increase the accuracy of predictions. Due to the traders' risk aversion the competitive environment in a rank-order tournament leads to the best prediction accuracy results.

However, the results of these empirical studies do not solve the main problem we identified when using experimental PM with non-monetary award, i.e. status quo bias.

3   PM Inflation as a Motivation Tool

In [18] the proposal and implementation of PM inflation was presented and the influence of PM inflation on behavior of participants was analyzed. Participants in experimental PM FreeMarket with non-monetary award, that were endowed with a start portfolio, may have a tendency to stick to the initial endowment ("status quo bias") or their willingness to accept greatly exceeds their willingness to pay ("endowment effect") [19]. As the solution of this problem the inflation factor was introduced. This factor would have to decrease the value of the free uninvested virtual money in the participants' accounts and encourages them invest their money on the market. The inflation rate depends on the ratio of the free and invested funds and motivates the participants to invest their funds into the market assets. This motivation incentive is called the PM inflation.

The implemented PM inflation decreases the daily nominal value of the free points of the traders. This process is illustrated by the following formula:

$$CM(t + 1) = CM(t) \times (1 - R_f/360) \qquad (1)$$

where

- $CM(t)$ - current free points on the participants' accounts before the adjustment of the PM inflation for the given period;
- $CM(t + 1)$ - points on the participants' accounts after the adjustment of the PM inflation for the given period;
- R_f - PM inflation ratio.

With regard to the described reasons and setting the sources of the PM inflation, the PM inflation ratio is constructed according to the following formula:

$$R_f = max(0, (TC - TD)/TS - 1) \qquad (2)$$

- TC - total volume of the funds on all accounts;
- TD - total demand = the number of demanded shares $\times$ the buying price;

- TS - total supply = total value of the newly issued shares (number × the current price) + total value of the shares kept (number × the current price).
- $TC - TD$ - totaql volume of **free** funds on all acounts. TC includes the funds that are blocked for active unfilled orders.

The defined ratio (2) was not too elastic especially at the start period of market trading, when the total supply TS of newly issued shares markedly exceeds funds TC of lower number of registered participants. The PM inflation ratio was zero and it didn't respond to low number of trading orders for demanded shares. By this way the ration didn't motivate participants to invest their free funds and protect the value of funds. The next construction of ratio solves the described problem.

$$R_f = \rho^{(TC-TD)/TS} - 1 \qquad (3)$$

where

ρ - PM inflation weight.

The PM inflation weight is a corrective parameter that is set to a constant value and it determines the rate of PM inflation increase. The weight is usually set to values between 1.1 and 1.5.

In [18] the comparison of two groups of participants was presented according to influence of PM inflation application. The PM inflation was applied only for one group of participants. The trading volumes of both groups were compared and the higher activity of group with PM inflation was detected. The deficiency of this experiment was the disparity of both groups with respect to market parameters. The main dissimilarity was in different markets, i.e. different shares on market and different trading processes according to different trading periods of both groups. This was corrected in a following experiment and the activities of groups on the same market were compared.

4   Comparison of Groups on the Same Market

4.1   Methodology of Experiment

First of all the signals system was implemented on experimental PM FreeMarket. This system allows to distribute different signals-messages to selected market participants and to switch PM inflation on or off with respect to participant groups.

The experiment participants were the students of one of the Financial mathematics course. They were divided into two groups according to the campus location (the campuses are located in two cities). The groups were designated by letters **P** and **C**. The experiment extended over four periods. In the first period the PM inflation was not applied to any group. It was switched on in the second period only for the group P and it was applied in the third period to the group C. In the fourth period the PM inflation was switched off in both groups. The comparison focused on the first three periods. The Table 1 summarizes the process of PM inflation application.

Table 1. Inflation application in groups and periods (Source: own)

Group/Period	1.period (14.2.-31.3.)	2.period (31.3.-24.4.)	3.period (24.4.-2.6.)	4.period (2.6.-20.6.)
Group P	No inflation - P1	Inflation - P2	Inflation - P3	No inflation - P4
Group C	No inflation - C1	No inflation - C2	Inflation - C3	No inflation - C4

The participants' activity in particular groups is compared. The activity is measured as the trading volume by group and period. The standard trading volumes (TV) per participant and per day are compared for the three periods with the help of statistical methods for parameters of groups (median, average). The equal number of means tests were performed. The mutual comparison of groups focuses mainly on groups P1, P2, P3 and on C1, C2, C3. The validity of the following hypotheses: *Trading volumes per participant and day in groups with PM inflation are higher than trading volumes in groups without PM inflation.* was evaluated.

That's why the P2 group trading volume would have to be higher than in groups P1 and C2. Similarly the C3 activity after PM inflation application would have to be higher than C2 group activity and at the same time it would have to be comparable with the group P3. Described hypotheses are summarized in the matrix in the Table 2. The symbol = denotes identical activities and symbols <,> denote lower/higher activity. Shaded cells indicate three main hypotheses that describe influence of the PM inflation.

Table 2. Hypotheses of activity comparison in particular groups (Source: own)

TV/TV	TV_{P1}	TV_{P2}	TV_{P3}	TV_{C1}	TV_{C2}	TV_{C3}
TV_{P1}	-	<	<	=	=	<
TV_{P2}	-	-	=	>	>	=
TV_{P3}	-	-	-	>	>	=
TV_{C1}	-	-	-	-	=	<
TV_{C2}	-	-	-	-	-	<
TV_{C3}	-	-	-	-	-	-

When comparing the groups we have to take into account the dependence or independence of particular groups in pairs. First of all we compare group pairs from different locations (P,C) irrespective of the period, e.g. P2-C1, P3-C2 etc. For these comparisons we use average daily TVs per participant to eliminate the effect of different participants number. These groups are considered as independent with the exception of groups in the same period and different locations (P1-C1, P2-C2, P3-C3). The exploration data analysis of basic characteristics and values recommended to use medians for statistical analysis and non-parametric tests were employed. Mann-Whitney test was used for equality of means tests. In the case of dependent groups P1-C1, P2-C2 a P3-C3 non-parametric Wilcoxon pair test was used for pairs of values in one day and in both locations (P,C).

A different view is used for comparison of the groups in the same locations and different period, e.g. P1-P2, P1-P3, C1-C2 etc. The members of these groups are the same participants in different time periods and that's why the groups are considered as dependent. The used Wilcoxon pair test compared activity in different time periods for every participants. The average daily trading volumes in the period for every participants are used to consider different lengths of periods in days.

4.2   The Results of Comparison

The results are summarized in Table 3. The cells show p-values and mutual position of medians of groups. Not proved hypotheses are colored in the table. In some cases the p-values indicate statistically significant differences in participants activity however these hypotheses can't be proved, because the result shows contrary mutual position of medians (e.g. activity in P2 group with PM inflation is lower than in P1 group without PM inflation and the equality of medians is rejected p-value = 0.0023).

Table 3. P-values of mutual tests of medians equivalence in particular groups (Source: own)

TV/TV	TV_{P1}	TV_{P2}	TV_{P3}	TV_{C1}	TV_{C2}	TV_{C3}
TV_{P1}	-	0.0023 (>)	0.0000 (>)	0.2387 (=)	0.0002 (>)	0.0001 (>)
TV_{P2}	-	-	0.0000 (>)	0.1078 (=)	0.0487 (>)	0.1902 (=)
TV_{P3}	-	-	-	0.0008 (<)	0.3787 (=)	0.7332 (=)
TV_{C1}	-	-	-	-	0.7548 (=)	0.0019 (>)
TV_{C2}	-	-	-	-	-	0.0014 (>)
TV_{C3}	-	-	-	-	-	-

The most of hypotheses were not proved (the cells are shaded). Only the main hypothesis P1-P2, P2-C2 ev. C2-C3 describe the PM inflation influence. These relations represent pair of one group with PM inflation and second one without PM inflation. The core of experiment is pair P2-C2 that compares P2 group with PM inflation and C2 group without PM inflation in the same period. Other factors, i.e. the number of newly issued shares, the number of actual shares and the prices of shares, are identical for both groups. The hypothesis of mutual relation was proved only for this pair P2-C2. The influence of PM inflation is supported also by not rejected equality of medians for pairs P1-C1 and P3-C3, i.e. similar activity in the same period and the same factors in groups on both locations (non PM inflation in both groups in 1. period and applied PM inflation in both groups in 3. period). On the other side the hypothesis of PM inflation influence on activity of the same participants in different periods P1-P2 a C2-C3 were not proved. With respect to these results we cannot state a definitive conclusion about the influence of PM inflation on participants activity.

The dependency graph of trading volume per participant on days offers another possible view. Figures 1 and 2 depict the progress of trading volume as a function of time with help of moving medians and averages. These figures present a more detailed overview of trading volume progress per participant also during particular periods.

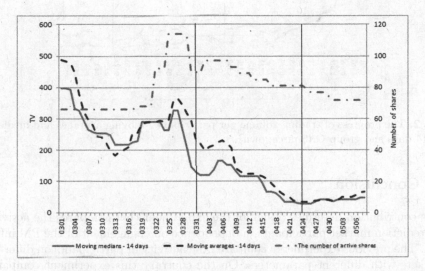

Fig. 1. The progress of trading volume per participant - moving averages and medians for 14 days and group P (Source: own)

On the figures the high TV at the start of experiment is detected and the decreasing of TV during experiment to the end. The initial high TVs distort averages and medians especially in P2 and P2 periods and by this way they distort comparison with TVs in other periods. Nevertheless the exclusion of these cases didn't affect the results in Table 3. At the same time the graphs show slight temporary TV increase in P group after 31.3. and in C group after 24.4. Next TVs increase follow increase of the active shares number after issue of new shares. According to this view the prevailing influence of these two factors is not purely clear. Only in C group the slight increase after 24.4. can be connected only with PM inflation according to decrease of active shares number.

The possible connection between daily TV and per participant and the number of active traded shares were analyzed with the help of Pearson and Spearman coefficients. According to these coefficients the TVs per day and participants and the number of active shares show weak negative dependency, eventually no dependency for the whole experiment period.

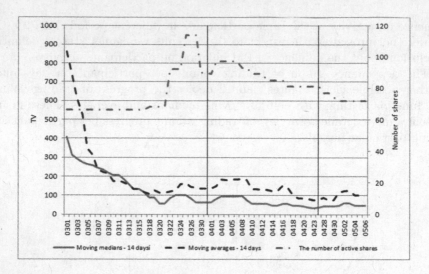

Fig. 2. The progress of trading volume per participant - moving averages and medians for 14 days and group C (Source: own)

5   Conclusion

This contribution describes design of the experiment that compares the activity of prediction market participants in dependence on engagement of the PM inflation. The previous experiments compared the activity of participants on different markets with different parameters. On the contrary this experiment compares participants activities on the same market in the same period, i.e. in the same conditions. The system of signals was implemented on the experimental prediction market and the participants in selected groups were informed about the PM inflation by this signal system. This system allowed to compare the group with PM inflation and no PM inflation group in the same period on the same market. The main goal of experiment was to prove or reject the hypothesis about the positive influence of the PM inflation on the increase of participants activity measured with trading volumes.

Following the results of comparison we can consider the short-term influence of the PM inflation on increase of the trading volume and on the participants activity. On the other side we cannot except the influence of other factors in full. One of them is the number of active shares on market. The impact of the increase of shares offers in short term after issue of new shares has to be taken into account together with PM inflation.

References

1. Drozdová, M.: New business model of educational institutions. E+M Ekonomie a Manage. **11**(1), 6–67 (2008)

2. Rosman, P.: M-learning - a paradigm of new forms in education. E+M Ekonomie a Manag. **11**(1), 119–125 (2008)
3. Skiera, B., Spann, M.: Opportunities of Virtual Stock Markets to Support New Product Development, chap. 13, pp. 227–242. Wiesbaden, Verlag Gabler (2004)
4. Forsythe, R., Nelson, F., Neumann, G., Wright, J.: Anatomy of an experimental political stock market. Am. Econ. Rev. **82**, 1142–1161 (1992)
5. Hanson, R.: Idea futures: Encouraging an honest consensus. Extropy **3**, 7–17 (1992)
6. Forsythe, R., Nelson, F., Neumann, G., Wright, J.: The 1992 Iowa political stock market: September forecasts. Polit. Methodologist **5**, 5–19 (1994)
7. Forsythe, R., Rietz, T., Ross, T.: Wishes, expecations and actions: a survey on price formation in election stock markets. J. Econ. Behav. Organ. **39**, 83–110 (1999)
8. Surowiecki, J.: Decisions, decisions. The New Yorker, 24 March 2003
9. Pennock, D.: A dynamic pari-mutuel market for hedging, wagering, and information aggregation. In: Proceedings of 5th ACM Conference on Electronic Commerce. pp. 170–179. ACM, New York (2004)
10. Hanson, R.: Combinatorial information market design. Inf. Syst. Frontiers **5**, 107–119 (2003)
11. Hanson, R.: Logarithmic market scoring rules for modular combinatorial information aggregation. J. Prediction Markets **1**, 3–15 (2007)
12. Hahn, R., Tetlock, P.: Information markets: A new way of making decisions. Washington D.C./Blue Ridge Summit PA: AEI-Brookings Joint Center for Regultory Studies (2006)
13. Ellis, C.M., Sami, R.: Learning political science with prediction markets: an experimental study. PS. Polit. Sci. Polit. **45**(2), 277–284 (2012)
14. Damjanovic, I., Faghihi, V., Scott, C., et al.: Educational prediction markets: construction project management case study. J. Prof. Issues Eng. Educ. Pract. **139**, 134–138 (2013)
15. Buckley, P., Garvey, J., McGrath, F.: A case study on using prediction markets as a rich environment for active learning. Comput. Educ. **56**, 418–428 (2011)
16. Passmore, D., Cebeci, E.: Potential of idea futures markets in organizational decision-making for educational technology. In: ITHET 2004: Proceedings of the fifth International Conference on Information Technology Based Higher Education and Training, pp. 550–555. IEEE, 345 E 47th St., New York, NY 10017 USA (2004)
17. Luckner, S., Weinhardt, C.: How to pay traders in information markets? Results from a field experiment. J. Prediction Markets **1**(2), 147–156 (2007)
18. Gangur, M., Plevný, M.: Tools for consumer rights protection in the prediction of electronic virtual market and technological changes. Amfiteatru Econ. **16**(36), 578–592 (2014)
19. Kahnemam, D., Knetsch, J., Thaler, R.: Anomalies: the endowment effect, loss aversion, and status quo bias. J. Econ. Prespectives **5**, 193–206 (1991)

Decision Processes in Smart Learning Environments

Peter Mikulecky[✉]

University of Hradec Kralove, 50003 Hradec Kralove, Czech Republic
peter.mikulecky@uhk.cz

Abstract. Smart learning environments can be naturally considered to be a new challenging step of computer enhanced learning evolution, offering many new interesting facilities. Smart environments in general as well as their special case smart learning environments can be studied as being based on a sophisticated multi-agent based architecture, which is behind all the decision processes that appear in the environment and enable the environment's functionality. The aim of this paper is to summarize recent state of the art in the area, with strong focus on decision processes in smart learning environments. We are not presenting any new algorithms for these processes, but we discuss several various possibilities and approaches instead. The paper intends to be a starting point in looking for new ways or approaches for creating decision processes by which smart learning environments cope with the problem of multiple residents in a smart (learning) environment.

Keywords: On-line learning · Smart environments · Multi-agent architectures · Decision processes

1 Introduction

We are witnessing now a big effort focused on research in the area of technological support of education. Most recent step in this effort resulted in the concept of smart learning environments, illustrated by the number of interesting and really implemented projects. Smart learning environments, based on approaches and technologies deployed in related areas of Ambient Intelligence and Smart Environments, certainly deserve considerable attention of the large community oriented on technology enhanced learning. Smart learning environments can be naturally considered to be a new challenging step of computer enhanced learning evolution, offering many new interesting facilities. Moreover, a number of new information technologies and approaches certainly will influence research in the area of smart learning environments. Let us mention cloud based architectures and applications as a representative.

As [1] pointed out, besides its technological perspective, social perspective, or ethical perspective, it is possible to study the area of Ambient Intelligence also from an educational perspective. This educational perspective deals with problems and challenges related to proper education in relevant areas.

N.T. Nguyen et al. (Eds.): ICCCI 2016, Part II, LNAI 9876, pp. 364–373, 2016.
DOI: 10.1007/978-3-319-45246-3_35

The famous ISTAG Report [2] started in 2001 a decade of various research initiatives in the rapidly growing area of ambient intelligence. Among the four basic scenarios described in the report, it introduced also a smart environment example in the form of the *Scenario 4: Annette and Solomon in the Ambient for Social Learning*. That was a vision of a learning environment, based on a position that learning is a social process.

According to the original description in the ISTAG report [2], *the Ambient for Social Learning (ASL) is an environment that supports and upgrades the roles of all the actors in the learning process, starting with the roles of the mentor and the students as most concerned parties. The systems that make up the ASL are capable of creating challenging and interacting learning situations that are co-designed by the mentor and students in real-time. Students are important producers of learning material and create input for the learning 'situations' of others. In other words, the ASL is both an environment for generating new knowledge for learning and a 'place' for learning about learning.*

Of course, the Ambient for Social Learning should be also a physical space (consisting of a room or even several rooms in a building) together with all of its ambient facilities, including many linkages with similar places. Its layout and furnishing has to be flexible and diverse, so that it can serve the learning purposes of many different kinds of groups and individuals. Such an Ambient for Social Learning could be an intelligent classroom supporting and upgrading roles of all the actors in the learning process, with a special accent on the roles of the mentor and students as the most important roles. The Ambient for Social Learning is conceived as a 'learning system' that is growing and improving simply by using it. Nevertheless, discussing the importance of the *ISTAG Annette and Solomon scenario*, Alsaif et al. [3] pointed out, that if we took into account this ISTAG scenario that could serve as an ideal case for a smart learning environment, the popular view of *anywhere and anytime learning* should be considered as impractically broad. We already discussed that in [4], where an agent-based architecture for the original ASL was proposed.

An attempt to specify smart learning environments in a slightly detailed way was published in [5]. A bit more general overview of the Ambient Intelligence possibilities in education brings our recently published paper [6]. The aim of the paper was to identify and analyze key aspects and possibilities of Ambient Intelligence applications in educational processes and institutions (universities), as well as to present a couple of possible visions for these applications. The conclusion of the presented research was that exploitation of Ambient Intelligence approaches and technologies in educational institutions was possible and could bring us new experiences utilizable in further development of various Ambient Intelligence applications.

In the scope of a recent research project we intend to go deeper into the nature and structure of decision processes conducted by autonomous multi-agent architectures that are behind intelligent environments of various kinds (e.g., intelligent offices, intelligent workplaces, smart learning environments, etc.), enabling their proper functionality. Structures of these decision processes are being investigated and new knowledge about their nature should be obtained. In this paper decision processes of autonomous multi-agent architectures of smart learning environments are studied, with a special focus on their specific problems.

2   Related Works

2.1   Smart Environments for Learning

According to [7], *a learning environment can be considered smart when the learner is supported through the use of adaptive and innovative technologies from childhood all the way through formal education, and continued during work and adult life where non-formal and informal learning approaches become primary means for learning.* That is, Kinshuk supports the meaning of smart learning environments as neither pure technology-based systems nor a particular pedagogical approach, but a subtle mixture of both.

In [8], *a smart environment for learning is defined as any space where ubiquitous technology informs the learning process in an unobtrusive, social or collaborative manner.* Alsaif et al. [3] pointed out, that a smart environment can be an *aware* room or building, capable of understanding something about the context of its inhabitants or workers; it can be a digitally enhanced outdoor space park, cityscape or rural environment; or it can be the environment created when peoples meetings or interactions are augmented by wearable devices.

Another interesting opinion can be found in [9]. They understand Smart Learning Environments as *systems that apply novel approaches and methods on the levels of learning design and instruction, learning management and organization, and technology to create a context for learning that provides learners with opportunities for individualized learning and reflection in a motivating way, and that allow teachers to facilitate learning, providing scaffolding and inspiration based on the learners needs and a careful observation of her learning activities.* As a conclusion, they pointed out that approaches in the direction of Smart Learning Environments cannot be restricted only on the technological level, but should involve also another dimensions.

According to [10], *a smart learning environment not only enables learners to access digital resources and interact with learning systems in any place and at any time, but also actively provides the necessary learning guidance, hints, supportive tools or learning suggestions to them in the right place, at the right time and in the right form.* The features just mentioned should be essential for smart learning environment. A smart learning system can be perceived as a technology-enhanced learning system that is capable of advising learners to learn in the real world with access to the digital world resources. A smart learning environment aims to help students gaining knowledge even when they are doing leisure activities. Hwang concludes that a simple incorporation of an intelligent tutoring system into a context-aware ubiquitous learning environment is not enough for obtaining a real smart learning environment.

2.2   Decision-Making in Smart Learning Environments

Kinshuk and his colleagues [11] stressed, that there are three major features of the development of smart learning environments that separates smart learning environments from other advances in learning technologies. The three features (or directions) are:

- full context awareness,
- big data and learning analytics,
- autonomous decision-making.

Another important feature of smart learning environments, which differ them from other learning environments, is their autonomous knowledge management capability that enables them to automatically collect individual learners life learning profiles. Collected learning profiles can track the learning progress of each individual learner over quite long periods and across the range of sources of evidence about the learners progress. According to Kinshuk et al. [11], based on individual life learning profiles and the techniques of big data and learning analytics, smart learning environments can precisely and autonomously analyze learners learning behaviors in order to decide in real time, for example,

- what interactions with the physical environment to recommend to the individual learners to undertake various learning activities, as well as the best location for those activities,
- which problems the learners should solve at any given moment,
- which online and physical learning objects are the most appropriate,
- which tasks are the best aligned with the individual learners cognitive and meta-cognitive abilities,
- what group composition will be the most effective for each group members learning process,
- and so on.

Such autonomous decision making and dynamic adaptivity has according to Kinshuk the potential to generalize and infer learners learning needs in order to provide them with suitable learning conditions. This capability of decision making and the capability of being dynamically adaptive to the learners' needs, based on learning analytics utilization, seems to be essential for a really successful smart learning environment.

As Spector [12] points out, it is necessary for a smart learning environment to autonomously provide *different learning situations and circumstances, as... a human teacher or tutor... to help learners become more organized and aware of their own learning goals, processes and outcomes.*

2.3   Decision-Making in Smart Environments

Following the concept of ubiquity, Ambient Intelligence is focused on technologies and approaches for development of intelligent environments aiming at supporting their users, that is, persons surrounded by these environments. Decision processes in intelligent environments are, for the sake of simplicity, usually studied in specific types of environments, like smart homes or smart offices.

Architectures of these environments are usually modeled by multi-agent systems, facilitating thus investigation of processes enabling functionality of the environments. Therefore, in general it is possible to study decision processes in smart environments as decision processes provided by the underlying multi-agent system, that are usually focused on fulfilling requirements or needs of users surrounded by the environment. The situation is relatively easy in the case of a single user, as it is no need to solve different, frequently also contradictory requirements of individuals, situated in the environment. The single user case can be taken into account in such situations, like, e.g., a senior is living in her smart household, or a smart hospital room is monitoring the only patient recovering herself in the room.

Of course, there are some interesting exceptions also in the case of one human situated in a smart room. As Becerra and Kremer presented in [13], also several other types of inhabitants could be taken into account, especially non-human, e.g., animals, plants, but even valuable paintings, or furniture requiring some specific treatment. If these non-human inhabitants are considered to be nearly equally important as their human house-mate, it would be necessary to take also their requirements into account and include them into decision processes of the respective smart environment. An example introduced in [13] could illustrate the case: *When the human inhabitant is not in the room, the environmental conditions in the room (e.g. light, temperature and humidity) are not monitored. This could have adverse effects on the other (non-human) inhabitants of the room. A dark room could prevent the plants from growing; a cold room would make the leather sofa uncomfortable to sit on when the human inhabitant returns; a warm, bright room could damage the valuable paintings.*

Just described case of one human and several non-human inhabitants of a smart environment can certainly be solved as a case of single resident, with several co-residents having just a restricted number of relatively standardized requirements. It could be clear from the fact, that those other inhabitants are important but are not able to interact with the environment. Their possible requirements could be in a sense inserted into the smart environment decision making facility. For instance, a plant do not like bright light and temperature below 20 C. The smart environment, when monitoring the daylight and temperature, could always take these restrictions into account.

The multiple resident case in intelligent environments, in contrary to the single resident case, is very interesting and far more difficult. As Cook and Das [14] pointed out, a lot of mobility tracking algorithms worked well just for single resident case. The multiple resident case still has been not solved satisfactorily. It was proven already that optimal location tracking of multiple residents is a NP-hard problem [15]. Nevertheless, as Cook and Das [14] argued further on, it can be supposed that each resident in an intelligent environment behaves selfishly in order to fulfill her/his own preferences or objectives and to maximize her/his utility. Therefore, the appearance and activities of multiple residents in an intelligent environment might lead to conflicting goals and serious problems if not deficiencies in decision-making behavior of the whole intelligent system. An intelligent environment must be intelligent enough in order to cope with such kind of problems by striking a balance between multiple preferences, in many cases having even contradictory nature. In [15], the problem of location tracking of multiple residents was investigated from the perspective of stochastic game theory; however, this solution seems to be just a theoretic result still without practical application.

3   Decision Processes

3.1   Introductory Remarks

Decision processes as such are studied broadly with a focus on their nature and usage in various organizations, see, e.g., [16] or [17], usually as Markov decision processes. Decision processes in multi-agent architectures are studied as well, see, e.g., [18, 19], or [20], usually with a focus on a particular application. Quite often approaches, as multi-agent reinforcement learning, are used [21]. There is also possible to use modeling and simulations of multi-agent systems for studying structure and actual impact of particular decision processes, see, e.g., [22].

Agent-based simulation is now a well established simulation modeling tool in academia and on the way to achieving the same recognition in industry, as Siebers and Aickelin [22] pointed out. Agent-based simulation is well suited to modeling systems with heterogeneous, autonomous and pro-active actors, such as human-centered systems. A special position among them have smart environments, which certainly are human-centered, their actors (agents) are autonomous up to a very high level and their important task is to be pro-active in serving the users residing in such an environment.

Luck et al. [23] made a distinction between two *Multi-Agent System* paradigms: *multi-agent decision systems* and*multi-agent simulation systems*. In multi-agent decision systems, agents participating in the system must make joint decisions as a group. Mechanisms for joint decision-making can be based on economic mechanisms, such as an auction, or alternative mechanisms, such as argumentation. Multi-agent simulation systems are used as models to simulate real-world domains where agent-based modeling is appropriate.

3.2   Decisions of Smart Learning Environments

First, let us go back to the ISTAG *Scenario 4: Annette and Solomon in the Ambient for Social Learning.* According to the ISTAG group, a number of specific technologies would be needed for implementation of this Smart Learning Environment, among others the following ones:

– Recognition (tracing and identification) of individuals, groups and objects.

- Interactive commitment aids for negotiating targets and challenges (goal synchronization).
- Natural language and speech interfaces and dialogue modeling.
- Projection facilities for light and soundfields (visualization, virtual reality and holographic representation), including perception based technologies such as psycho-acoustics.
- Tangible/tactile and sensorial interfacing (including direct brain interfaces).
- Reflexive learning systems (adaptable, customisable) to build aids for reviewing experiences.
- Content design facilities, simulation and visualization aids.
- Knowledge management tools to build community memory.

An important problem in each smart environment is the problem of how the environment evaluates users needs and how it assigns preferences to them. Actually, when many users are involved in a ubiquitous environment, the decisions of one user can be affected by the desires of others. This makes learning and prediction of user preference difficult. To address the issue, Hasan et al. [24] propose an approach of user preference learning which can be used widely in context-aware systems. The approach based on Bayesian RN-Metanetwork, a multilevel Bayesian network to model user preference and priority is used there.

According to [24], a real smart system should have the following three capabilities:

1. A smart system should be able to do inference.
2. A smart system should be able to learn by itself. User and developers can act like teachers, but the knowledge should be improved incrementally.
3. A smart system should be able to solve some difficult problems, such as the conflict among the users.

When applying the first capability on the case of smart learning environments, the environment should monitor the users (in this case the learners making use of the environment), and on the basis of collected data and using inferences from the data the smart learning environment should decide about such matters as what learning activities could be recommended to individual learners, or what are the problems the learner should solve in the given moment. Certainly this capability is useful in the case when assessing learners' skill should be provided by the smart learning environment itself [25]. In this case decisions of the environment could influence heavily the further study path of each individual learner.

If we think about the second capability, certainly it is expected that a really smart learning environment has to learn important facts related to the learners' progress by itself. If this is possible, then the environment would be able to decide, e.g., what group composition will be the most effective for each group members learning process (cf. [11]). Such a learning of the smart learning environment could be achieved by using various modern knowledge management approaches, see, e.g., [1, 26], or [27].

The third capability seems to be most important from the research behind this paper point of view. Solving difficult problems, as conflict among the users of the smart learning environment, is certainly a very difficult task. These conflicts usually are based on users' preferences that are changing over the time. It is clear that always there is certain set of preferences of the particular user, which are nearly static, not changing in a long term period. For instance, if a person does not like to strong light, it is unlikely, that a day after this preference will change. Such preferences can be evaluated by the environment quickly and possible conflicts with others could be solved simply by comparing the preferences and using some of simple comparative algorithms (the biggest one is the winner, or something like that). However, usually this is not the case of a group of learners residing in a smart learning environments and having very often contradictory requirements as to use some resources, or collaborating mutually on a project, which can be a source of numerous disputes and disagreements. There it is still lack of methods for solving such conflicts, and new approaches should be found and applied. According to [24], *the preference of user changes over time or based on situation. It makes online learning (or adaptation) a crucial requirement. Sometimes there is uncertainty in users temporary preference. User does not always select the most weighted choice. Again, when there are many users in the smart environment, the action of one user can affect others choice. It raises the challenges of distinguishing the preference of each user as well as resolving the conflicts among different user preferences. The introduction of probabilistic model can handle these uncertainty and adaptive prioritization of users.* This adaptive prioritization of users seems to be a promising way for handling the users' preferences successfully, however, this is still not solved satisfactorily.

4   Conclusions

Studying decision processes in smart learning environments can be one of ways how to understand their functionality and adjust the underlying multi-agent based architecture accordingly. It seems, that one of the most important points here lies in investigating various methods how the users' preferences can be represented, evaluated, and compared mutually aiming to give the smart environment a possibility of as proper decisions about its future actions as it is possible. Probabilistic approaches, as the one by [24] could be promising, however, it is necessary to elaborate them further on. There could be some promising directions based on ontologies, see, e.g., [28], nevertheless, the referenced paper was not written with that aim. So, a lot of further research seems to be ahead. The acknowledged project DEPIES is one of such initiatives aiming to contribute to optimization and better coordination of decision processes oriented on smart environment activities focused on the multiple residents case. And smart learning environments are typically used by a group of residents, so they are a good target for focused research with the described orientation. In the project, new approaches and algorithms for decision processes in suitable multi-agent architectures are investigated using the multi-agent modeling and simulation of these architectures. In order to enable proper investigation of various decision processes in a typical multi-agent based architecture of an intelligent environment, the modeling and simulation environment AnyLogic [29] is used broadly.

Some most recent results in this direction have been published already in [30]. We believe, that agent-based simulation will be a strong tool in our way towards deeper understanding of decision processes in smart environments, with a special accent on smart learning environments.

Acknowledgment. The support of Czech Science Foundation GAČR #15-11724S DEPIES is gratefully acknowledged.

References

1. Bureš, V., Čech, P., Mls, K.: Educational possibilities in the development of the ambient intelligence concept. Problems Educ. 21st Century **13**, 25–31 (2009). ISSN 1822-7864
2. Ducatel, K., Bogdanowicz, M., Scapolo, F., Leijten, J., Burgelman, J.C.: Scenarios for ambient intelligence 2010, ISTAG report, European commission. Institute for Prospective Technological Studies, Seville, November 2001. ftp://ftp.cordis.lu/pub/ist/docs/istagscenarios2010.pdf
3. Alsaif, F., et al.: Determination of smart system model characteristics for learning process. In: International Conference on Convergence Technology, vol. 4, pp. 881–885 (2014)
4. Mikulecký, P.: Smart learning environments - a multi-agent architecture proposal. In: 10th International Scientific Conference on Distance Learning in Applied Informatics (DIVAI 2014), Wolters Kluwer (2014)
5. Mikulecký, P.: Smart environments for smart learning. In: DIVAI 2012 (2012)
6. Bureš, V., Tučník, P., Mikulecký, P., Mls, K., Blecha, P.: Application of ambient intelligence in educational institutions: visions and architectures. Int. J. Ambient Comput. Intell. (IJACI) **7**(1), 94–120 (2016)
7. Kinshuk: Roadmap for adaptive and personalized learning in ubiquitous environments. In: Kinshuk, Huang, R. (eds.) Ubiquitous Learning Environments and Technologies. Lecture Notes in Educational Technology, pp. 1–13. Springer, Heidelberg (2015)
8. Winters, N., Walker, K., Rousos, D.: Facilitating learning in an intelligent environment. In: The IEE International Workshop on Intelligent Environments, pp. 74–79. Institute of Electrical Engineers, London (2005)
9. Libbrecht, P., Müller, W., Rebholz, S.: Smart learner support through semi-automatic feedback. In: Chang, M., Li, Y. (eds.) Smart Learning Environments. Lecture Notes in Educational Technology, pp. 129–157. Springer, Heidelberg (2015)
10. Hwang, G.J.: Definition, framework and research issues of smart learning environments-a context-aware ubiquitous learning perspective. Smart Learn. Environ. **1**(1), 1–14 (2014)
11. Kinshuk, Chen, N.S., Cheng, I.L., Chew, S.W.: Evolution is not enough: revolutionizing current learning environments to smart learning environments. Int. J. Artif. Intell. Educ. **26**(2), 561–581 (2016)
12. Spector, J.M.: Conceptualizing the emerging field of smart learning environments. Smart Learn. Environ. **1**(1), 1–10 (2014)
13. Becerra, G., Kremer, R.: Ambient intelligent environments and environmental decisions via agent-based systems. J. Ambient Intell. Human. Comput. **2**(3), 185–200 (2011)

14. Cook, D.J., Das, S.K.: How smart are our environments? An updated look at the state of the art. Pervasive Mob. Comput. **3**(2), 53–73 (2007)
15. Das, S.K., Roy, N., Roy, A.: Context-aware resource management in multi-inhabitant smart homes: a framework based on nash H-learning. Pervasive Mob. Comput. **2**(4), 372–404 (2006)
16. Pettigrew, A.M.: The Politics of Organizational Decision-Making. Routledge, New York (2014)
17. Sigaud, O., Buffet, O.: Markov Decision Processes in Artificial Intelligence. Wiley, New York (2013)
18. Yang, X., Yao, J.: Modelling multi-agent three-way decisions with decision-theoretic rough sets. Fundam. Inf. **115**(2–3), 157–171 (2012)
19. Burnett, C., Norman, C., Sycara, K.: Trust decision-making in multi-agent systems. In: Proceedings of the 22nd International Joint Conference on Artificial Intelligence, IJCAI 2011, pp. 115–120 (2011)
20. Yu, C.H., Werfel, J., Nagpal, R.: Collective decision-making in multi-agent systems by implicit leadership. In: Proceedings of the 9th International Conference on Autonomous Agents and Multiagent Systems, vol. 3, pp. 1189–1196. International Foundation for Autonomous Agents and Multiagent Systems (2010)
21. Wu, J., Xu, X., Zhang, P., Liu, C.: A novel multi-agent reinforcement learning approach for job scheduling in grid computing. Future Gener. Comput. Syst. **27**(5), 430–439 (2011)
22. Siebers, P.O., Aickelin, U.: Introduction to multi-agent simulation. arXiv preprint (2008). http://arxiv.org/abs/0803.3905
23. Luck, M., McBurney, P., Shehory, O., Willmott, S.: Agent technology: computing as interaction (a roadmap for agent based computing). Technical report, University of Southampton, UK (2005)
24. Hasan, M.K., Ánh, K., Mehedy, L., Lee, Y.-K., Lee, S.-Y.: Conflict resolution and preference learning in ubiquitous environment. In: Huang, D.-S., Li, K., Irwin, G.W. (eds.) ICIC 2006. LNCS (LNAI), vol. 4114, pp. 355–366. Springer, Heidelberg (2006)
25. Klimova, B.: Assessment in smart learning environment – a case study approach. In: Uskov, V.L., Howlett, R.J., Jain, L.C. (eds.) Smart Education and Smart e-Learning. Smart Innovation, Systems and Technologies, vol. 41, pp. 15–24. Springer, Heidelberg (2015)
26. Mikulecky, P.: User adaptivity in smart workplaces. In: Pan, J.-S., Chen, S.-M., Nguyen, N.T. (eds.) ACIIDS 2012, Part II. LNCS, vol. 7197, pp. 401–410. Springer, Heidelberg (2012)
27. Mikulecký, P.: Learning in smart environments - from here to there. In: 10th European Conference on e-Learning, pp. 479–484. ACL, Reading (2011)
28. Uskov, V.L., Bakken, J.P., Pandey, A.: The ontology of next generation smart classrooms. In: Uskov, V.L., Howlett, R.J., Jain, L.C. (eds.) Smart Education and Smart e-Learning. Smart Innovation, Systems and Technologies, vol. 41, pp. 3–14. Springer, Heidelberg (2015)
29. Tučník, P., Bureš, V.: Inclusion of complexity: modelling enterprise business environment by means of agent based simulation. Int. Rev. Model. Simul. (IREMOS) **6**(5), 1709–1717 (2013)
30. Mis, K., Cimler, R., Mikulecky, P.: Agent-based simulation for identifying the key advantages of intelligent environments for inhabitants with special needs. In: Sulaiman, H.A., Othman, M.A., Othman, M.F.I., Rahim, Y.A., Pee, N.C. (eds.) Advanced Computer and Communication Engineering Technology: Proceedings of ICOCOE 2015. Lecture Notes in Electrical Engineering, vol. 362, pp. 1031–1041. Springer, Heidelberg (2016)

Big Data Mining and Searching

Upgrading Event and Pattern Detection to Big Data

Soumaya Cherichi[1(✉)] and Rim Faiz[2]

[1] LARODEC, ISG Tunis, Bardo, Tunisia
soumayacherichi@gmail.com
[2] LARODEC, IHEC Carthage, Carthage Presidency, Carthage, Tunisia
Rim.Faiz@ihec.rnu.tn

Abstract. Social mediating technologies have engendered radically new ways of information and communication, particularly during events; in case of natural disaster like earthquakes tsunami and American presidential election. The growing complexity of these social mediating technologies in terms of size, number of users, and variety of bloggers relationships have generated a big data which requires innovative approaches in order to analyse, extract and detect non-obvious and popular events. This paper is based on data obtained from Twitter because of its popularity and sheer data volume. This content can be combined and processed to detect events, entities and popular moods to feed various new large-scale data-analysis applications. On the downside, these content items are very noisy and highly informal, making it difficult to extract sense out of the stream. Taking to account all the difficulties, we propose a new event detection approach combining linguistic features and Twitter features. Finally, we present our event detection system from microblogs that aims (1) detect new events, (2) to recognize temporal markers pattern of an event, (3) and to classify important events according to thematic pertinence, author pertinence and tweet volume.

Keywords: Big data · Microblogs · Event detection · Temporal markers · Patterns · Social network analysis

1 Introduction

Recent years have revealed an important increase of online social networks and social media platforms, which gave birth to a huge volume of data in blogs and more precisely microblogs. The importance of social media comes from the fact that each user is henceforth a potential author and the language is closer to the reality than any linguistic norm. However, this special kind of chaotic data creates an opportunity to improve effective event detection models by taking advantage of patterns that is created from big data analysis. During the « Arab Spring Movement », Twitter is considered as an important information source to coordinate protests and to bring awareness to the atrocities [1]. In recent world events, social media data has been shown to be effective in detecting earthquakes [2]; rumors [3]; crisis and disaster [4], spam [5], and identifying characteristics of information propagation [6, 7].

© Springer International Publishing Switzerland 2016
N.T. Nguyen et al. (Eds.): ICCCI 2016, Part II, LNAI 9876, pp. 377–386, 2016.
DOI: 10.1007/978-3-319-45246-3_36

A system that can extract this information from twitter big data and present an overview of upcoming popular events, such as sports matches, national holidays, and public demonstrations, is of potentially high value. This functionality may not only be relevant for people interested in attending an event or learning about an event; it may also be relevant in situations requiring decision support to activate others to handle upcoming events, possibly with a commercial, safety, or security goal.

Nowadays, people often attempt to search for trending news and hot topics in real time from microblogging messages to satisfy their information needs. Under such a circumstance, a real demand is to find a way to allow users to organize a large number of microblogging messages into understandable events.

Recently, several systems (e.g., [2]) have been proposed to detect events from tweets, but most of them are missing the analysis component. In the literature, several systems (e.g., [8, 9]) are proposed to analyze events from blogs, but they may fail in processing tweets, which are short and noisy, and do not explore rich information (e.g., users's network) in Twitter. This incites us to study the problem of event detection, which is an interesting and important task in such circumstances.

While most existing work either ignore structured aspects of the information present in Twitter, or transpose traditional approaches of NLP to extract these structures (such as parsing), a peculiarity of this work is rather to make the maximum specificities of Twitter (explicit and implicit links between tweets, redundant information, temporal and spatial co-occurrence, metadata…) to rebuild these structures. So while very many tweets cannot be the subject of parsing, because of their ungrammaticality, a structure linking the events and entities mentioned in the tweet can often be still inferred through the correlation between this particular tweet and others in the same field or a related event. Making sense of social media content is not trivial. Big Data streams from social media platforms usually contain much:

- **Informal use of language:** Twitter users produce and consume information in a very informal manner compared with traditional media. Tweets are 140 characters in length, forcing users to use short forms to convey their message. Many routine words are shortened such as "pls" for "please", "forgt" for "forgot", also the use of slang words, abbreviations and compound hashtags.
- **Noisy information:** While traditional event detection approaches assume that all documents are relevant, Twitter data typically contains a vast amount of noise and not every tweet is related to an event.

Social media platforms have empowered users to collaboratively create, distribute and exploit information as soon as a real-world event occurs in a wisdom of the crowd fashion. However, finding relevant events from an amount of noise and inter-personal communication can be a challenging task, since the relevance of a post is dependent both on its linguistic features and Twitter features. In this paper, we propose a novel system that combines a big data analytics environment with Twitter analytics to semantically analyze, detect and categorize events in real time by exploring rich information from Twitter. It fully supports the four functions proposed above: (1) handle the informality of language in a Twitter stream with a clustering stage to decrease noisy information and a procedure to rank tweets that describe an event best by their informativeness using NLP tools, (2) generating temporal patterns and cause consequence verbs for events, (3) detecting new events not precedently known, and (4) categorising and ranking events according to their importance.

The remainder of this paper is organized as follows. In Sect. 2, we give an overview of related works. In Sect. 3, we present our system outline and discuss experiments and obtained results in Sect. 4. Finally, Sect. 5 concludes this paper and outlines future work.

2 Related Works

By definition, "Big data refers to data sets whose size is beyond the ability of typical database software tools to capture, store, manage and analyse" [10]. In fact, nowadays, we are in the dawn of the big data era, where people are concerned with how to rapidly get the desired information, and Big Data analytics present suitable methods and techniques to extract key information from massive data. Indeed, big data analytics is where advanced analytic techniques are applied on big data sets [11]. Thence, those analytics have as a goal to discover associations and understand patterns and trends within a huge volume of data. Big data analytics are described by three primary characteristics: volume, velocity and variety and it has the potential to take advantage of the explosion in data to extract insights for making better informed decisions and bring values for enterprises and individuals.

Previous work on event extraction [12, 13] have focused largely on news articles, as historically this genre of text has been the main source of information on current events. In the meantime, social networking sites such as Facebook and Twitter have become an important complementary source of such information. The problem of event identification in social media was introduced by Becker et al. [14] who presented an incremental clustering algorithm that classifies social media documents into a growing set of events.

Few existing approaches are designed for streaming Twitter data and even fewer are scalable to real-time streams [16]. And most of these works or neglect structured aspects of the information present in Twitter, or transpose traditional approaches of NLP to extract these structures (such as parsing), a feature of this work is rather to exploit a maximum of specificities of Twitter (explicit and implicit links between tweets, redundant information, temporal and spatial cooccurrence, metadata ...) to rebuild these structures. So while very many tweets can not be the subject of parsing, because of their ungrammaticality, a structure linking the events and entities mentioned in the tweet can often be still inferred through the correlation between this particular tweet and others in the same field or a related event. To achieve this, the use of sequence modules (e.g. morphosyntactic analysis and syntactic and semantic roles in ...) will not suffice, and theoretical models well founded and optimized globally (e.g. algortithme of machine learning) will be offered. To achieve these objectives, we propose an approach that examines and highlights the poor performance of conventional NLP tools when applied to microblogs, and we are developing a series of tools for this type of text, which have very particular characteristics compared to other types of classical texts, especially in terms of redundancy of information, new types of structures (retweets, mentions, hashtags ...) and new lexical conventions (abbrévations, emoticons ...) which provide additional information to the traditional view of NLP and can contribute significantly to compensate for deficiencies of traditional information sources, i.e. grammatical and lexical in the context of microblogs. So, our goal is to extract and detect complete structures in tweets, unlike Ritter in [6, 17] that operates this formalism to classify types of events previously detected.

Processing within our system is divided in four stages. The first is tweet processing, during which hashtags are normalized then gathered. The second stage is Tweet Clustering, during which tweets are divided to two clusters: Cluster EVENT and cluster NOT EVENT according to a set of features selection. The third stage is Tweet Annotation during which potential key event information, temporal markers, and event terms (cause effect verbs), are extracted from single tweets in the Twitter stream. The fourth stage is event extraction, during which the strongest pairs of dates and event terms which match with patterns are extracted as events. The fifth and final stage is event presentation, during which additional event terms are extracted, the final set of event terms is selected and ordered, and tweets that mention an event are ordered. We describe and motivate the different components below. A separate evaluation of the most important components is presented in Sect. 4.

3   System Outline

TEXEV extracts a 4-tuple representation of events which includes a named entity, temporal markers such as calendar date, cause effect verbs and co-occurent hashtag frequency. This representation was chosen to closely match the way important events are typically mentioned in Twitter. An overview of the various components of our system for extracting events from Twitter is presented in Fig. 1. Given a raw stream of

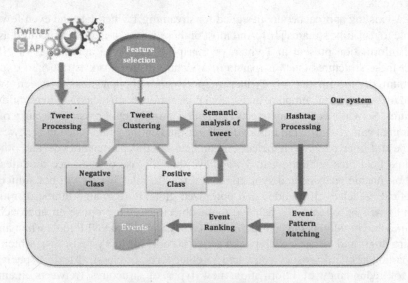

Fig. 1. Our system achitecture

tweets, our system extracts hashtags in order to normalize and group them in a set of clusters then our system gathered the stream of tweets in two clusters Positive Class: tweet contain event and Negative class if not, according to a set o features selection. Once we selected tweets containing event, we extracts named entities in association with event phrases, cause effect verb and temporal markers which are involved in significant events. Fourth step during which the extracted events are categorized into types. We extend the approach with a clustering stage to decrease duplicate output and a procedure to rank tweets that describe an event best by their informativeness. We implemented TEXEV based on Java and PHP with support of MySQL, Twitter4J.

3.1 Tweet Preprocessing

Information shared in social media, especially in discussion forums, blogs and tweets, is rich and vibrant. The application of the usual methods of NLP in this context, is not done without difficulty because of the noise and the "unusual" spelling. Blogs, tweets and statuses updates are written informally and look more like a "state of mind". This informality creates different challenges to the field of NLP.

Despite our filtered and focused crawling, many users use the hashtags popular and keywords to spam the stream to get attention. Including these tweets due to the mere presence of hashtags or keywords may bias the analysis, so a further round of denoising is performed following a few simple heuristics as described below:

- Tweets which contain hashtags.
- Tweets with similar content but have different user names and with the same timestamp are considered as multiple accounts.
- Tweets with same account and same content are considered as duplicate tweets.
- Tweets with same account and same content at multiple times are considered as spam tweets.

3.2 Hashtag Processing

As we have argued in the introduction, the problem of classifying a stream of social media data into events can be seen as an instance of stream data classification where the set of classes is constantly growing and evolving. We consider an event as something that happens in one day and in one place, such as event, or that spans several days or locations such as epidemic. An event will be represented by a set of terms whose frequency increases sharply to one or more times during the period analyzed. As hashtags used to give a general idea about the topics discussed in a tweet, one of our methods uses these elements to identify salient topics. Our first task is normalizing and grouping hashtags to use them as an event indicator.

Normalization of Hashtags. In this section, we present the techniques that we have applied to normalize terms. The microblogs users often make spelling mistakes creating several variants for the same term. This problem was already addressed by spelling systems using phonetic algorithms. These algorithms will index the words according to their pronunciation. The principle is to use the pronunciation of a misspelled word to predict the correct word with the same pronunciation, which corresponds to it. We used the algorithm of Soundex to normalize the terms that have a similar pronunciation.

Grouping of Hashtags. Our method is based on terms found in tweets to determine the salient topics; whereas, a subject is often represented by more than one term. Therefore it is important to group the terms referring same subject, if each term in the corpus represent a different subject. The semantic relationship between two terms is used to identify their degree of association. This information plays an important role in several areas of NLP such as the automatic construction of thesaurus, information retrieval... For example, it is useful to use terms similar to those specified in the user's query to retrieve relevant documents. Several studies have relied on bases manually constructed by linguists (e.g. WordNet) to determine the semantic relationships between terms. These databases contain information indicating the type of relationship (synonym, antonym, hypernym...) between terms. However, they do not cover the dialects or the writing mode (errors, abbreviations) used in social media. Furthermore, the language used in social media is frequently enriched with new terms invented by users (e.g., products, people, political party...).

We implemented a system based on co-occurrence hashtags to group a common topic. Since the tweets are short, we considered the whole tweet like window. We found that hashtags that frequently co-occur are similar or relate to the same subject. To measure the degree of relationship between two hashtags, we used the measure Pointwise Mutual Information (PMI). PMI measures the amount of information provided to the simultaneous presence of a pair of terms in our case the Hi hashtags.

$$\mathrm{PMI(Hi, Hj)} = \log\left(\frac{P(Hi\&Hj)}{P(Hi)P(Hj)}\right) = \log\left(\frac{N*a}{(a+b)*(a+c)}\right) \qquad (1)$$

P (Hf&Hj) is the probability that Hi and Hj appear together in one tweet, P (Hi) P (Hj) is the probability that Hi and Hj appear together, if they are statistically independent. Ratio P(Hi & Hj) and P (Hi) P (Hj) measure the degree of dependence between Hi and Hj. PMI is maximized when Hi and Hj are perfectly associated. N is the number of tweets considered; a is the number of times Hi and Hj appear together, b the number of times that Hi is present but Hj is absent while c is the number of times Hj is present, but where Hi is absent. In our case, the a priori determination of the number of clusters is hardly en-visageable since each cluster represents a subject, which we do not know the number in the corpus. Therefore we normalize PMI values using the method.

$$\mathrm{NPMI(ei, ej)} = \log\left(\frac{P(ei\&ej)}{P(ei)P(ej)}\right)/-\log P(ei\&ej) = \mathrm{PMI}/-\log P(ei\&ej) \qquad (2)$$

P (ei & ej) is the probability that ei and ej appear together. The value of NPMI is comprised within the range [−1, 1]. NPMI (ei, ej) is 1 when ei and ej are completely dependent and −1 if they are completely independent.

3.3   Semantic Analysis of Tweet

Tweet Clustering. For most clustering algorithms, the user must have sufficient knowledge of the data to determine the number of clusters. For example, the k-mean algorithm requires specifying in advance the number k of clusters to use. Therefore, we used k-means to classify tweets into two clusters: Positive class: if a user makes a tweet about a target event and Negative class: if a user doesn't make a tweet about a target event. As features for this task we use the presence of date, of named entities and of event terms by using dictionaries of event terms gathered from WordNet.

Pattern Matching. In order to extract event mentions such as temporal markers and cause effect verb from Twitter's noisy text, we first annotate a corpus of tweets, which is then used to train sequence models to extract events.

Our pattern acquisition approach involves multiple consequtive iterations of ML followed by manual validation. Learning patterns for each event-specific semantic role requires a separate cycle of learning iterations.cf

After analyzing 100000 tweets; we noticed that they might have one of the following patterns:

(1) Calendar term followed by an event.
(2) Preposition followed by a calendar term.
(3) Event followed by a calendar term.
(4) Subject followed by a relative pronoun, followed by a verb cause-consequence, and followed by event.
(5) Subject followed by a verb cause-consequence, followed by event.
(6) Subject followed by a verb cause-consequence, followed by event.

We sequence these **linguistic markers** according to their types.

(1) The calendar term class

- propo-num stands for preposition + number.
- Cal-num stands for: calendar + number.
- Prepo stands for preposition.
- Num-cal-num stands for number + calendar + number.

(2) The occurrence indicator class

- Adj_occ stands for adjective + occurrence.
- Adt_det_occ stands for tense adverb + determiner + occurrence.

(3) The relative pronoun class

- Prr_aux_ppa: relative pronoun + auxiliary + past participle.
- Prr_aux_adv_ppa stands for relative pronoun + auxiliary + adverb + past participle.

(4) The cause-consequence verb

- Verbconsq_subject: event + verb + event.
- Verbconsq_argument: subject + verb + event.

Tweet Ranking. The information to be extracted can target a specific event, or it can involve a discovery process in which latent events and their associated information are detected. We develop our ranking model based on our previous metrci measure for tweet ranking that integrates different criteria namely the social authority of micro-bloggers, the content relevance, the tweeting features as well as the hashtag's presence. [18] Among the most important tasks for a ranking system tweet is the selection of features set. We offer three types of features to rank tweets:

(1) Content features refer to those features which describe the content relevance between events and tweets namely how many tweets are similar to the tweet using TF-IDF Formulas and after how much time was retweeted several times.
(2) Tweet features refer to those features, which represent the particular characteristics of tweets, as URLs and hashtags in tweet. A tweet that contains a link or a hash tag may have a broad coverage, which may be also important. Also we capture the spread of a tweet through its re-tweet number and the number of favorites.
(3) Author features refer to those features which represent the authority of authors of the tweets in Twitter. According to our previous work a tweet from an authority is more significant than the one from a user with less credibility. Several signals can indicate the credibility of a user such as whether a user is a verified account (e.g., a news agent or a police department), how many followers a user has, the age of the account, and the number of tweets that a user has.

Finally, we use a linear equation to estimate the importance of a tweet based on the above features.

4 Experimental Evaluation

4.1 Data and Baseline

For evaluation purposes, we gathered roughly the 10 million most recent tweets on February 2nd 2016 to March 20th 2016 collected using a Java program that used the Twitter4J library. This library provides access to data (tweets, user information...) Twitter via its programming interface, Twitter API. We mainly studied the content of tweets (their sizes, the most frequents words,...), the preoccupations users based on hashtags used, the behavior of users. ... To demonstrate the importance of natural language processing and information extraction techniques in extracting informative events, We compare against a simple ngram baseline which does not make use of our NLP tools. Note that the ngram baseline is only comparable to the entity + date precision since it does not include event phrases or types.

4.2 Results

To quantify the effectiveness of our approach in detecting events, we compute the F1 score which captures both the Precision and Recall.Precision is computed as the number of detected events that match the ground truth including sub-events. Recall is

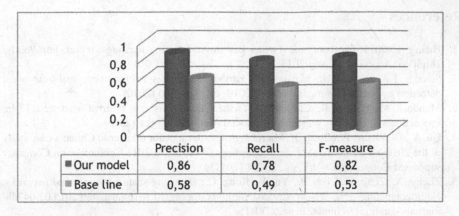

Fig. 2. System evaluation

	Precision	Recall	F-measure
Our model	0,86	0,78	0,82
Base line	0,58	0,49	0,53

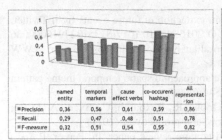

	named entity	temporal markers	cause effect verbs	co-occurent hashtag	All representation
Precision	0,36	0,56	0,61	0,59	0,86
Recall	0,29	0,47	0,48	0,51	0,78
F-measure	0,32	0,91	0,54	0,55	0,82

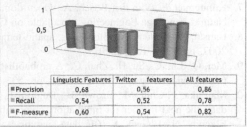

	Linguistic Features	Twitter features	All features
Precision	0,68	0,56	0,86
Recall	0,54	0,52	0,78
F-measure	0,60	0,54	0,82

Fig. 3. Individual feature performance   **Fig. 4.** Individual marker performance

computed as the number of events from the ground truth which were successfully detected. The F1 Score for our dataset was 0.82 (Fig. 2).

However, when we included Twitter features, both precision and recall further increased to 86 % and 78 % respectively, as shown in Fig. 3.

We analyzed each marker individually, to evaluate its performance, when we included all markers both precision and recall further increased to 86 % and 78 % respectively, as shown in Fig. 4.

5 Conclusion

In this work, we presented our novel approach and introduced our system TEXEV to detect and extract real time events in informal and high volume Twitter streams. The results demonstrate that the proposed approach can handle the informality of language in Twitter streams, through the use of our NLP method to extract event markers. Through experiments we demonstrated that the proposed approach is efficient and is able to capture reasonable events in topic streams and random streams on Twitter. Looking ahead, we intend to expand our system to categorize events of various kinds using Twitter according to their topic and to propose a summary of each event.

References

1. Huang, C. 2011. Facebook and Twitter key to Arab Spring uprisings: report. http://bit.ly/1bh6jV6. Accessed 28 Aug 2013
2. Sakaki, T., Okazaki, M., Matsuo, Y.: Earthquake shakes twitter users: real-time event detection by social sensors. In: WWW 2010, pp. 851–860 (2010)
3. Mendoza, M., Poblete, B., Castillo, C.: Twitter under crisis: can we trust what we RT? In: Proceedings of the First Workshop on Social Media Analytics (2010)
4. Qu, Y., C., Zhang, P., Zhang, J.: Microblogging after a major disaster in China: a case study of the 2010 Yushu Earthquake. In: Proceedings of the ACM 2011 Conference on Computer supported Cooperative Work, pp. 25–34 (2011)
5. Zheng, X., Zeng, Z., Chen, Z., Yu, Y., Rong, C.: Detecting spammers on social networks. Neurocomputing 159, 27–34 (2015). http://dx.doi.org/10.1016/j.neucom.2015.02.047db/journals/ijon/ijon159.html#ZhengZCYR15
6. Ritter, S., Mausam, C., Etzioni, O.: Named entity recognition in tweets: an experimental study. In: Proceedings of the Conference Empirical Methods in Natural Language Processing 2011 (2011)
7. Kunneman, F., Van den Bosch, A.: Open-domain extraction of future events from Twitter. Natural Language Engineering, Available on CJO 2016. doi:10.1017/S1351324916000036
8. Bansal, N., Koudas, N.: Blogscope: spatio-temporal analysis of the blogosphere. In: WWW 2007, pp. 1269–1270 (2007)
9. Mei, Q., Liu, C., Su, H., Zhai, C.: A probabilistic approach to spatio temporal theme pattern mining on weblogs. In: WWW 2006, pp. 533–542 (2006)
10. Manyika, J., Chui, M., Brown, B., Bughin, J., Dobbs, R., Roxburgh, C., Byers, A.H.: Big data: the next frontier for innovation, competition, and productivity (2011)
11. Elgendy, N., Elragal, A.: Big data analytics: a literature review paper. In: Perner, P. (ed.) ICDM 2014. LNCS, vol. 8557, pp. 214–227. Springer, Heidelberg (2014)
12. Sankaranarayanan, J., Samet, H., Teitler, B., Lieberman, M., Sperling, J.: Twitterstand: news in tweets. In: Proceedings of the 17th ACM SIGSPATIAL International Conference on Advances in Geographic Information Systems, pp. 42–51, Seattle, WA, USA, November 2009
13. Faiz R.: Identifying relevant sentences in news articles for event information extraction. Int. J. Comput. Process. Orient. Lang. (IJCPOL), World Sci. 19(1), 1–19 (2006)
14. Becker, H., Iter, D., Naaman, M., Gravano, L.: Identifying content for planned events across social media sites. Proceedings of the 5th ACM International Conference on Web Search and Data Mining, pp. 533–542. ACM, New York (2012)
15. Doan, S., Vo, B.K.H., Collier, N.: An analysis of Twitter messages in the 2011 Toho earthquake. Arxiv preprint arXiv:1109.1618 (2011)
16. Cherichi, S., Faiz, R.: Analyzing the behavior and text posted by users to extract knowledge. In: Hwang, D., Jung, J.J., Nguyen, N.-T. (eds.) ICCCI 2014. LNCS, vol. 8733, pp. 524–533. Springer, Heidelberg (2014)
17. Soumaya, C., Rim, F.: Big data analysis for event detection in microblogs. In: Król, D., et al. (eds.) Recent Developments in Intelligent Information and Database Systems. SCI, vol. 642, pp. 309–319, Springer, Heidelberg (2016)

A Graph-Path Counting Approach for Learning Head Output Connected Relations

Nuran Peker and Alev Mutlu[(⊠)]

Computer Engineering Department, Kocaeli University, Kocaeli, Turkey
{145112003,alev.mutlu}@kocaeli.edu.tr

Abstract. Concept discovery is a multi-relational data mining task where the problem is inducing definitions of a relation in terms of other relations. In this paper, we propose a graph-based concept discovery system to learn definitions of head output connected relations. It inputs the data in relational format, converts it into graphs, and induces concept definitions from graphs' paths. The proposed method can handle n-ary relations and induce recursive concept definitions. Path frequencies are used to calculate the quality of the induced concept descriptors. The experimental results show that results obtained are comparable to those reported in literature in terms of running time and coverage; and is superior over some methods as it can induce shorter concept descriptors with the same coverage.

Keywords: Concept discovery · HOC · Graph path counting · Support · Confidence

1 Introduction

Multi-relational data mining (MRDM) [1] is concerned with inducing patterns hidden in multiple relations. One of the most commonly addressed tasks in MRDM is concept discovery. Given a set of target instances and a set of related facts, concept discovery aims to discover logical definitions of the target instances in terms of related facts.

The concept discovery problem has long been studied under Inductive Logic Programming (ILP) [2] research and promising results in various domains are reported [3]. ILP-based concept discovery systems are vulnerable to the local minima problem, where refine by one-literal-at-a-time operators of ILP fail to improve concept descriptors quality. Graph-based approaches in concept discovery were first proposed to overcome such situations [4]. Graph-based approaches in concept discovery employ graph-theoretic approaches such as substructure discovery and path finding to induce concept descriptors.

In this paper we present a path finding method for concept discovery for head output connected relations. It inputs a relational database, represents each target instance and its related facts as an individual graph and discovers paths that frequently appear the graphs. Solution clauses are those concept descriptors that

© Springer International Publishing Switzerland 2016
N.T. Nguyen et al. (Eds.): ICCCI 2016, Part II, LNAI 9876, pp. 387–396, 2016.
DOI: 10.1007/978-3-319-45246-3_37

satisfy user provided minimum support and minimum confidence. This paper introduces a novel approach to calculate support and confidence of a graph path. In the proposed support/confidence calculation method, target instances that are modeled by a path contribute to the support of the path while instances, either from target relation or background knowledge, that are modeled by the path contribute to confidence of the path. The proposed method works on n-ary relations and is capable of inducing recursive concept descriptors.

To evaluate the performance of the proposed method, we conducted experiments on data sets that belong to the different learning problems. The experimental results show that the proposed system can learn definitions of relations that belong to different learning problems, and is compatible to state-of-the-art systems in terms of running time and coverage.

The rest of the paper is organized as follows: in Sect. 2 we define the concept discovery problem and provide an overview of various graph-based concept discovery systems. In Sect. 3 we introduce the proposed method. In Sect. 4 we present the experimental results and compare the obtained results to several state-of-the-art methods. The last section concludes the paper with future directions.

2 Background

Multi-relational data mining (MRDM) is concerned with discovering patterns hidden in multiple relations. One of the most commonly addressed tasks in MRDM is concept discovery. Given finite number of concept instances and related facts, concept discovery is the problem of inducing logical definitions of the concept relation, called *target relation*, in terms of other relations, called *background knowledge*. More formally, the target relation, $E = E^+ \cup E^-$, is a set of positive and negative instances generally expressed as ground facts; background knowledge, β, is intensionally or extensionally expressed observations related to target instances; and $\mathscr{L}$ is a language bias, the problem is inducing finite length Horn clauses where, in its disjunctive form, the positive literal is from E, negated literals are from β and the clause grammatically satisfies $\mathscr{L}$.

Completeness and consistency are two metrics to evaluate the performance of concept discovery systems. A concept discovery system is called complete if it covers all of the positive target instances, and consistent if it covers none of the negative target instances. A concept descriptor, H, is said to cover a target instance, e, if, with respect to background knowledge, it entails the target instance, $\beta \cup H \vDash e$. As real world data is generally noisy, completeness and consistency constraints are extended, respectively, to cover as many positive target instances as possible and as few negative target instances as possible. Concept discovery systems proceed in an iterative manner, at each iteration they discover some concept descriptors that explain limited number of target instances, and restart another iteration to discover concept descriptors to explain the remaining target instances. The process terminates whenever (i) all target instances are covered, (ii) number of uncovered target instances drop below a certain threshold, (iii) no more concept descriptors can be induced from the remaining target instances.

Concept discovery problem has long been studied under ILP research. Such systems represent the relational data within first order logic framework and utilize logic operators to induce concept descriptors. Operators of ILP refine concept descriptors by one-literal-at-a-time and fail to do so when concept descriptors should be refined by two or more literals simultaneously. Graph-based approaches for concept discovery were first proposed to overcome such situations.

Graph-based concept discovery systems are classified as substructure and path finding-based approaches. Assumption behind the substructure-based systems is that concept definitions should be frequently appearing substructures in the graph. The assumption behind the path finding-based approaches is that concept descriptors should be frequently appearing paths that connect certain nodes. Graphs provide a flexible framework for relational data representation. In methods such as [5,6] vertices represent facts and edges connect vertices that have some arguments in common. In [4,7] vertices represent fact arguments and edges connect vertices that form a fact. Several graph traversal methods have been employed in path finding-based concept discovery systems. [4] employs bidirectional search, while [7] performs beam search, [8] performs depth first search, and [6] traverses the graph in depth-limited manner. Graph-based approaches in concept discovery also differ in the manner they evaluate the concept descriptors. Substructure-based approaches generally follow graph compression ratio to evaluate the quality of substructures [7,9]. [10] follows support and confidence metrics to evaluate the quality of the induced concept descriptors. In [4,7] concept descriptors with the highest accuracies are selected as solutions.

A definite clause, $h \leftarrow b_1, b_2, \ldots, b_n$, is called Head Output Connected (HOC) if at least one argument of each body literal is an argument of the head literal or is instantiated in a preceding body literal. [11] is an ILP-based concept discovery system particularly designed for such learning problems.

The method proposed in this study is a path finding method for HOC type of learning problems. It distinguishes from the state-of-the-art path finding methods by

(a) generating candidate concept descriptors while building the graph,
(b) modifying definitions of support and confidence to be applicable to graphs,
(c) handling n-ary relations and being capable of inducing recursive concept descriptors,
(d) no need to represent the entire data as a graph,
(e) considering all of the target instances at once, hence avoiding concept descriptors shadowing discovery of other concept descriptors.

3   The Proposed Method

The proposed method is a path finding approach for concept discovery in graphs for head output connected of relations. It inputs a relational database, minimum support, *min_sup*, minimum confidence, *min_conf*, and maximum concept descriptor length, *dmax*; outputs head output connected concept descriptors of at most *dmax* literals and satisfy the *min_sup* and *min_conf* constraints. In the proposed approach, relational data is represented as a labeled directed graph, where vertices represent facts and edges connect those vertices that have at least one argument in common. Support and confidence of concept descriptors are calculated by counting paths in the graph. The proposed method consists of three main steps, namely *graph construction*, *concept descriptor extraction*, and *concept descriptor evaluation*.

– **Graph Construction:** In this step a graph for each target instance is constructed, with the target instance being the initial vertex of the graph. Starting from the initial vertex, the graph is constructed by adding vertices that represent facts related to the those relations represented by vertices already in the graph. Two facts are called related if they have at least one argument value in common. Algorithm 1 outlines the graph construction step.

Algorithm 1. Graph construction

Require: Relational database, D
Ensure: D as graphs, G_i
1: **for** (each target instance t_i in D) **do**
2:    v = addVertex(G_i)
3:    v.rn = t_i
4:    v.path.push(v)
5:    toExpand.push(v)
6:    **for** (d = 1; d<dmax; d++) **do**
7:      **while** (!toExpand.empty()) **do**
8:       v = toExpand.pop()
9:       r = getRelatedFacts(v.rn)
10:       **for** (i = 0; i<r.size(); i++) **do**
11:        it = find(G_i, r[i])
12:        **if** (it != NULL) **then**
13:         it.path=v.path
14:         it.path.push(t)
15:         addEdge(G_i, v, it)

16:         nodesAtLevel[i][d].push(it)
17:        **else**
18:         t = addVertex(G_i)
19:         t.rn = rfs[i]
20:         t.path=v.path
21:         t.path.push(t)
22:         addEdge(G_i, v, t)
23:         nodesAtLevel[i][d].push(v)
24:         toExpandTmp.push(t)
25:        **end if**
26:       **end for**
27:      toExpand = toExpandTmp
28:      toExpandTmp.clear()
29:     **end while**
30:    **end for**
31: **end for**

In the proposed method vertices are implemented with bundled properties. The *rn* property stores the relation the vertex represents. The *path* property stores the path to be followed to reach that particular vertex from the initial vertex of the graph. The *addVertex()* function adds a new vertex to the graph. The *find()* function returns a vertex descriptor if its second argument, a relation, is represented by a vertex in the graph. The *toExpand* list keeps a track of nodes that need to be expanded at the current iteration, and *toExpandTmp* stores the vertices that will be expanded in the next iteration. For a related fact, a vertex representing it may or may not be present in the graph. If such a vertex exists in the graph, then a directed edge from the currently expanded vertex to that vertex is added, else a new vertex is created and its properties are set. If the expansion introduces a new vertex to the graph, the newly added vertex is added to a list called *toExpandTmp*. Once all the nodes in the *toExpand* list are examined and the graph is adjusted accordingly, nodes in the *toExpandTmp* are copied into *toExpand* list and a new expansion iteration starts. This expansion is performed *dmax* times as more iterations will produce paths of length longer that *dmax* which are out of interest. At each iteration an array called *nodesAtLevel* is maintained to keep vertices that are reachable at number of hopes indicated by the index of the second dimension from the initial node. Due to the limitation on the maximum rule length, this step does not convert the entire data set into a graph, but only the part of the data set that will generate concept descriptors with desired length properties.

- **Concept Descriptor Extraction:** Algorithm 2 outlines this step. Firstly, paths of length 2, 3, ..., *dmax* are extracted from the graphs. This is a straightforward process as *nodesAtLevel* list keeps vertices reachable at each level and vertices store their path information in the *path* property. Secondly, the extracted paths are generalized by substituting constant arguments with variables. In this step a hash table is constructed to store generalized paths and vertices whose generalization map to that generalized path is constructed.

Algorithm 2. Concept Descriptor Extraction

Require: Nodes reachable from initial node at most dmax hops, $nodesAtLevel[][]$
Ensure: Generalized HOC concept descriptors
 1: **for** (i = 0; i < #TargetInstances; i++) **do**
 2:   **for** (j = 0; j < dmax; j++) **do**
 3:     **for** (k = 0; k < nodesAtLevel[i][j].size(); k++) **do**
 4:       s = generalizePath(G_i[$nodesAtLevel[i][j][k]$])
 5:       **if** (isHOC(s)) **then**
 6:         it = map.find(s)
 7:         **if** (it == NULL) **then**
 8:           map.insert(s, <i, nodesAtLevel[i][j][k]>)
 9:         **else**
10:           map[i]→second.insert(s, <i, nodesAtLevel[i][j][k]>)
11:         **end if**
12:       **end if**
13:     **end for**
14:   **end for**
15: **end for**

In this step, generalized paths are also tested for being head output connected. Algorithm 3 outlines this process. It follows a set theoretic approach for HOC testing. For a concept descriptor to be HOC, at least on argument of a body literal should be a subset of the head literals or be instantiated in a preceding body literal. As concept descriptors induced by the proposed method are paths, at least one argument of a body literal is instantiated in its immediate predecessor. The only case that may violate HOC property in our system is introduction of a new variable in the last literal of the path. Hence, the arguments of the last body literals should be those that are introduced in the proceeding literals and never used again. As an example consider the concept descriptor elti(A,B):-wife(A,C), brother(C,D), husband(D,B). Following the Algorithm 3 s changes as follows: literal examined elti(A,B) and $s = \{A,B\}$, literal examined wife(A,C) and $s = \{B,C\}$, literal examined brother(C,D) and $s = \{B,D\}$. Final literal is husband(D,B) and s is $s = \{B,D\}$, hence the clause HOC.

Algorithm 3. HOC test

Require: Path, p
Ensure: True / False
 1: **for** (i = 0; i < p[0].arguments.size(); i++) **do**
 2:     s.insert(p[0].arguments[i].generalizedValue)
 3: **end for**
 4: **for** (i = 1; p.size() - 1; i++) **do**
 5:     **for** (j = 0; j < p[i].arguments.size(); j++) **do**
 6:         s1.insert(p[i].arguments[j].generalizedValue)
 7:     **end for**
 8:     s = s\s1 ∪ s1\s
 9:     s1.clear()
10: **end for**
11: **for** (i = 0; i < p[p.size()].arguments.size(); i++) **do**
12:     s1.insert(p[p.size()].arguments[i].generalizedValue)
13: **end for**
14: **if** s\s1 == ∅ **then**
15:     **return** true
16: **else**
17:     **return** false
18: **end if**

- **Concept Descriptor Evaluation:** In this step support and confidence of the paths are calculated. Number of target instances explained by a path constitute the support set of a path. Number of facts, either from target relation or background knowledge, contribute the confidence set of the path.

To calculate the support of a path, hash table constructed in step 2 is used. To find the support set of a generalized path, for each vertex, v, in the paths, distinct $G_i[v].path[0]$ are counted. Support a path is calculated by dividing the size of support set by the number of target instances.

To find confidence of a path, a method similar to support calculation is followed. Vertices that contribute to a certain generalized path are retrieved from the $< generalizedPath, paths >$ hash table. For each such vertex v, vertices connected to v are extracted from the graph. Arguments of $G_i[v].path[0]$ are replaced with the arguments of the arguments of the relation represented by the related vertex. This modified path is generalized and is checked against the

---

Algorithm 4. Concept Descriptor Evaluation

Require: Vector of $<generalized, paths>$, paths
Ensure: Frequent and strong concept descriptors
1: **for** (i = 0 i < paths.size(); i++) **do**
2:     p = paths[i]→second
3:     **for** (j = 0; j < p.size(); j++) **do**
4:       supSet.insert($G_{p[j] \to first}[p[j] \to second]$.rn)
5:       confSet.insert($G_{p[j] \to first}[p[j] \to second]$.rn)
6:       vC = getConnectedVertices()
7:       **for** (k = 0; k < vC.size(); k++) **do**
8:         $G_{p[j] \to first}[p[0]]$.rn.arguments = $G_{p[j] \to first}[vC[k]]$.rn.arguments
9:         **if** (paths[i]→first == generalize($G_{p[j] \to first}[vC[k]]$)) **then**
10:          confSet.insert($G_{p[j] \to first}[vC[k]]$.rn)
11:        **end if**
12:       **end for**
13:     **end for**
14:     support = $\frac{supSet.size()}{\#TargetInstances}$
15:     confidence = $\frac{supSet.size()}{\#ConfSet.size()}$
16:     **if** (support ≥ min_sup && confidence ≥ min_conf) **then**
17:       solution.push(paths[i]→first)
18:     **end if**
19: **end for**

---

original generalized path. Exact match of these two generalized paths means that the path also explains the related fact and this fact contributes to the confidence set of the path. The confidence of a path is calculated by dividing the size of the support set by the size of the confidence set. Algorithm 4 given below outlines the support and confidence calculation step.

4 Experimental Results

In order to evaluate performance of the proposed method we conducted experiments on three data sets. The first data *elti* consists of binary relations. The second data set is Mooney's kinship data set [12], which contains binary kinships. The third data set consists of facts that define Fibonacci numbers. It includes plus/3 and predecessor/2 relations as background knowledge, and fibonacci/2 as the target relation. The properties of the data sets are given in Table 1. Minimum support, minimum confidence and maximum concept descriptor length parameters for elti data set are obtained from [13], for the family data set from [14] and for the Fibonacci data set from [11].

The first data set is used to evaluate the performance of the proposed method that includes facts indirectly related to the target instances. The second data set is highly connected. The third data set is used to evaluate the performance of the proposed method on n-ary relations with recursive definitions. All relations in the data sets are from the categorical domain.

Table 1. Data sets and experimental settings

Name	# Relations	# Instances	Min. Sup.	Min. Conf.	Rule Length
Elti	9	224	0.2	0.6	3
Family	12	744	0.1	0.7	9
Fibanocci	5	30	1	1	5

In Table 2 we compare the results obtained for the Family data set to those reported for RPBL [5] and a Hybrid Graph-based Concept Discovery System [14]. The $\#C.D$ column lists the number of concept descriptors discovered, the $Av.\ L.$ column lists the average length of the concept descriptors discovered, the $Covr.$ column lists the coverage of the concept descriptors, and lastly the $R.T.$ column lists the running time.

Table 2. Comparison of the Family Data set

Data Set	Proposed M.				RPBL			Hybrid-GCD		
	#C.D	Av.L	Covr	R.T	#C.D	Av.L	Covr	#C.D	Av.L	Covr
Brother	18	2	100	1.19	2	6	95	2	2.5	100
Niece	4	2	100	1.06	2	6	95	7	2	100
Nephew	4	2	100	1.06	2	6	95	7	2	100
Mother	4	2	100	1.44	2	6	95	7	2	100
Daughter	4	2	100	1.45	2	6	95	7	2	100
Wife	7	2	100	1.12	2	6	95	7	2	100
Uncle	4	2	100	1.14	2	6	100	9	2	100
Aunt	4	2	100	1.14	2	6	100	10	2	100
Son	4	2	100	1.45	2	6	100	10	2	100
Father	4	2	100	1.45	2	6	100	10	2	100
Sister	10	2	100	1.18	2	6	100	10	2	100
Husband	50	2	100	1.09	2	6	100	10	2	100

The experimental results show that when compared to Hybrid-GCD, the proposed method discovers rules of almost the same length, while number of concept descriptors found vary. This is due to the support and confidence calculation method. Assume that $p/2$ is the target relation and $ar1$, $ar2$ are two arbitrary argument values in the data set. In Hybrid-GCD replacing arguments of the target relation with $ar1$ and $ar2$ is allowed while in the proposed method such a replacement is allowed if and only if there exists a fact with these two values as its arguments. When compared to RPBL, the proposed method induces shorter concept descriptors. Concept descriptors with large number of literals may be hard to interpret and such concept descriptors are subject to the over-fitting problem. When compared by means of number of concept descriptors discovered, the proposed method discovers larger number of concept descriptors. RPBL performs search for predefined number of target instances and if a concept descriptor is found to explain a target relation, this target relation is removed from the instance set and new concept descriptors are sought for remaining target instances. In such an approach, (i) ordering of target instances alters the final solution set, (ii) removal of a target instance may prevent discovery of some other concept descriptors due to support and confidence values. As the proposed method considers all the target instances at once, it is possible to discover concept descriptors supported by the same target instances. The proposed approach is capable of inducing much simpler concept descriptors with almost the same precision and recall values. When systems are compared with respect to running time, the proposed method induces the final concept descriptors around 1 s, while it is reported that RPBL induces the concept descriptors in 0.027 on the average and around 1 s for Hybrid-GCD.

For the Elti data set we compared the results obtained to those of Hybrid-GCD. The proposed method found exactly the same solution clauses with full coverage around 2 s, while running time of Hybrid-GCD is reported 0.51 s. This is due to the fact that, Hybrid-GCD builds a single graph for the entire data set while the proposed method builds a distinct graph for each data set. Although this may seem to be a downside of the proposed system, such a design model allows system to be parallelized.

To test the applicability of the proposed system on n-ary relations and relations with recursive definitions, we conducted experiments on manually generated Fibonacci data set. The proposed method discovered the following rule around 1 s. This rule is the same rule reported in [11].

$$\text{fib(A,B):-pred(C,A)fib(C,D)pred(E,C)fib(E,F)plus(F,D,B)}$$

5 Conclusion

In this work we propose a path finding method for concept discovery in graphs for head output connected relations. It inputs a relational database, creates as many graphs as the number of target instances and outputs those paths that satisfy user defined minimum support and minimum confidence values. Number of target instances modeled by the path contribute to the support of the path, and the number of facts modeled by the path contribute to the confidence of the path. The method can handle relations with arbitrary number of arguments and can learn recursive and non-recursive concept descriptors.

A major limitation of the proposed method is that it can not handle arguments from continues domain. A future work includes implementation of a discretization method to handle such data sets. Another future work is implementing parallel version of the proposed method. The proposed method is well-suited for parallelization as operations performed on graphs are independent.

References

1. Dzeroski, S.: Multi-relational data mining: an introduction. SIGKDD Explor. **5**(1), 1–16 (2003)
2. Muggleton, S.: Inductive logic programming. New Gener. Comput. **8**(4), 295–318 (1991)
3. Bratko, I., Muggleton, S.: Applications of inductive logic programming. Communications of the ACM **38**(11), 65–70 (1995)
4. Richards, B.L., Mooney, R.J.: Learning relations by pathfinding. In: Proceedings of the 10th National Conference on Artificial Intelligence, San Jose, CA, 12–16 July 1992, pp. 50–55 (1992)
5. Gao, Z., Zhang, Z., Huang, Z.: Extensions to the relational paths based learning approach RPBL. In: First Asian Conference on Intelligent Information and Database Systems, ACIIDS 2009, Dong hoi, Quang binh, Vietnam, 1–3 April 2009, pp. 214–219 (2009)
6. Abay, N.C., Mutlu, A., Karagoz, P.: A graph-based concept discovery method for n-Ary relations. In: Madria, S., Hara, T. (eds.) DaWaK 2015. LNCS, vol. 9263, pp. 391–402. Springer, Heidelberg (2015)
7. Gonzalez, J., Holder, L., Cook, D.J.: Application of graph-based concept learning to the predictive toxicology domain. In: Proceedings of the Predictive Toxicology Challenge Workshop (2001)
8. Yan, X., Han, J.: gSpan: Graph-based substructure pattern mining. In: Proceedings of the 2002 IEEE International Conference on Data Mining (ICDM 2002), 9–12 December 2002, Maebashi City, Japan, pp. 721–724 (2002)
9. Yoshida, K., Motoda, H., Indurkhya, N.: Graph-based induction as a unified learning framework. Appl. Intell. **4**(3), 297–316 (1994)
10. Inokuchi, A., Washio, T., Motoda, H.: An apriori-based algorithm for mining frequent substructures from graph data. In: Zighed, D.A., Komorowski, J., Żytkow, J.M. (eds.) PKDD 2000. LNCS (LNAI), vol. 1910, pp. 13–23. Springer, Heidelberg (2000)
11. Santos, J.C.A., Tamaddoni-Nezhad, A., Muggleton, S.: An ILP system for learning head output connected predicates. In: Lopes, L.S., Lau, N., Mariano, P., Rocha, L.M. (eds.) EPIA 2009. LNCS, vol. 5816, pp. 150–159. Springer, Heidelberg (2009)
12. Hinton, G.: UCI machine learning repository (1990)
13. Kavurucu, Y., Senkul, P., Toroslu, I.H.: Concept discovery on relational databases: new techniques for search space pruning and rule quality improvement. Knowl.-Based Syst. **23**(8), 743–756 (2010)
14. Mutlu, A., Karagoz, P.: A hybrid graph-based method for concept rule discovery. In: Bellatreche, L., Mohania, M.K. (eds.) DaWaK 2013. LNCS, vol. 8057, pp. 327–338. Springer, Heidelberg (2013)

A Framework for Analysis of Ontology-Based Data Access

Agnieszka Konys[(⊠)]

Faculty of Computer Science and Information Technology, West Pomeranian
University of Technology in Szczecin, Żołnierska 49, 71-210 Szczecin, Poland
akonys@wi.zut.edu.pl

Abstract. In last decade the terms related to Semantic Web become significant
elements in the efficient way of information retrieval, processing and supporting
availability of machine readable data. An ontology offers a wide spectrum of its
application for data access. Ontology-Based Data Access is regarded as a key
ingredient for the new generation of information systems, especially for
Semantic Web applications that involve large amounts of data. OBDA system
uses an ontology as a conceptual schema of the subject domain, and as a basis of
the user interface for SQL database systems. This paper presents the complex
overview of OBDA systems and a framework for analysis of selected tools for
OBDA system creation.

Keywords: OBDA · Ontology · Data access · Large data sets · OBDA
framework

1 Introduction

A huge amount of data is being generated every day, and it is an increasing trend. As a
consequence, it leads to information overload. Large amounts of data have been
accumulated by entities from a large variety of sources and in many different formats.
Data access is one of the determining factors for the potential of value creation processes. It directly influences on decision-making processes, analysis and effective
exploitation of the data. The process of gathering data over disparate sources is
a time-consuming task. However, data access is still a major bottleneck for many
entities. The problem is how to make use of knowledge and other technical skills to
extract data from different sources. Another problem concerns the interpretation and the
practical usage of the data. In real life, decisions must often be made quickly, on base
of current and relevant information, without wasting time for analyzing unnecessary
data. The heterogeneity of documents published on the Web may provides some
obstacles. The existence of diverse input sources both structured, semi-structured and
unstructured poses new challenges for handling data. Another problem concerns the
linking between different types of documents available on the Web [2]. Furthermore,
the high importance is assigned to the data storage in triples independently of physical
scheme. The data are stored in heterogeneous systems, thus the access (by posing
queries over the data) is undoubtedly a challenging task. There is no simple way to
avoid heterogeneous data sources.

© Springer International Publishing Switzerland 2016
N.T. Nguyen et al. (Eds.): ICCCI 2016, Part II, LNAI 9876, pp. 397–408, 2016.
DOI: 10.1007/978-3-319-45246-3_38

The paper describes available ontology-based solutions to knowledge extraction from different data sources. Moreover, the paper presents the complex overview of OBDA systems and a framework for analysis of selected tools for OBDA system creation. It considers the main features of OBDA tools, especially it describes drivers, implementation, reasoning, used DL knowledge base, provided interface, compatibility with ontology editors, extensions, mapping and others. Altogether, it contains 9 main criteria and 30 sub-criteria. It is possible to add more information or update existing ones.

2 Ontology-Based Solutions to Knowledge Extraction from Different Data Sources

In last decade the terms related to Semantic Web become significant elements in the efficient way of information retrieval, processing and supporting availability of machine readable data. A close relation between ontologies and the Semantic Web is noticeable, hence the role of ontologies and their application in knowledge extraction is significant. Nowadays a number of ontology-based solutions is still rising up. It is possible to indicate many methods, approaches, and tools supporting knowledge extraction (e.g. automatic ontology construction from different types of input sources: structured, semi-structured, unstructured, Question Answering Systems (QAS), Ontology-Based Information Extraction (OBIE) systems, Ontology-Based Data Access (OBDA), and different mining techniques) [2].

Ontology-based approaches provide a practical framework to address the semantic challenges presented by distributed and large data sets. Moreover, they allow to add metadata ontology for annotating data, and they bring benefits for search and retrieval information. The key element is that they offer reasoning processes and understanding of the obtained results. A significant role is assigned to mapping content to defined ontology, and then convert content to triples according to defined ontology. The process of combining the reasoning with large amount of data with ontological knowledge has made a significant meaning. The main problem concerns the way of ontology storage and effective reasoning, without losing out of sight the need of scalability [17]. The fundamental issue is to ensure scalable reasoners for a huge amount of collected data.

Ontology-based solutions to knowledge extraction support the process of dealing with heterogeneous data, and the access to relevant data, with regard to increasing number of it [9]. Generally, an ontology offers a wide spectrum of its application in Big Data context. It can be assumed that the combination of ontologies and Big Data can solve some of the problems identified for large data sets [17]. The use of ontologies for accessing data is one of the most exciting new applications of description logics in databases and other information systems. The aim is to limit or reduce querying relational databases. One of the most interesting usages of shared conceptualizations is Ontology-Based Data Access (OBDA). OBDA provides a convenient way to deal with large amounts of data spread over heterogeneous data sources [11]. OBDA allows users to formulate queries in a single user-friendly ontology language. These queries are then unfolded and executed on the data sources. It is a part of Ontology-Based Data Management (OBDM).

OBDA solutions provide a lot of opportunities to connect ontology with database, but it is not only one way (Fig. 1). Interesting possibility is to convert database (e.g. RDB) to ontology-based (Resource Description Format) database. RDF is the native language of linked data. It is based on triples {Subject, Property, Object}, and it is machine-readable. Then, RDF Schema provides the vocabulary for RDF, and it allows to create class hierarchy and properties. Moreover, OWL format can be easily converted into RDF, XML or other syntax. The RDF and OWL are used to integrate all data formats and standardise existing ontologies.

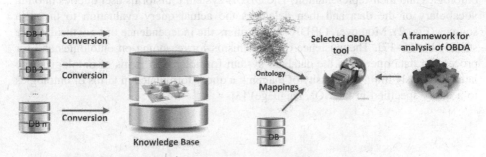

Fig. 1. Different ways to knowledge extraction using databases and ontologies

The queries may be executed by SPARQL (SPARQL Protocol And RDF Query Language) as RDF query language. SPARQL may be used to extract information from databases as the semantic query language. It allows to retrieve and manipulate data stored in RDF format. Moreover, more expressive languages exist, e.g.: RIF (Rule Interchange Format) and SWRL (Semantic Web Rule Language). They provide support rules, and distinguish relations which are not directly described by description logic in OWL.

2.1 Ontology-Based Data Management

The general aim of Ontology-Based Data Management (OBDM) is to separate the conceptual model of data from the way of data storage. The architecture of OBDM is composed of the three layers: data source on the bottom, and the ontology on the top, mapping between queries over the sources and queries in the ontology language. OBDM offers a support in reading the source data. Instead, OBDA is sufficient in applications where the data sources located at the bottom layer are maintained independently of the query framework at the top level [15].

2.2 Ontology-Based Data Access

OBDA is regarded as a key ingredient for the new generation of information systems, especially for Semantic Web applications that involve large amounts of data. The underlying idea is to facilitate access to data by separating the user from the raw data sources using an ontology that provides a user-oriented view of the data and makes it accessible via queries formulated solely in the language of the ontology without any knowledge of the actual structure of the data.

OBDA system uses an ontology as a conceptual schema of the subject domain, and as a basis of the user interface for SQL database systems. OBDA offers the direct access to data source [15]. The ontology defines a high-level global schema and provides a vocabulary for user queries. Moreover, the ontology gives the possibility to hide details. It makes use of database expert knowledge when mapping relationships. The queries are executed by mappings describing relationship between concepts in the ontologies and their representation. The OBDA system transforms user queries into the vocabulary of the data and then delegates the actual query evaluation to the data sources (Fig. 2). Moreover, OBDA system offers the independence of how the data are stored [11, 15–17]. The existence of mechanisms for reasoning on ontologies and for processing data queries in the database system formulated in terms of ontologies (the latter obviously implies necessity of mapping a query formulated in terms of ontologies to a query specified in the SQL language) [3].

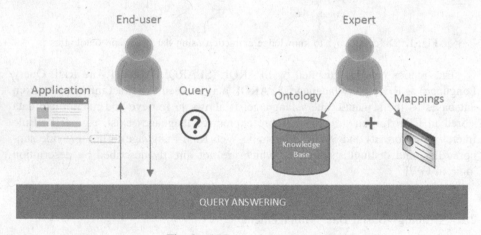

Fig. 2. OBDA main processes

The OBDA is not deprived of disadvantages. Many efforts have been put in defining the appropriate language for the semantic layer, defining the structure and language of the mappings used to link the data and the semantic layer, studying the complexity of offering a set of useful services such as query answering, database schema extraction, inconsistency management, etc. Moreover, ontology and mappings are expensive and take some time [4]. The efficiency for translation process and execution queries might be improved.

3 OBDA Specification and Query Answering

OBDA specification is composed of the following elements: the ontology, the knowledge base, the source schema and the mappings.

3.1 Ontology

An ontology can be defined as an explicit representation of a shared conceptualization. It enables defining concepts, relations and instances [21]. The one of the advantages of an ontology is that it can cope with the real world complexity and adapt with the changes. The common standard is OWL (Ontology Web Language). Description Logic (DL) is a theoretical base for OWL language for expressing and modeling ontologies. It was created to help in achieving an acceptable compromise between its expressiveness, which remains sufficient for many applications, and computational complexity of reasoning on ontologies and processing queries to data stored in large databases. The general idea of creation and using logical languages is to produce those semantic layers [17]. Development of standards of ontology description languages and Semantic Web indicates a new activity wave in developing tools for systems of semantic access to databases and a new class of database systems, called Ontology Based Data Access (OBDA) systems [15].

3.2 Knowledge Base

Description Logic is commonly used to describe ontologies. It allows to express axioms about concepts (unary properties), roles (binary relations) and individuals (constants). DL offers a wide spectrum of expressive power and computational complexity of complete reasoning [2]. A DL-Lite knowledge base (KB) is composed of two components: an intensional level (TBox), used to model the concepts and the relations (roles) of the ontologies, and an extensional level (ABox), used to represent instances of concepts and roles. DL ontology (O) is a pair (T, A), where T is the TBox, and A is the ABox. The aim of reasoning in DL-Lite is to provide computing subsumption between concepts, and to check satisfiability of the knowledge base. DL used in OBDA systems as a language of conceptual modeling of subject domains came to existence as a result of synthesis of expressive means of the first order logic and structural languages of knowledge representation (semantic networks and frames) [3, 18].

3.3 The Source Schema and Mappings

The source schema is a schema for the source databases. Then, the mapping links the terminology of the ontology to queries over the source schema. A mapping assertion m from a source schema S to a TBox T has the form $\varphi(x) \sim> \psi(x)$, where $\varphi(x)$, called the body of m, and $\psi(x)$, called the head of m, are queries over S and T, respectively, both with free variables x, which are called the frontier variables. The number of variables in x is the arity of the mapping assertion. In principle, $\varphi(x)$ and $\psi(x)$ can be specified in generic query languages. The literature on data integration and OBDA has mainly considered $\varphi(x)$ expressed in a (fragment of) first-order logic, and $\psi(x)$ expressed as a CQ - Conjunctive Query [16, 17, 19].

3.4 OBDA Formalization

From the formal point of view, a traditional OBDA system (O) can be viewed as a triplet of the form: (T, M, D), where T is the TBox of the used ontology (a set of OWL 2 QL axioms [14]), D is a relational database which is its ABox, and M is a set of assertions of mappings between T and D. A mapping is a set of assertions, each one associating a query $\varphi(x)$ over the source schema with a query $\psi(x)$ over the ontology. The intuitive meaning of a mapping assertion is that all the tuples satisfying the query $\varphi(x)$ also satisfy the query $\psi(x)$ [16]. The OWL 2 QL profile of OWL 2 ensures that queries formulated over the T can be rewritten into SQL [15].

3.5 Query Answering

The process of query answering in OBDA is divided into three stages. In the first stage called ontology rewriting, it is possible to rewrite an initial ontology query. The query considers the set of data or data a user is looking for. The knowledge base is used in this rewriting process. At the end, a collection of queries is provided. The next stage, unfolding, allows user to replace the ontology vocabulary with database queries. A base for unfolding is mapping. Finally, a collection of queries over the database schema is provided. After the execution (the third stage) of the final queries it is possible to quit DBMS. In literature, a number of rewriting techniques and algorithms exists. It is possible to indicate the following solutions: CGLLR, Rapid, REQUIEM (REsolution-based QUery rewrIting for Expressive Models), PerfectRef, Presto, the tree-witness rewriting and its extensions [1, 16, 18, 20].

4 A Framework for Ontology-Based Data Access Analysis

4.1 Selected Tools for Creating OBDA Systems

The analysis presents the description of each of selected OBDA tools, including the information about drivers, implementation, reasoning, used DL knowledge base, provided interface, compatibility with ontology editors, extensions, mapping and others. This analysis encompasses different criteria, but it is worth to notice that they are not equal for all of analyzed solutions.

4.2 A Framework for Analysis of Selected Tools for OBDA System Creation

The necessity of framework construction of OBDA has received a lot of attention in the last years. In literature, many theoretical studies have been taken for OBDA systems [3, 4, 16], and the development of OBDA projects in various domains (e.g. Optique

Table 1. The analysis of selected tools for OBDA systems [1, 3, 5–8, 10–15, 20]

Name	Criteria
QuOnto (QUerying ONTOlogies) system	Drivers: Oracle, DB2, SQL Server, MySQL, Postgres (and others drivers); Implementation: Java; Reasoner: A reasoner for the DL (DL Lite family); implemented query algorithm to check consistency; data processing for large volume ABox and ensures compatibility with OWL2 and SPARQL; Interface: Extended DIG interface; DL knowledge base: The intensional level of the knowledge base is simply a DL-Lite TBox. The ABox is stored under the control of a DBMS. The system allows to efficiently answer complex conjunctive queries
DIG QuOnto server	Implementation: An implementation of an extension of DIG interface for the QuOnto system; Reasoning: Performance of reasoning by means of offered reasoner on ontologies, and rewriting data queries into a union of conjunctive SQL queries; Mapping: Ontology mapping is provided by widely accepted SQL DBMSs; Extensions: Required OBDA extension for DIG for the client-server architecture
QToolKit	Interface: Graphic user interface for system QuOnto; Reasoning: QuOnto reasoner; Features The internal data repository of QuOnto is used for storing the ABox; it allows to represent DL Lite ontologies
DIG interface	Description Logic: The descriptive logic SHOIQDn; it allows a minimal set of messages that the reasoners equipped with DIG interfaces exchange with their clients; Reasoner: CEL, FaCT ++, Pellet, and RacerPro; a standardized HTTP/XML interface to Description Logics (DL) reasoners (developed by the DL Implementation Group); Interface: Specification of the standard interface for interaction of various applications with reasoners on the descriptive logics in a network environment, created by the DL Implementation Group (DIG). An interface is supported by the majority of ontology editors and reasoners; Ontology editors: Compatibility with the ontology editors (e.g. Protege, SWOOP) and the Jena framework
OBDA extension for DIG	OBDA extension for DIG: It extends functionality of the DIG interface by possibilities of specification of the data source parameters and mapping queries to ontologies onto queries to data sources for reasoners of the OBDA systems
OBDA Plugin for Protégé	Main features: It allows to describe the mappings connecting the data source and the entities of the ontology; it enables users describing the data sources of the OBDA system, and it sends the descriptions of these components to an OBDA enabled reasoner; Type: An open code plugin for the ontology editor Protégé; Extensions: An extension of capabilities of Protégé

(Continued)

Table 1. (*Continued*)

Name	Criteria
Ontop	Connections with OBDA: Ontop starts its work by constructing the semantic-based tree-witness rewriting of q and T over H-complete ABoxes; It is available as a plugin for Protege 4 and upper, SPARQL end-point and OWLAPI and Sesame libraries
ROWLkit	Main features: It uses QuOnto services augmented by means required for operating by the OWL2 QL ontologies; it can perceive OWL2 QL ontologies as input data owing to the interface OWL API; it can operate by data stored in the external memory; Implementation: The implementation of the OWL2 QL profile; Reasoning: Reasoning on the ontologies described by means of OWL2 QL and their verification, as well as data query processing in ontology terms; Extensions: A graphic user interface is provided; Implementation: Java; it uses in Memory DBMS H2 Java to store ABox
MASTRO	Main features: The low computational complexity of MASTRO, DIG-MASTRO is an attractive component of an OBDA system, where the amount of data stored in the sources is large, and efficient reasoning is mandatory; Implementation: Java; Components: The OBDA-enabled reasoner - DIG-MASTRO and the OBDA plugin for the standard ontology editing tool Protege; Extensions: A functional extension of the QuOnto system; A version with the DIG interface - DIG MASTRO; together with the OBDA Plugin for Protégé, it offers a complete cycle of the OBDA system development and operation; Mapping: DIG-MASTRO enables MASTRO to interact with any client conforming to the OBDA extension of DIG. It provides facilities to specify data sources and mappings; DL: Ontologies in MASTRO are specified by means of the DL LiteA, logic of the DL Lite family; Reasoning: It is possible to specify and verify ontologies and to perform data queries and reasoning on the ontologies

project, 7th Framework Programme). The deep analysis of selected OBDA tools allows to create a framework for analysis. It includes the common features of analyzed set of OBDA. The key features were indicated to compare the selected OBDA tools and integrated algorithms supporting them. The set of criteria is reduced to the most important ones, and on base on it, the main establishments of proposed framework are created. The access to the selected OBDA tools is seamless. Most of presented tools for OBDA systems is available on the Web. The main obstacles concern the proper connection and integration data with ontology and creating mappings.

The proposed framework includes selected tools and extensions for OBDA system creation. (DIG Interface, DIG QuOnto server, Ontop, QToolKit, ROWLkit, Quest, MASTRO, OBDA extension for DIG, QuOnto, OBDA plugin for Protege). The set of criteria was build on base of the information presented in the Table 1. The 9 main criteria (Description Logic, Mapping, Rewriting, Implementation, Reasoning process, Ontology Editors, Interface, Extensions, Drivers) and 30 sub-criteria were defined (Description Logic: DL Lite Family, DL LiteA, SHOIQDn; Drivers: DB, Oracle, Postgres, MySQL, H2, SQL Server; Extensions: extension for QuOnto system, OBDA extension for DIG, OBDA Plugin for Protégé; Implementation: Implementation OWL2 QL, Implementation Java; Interface: DIG interface, Extended DIG interface, user interface for QuOnto, OWL API; Mapping: SQL DBMSs, DIG MASTRO; Ontology Editors: Jena Framework, SWOOP, Protege; Reasoning Process: DIG-MASTRO, CEL, FaCT++, QuOnto, Pellet, RacerPro; Rewriting: Semantic-based tree-witness rewriting). Due to limited space of the paper it is not possible to include the detailed table with the set of criteria and sub-criteria and their fulfillment by a given OBDA tool.

The framework was written in OWL/XML language. It exploits the ontology. The ontology was created manually, on base of the data included in the Table 1. Protégé software was used to support the process of the ontology construction. The procedure to ontology construction was composed of the 8 steps: (1) defining a set of criteria, (2) taxonomy construction, (3) ontology construction, (4) formal description, (5) defined classes creation, (6) reasoning process, (7) consistency verification, (8) a set of results. A domain of modeling encompasses the set of ODBA tools. The general aim of the proposed framework is to provide the knowledge of selected OBDA tools and to ensure a specified guideline supporting OBDA tool classification and selection processes. A user interface is prepared. Ultimately, it is a part of the elaborated procedure supporting ontology-based knowledge discovery (Fig. 3).

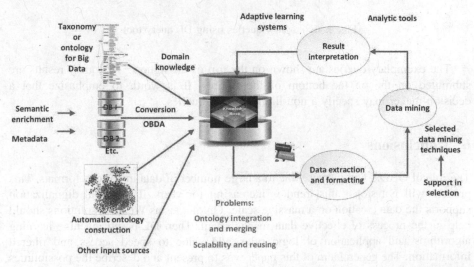

Fig. 3. Ontology-based knowledge discovery - a proposal of a procedure

5 Case Studies: Practical Examples of an Application of the Proposed Framework

The case study presents practical examples of the framework for analysis of OBDA. DL query is used to extract the answers. It is supposed that a decision-maker is looking for the OBDA tool that fulfils a set of pre-defined requirements. It is assumed that preferable OBDA tool should have not all of the following requirements: (the first case study) OBDA extension for DIG, or DIG-MASTRO, or Extended DIG interface. The application of the reasoning mechanism provides a set of results (OBDA tools) with regard to the pre-defined requirements. These requirements are fulfilled by 4 OBDA tools: DIG QuOnto Server, MASTRO, OBDA Extension for DIG, and QuOnto. In the second and third case studies, the set of criteria has been changes subtly. After the reasoning process, the set of results includes: DIG QuOnto Server, MASTRO, Ontop and QToolKit in the second case study, and DIG Interface in the third case study (Fig. 4).

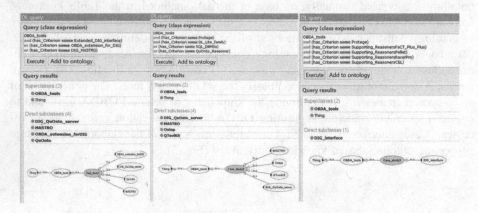

Fig. 4. Exemplary queries using DL query tool

The exemplary queries are shown on the top of the figures. The query results are submitted on the on the bottom of the figures. It is worth to emphasize that a decision-maker may specify a non-limited set of queries.

6 Conclusions

The role of OBDA will increase due to a large number of data in various formats. This process will not stop - this trend is increasing for years. The global digitalization supports the data creation on a massive scale. Ontologies as Web 3.0 solutions should help in the process of effective data management. Then DL-based machine learning algorithms and application of logic rules add value to speed access and filtered information. The general aim of this paper was to present and describe the possibilities offered by OBDA solutions. Apart from that, the analysis of selected tools for OBDA system creation was attached. Based on it, the framework for analysis OBDA was constructed. The aim of it was to collate the tools for OBDA and to extract the preferred solutions with regard to a pre-defined set of criteria. The framework was written in OWL/XML language and it exploited the ontology. The queries were posed using DL query. It has contained 9 main criteria and 30 sub-criteria. It is possible to add more information or update existing ones. The aim is to provide necessary information of OBDA tool for the end-user, especially with regard to his preferences. It is worth to emphasize that the proposed framework for analysis of OBDA is a part of the more complex procedure supporting ontology-based knowledge discovery.

References

1. Calvanese, D., De Giacomo, G., Lembo, D., Lenzerini, M., Rosati, R.: Tractable reasoning and efficient query answering in description logics: the DL-Lite family. J. Autom. Reasoning **39**, 385–429 (2007)
2. Konys, A.: Knowledge-based approach to question answering system selection. In: Núñez, M., et al. (eds.) ICCCI 2015. LNCS, vol. 9329, pp. 361–370. Springer, Heidelberg (2015). doi:10.1007/978-3-319-24069-5_34
3. Kogalovsky, M.R.: OntologyBased data access systems. Program. Comput. Softw. 38(4), 167–182 (2012). Pleiades Publishing, ISSN 03617688
4. Savo, D., et al.: MASTRO at work: experiences on ontology-based data access. In: Proceedings of the 23rd International Workshop on Description Logics, CEUR WS 573, Waterloo, Canada (2010)
5. Acciarri, A., Calvanese, D., De Giacomo, G., Lembo, D., Lenzerini, M., Palmieri, M., Rosati, R.: QUONTO: QUerying ONTOlogies. In: Proceedings of the 20th National Conference on Artificial Intelligence, Pittsburgh, Pennsylvania, vol. 4, pp. 1670–1671 (2005)
6. Calvanese, D., De Giacomo, G., Lembo, D., Lenzerini, M., Poggi, A., Rodriguez-Muro, M., Rosati, R.: Ontologies and databases: the *DL-Lite* approach. In: Tessaris, S., Franconi, E., Eiter, T., Gutierrez, C., Handschuh, S., Rousset, M.-C., Schmidt, R.A. (eds.) Reasoning Web. LNCS, vol. 5689, pp. 255–356. Springer, Heidelberg (2009)
7. Description Logic Implementation Group. http://dl.kr.org/dig/
8. Turhan, A., et al.: DIG 2.0—Towards a Flexible Interface for Description Logic Reasoners. http://www.owllink.org/publications/dig20_OWLED06.pdf
9. Calvanese, D., et al.: The mastro system for ontology-based data access. Semant. Web J. 2(1), 43–53 (2011)
10. The DIG interface standard (DIG 2.0). http://dl.kr.org/dig/interface.html
11. Ontology-Based Data Access. http://obda.inf.unibz.it/dig_11_obda/
12. Kalichenko, L.A.: SINTHESIS: Definition, Design, and Programming Language for Interoperable Environments of Heterogeneous Information Resources, 2nd edn. Institute of Informatics Problems of Russian Academy of Sciences (1993)
13. Kifer, M., Lausen, G.: F-Logic: a higer-order language for reasoning about objects, inheritance, and schema. SIGMOD Rec. **18**(2), 134–146 (1989)
14. Motik, B., Cuenca Grau, B., Horrocks, I., Wu, Z., Fokoue, A., Lutz, C.: OWL 2Web Ontology Language profiles, 2nd edn. W3C Recommendation, W3C (2012)

15. Rodríguez-Muro, M., Kontchakov, R., Zakharyaschev, M.: Ontology-based data access: *Ontop* of databases. In: Alani, H., Kagal, L., Fokoue, A., Groth, P., Biemann, C., Parreira, J. X., Aroyo, L., Noy, N., Welty, C., Janowicz, K. (eds.) ISWC 2013, Part I. LNCS, vol. 8218, pp. 558–573. Springer, Heidelberg (2013)
16. Lembo, D., Mora, J., Rosati, R., Savo, D.F., Thorstensen, E.: Mapping analysis in ontology-based data access: algorithms and complexity. In: Arenas, M., et al. (eds.) ISWC 2015. LNCS, vol. 9366, pp. 217–234. Springer, Heidelberg (2015). doi:10.1007/978-3-319-25007-6_13
17. Heymans, S., et al.: Ontology Reasoning with Large Data Repositories, Ontology Management. Computing for Human Experience, vol. 7, pp. 89–128 (2008)
18. Calvanese, D., De Giacomo, G., Lembo, D., Lenzerini, M., Poggi, A., Rosati, R.: Linking data to ontologies: The description logic DL-Lite A, vol. 216 (2006)
19. Rodriguez-Muro, M., Kontchakov, R., Zakharyaschev, M.: Query rewriting and optimisation with database dependencies in ontop, In: Eiter T., et al. (eds.) Informal Proceedings of DL, vol. 1014, pp. 917–929. CEUR Workshop Proceedings (2013)
20. Gruber, T.: Toward principles for the design of ontologies used for knowledge sharing. Int. J. Hum. Comput. Stud. **43**(5–6), 907–928 (1995)

Towards a Real-Time Big GeoData Geolocation System Based on Visual and Textual Features

Sarra Hasni[1,2(✉)] and Sami Faiz[1,3]

[1] Laboratory of Remote Sensing and Spatial Information Systems,
University of Al Manar, Tunis, Tunisia
hasni.sarra@gmail.com, sami.faiz@insat.rnu.tn
[2] Tunisia Polytechnic School, University of Carthage, La Marsa, Tunisia
[3] Higher Institute of Multimedia Arts of Manouba, University of Manouba,
La Manouba, Tunisia

Abstract. During the recent years, new sensors and new methods of collecting geospatial information emerged. This technological advancement is behind the apparition of Big GeoData concept. The exploitation of such concept demonstrates its effectiveness in various activity sectors. However, it poses new challenges that cannot be overcomed using traditional solutions. We believe that a good exploitation of Big GeoData can be guaranteed by establishing, first of all, a good placing strategy yet called geolocation estimation activity. In fact, generating efficient geospatial knowledge is closely linked to positional accuracy. Since that traditional geolocation estimation methods are not able to handle Big GeoData variety, a new solution has to be proposed. This is the object of the present article where the architecture of a real-time multimode approach is described.

Keywords: Big GeoData · VGI · Geolocation estimation · Visual features · Textual features

1 Introduction

In the literature, several studies emphasize the particularity of geospatial information. For example, [1] considers that "space is special". We approve these statements as we believe that human activities are often linked to the space whether it is private or public. These reasons highlight the need of determining geospatial aspects of objects to understand human behaviors and then the real-world alterations. [2] thinks that positional accuracy is one of the most important quality standards of geospatial information. Since then, the task of "geolocation estimation", yet called "placing" can be a good solution for inferring unspecified object location. Much efforts were basically dedicated for text or image geolocation. In other words, each of the already developed work is focused on a single modality [3]. Nowadays, the geolocation estimation task becomes more and more complex. In fact, new sources of data have become available.

N.T. Nguyen et al. (Eds.): ICCCI 2016, Part II, LNAI 9876, pp. 409–417, 2016.
DOI: 10.1007/978-3-319-45246-3_39

The unprecedented amount of already geolocated data is generated in continuous and exponential way giving rise to a new phenomenon: Big GeoData. A large part of this trend consists of Volunteered Geographic Information (VGI).

[4] defines VGI as generated geographic information by volunteer contributors and crowd-sourced data obtained essentially from social media and Web-based platforms. The exploitation of such data is useful especially for geospatial applications. For example, an elaborated study by McKinsey Global Institute (MGI) has shown the effect of Big Geodata on economics. In fact, when exploiting the personal location data, it is possible that consumers will save about $600 billion annually by 2020 [5]. As far as the health sector is concerned, Big GeoData analysis presents a considerable asset. When Ebola outburst, it was possible to predict the development of the areas of contagion through the analysis of flows and displacement of populations [6].

However, another big part of data shared by social media and Web-based platforms has no specified location. So they are less useful for geospatial application. In order to ensure an effective generation of geographical knowledge, the preservation of the spatial dimension of shared data is a primary need to create Big GeoData and VGI.

In this paper we propose a new multimode placing approach for shared data on social media and Web-based platforms. This approach can operate on real-time and support diverse VGI sources. The remainder of this article is organized as follows: some related works to placing task are discussed in Sect. 2. In Sect. 3, we introduce the Big GeoData phenomenon. Then, in Sect. 4 we propose the architecture of our geolocation estimation approach. Finally, a conclusion and future works are presented in Sect. 5.

2 Overview of Geolocation Task

According to [7], placing task is vital for geospatial applications. In the following, we describe recent works related to this task. These works broadly fall into two basic models: Text-based model and image-based model geolocation.

2.1 Text-Based Geolocation Methods

The volume of user-generated text shared, especially on social media, can be exploited for many purposes (e.g. opinion analysis, targeted advertising and natural phenomena detection). However user's position is not declared in most cases. For example, it is estimated that only 1 % to 3 % of tweets are geotagged. Thus, text-based geolocation becomes of increasing interest [7]. A recent spike in this interest results in developing several approaches. For example, [8] emphasizes the role of geoparsing techniques in constructing gazetted expressions. Where, [9] creates a local lexicon of geographic locations based on Wikipedia. These solutions were used in recent works and have shown their sparsity in coverage. [7] proposes a new framework for tweets geolocation. He starts by evaluating a subset of feature selection techniques in order to extract LIW (Location Indicative Words). Then, he studies factors which impact placing accuracy (e.g. temporal variance, user geolocability, non-geotagged tweets). In the same context, another approach is suggested by [10]. This approach is based on network prediction model and unified text. For more precision, it consists of a label propagation approach for the geolocation task based on "modified Adsorption".

2.2 Image-Based Geolocation Methods

The emergence of advanced technologies leads to an exponential increase of imagery shared on the Web. However, the amount of images with defined geo-coordinates is estimated at only 5 % [11]. Thus, the need to estimate geolocation of images is in high demand and is addressed by recent works. One of the most studied solutions is to estimate a distribution over geographic locations from an image. This solution is basically a purely data-driven scene matching approach. In this context, the Im2GPS system is able to estimate image geolocation as a probability distribution over the Earth's surface [12]. By evaluating Im2GPS, [13] notes that this system has some weaknesses. It cannot achieve a high accuracy. In fact, it is incredibly complex to explicitly estimate images positions from only visual features unless they are characterized by unique feature or they contain an explicit landmark. Less general approaches are proposed by [14, 15]. They consist of full 3D models of landmarks which aim to geolocate images with more precision. However, these approaches necessitate a large set of training images per place. So they are only useful around popular landmarks.

2.3 Discussion

By studying aforementioned approaches, we appreciate dedicated efforts for the geolocation estimation task. Nevertheless, we deduct their drawbacks which limit their effectiveness. Principally, these approaches operate on batch mode. So, they are not able to support current flows of objects. In addition, most described text and images geolocation methods operate on objects from definite targets (respectively Twitter and Flickr). However, other sources have demonstrated their richness and are able to generate effective geographic knowledge. Note also that additional techniques, poorly adopted by these works, must be applied to identify noisy tags (e.g., entirely non-local indicative words (hello, summer, home, etc.), abbreviations, etc.) before applying text placing solutions.

3   Era of Big GeoData

With the Big Data revolution, large quantities of geospatial data are generated. The Big Data concept and geospatial data union are often denoted by: Big GeoData.

3.1   Definition and Features of Big GeoData

The Big GeoData can be defined as a set of spatial data that exceed the capacity of traditional computing systems. Other definitions are even proposed, to express new viewpoints on this phenomenon in a more specified frame. [16] thinks that the Big GeoData is not only a "business" but also a "big" science. [17] thinks in turn, that the Big GeoData is a social opportunity. In fact, human beings are faced with new challenges (e.g. health issues, climate change, energy management, etc.). So, they can take benefit from this phenomenon to resolve them. Thenceforth, we can consider the Big GeoData as a multidisciplinary concept. The Big GeoData inherits Big Data features. Obviously, size is the first thing that comes to mind when asking about the definition of Big GeoData. However, other features have emerged recently to describe this phenomenon. In fact, studying the expression "4 V" is jointly acknowledged:

- Volume: Much of Big Data is composed of location data and the rate of recent increases is by 20 % each year [18].
- Velocity: It is estimated that 500 million tweets are daily generated. From 1–5 % of this quantity is produced with explicit geographical information [19].
- Variety: The variety of sources introduces a variety of types of the generated data. For example, tags on OpenStreetMap can be structured or unstructured and twitters can include multilingual text, hashtags and URLs [19].
- Veracity: Much of Big GeoData is accumulated from unverified sources which are characterized by a low or an unknown level of accuracy [20].

3.2   Relative Technologies to Big GeoData

In order to take advantages and to fully exploit Big GeoData as cited previously, some motivations occurred. These motivations aim to extend emerging technologies for Big Data and to propose new solutions which support geographical dimensions. For example, SpatialHadoop [21] and GeoSpark [22] are respectively instance of Hadoop and Spark. Furthemore, GeoMesa [23] is an hybrid approach which uses Hadoop and cloud computing technologies at once. More precisely, Spatial analytics in GeoMesa can leverage Hadoop to perform computations in parallel on a cloud [24].

Being an open source document database which belongs to NoSQL family, MongoDB is also able to handle geographic information.

Alongside the studies that we carried out on these frameworks, we discovered that they share a common flaw. They are devoid of geolocation mechanisms. In fact, they operate only on data which are already geolocated as they offer to their users a defined set of operations (e.g., KNN, Spatial join, etc.). Thus neglecting such a treatment can lead to a lack of manipulation and analysis accuracy.

4   A Multimode Approach for Big GeoData Placing

In order to overcome limitations of previous Big GeoData systems, we propose a new approach for Big GeoData geolocation estimation. We take benefit from both visual and textual information to guarantee more accuracy. The planned architecture of this system is described on Fig. 1.

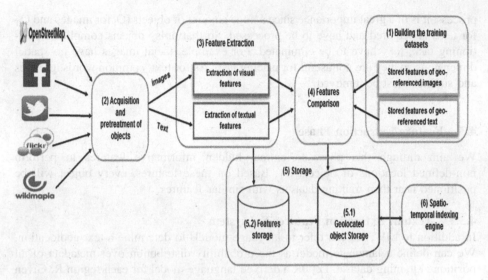

Fig. 1. The global architecture of our project

4.1 Building the Training Datasets

We start by selecting a large number of objects T (images and texts) whose geospatial dimension is defined. Since these objects can be labeled with additional text (metadata or keywords), establishing a connection among words and regions can be helpful for the geolocation estimation task. In this context, extracting LIW (Location Indicative Words) proves a good solution. We propose in the following, a simple solution to extract LIW. This heuristic-based solution is derived from TF-IDF (Term Frequency-Inverse Document Frequency) method. Let:

$$\text{Weight}_w = \text{WF}_w * \text{IRF}_w. \tag{1}$$

Where: WF_w designs the Word Frequency and IRF_w presents the inverse Region Frequency. Note that IRf_w is calculated in terms of:

$$\text{IRF}_w = (\log|R|) / |\{r_j : w_i \in r_j\}|. \tag{2}$$

Where: R designs the number of regions and r_j is the number of regions where users use w in the training set. In other words, where $n_{ij} \neq 0$.

4.2 Acquisition and Pretreatment Phase

We think that a real-time acquisition of Big GeoData is of a great importance especially to detect real phenomena. To do this Apache solutions like Storm and Kafka may be a good candidate to process data from diverse sources. They consist of free and open sources distributed computation systems. Note that additional sources can be supported by such frameworks and then a larger range of data can be exploited. As for pretreatment process, it is of a great importance since a huge amount of objects (O_i for images and O_T for text) is acquired and have to be processed. So that noisy objects complicate positioning task, they have to be eliminated. For example, facial images have no spatial dimensions so they are not useful for placing task. As for text, common words, prefixes and suffixes must be removed.

4.3 Feature Extraction Phase

We aim through this phase to extract hidden informative features to perform non-defined locations of objects O_T. Based on these features, every object will be positioned near than training datasets with similar features.

4.3.1 Textual Extraction Features Subsystem

In addition to LIW, we can refer to language models to determine a text geolocation. We can define a language model as the probability distribution over metadata of all positioned training dataset. Let θR a derived language model for each region R. Given an acquired object O_T (e.g. commentaries on blogs, tweets, etc.) with words $W_o = \{w_1,..,w_n\}$, θR will be ranked based on their probability to generate W_o.

$$P(\theta R|\ W_o) = ((P\ (W_o|\ \theta R\ )\ P(\theta R))\ /\ P\ (W_o)) \propto P\ (\theta R) * \prod_{i=1}^{n} P\ (w_i|\ \theta R). \quad (3)$$

$P(w_i|\theta R)$ is the maximum likelihood probability of generating w_i from θR [25]. It is smoothed by calculating the maximum likelihood of generating w_i from the background language model. This latter is generated over all objects in the training dataset.

4.3.2 Visual Extraction Features Subsystem

We adopt the same strategy of Im2GPS system [12] to geolocate image objects O_i. So the following features have to be extracted: colour histograms, texton histograms, line features and Gist Descriptor. In addition to color histograms and line features, we recommend the extraction of feature vectors for object detection in images. In fact, they allow us to determine easily object shapes and types (built or natural) and to ameliorate the geolocation task accuracy. For this purpose, we suggest the following additional features to enhance Im2GPS performance:

- Grey-Level Co-occurrence Matrix (GLCM): It is also called "the gray-level spatial dependence matrix". It measures the occurrence of pairs of pixels with particular values to characterize the texture in an image.
- Histogram of Oriented Gradients (HOG): It consists of a feature descriptor frequently used in image processing to detect objects. It measures occurrences of gradient orientation in localized parts of an image. Thus, each object appearance and shape in an image can be measured in terms of distribution of intensity gradients.

Note that we can also take benefit from textual features to geolocate O_i since they can be accompanied with textual data (metadata keywords, etc.).

4.4   Features Comparison

Given an acquired object A, objects T from the training datasets are ranked in function of their scores S(A, T). The S(A, T) measures the similarity between A and T [26]. Then, it is estimated that an object T which is characterized by a high score, shares a close geolocation in the space with A. Let:

$$S(A, T) = e^{-\gamma D(A,T)}. \tag{4}$$

Where: $\gamma = 1.0$ and $D(A,T)$ is a weighted sum of distances between extracted features:

$$D(A, T) = \sum_{n=1}^{N} w_n * d_n(A_n, T_n). \tag{5}$$

Note that A_n, T_n, w_n and d_n design respectively: the nth feature of the acquired object A, the nth feature of the training object T, the weighting coefficient and the corresponding distance. We can apply the Maximum Entropy (ME) approach to obtain appropriated feature weights. We think that this approach is more suitable than others with the same finality. In fact, we deal with a variety of features and such an approach supports independencies between features. This is useful in our case, especially for shared images with additional textual components. We adopt the same strategy deployed by [26] to measure weights of features based on ME. Given an acquired object A, the objects T have to be classified to: relevant objects (RO) and irrelevant objects (IO). The distances between the ith feature of A and T:

$$f_i(A, T) = d_i(A_i, T_i). \tag{6}$$

We include a constant feature $f_{i=0}(A, T) = 1$ to guarantee a prior probabilities. Then, the scores S(A, T) can be replaced by the posterior probability for the RO class and the ranking and combination of several objects A is done as the following:

$$S(A, T) = P (RO|A, T)$$

$$= \left(\exp\left[\sum_i \lambda_{Roi} f_i(A, T) \right] \right) / \sum_{k \in \{RO,IO\}} \exp\left[\sum_i \lambda_{ki} f_i(A, T) \right]. \tag{7}$$

4.5   Storage Phase

Geolocated objects are transferred to the storage support which must be selected taking into account Big GeoData features. In this context, we think that data centers can be good candidates. Thanks to their interactivity, they can deal flexibly with an Apache framework which is exploited to acquire Big GeoData streams. We aim by this integration to take advantage from both: long term storage and real-time accumulation of data. This is useful for future use including detecting events and studying their evolution over the time. For this purpose, we propose a spatiotemporal indexing mechanism.

It consists of a simple method for encoding geo-time coordinates of objects into an index entry. Note that geo-coordinates can results from features comparison phase while timestamp can be determined as the moment of objects acquisition.

5 Conclusion

In this paper, we examined several methods for placing task. These methods demonstrated their effectiveness for inferring geolocation for images or text. Despite of emphasizing the power of location, they are not able to handle large amounts of multi-types objects generated with high velocity. Then, they are not able to handle Big GeoData. In our new research project, we aim to propose a multimode system for Big GeoData positioning based on both visual and textual features. We aim to estimate the geolocation of shared VGI on social networks and Web-based platforms whose power still partially revealed using traditional solutions. We focus our future works to support additional objects types like videos which can be treated as sequence of images. The design of this system aims to fully take advantage from the placing task: Since real-time objects features can be determined, we can go further and exploit them for additional activity such as real phenomena detection. Finally, to evaluate our approach, future efforts will be conducted to practically measure its effectiveness. It can be also the object of some enhancements to guarantee a higher accuracy.

References

1. Anselin, L.: Some robust approaches to testing and estimation in spatial econometrics. Reg. Sci. Urban Econo. **20**(2), 141–163 (1990)
2. Haklay, M.: How good is volunteered geographical information? a comparative study of OpenStreetMap and Ordnance Survey Datasets. Environ. Plan. B: Plan. Des. **37**, 682–703 (2010)
3. Han, B., Cook, P., Baldwin, T.: Text-based twitter user geolocation prediction. Artif. Intell. Res. **49**, 451–500 (2014)
4. Goodchild, M.F.: Citizens as sensors: the world of volunteered geography. GeoJournal **69**(4), 211–221 (2007)
5. Manyika, J., Chui, M., Brown, B., Bughin, J., Dobbs, R., Roxburgh, C., Byers, A.H.: Big data: the next frontier for innovation, competition, and productivity. McKinsey Global Institute (2011)
6. Hunting Down Ebola with Big Data. http://www.datanami.com/2014/11/03/hunting-ebola-big-data/
7. Choi, J., Friedland, J.: Multimodal Location Estimation of Videos and Images. Springer (2014)
8. Leidner, J L., Lieberman, M D.: Detecting geographical references in the form of place names and associated spatial natural language. SIGSPATIAL Spec. 3(2), 511 (2011). ACM, New York

9. Quercini, G. Samet, H. Sankaranarayanan, J., Lieberman, M.D.: Determining the spatial reader scopes of news sources using local lexicons. In: Proceedings of the 18th SIGSPATIAL International Conference on Advances in Geographic Information Systems, pp. 43–52. ACM, New York (2010)

10. Rahimi, A., Cohn, T., Baldwin, T.: Twitter user geolocation using a unified text and network prediction model. In Proceedings of the 53rd Annual Meeting of the Association for Computational Linguistics and the 7th International Joint Conference on Natural Language Processing, pp. 630–636. Cornell University Library (2015)

11. Friedland, G., Vinyals, O., Darrell, T.: Multimodal location estimation. In: ACM Multimedia (2010)

12. Hays, J., Efros, A.A.: IM2GPS: estimating geographic information from a single image. In: Proceedings of IEEE Conference on computer Vision and Pattern Recognition, pp. 1–8 (2008)

13. Hare, J., Davies, J., Samangooei, S., Lewis, P.: Placing photos with a multimodal probability density function. In: Proceedings of International Conference on Multimedia Retrieval, Glasgow, GB (2014)

14. Crandall, D., Owens, A., Snavely, N., Huttenlocher, D.: SfM with MRFs: Discrete-continuous optimization for large-scale structure from motion. IEEE Trans. Pattern Anal. Mach. Intell. **35**(12), 2841–2853 (2013)

15. Snavely, N., Seitz, S M., Szeliski, R.: Modeling the world from internet photo collections. IJCV 80(2), 189–210 (2008)

16. Dalton, C.M., Thatcher, J.: Inflated granularity: Spatial "Big Data" and geodemographics. Big Data Soc. **2**(2), 1–15 (2015)

17. Eagle, N., Greene, K.: Reality Mining: Using Big Data to Engineer a Better World, 1st edn. The MIT Press, Cambridge (2014)

18. Big data: The future is in analytics. http://geospatialworld.net/Magazine/MArticleView.aspx?aid=30512

19. Mooney, P., Winstanley, A. C.: Is VGI big data? In: GISRUK UK (2015)

20. Li, S., Dragicevic, S., Anton, F., Sester, M., Winter, S., Coltekin, A., Pettit, C., Jiang, B., Haworth, J., Stein, A., Cheng, T.: Geospatial big data handling theory and methods: a review and research challenges. ISPRS J. Photogrammetry Remote Sens. **115**, 119–133 (2015)

21. Eldawy, A., Mokbel, M F.:The ecosystem of SpatialHadoop. In: ACM SIGSPATIAL Special (2014)

22. Yu, J., Wu, J, Sarwat, M.: A demonstration of GeoSpark: a cluster computing framework for processing big spatial data. In: Proceedings of IEEE International Conference on Data Engineering, Helsinki, Finaland (2016)

23. GeoMesa. http://www.ccri.com/case-studies/geomesa/

24. GeoMesa: Scalable Geospatial Analytics on Accumulo. https://www.locationtech.org/content/geomesa-scalable-geospatial-analytics-accumulo-0

25. Hauff, C., Houben, G.-J.: Geo-location estimation of flickr images: social web based enrichment. In: Baeza-Yates, R., de Vries, A.P., Zaragoza, H., Cambazoglu, B., Murdock, V., Lempel, R., Silvestri, F. (eds.) ECIR 2012. LNCS, vol. 7224, pp. 85–96. Springer, Heidelberg (2012)

26. Deselaers, T., Weyand, T., Ney, H.: Image retrieval and annotation using maximum entropy. In: Peters, C., Clough, P., Gey, F.C., Karlgren, J., Magnini, B., Oard, D.W., de Rijke, M., Stempfhuber, M. (eds.) CLEF 2006. LNCS, vol. 4730, pp. 725–734. Springer, Heidelberg (2007)

A Parallel Implementation of Relief Algorithm Using Mapreduce Paradigm

Jamila Yazidi[✉], Waad Bouaguel, and Nadia Essoussi

LARODEC, ISG, University of Tunis, Tunis, Tunisia
yazidi.jamila@gmail.com, bouaguelwaad@mailpost.tn,
nadia.essoussi@isg.rnu.tn

Abstract. Feature selection is an important research topic in machine learning and pattern recognition. In recent years, data has become increasingly larger in both number of instances and number of features. In fact the number of features that can be contained in a Big Data is hard to deal with. Unfortunately, the number of features that can be processed by most classification algorithms is considerably less. As a result, it is important to develop techniques for selecting features from very large data sets. However the efficiency of existing feature selection algorithms significantly downgrades, if not totally inapplicable, when data size exceeds hundreds of gigabytes. Traditional methods like Filters, Wrappers and Embedded methods lack enough scalability to cope with datasets of millions of instances and extract successful results in a finite time. Therefore, the main purpose of this paper is to propose a new parallel feature selection framework that enable the use of feature selection methods in large datasets.

Keywords: Feature selection · MapReduce · Parallel computing

1 Introduction

Over the last few years, dimensionality become a serious problem since data has become increasingly larger in both number of instances and number of features. Unbrokenly, existing algorithms of classification and clustering need less number of features to work efficiently without a hitch [1]. Thus, data reduction is proposed to reduce the number of features and increase the learning performance. Two ways to reduce the number of variables exist: feature extraction and feature selection.

In feature extraction we are interested in looking for new features that are the result of transformation of the original data by using some transformation functions [1]. In feature selection the goal is finding a minimum subset of non redundant and relevant features from the original set. Although efficient, dimensionality reduction methods are used in the absence of Big Data which means that existing algorithms of both feature extraction and feature selection can be used with large data sets but cannot deal with very large data.

© Springer International Publishing Switzerland 2016
N.T. Nguyen et al. (Eds.): ICCCI 2016, Part II, LNAI 9876, pp. 418–425, 2016.
DOI: 10.1007/978-3-319-45246-3_40

The existing approaches of feature selection are grouped into three categories: filter, wrapper and embedded methods. Filters concept consist of filtering the undesirable features out before learning [1] which make them fast and independent of the learning process. The goal of a wrapper approach is straightforward; to achieve the highest accuracy rate by selecting features that accomplish that for a specific algorithm. Although efficient, when the size of the data is very large the wrapper approach become not suitable for Big Data. Embedded methods are similar to wrappers since the features are selected during the learning process [2]. Based on the comparison between filters and wrappers and our goal which is the scalability of those algorithms for Big Data, we decided to work with filter method using a univariate approach.

Some algorithms of feature selection that exist are considered as exhaustive methods and cover the whole search space as Focus algorithm, Branch and Bound, Best First Search [3,4]. Other algorithms are considered as heuristic methods such as Set Cover and Mutual Information [3,4]. A third category of algorithms are considered as nondeterministic methods known as stochastic methods allowing a generation randomly of subsets. They uses several techniques like genetic algorithm and simulated annealing for instance: Las Vegas Filter (LVF) and Las Vegas Wrapper (LVW) [3,4]. For this paper, we focus on Relief algorithm which belongs to a fourth category named Feature Weighting Methods. Previous works have shown that it is effective with large datasets with both high number of instances and of features.

It is obvious from literature that the performance of feature selection techniques downgrade simultaneously with the increase of volume of data. The number of features and the number of instances affect the complexity of feature selection algorithms so that for very large data sets with thousands of features or instances, the efficiency can be reduced totally. For big data these traditional algorithms are inapplicable too. This is due to the existence of all data in a single memory that cannot be supported.

Machines are not able to process this huge amount of data and parallel solutions are considered suitable mechanism for dealing with. We propose in this paper using a parallel paradigm which allows to reduce the execution time of feature selection by implementing Relief algorithm. The rest of the paper is organized as follows. Section 2 provides some background about MapReduce and the state of the art about feature selection algorithms using the MapReduce programming model. Section 3 describes the proposed approach for the Relief algorithm using the MapReduce. The results are discussed and analyzed in Sect. 4. Finally the summary of the paper is presented in the conclusion in Sect. 5.

2   Feature Selection Algorithms Using the MapReduce: The State of the Art

[5] describes Big Data as having three main characteristics: Volume, Variety and Velocity. This is the most known definition of Big Data, although many authors add other V's such as Value and Veracity.

The huge volume of Big Data makes the processing with a centralized solution in a traditional way an impossible mission, that is why many distributed approaches have been proposed in literature: Message Passing (MPI), threads, workflow and MapReduce. The MapReduce is proposed by Google as a programming model to process by parallel and distributed algorithms, large and scalable datasets [6]. Those algorithms are distributed on a cluster which can contain thousands of machines with reliable and fault tolerance manner. The job of the MapReduce consists of three principle steps which are Mapper procedure, Shuffle procedure and Reducer procedure. However, only the mapper and reducer procedures need to be implemented since Shuffle is executed automatically by the computer [6].

The input of the MapReduce is split into many blocks. Hence, the first step is to process each block with a map. As a consequence the whole input is processed in a parallel way and the output of the map procedure is a set of (Key, value) pairs. After that, the shuffle procedure works automatically in order to group the data with same key. The last step is the reducer procedure, and as output of the shuffle we have several unique keys where each key has many values. The goal of the reducer step is to obtain the best value according some criteria for each key.

[7] proposed feature subset selection problem using wrapper approach in supervised learning. It was a new Wrapper approach using Genetic Algorithm (GA) as random search technique. [8] proposed a new parallel feature selection model based on MapReduce which is called Twister. By using this model, The iterative MapReduce computation is useful to implement many parallel algorithms with simple iterative structure. The client controls the program of MapReduce assigning methods to the MR job, preparing keyValue pairs, preparing statistic data to the MR tasks through the partition file if required.

[9] proposed a parallel feature selection using positive approximation (FSPA) based on MapReduce. In fact this proposition is an improvement for a previous contribution which is a combination between rough set based feature selection method and positive approximation. [10] proposed an iterative MapReduce for feature selection. A method which consists in using parallel Kmeans on MapReduce for clustering.

All the previous approaches are models or framework that have not been implemented yet. However, only [11] implemented two filter algorithms: Ranking based on Pearson correlation coefficient and mRMR.

mRMR is a feature selection algorithm which belongs to filter category and based on Mutual Information measure for Maximum Relevance or Minimum Redundancy. In one hand, the Maximum Relevance aims to maximize the relevance between a feature and the target class. For this step features are sorted in a descent way according to their mutual information. On the other hand, the Minimum Redundancy tends to minimize the repeat between features by maximizing the mean of all mutual information values (between each individual feature with the target class).

Two Map Reduce jobs are required: The first one calculates the mutual information for each candidate feature. For this, the Map step achieve the parallelism over records and the reduce step is for the parallelism over features.

The second job aims to select the best subset of features that have the highest mutual information already done in the first job.

Ranking based on Pearson correlation coefficient is an algorithm of feature selection that belongs to filter category too. Its principle consists in computing a coefficient based on Pearson Correlation between each feature and the target class. Thus best features are those with the highest coefficient. For this only one job Map Reduce is required: The map step iterates over the training set records to achieve the parallelism over records. As output of the map, the product of each feature and the target class value constitute the value while the feature index is the key. The reduce step is responsible of the parallelism over features by summing up the values of map output. Table 1 resume the two existing algorithms of feature selection that have been implemented where n is number of instances, m the number of features, s the number of features to select and b the number of nodes in a cluster.

Table 1. Summary of implemented algorithms of feature selection using Map Reduce

Algorithm	Nb jobs MapReduce	Post-processing	Parallelism	Complexity
mRMR	2	No	Records-features	$O(\frac{n*m}{b})$
Ranking	1	Yes	Records-features	$O(\frac{n*m''*s^2}{b})$

3   Proposed Approach

Our proposed solution consists in implementing a Relief feature selection algorithm using MapReduce. Relief algorithm is based on sampling the whole set of instances. This is inspired from the idea that relevant features are those whose values can distinguish from close instances to each other. Thus, for each instance I, the algorithm find the two nearest neighbors which are called near hit H and near miss J, belonging to two different classes (For binary classification). By using Relief, a feature is considered as relevant if its values are the same between I and H and different between I and J. This relevance is computed through distance. This distance between the instance I and each feature is accumulated in a weight vector w which has the same number as the number of dimensions. Only features which have a weight higher than a relevance threshold τ are selected. τ can be statistically estimated. The samples have a size m. More the samples are larger, more the approximation I is reliable. As mentioned above, our solution is to implement this algorithm using the parallel paradigm MapReduce. The following scheme explains the architecture of this programming model.

The following steps represent the details of the solution: First we split the original set of features into many subsets of features. Then, many smaller

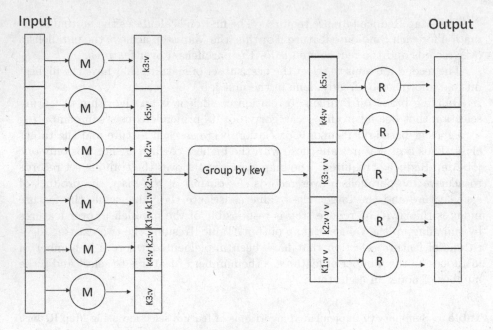

Fig. 1. Mechanism of MapReduce

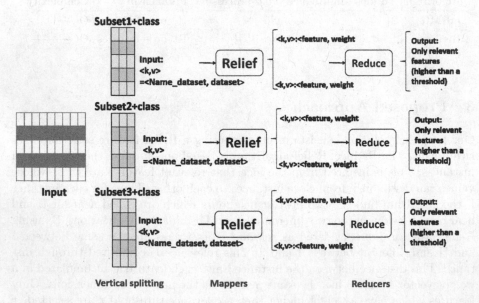

Fig. 2. Proposed approach

datasets from the original one are formed according to the created subsets of features with adding the attribute class for each subset. In other words, the generated datasets are the result of a vertical splitting of the big dataset.

Secondly, for the map step, the generated datasets are the input value with an input key which refers to the name of the generated dataset. Relief algorithm is applied for each one of the new datasets. As output of the map, for each treated dataset of m features we have m pair of $<k, v>$ refer to the $<feature, weight>$.

Finally for the reduce step the weights are compared to an empirical threshold τ in order to keep only features with a weight higher than τ. Thus, only relevant features and their weights are generated from the reduce step as $<k, v>$.

Hence, we can handle the large number of features thanks to the vertical splitting of the dataset. Moreover, each treated file for the map step is splitted in an horizontal way by default. This allows to deal with the big number of instances.

4   Results and Discussion

The experiments were conducted on two databases which are available at the UCI machine learning repository. The first one is the Amazon Commerce reviews dataset which is a large dataset derived from the customer reviews in Amazon Commerce Website for authorship identification. This data set includes 1500 instances described by 10000 features. We consider also the Dexter dataset that consists of 2600 instance which include 20000 features presenting frequencies of occurrence of word stems in text. Table 2 displays the characteristics of the datasets that have been used for evaluation.

Table 2. Datasets summary

Name of dataset	Amazon	Dexter
Number of features	10000	20000
Number of instances	1500	2600
Missing value	Yes	Yes
Type of variables	Real	Integer

Our algorithm was implemented with the MapReduce on Hadoop using R language by Rstudio. Actually, using R with Hadoop is possible thanks to a component of Hadoop which is Hadoop streaming. Thus, many packages in R are required to implement the MapReduce with R: RMR package to execute the job MapReduce and RHDFS package to enable the use of HDFS in Hadoop. The feature selection algorithm Relief is already implemented in R available from the package FSelector. Big memory package was also used in order to read large data files. These are the main used packages.

The results in Tables 3 and 4 show that the running time of Relief algorithm using our proposed approach decreases considerably compared to non parallel results. Moreover, using parallel MapReduce depends on the number of used nodes: Increasing the number of nodes decrease the running time.

Table 3. Results given by Dexter dataset

	Running time (s)	Number of selected features
Non parallel (Weka)	11670	565
Parallel (MapReduce single node)	3580	570
Parallel (MapReduce two nodes)	2633	572

Threshold of Relief $\tau = 0.0$

Table 4. Results given by Amazon dataset

	Running time (s)	Number of selected features
Non parallel (Weka)	3450	1198
Parallel (MapReduce single node)	2519	1196
Parallel (MapReduce two nodes)	2340	1205

Threshold of Relief $\tau = 0.5$

The difference in the number of selected features (which is between 565 and 572 for Dexter and between 1196 and 1205 for Amazon) is explained by the samples considered in Relief algorithm.

Tables 5 and 6 show the role of selecting features using MapReduce in improving the results of classification algorithms. Table 5 show that some algorithms of classification like Bayesian Network are not applicable without selecting features. This is due to the huge number of features that influence the scalability of those algorithms. Here, feature selection is a crucial task for the classification algorithm.

Table 6 shows that even when a classification algorithm is applicable for a whole dataset, selecting features using MapReduce reduces the time for building models. It decreases from 36.44 s to 1.92 s. Moreover, our proposed approach helps to improve the percentage of correct instances, Precision, Recall and F-measure.

Table 5. Comparison of classification results for Dexter dataset using Bayesian Network

	Time for building model	correct instances	Precision	Recall	F-measure
All features	*	*	*	*	*
570 features	0.67	95.23	0.953	0.952	0.952

*model couldn't be built

Table 6. Comparison of classification results for Dexter dataset using Decision Stump

	Time for building model (s)	% correct instances	Precision	Recall	F-measure
All features	36.44	96.1	0.796	0.633	0.674
570 features	1.92	97.2	0.971	0.702	0.971

Selecting features using Relief on MapReduce, reduces the time of Relief algorithm on large datasets. In addition, it improves the quality of classification by improving the main measures cited above.

5   Conclusion

In this paper we presented a new approach dealing with Big datasets using MapReduce and consists in a parallel implementation of Relief algorithm. The proposed approach was presented on two steps. The first one is a split phase in order to reduce the complexity of large datasets. The second one is the parallelism over records in the map step to apply Relief algorithm. Experimentations show that the idea is applicable and gives good results for large datasets. Moreover, we proved that we can improve the quality of classification algorithms either for inapplicable ones on high dimensional data or for improving measures. For future work, we aim to experiment our approach on massive data, and we want to extend the number of nodes.

References

1. García, S., Luengo, J., Herrera, F.: Feature selection. In: García, S., Luengo, J., Herrera, F. (eds.) Data Preprocessing in Data Mining, pp. 163–193. Springer, Heidelberg (2015)
2. Guyon, I., Elisseeff, A.: An introduction to variable and feature selection. J. Mach. Learn. Res. **3**, 1157–1182 (2003)
3. Arauzo-Azofra, A., Benitez, J.M., Castro, J.L.: Consistency measures for feature selection. J. Intell. Inf. Syst. **30**(3), 273–292 (2008)
4. Almuallim, H., Dietterich, T.G.: Learning with many irrelevant features. In: AAAI, vol. 91, pp. 547–552, Citeseer (1991)
5. Kaisler, S., Armour, F., Espinosa, J.A., Money, W.: Big data: Issues and challenges moving forward. In: 2013 46th Hawaii International Conference on System Sciences (HICSS), pp. 995–1004. IEEE (2013)
6. HajKacem, M.A.B., N'cir, C.B., Essoussi, N.: Mapreduce-based k-prototypes clustering method for big data. In: 2015 IEEE International Conference on Data Science and Advanced Analytics, DSAA 2015, Campus des Cordeliers, Paris, France, 19–21 October 2015, pp. 1–7 (2015)
7. Karegowda, A.G., Jayaram, M., Manjunath, A.: Feature subset selection problem using wrapper approach in supervised learning. Int. J. Comput. Appl. **1**(7), 13–17 (2010)
8. Sun, Z.: Parallel feature selection based on mapreduce. In: Wong, W.E., Zhu, T. (eds.) Computer Engineering and Networking, pp. 299–306. Springer, Heidelberg (2014)
9. He, Q., Cheng, X., Zhuang, F., Shi, Z.: Parallel feature selection using positive approximation based on mapreduce. In: 2014 11th International Conference on Fuzzy Systems and Knowledge Discovery (FSKD), pp. 397–402. IEEE (2014)
10. Kourid, A.: Iterative mapreduce for feature selection. Int. J. Eng. Res. Technol. **3** (2014). ESRSA Publications
11. Reggiani, C.: Scaling feature selection algorithms using mapreduce on apache hadoop (2013)

Machine Learning in Medicine
and Biometrics

Rational Discovery of GSK3-Beta Modulators Aided by Protein Pocket Prediction and High-Throughput Molecular Docking

Rafael Dolezal[1,2(✉)], Michaela Melikova[1,2], Jakub Mesicek[1,2], and Kamil Kuca[1,2]

[1] Center for Basic and Applied Research, Faculty of Informatics
and Management, University of Hradec Kralove, Rokitanskeho 62,
50003 Hradec Kralove, Czech Republic
{rafael.dolezal,michaela.melikova,
jakub.mesicek,kamil.kuca}@uhk.cz
[2] Biomedical Research Center, University Hospital Hradec Kralove,
Sokolska 581, 500 05 Hradec Kralove, Czech Republic

Abstract. Over the last three decades, computer-aided drug design (CADD) methods have attracted increasing attention of medicinal chemists especially due to their potential to penetrate into the molecular level of drugs' mechanism of action. So far, CADD techniques have significantly contributed to rational development of more than two hundred novel drugs. Brute force of supercomputers has enabled chemists to screen virtual ligand libraries of millions of chemical structures in affordable time span while indicating which compounds should be prioritized in further preclinical research and which can be eliminated *a priori*. A prominent position in CADD is held by structure-based methods that analyze 3D structures of biological targets to find optimal small binding molecules modulating the target's bioactivity. In the current work, we have performed a protein binding pocket screening on an X-ray model of human Glycogen Synthase Kinase 3 beta (GSK3β) employing an algorithm based on Voronoi tessellation. The found binding sites were analyzed and compared with the results of surface screening of the GSK3β model by molecular docking based calculations. Finally, the revealed binding sites were exploited in a structure-based virtual screening supported by pleasingly parallelized calculations on a peta-flops-scale supercomputer. The most promising GSK3β modulators resulting from the *in silico* screening have been proposed for *in vitro* testing.

Keywords: Computational biology · Drug discovery · GSK3-beta · Virtual screening · Protein binding pocket prediction

1 Introduction

Computer-aided drug design (CADD) methods have become an important strategy in medicinal chemistry and pharmaceutical research since they allow in many cases to notably reduce time and cost required for drug lead discovery and optimization [1].

© Springer International Publishing Switzerland 2016
N.T. Nguyen et al. (Eds.): ICCCI 2016, Part II, LNAI 9876, pp. 429–439, 2016.
DOI: 10.1007/978-3-319-45246-3_41

Classically, new drugs are developed in a tedious process starting with wet-chemistry screening for unknown chemical compounds exhibiting high *in vitro* affinity towards a selected biological target (e.g. enzyme, receptor, nucleic acid, cells, etc.). Consequently, the identified hit compounds are chemically optimized by organic synthesis methods in a trial-error manner until a chemical structure with optimal pharmacodynamic and pharmacokinetic properties is obtained. Generally lacking sufficient logic, such an approach can consume more than 1 billion dollars and take over 15 years on average to introduce a novel drug into clinical practice, although sometimes luckier drug discovery stories happen to chosen researchers [2]. Fortunately, due to the massive development of computer technologies in last decades, *in silico* drug discovery and design techniques can considerably enhance, rationalize and streamline these classical laboratory methods, as has been demonstrated, to mention a few, in the cases of Dorzolamid, Captopril, or Saquinavir [3].

A specific tactic is employed in structure-based (SB) CADD methods where the main task is to analyze 3D models of biological targets [4]. Novel drug candidates are proposed depending on their ability to interact with the biological target model and to elicit the desirable conformational changes from it. Generally, flexible molecular docking, molecular dynamics or even hybridized quantum chemical and molecular mechanical calculations (QM/MM) are employed in such simulations [5]. For SB methods, both the ligand and the target structures are imperative. At present, several approaches based on X-ray crystallography, nuclear magnetic resonance (NMR) or cryo-electron microscopy are applicable to experimentally determine the active sites. Once the 3D protein structure has been elucidated, numerous computational chemistry analyses can be conducted to properly exploit the chemical information [2]. Especially, looking for increased specificity of ligand-enzyme/receptor interactions due to binding in allosteric active sites can bring desirable innovation in drug design towards more potent and selective therapeutics.

Therefore, we focused in the study on computational analyses of a 3D model of glycogen synthase-kinase 3 beta (GSK3β) enzyme with the aim to find its potential binding sites. Herein, two approaches were employed: (1) geometric analysis of the protein surface based on Voronoi tessellation, and (2) a molecular-docking based method. Both approaches are discussed, compared, and eventually utilized in a high throughput molecular docking for novel GSK3β ligands as candidates for anti-Alzheimer's disease drugs.

2   Problem Definition

In the current work, we have designed a computational analysis of human GSK3β which is involved in several pathological processes such as *diabetes mellitus* (type 2), Alzheimer's disease, chronic inflammatory disorders, etc. Although the orthosteric binding site of GSK3β is well-known from X-ray crystallography, there is an urgent need for development of novel allosteric inhibitors binding to different parts of the enzyme. Thus, the objective of the present study is to show how to identify other (i.e. allosteric) binding sites in an X-ray model of GSK3β by means of geometrical and molecular mechanical analysis of the enzyme's 3D structure. The protein pocket binding prediction is performed by a Voronoi tessellation algorithm in Fpocket program and the obtained results are compared with an energy-based active site search by molecular docking in AutoDock Vina. The predicted binding protein pockets' locations are finally introduced as key parameters into SB virtual screening for novel GSK3β modulators. Since the proposed virtual screening by high throughput flexible molecular docking is highly time-expensive, we utilized a supercomputer of 2 PFlops performance with 1008 compute nodes and 24 192 CPUs to perform the calculation in parallel. The resulting top-scoring candidates for GSK3β modulators are suggested for further *in vitro* investigations.

3 State of the Art

3.1 Pocket Binding Prediction Within Rational Drug Design

In order to discover novel lead structures *in silico*, 3D structure of binding sites in receptors/enzymes are necessary to be known. Accordingly, co-crystallization of a ligand with the target and a subsequent X-ray, NMR, or cryo-electron microscopy analysis generally has to be carried out to reveal the active site as well as the target's whole structure. However, further effort needs to be made if new binding sites are to be discovered. In this regard, computational methods are often used for identification of allosteric binding sites on experimentally determined 3D target models [6]. Currently, four main methods are used to determine and characterize protein binding sites: (1) geometric analyses of the target's protein surface to find concave pockets; (2) computational chemistry screening of the target's protein surface by a probe ligand to reveal high binding energy regions; (3) molecular dynamic studies simulating ligand-enzyme/receptor trajectories as a function of time; (4) protein sequence or phylogenetic analyses to identify homologous parts with active site templates in a profile library. The mainstream techniques are represented by geometric and energy-based methods. Demanding molecular dynamics studies are generally used for finding the thermodynamic equilibrium of solvated states when the binding sites have already been localized by molecular docking. Finally, bioinformatics-based methods rely on the databases of known binding sites' sequences and similarity analyses and, thus, they are not universally applicable for novel active sites discovery. For these reasons, we hold forth only on the first two methods mentioned above.

Geometric Approach. As the ligand binding sites mostly form the largest pockets, cavities or tunnels in the protein surface, geometrical methods for the active sites prediction frequently utilize algorithms splitting the protein into regular segments with a 3D grid box and finding the grid points at which a probe sphere does not coincide with any protein atoms. This simple search protocol is employed, for instance, by Pocket program. Since such algorithm for protein pocket detection is strongly dependent on the protein input orientation relative to the coordinate reference frame, further improvements have been introduced to the calculations. Briefly, Ligsite program scans the protein structure in seven directions; Delaney program is optimized to form a monolayer of particles covering the protein surface which is followed by cyclic surface expansions and contractions to concentrate the probes in protein cavities; Pass program designates cavities by cumulating such probe spheres that are tangentially connected with three atoms assigned to the protein surface, the algorithm of Del Carpio program searches cavities in a growing process starting from the center of gravity of the protein identifying atoms integrated within a concave surface, Apropos algorithm represents the protein by α-shape triangle bodies (i.e. Delaunay models) to find the cavities as steric differences to the convex hull formed when the α parameter approaches infinity. Similarly, Cast program utilizes the Apropos's algorithm which is extended by a flow theory approximation to determine accessible protein pockets [7]. A relatively novel algorithm which attempts to face the problem is based on building α-spheres and application of Voronoi tessellation. In the present work, we have focused on investigation of this algorithm as it has been implemented in the open source program Fpocket. Basic principles of Voronoi tessellation are introduced in the next chapter.

Energetic Approach. The other approach which has found its way into protein pocket predictions is based on molecular docking. The main objective of the technique is to find the global minimum on the potential energy hypersurface representing the Cartesian coordinates of atoms in a ligand-protein complex. Usually, molecular docking calculations utilize experimentally determined 3D protein structures trying to substitute the co-crystalized ligand with other compounds to estimate the best alignment in the active site and the resulting binding affinity or even Gibbs free energy. Since the binding energy is closely related with drugs' inhibition potencies towards specific enzymes, molecular docking as an important part of CADD and lead optimization can be employed to evaluate the binding properties of several drug candidates or to screen extensive virtual ligand databases in high-throughput SB virtual screening. Eventually, the process of molecular docking may easily be extended to search for other binding sites on the whole protein surface by means of a suitable probe ligand molecule. The only burden of the approach is higher computational strenuosity which requires high performance computing (HPC) clusters or supercomputers.

Being developed from classical mechanics, the interacting molecular system handled by molecular docking is approximated by rigid spheres and elastic bonds, the positions (R: distance; θ, φ: angles) of which are governed by attractive and repulsive forces.

$$E_s = \sum_{i=1}^{n} K_s \left(R_i^a - R_i^0\right)^2 ; E_b = \sum_{i=1}^{n} K_b \left(\theta_i^a - \theta_i^0\right)^2 ; E_t = \sum_{i=1}^{n} \frac{V_n}{2} \left(1 + \cos\left(n\varphi_i^a - \varphi_i^0\right)\right) ;$$

$$E_{vdW} = \sum_{i,j \in vdW; i \neq j}^{n} \left[\frac{A_{ij}}{R_{ij}^{12}} - \frac{B_{ij}}{R_{ij}^{6}}\right] ; E_{hb} = \sum_{i,j \in hb; i \neq j}^{n} \left[\frac{C_{ij}}{R_{ij}^{12}} - \frac{D_{ij}}{R_{ij}^{10}}\right] ; E_e = \sum_{i,j \in e; i \neq j}^{n} \left[\frac{q_i q_j}{\varepsilon R_{ij}}\right] . \tag{1}$$

Given the high number of atoms, especially in proteins, and many degrees of freedom, the system is often split in to a rigid part, which is not allowed to move, and a flexible part, which can change its position and conformation during the simulation. In molecular docking programs, the total potential energy is approximated by a scoring function consisting of several components responsible for: (1) bond stretching (E_s); (2) bond bending (E_b); (3) dihedral torsion (E_t); (4) van der Waals interactions (E_{vdW}); (5) hydrogen bonding (E_{hb}); (6) electrostatic interactions (E_e) (Eq. 1) [5].

As a systematic sampling of the combinatorial space is very computationally demanding, various stochastic approaches, Monte Carlo based methods or genetic algorithms are often used for minimization of scoring functions. However, development of accurate scoring function is an active area of research and, thus, the results of molecular docking still have to be accepted cautiously. So far, several dozen of molecular docking program with different scoring functions and search algorithms have been developed (e.g. Dock, Gold, Glide, Icm, Molegro and FlexX).

In the present study, we have decided to utilize an open-source program AutoDock Vina (shortened as Vina) to predict novel/allosteric binding sites on a 3D model of GSK3β enzyme as well as to perform subsequent SB virtual screening aimed at discovery of novel inhibitors of the enzyme. Among flexible molecular docking programs, Vina is one of the favorite ones since it can employ multithreading on multicore computers and provides relatively good ligand binding pose predictions in terms of reproducing the observations determined by X-ray or NMR. Utilizing principles of pleasing/embarrassing parallelization, Vina is ideally robust to serve as a molecular docking engine in high-throughput screening in supercomputers. The program relies on a memetic algorithm that interleaves global and local searches by stochastic and deterministic techniques. Briefly, the global optimizing algorithm is based on Markov chain Monte Carlo algorithm with restart disturbed by random mutation of the current solution. The result of local search is accepted if it satisfies Metropolis criterion. In the local search, Vina employs Broyden-Fletcher-Goldfard-Shanno (BFGS) algorithm aimed at nonlinear unconstrained optimization. For further information on Vina program, the reader is kindly referred to the literature [5].

3.2 GSK3β as a Target in CADD

Glycogen synthase kinase 3 beta (GSK3β; EC: 2.7.11.26) is an essential enzyme present within nearly all human cells (i.e. in cytoplasm, nucleus, cell membrane) and as such it is involved in the control of glucose metabolism, Wnt signal pathways, DNA transcription processes and microtubules maintenance. The constitutive mechanism of GSK3β action consists in serine/threonine phosphorylation of several enzymes like glycogen synthase (GYS1, GYS2), CTNNB1/beta-catenin, APC, NFATC1/NFATC, MAPT/TAU and MACF1. For example, by phosphorylation of GYS1, GSK3β inhibits glycogen synthesis in skeletal muscles, by phosphorylation of MAPT/TAU on threonine 548 it suppresses the ability of MAPT/TAU to stabilize microtubules, through phosphorylation of MACF1 it can restrict bulge stem cell migration, etc. Thus, GSK3β exhibits pleiotropic activity which has made it a desirable target for the treatment of diseases such as *diabetes mellitus* 2, Alzheimer's disease (AD), manic depressive disorders, neurodegenerative disorders, chronic inflammatory disorders. Especially, a nexus between GSK3β and AD has recently gained growing attention in drug design. According to current hypotheses, increased activity of GSK3β contributes to hyperphosphorylation of tau proteins and the formation of neurofibrillary tangles. Additionally, an up-regulated level of GSK3β has been suggested to be responsible for enhanced triggering of β-amyloidogenic processes *via* modulating γ-secreatase. As hyperphosphorylated tau proteins and β-amyloid neural plaques are the main hallmarks of AD pathology, modulators of GSK3β represent important drug candidates for causal treatment of AD. Although human organism is capable to restore declined GSK3β activity, it is not able to downregulate increased level of this enzyme in pathological states.

At present, several dozen X-ray structures of GSK3β with co-crystalized inhibitors are available in Protein Data Bank (a free online database) for computational chemistry and biology analyses. In the present study, a model of GSK3β complexed with NMS-869553A compound (PDB ID: 3DU8, resolution: 2.20 Å) was chosen for *in silico* investigations in Fpocket and AutoDock Vina program. Finally, this model was used for virtual screening.

4   Experimental Setup

4.1   Pocket Binding Search in 3D Model of GSK3β

Voronoi Tessellation. Fpocket is an open-source program which has been designed for geometry-based search for pockets in 3D protein models [8]. Generally, the algorithm is based on Voronoi tessellation and may be applied to spatial analysis of X-ray as well as homology protein models. In simple terms, a Voronoi region V_i of an atom z_i can be defined as a domain of all points x which have shorter Euclidian distance in a metric subspace $\Omega \in R^N$ to that atom than to other atoms $z_{j...k}$ (Eq. 2):

$$V_i = \left\{ x \in \Omega \,||\, x - z_i| \leq |x - z_j| \,\text{for}\, j = 1, \ldots, k, j \neq i \right\}. \tag{2}$$

If applied on all the protein atoms, the 3D space delimited by the protein structure can be decomposed into a complex of Voronoi objects. Interestingly, the Voronoi region of an atom includes the geometrical centers of all possible α-spheres that contact the atom. As for α-spheres, they are defined as spheres touching at least 3 different atoms of proteins and containing no atoms inside. The distribution of α-spheres depends on: (1) the α-sphere radius, and (2) atom coordinates of the proteins. Large spheres cover the external shape, while smaller spheres concentrate in densely populated protein pocket and cavities. Therefore, location of α-spheres with a suitable radius leads to identification of potential active sites. Practically, the task of α-spheres distribution is solved through Voronoi tessellation of 3D space into Voronoi regions, each of them being delimited by edges and vertices. Intersects of 4 Voronoi regions, which are called Voronoi vertices, are then used as the centers of α-spheres. Generally, perpendicular distances between atoms and α-spheres are determined instead of tangential distances to make the calculation easier.

The solution of 3D space Voronoi decomposition, that is finding the Voronoi vertices, is similar to the Delaunay triangulation. In case where the perpendicular distance approximation is assumed, atoms which can be interconnected by a Voronoi edge are taken as a Delaunay group that is consequently projected on a 4D paraboloid hypersurface. Finally, the Voronoi vertices are obtained as convex hulls of each Delaunay group, which corresponds to finding a set of atoms coinciding with the intersection of all half-spaces that involve the atoms [9]. To solve the convex hull problem, Fpocket program utilizes Qhull source code, which is based on determining the extreme points in the 4D paraboloid hypersurface fitted on the Voronoi intersections (Fig. 1) [8].

In order to find binding pockets in 3DU8 model of GSK3β, several parameters had to be tuned up. At first, ligands and waters molecules were removed from the input file. Then, pruning α-spheres to eliminate solvent inaccessible areas was done by adjusting the sphere radius, affinity of α-spheres to protein pockets was increased through setting condition on tight clustering of the spheres.

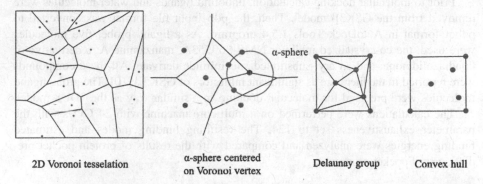

2D Voronoi tesselation   α-sphere centered
                              on Voronoi vertex   Delaunay group   Convex hull

Fig. 1. Voronoi tessellation for point generators in 2D space. To find the Voronoi vertices to center α-spheres, Delaunay groups need to be defined and solved as convex hulls.

Thus, after scanning mutual contacts between α-spheres, relatively isolated clusters are removed while larger complexes are aggregated into a single cluster. Additionally, Fpocket program can also evaluate which atoms are touching α-spheres in terms of differentiating the environment polarity. Close contacts with carbon and sulphur atoms are registered as a non-polar surrounding, while contacts with oxygens and nitrogens are interpreted as a polar area. Based on this simple chemical analysis, Fpocket eventually ranks the found protein pockets according to their potency to bind small drug-like molecules. Finally, an output with top-scoring binding sites is prepared (Fig. 2).

Molecular Docking. Employing AutoDock Vina program, protein pockets can be found when the gridbox for global search of the lowest potential energy binding mode is set up to enclose the whole enzyme structure. As already mentioned above, a Monte Carlo algorithm can evaluate the interactions between a probe ligand and the enzyme throughout the complete protein surface including pockets, cavities and tunnels. In order to accomplish such molecular docking task in reasonable time, common setting of flexible aminoacid residues had to be omitted. Only molecular docking with flexible ligand probe and rigid GSK3β model (i.e. PDB ID: 3DU8) could be performed.

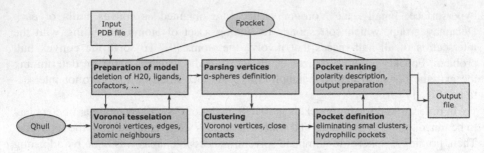

Fig. 2. Workflow of protein pocket prediction in Fpocket program [8].

Prior to molecular docking calculation, unbound ligands and water molecules were removed from the GSK3β model. Then, the pdb input file format was converted to pdbqt format in AutoDock Tools 1.5.4 program. As a ligand probe, five molecules were used: the co-crystalized inhibitor NMS-869553A, manzamine A, a derivative of thiadiazolidinone, and an aryl-substituted succinimide derivate. All these compounds were reported in the literature as significant inhibitors of GSK3β [10]. The probe ligand molecules were prepared for molecular docking in a similar way as the 3DU8 model.

The calculations were performed on a multicore machine with 24 CPUs with the parameter exhaustiveness set to 124. The resulting binding modes and estimated binding energies were analyzed and compared with the results of protein pocket prediction by Fpocket program.

4.2 Virtual Screening

After completing the protein pocket predictions and molecular docking analyses of the GSK3β model, a compromise between both solutions was proposed to define the top-scoring active site of the enzyme. Unlike molecular docking search for potential binding sites, the settings for virtual screening (VS) involved only a part of the enzyme which was used as the target site for virtual screening of virtual ligand library by flexible molecular docking on a Pflops-scale supercomputer Salomon in the Czech Republic. The supercomputer employs approximately 1000 powerful compute nodes with about 25 000 CPUs in total. For the virtual screening, we utilized online library ZINC from which 5 000 chemical compounds were selected based on the structure similarity indices with respect to the four ligand probes used for screening the GSK3β model by molecular docking. Thus, the virtual ligand library was composed of molecules sharing > 80 % structural similarity with the templates. The virtual ligand library was prepared for molecular docking with AutoDock Tools 1.5.4 program.

Regarding the GSK3β model (PDB ID: 3DU8), 30 aminoacid residues surrounding approximately the designed active site were set as flexible. The gridbox of $V \approx 26$ 000Å^3 was centered at x = 25.324, y = −10.807, and z = −9.471, which allowed to encompass all flexible residues. Execution of 5 000 jobs was performed distributively using 24 CPUs per node by our in-house shell program, which we described previously

[2]. The resulting binding energies were ranked and the top-scoring molecules were considered for further investigations, for instance, in biochemical experiments.

5 Results and Discussion

Calculations in Fpocket were set to use α-spheres of R = 3.0–6.0Å, touching at least 3 apolar neighbours with maximal distance for sphere clustering of 1.73Å. All other parameters were left in default mode. As a result, 23 potential protein pockets were predicted on the GSK3β model, the top scoring one having the overall score of 44.32 and being composed of relatively polar aminoacids. According to literature, the pocket no. 1. corresponds to the main ATP binding active site. The other pockets were characterized by considerably lower scores. These secondary pockets may be considered as potential allosteric sites of GSK3β. The localization and basic properties of the top 5 pockets are given in Fig. 3 and Table 1.

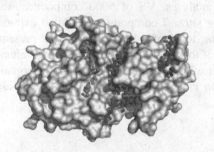 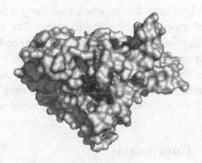

Fig. 3. 5 top-scoring protein pockets detected in GSK3β model (PDB ID: 3DU8) by Fpocket. (first top – red, second – green, third – blue, forth – yellow, fifth – magenta) (Color figure online)

Table 1. Properties of 5 top scoring protein pocket detected in GSK3β model (PDB ID: 3DU8).

No.	Score	Volume Å	Hydrophobicity score	Charge score	Polarity score
1.	44.33	5352.44	23.95	8.00	25.00
2.	19.36	3732.92	47.39	1.00	10.00
3.	17.63	2121.04	21.53	−2.00	11.00
4.	11.82	2640.69	17.60	−1.00	10.00
5.	11.02	2537.91	20.86	3.00	10.00

Searching for potential binding sites by molecular docking provided similar results as the protein pocket prediction. Probing the GSK3β protein surface by the succinimide derivative revealed only 1 site (−7.9 kcal/mol), manzamine A showed three binding sites (−11.2; −9.6; −9.4 kcal/mol), NMS-869553A, two sites (−8.1; −7.2 kcal/mol), and the thiadiazolidinone derivative, two sites (−5.6; −5.4 kcal/mol). After proper analysis of the resulting modes, it can be concluded that in the case of GSK3β model, geometric and energetic approaches identified the same binding sites. In both methods, the ATP binding site resulted as the best one. However, molecular docking seems to be more accurate since it can evaluate intermolecular interactions and indicate the preferred binding mode for a particular ligand (Fig. 4).

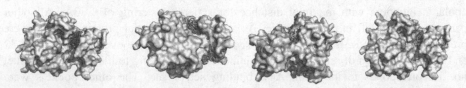

Fig. 4. Binding sites identified by molecular docking with four different probes. From left to right: succinamide derivate, manzamine A, NMS-869553A, and thiadiazolidinone derivate.

Utilizing the results of abovementioned analyses, VS of 5000 compounds was performed targeting the localized five binding sites. 7 compounds exhibited extraordinary *in silico* binding affinity exceeding the level of −15.0 kcal/mol. At present, further computational studies at a higher level of theory are being conducted to evaluate the obtained results. With respect to the high affinity estimates, purchase or organic synthesis of the top scoring compounds and *in vitro* tests for activity against GSK3β will be suggested to appropriate laboratories.

6   Conclusions

Voronoi tessellation represents a straightforward technique for detecting pockets and cavities on 3D protein surfaces. Such a geometrical approach is considerably less time consuming than energy-based molecular docking which needs to evaluate multiple interatomic forces within a large conformational space. From this study on GSK3β it has become evident that the primary determinant of ligand binding is pocket geometry. That's the reason why geometric and energetic approaches provided similar results herein, in case molecular dynamic (MD) forces were not taken into account. However, it is a matter of further research whether demanding MD simulations can bring more accurate results into this area. Being optimistic, we revealed 7 promising candidates for GSK3β activity modulation which should shed more light on the quality of the protein pocket predictions after *in vitro* experiments have been performed.

Acknowledgements.   The support of the Specific research project at FIM UHK is gratefully acknowledged. This work was also supported by long-term development plan of UHHK, by the IT4Innovations Centre of Excellence project (CZ.1.05/1.1.00/02.0070), and Czech Ministry of Education, Youth and Sports project (LM2011033).

References

1. Sliwoski, G., Kothiwale, S., Meiler, J., Lowe Jr., E.W.: Computational methods in drug discovery. Pharmacol. Rev. **66**, 334–395 (2014)
2. Dolezal, R., Sobeslav, V., Hornig, O., Balik, L., Korabecny, J., Kuca, K.: HPC cloud technologies for virtual screening in drug discovery. In: Nguyen, N.T., Trawiński, B., Kosala, R. (eds.) ACIIDS 2015. LNCS, vol. 9012, pp. 440–449. Springer, Heidelberg (2015)
3. Kubinyi, H.: Success Stories of Computer-Aided Design. Computer Applications in Pharmaceutical Research and Development. John Wiley & Sons Inc., New York (2006)
4. Lyne, P.D.: Structure-based virtual screening: an overview. Drug. Discov. Today **7**, 1047–1055 (2002)
5. Dolezal, R., Ramalho, T.C., França, T.C., Kuca, K.: Parallel flexible molecular docking in computational chemistry on high performance computing clusters. In: Núñez, M., et al. (eds.) ICCCI 2015. LNCS, vol. 9330, pp. 418–427. Springer, Heidelberg (2015). doi:10.1007/978-3-319-24306-1_41
6. Campbell, S.J., Gold, N.D., Jackson, R.M., Westhead, D.R.: Ligand binding: functional site location, similarity and docking. Curr. Opin. Struct. Biol. **13**, 389–395 (2003)
7. Laurie, A.T., Jackson, R.M.: Methods for the prediction of protein-ligand binding sites for structure-based drug design and virtual ligand screening. Curr. Protein Pept. Sci. **7**, 395–406 (2006)
8. Le Guilloux, V., Schmidtke, P., Tuffery, P.: Fpocket: an open source platform for ligand pocket detection. BMC Bioinform. **10**, 168 (2009)
9. Labute, P., Santavy, M.: Locating binding sites in protein structures. J. Chem. Comput. Group (2007)
10. Palomo, V., Soteras, I., Perez, D.I., Perez, C., Gil, C., Campillo, N.E., Martinez, A.: Exploring the binding sites of glycogen synthase kinase 3. Identification and characterization of allosteric modulation cavities. J. Med. Chem. **54**, 8461–8470 (2011)

A Machine-Learning Approach to the Automated Assessment of Joint Synovitis Activity

Konrad Wojciechowski[1], Bogdan Smolka[2], Rafal Cupek[2(✉)], Adam Ziebinski[2], Karolina Nurzynska[2], Marek Kulbacki[1], Jakub Segen[1], Marcin Fojcik[3], Pawel Mielnik[4], and Sebastian Hein[5]

[1] Polish-Japanese Institute of Information Technology, Warsaw, Poland
{kulbacki,segen}@pjwstk.edu.pl, kwojciechowski@polsl.pl
[2] Silesian University of Technology, Gliwice, Poland
{bsmolka,rcupek,aziebinski,knurzynska}@polsl.pl
[3] Høgskulen i Sogn og Fjordane, Sogndal, Norway
Marcin.Fojcik@hisf.no
[4] Helse Førde, Førde, Norway
pawel.franciszek.mielnik@helse-forde.no
[5] Medical Technology and Equipment Institute, Zabrze, Poland
sebastian@itam.zabrze.pl

Abstract. Medical ultrasound imaging is an important tool in diagnosing and monitoring synovitis, which is an inflammation of the synovial membrane that surrounds a joint. Ultrasound images are examined by medical experts to assess the presence and progression of synovitis. Automating image analysis reduces the costs and increases the availability of the ultrasound diagnosis of synovitis and diminishes or eliminates subjective discrepancies. This article describes research that is concerned with the problem of the automatic estimation of the state of the activity of finger joint inflammation using the information that is present in ultrasonography imaging.

1 Introduction

Chronic arthritis is a heterogeneous group of diseases that are characterised by the long-lasting inflammation of joints, which can influence the general condition of patients and which may involve other organs besides the joints. Chronic arthritis is estimated to affect up to 1.5 % of the population. Rheumatoid arthritis is the most frequent and has an estimated prevalence from 0.5 to 1.0 % of the population. The accurate measurement of disease activity is crucial to providing adequate treatment and care to patients and a medical ultrasound examination can provide useful information regarding this activity.

A medical ultrasound examination is a method for visualising the human body structures with high frequency acoustic waves. The power Doppler method uses the Doppler effect to measure and visualise the circulation of blood in tissues. Ultrasound examinations using the power Doppler method is a validated

© Springer International Publishing Switzerland 2016
N.T. Nguyen et al. (Eds.): ICCCI 2016, Part II, LNAI 9876, pp. 440–450, 2016.
DOI: 10.1007/978-3-319-45246-3_42

method for assessing joint inflammation [1]. This procedure, in which high frequency transducers (12.5 MHz and above) are used, is standardised. Different projections can be exploited to visualise the joints and joint inflammation, but the dorsal medial line in the joints of the hand is used most frequently. Both quantitative and semi-quantitative methods of activity measurement have been evaluated in clinical praxis [2,3]. Both are considered to be reliable and repeatable. When quantitative methods directly measure different parameters such as the intensity of a Doppler signal, the semi-quantitative method evaluates the ultrasound image. The second one is performed by a human examiner, who estimates the synovitis activity based on their own experience or standardised atlases. The result is registered as a number from 0 to 3, where 0 means no inflammation and 3 represents the highest possible inflammation activity. The semi-quantitative method is used more often in clinical praxis and scientific studies [2]. Possible discrepancies between different examiners and different examinations can influence the results in this method [4].

Developing a software system to automate these assessments can reduce the number of human-dependent discrepancies in the evaluation of synovitis. Such systems can be used in large clinical trials as well as in everyday clinical practice. Moreover, they can improve the quality of the results from large multicenter studies in which comparability of assessments from different sources is crucial. To the best of our knowledge, such a system does not exist and we have not found any indications that one is being developed. Therefore, this work presents studies concerning the development of a system to support the automatic evaluation of joint synovitis activity. Ultrasound images that were obtained using the standard procedure were analysed digitally to assess the stage of synovitis activity. To reduce any bias associated with different joint anatomies, we limited the assessments to the metacarpophalangeal (MCP) and proximal intra-phalangeal joints (PIP).

2   State of the Art

Brightness mode Doppler images together with colour and power mode imaging techniques are affected by various kinds of noise, such as electronic noise, which originates in the electronics of a system; speckle noise, which is a random interference pattern in an image that is formed through the coherent radiation of a medium that contains many sub-resolution scattering objects; noise that is introduced by vibrations that result from moving tissues; audio sound disturbances that are generated within the human body, which are indistinguishable from the Doppler shift that is produced from the blood and tissues and flash artefacts that are generated by the movement of the transducer with respect to the tissue [5].

Within the proposed research project, we intend to develop a family of novel image enhancement techniques, whose goal is noise suppression and are aimed at increasing the ability to extract relevant image features. The new noise suppression algorithms, which will be developed within the research project, will be tuned to the noise characteristics that are found in ultrasound images and will be based on the concepts of:

- Non-Local-Means, which are based on averaging the image pixels in such a way that the new value of the restored image at a given location is estimated as the weighted average of the intensities of all of the image pixels, whose local filtering windows show a high degree of similarity [6];
- Total-Variation Regularisation inspired approaches, which minimise the image variation, but which are subject to constraints that involve the noise statistics [7];
- Anisotropic Diffusion, which reduces the noise component while preserving the image details [8];
- Digital Paths extracting the features of the local image using digital paths to explore the extended neighbourhood of the pixels being processed to determine the optimal connections between the central pixel of the filtering window and its neighbours [9,10];
- Myriad Filters, which are able to suppress the noise modelled by alpha-stable distributions, whose interesting feature is that it is possible to model the Gaussian and heavy tailed distributions by changing their main coefficient [11];
- Bilateral Filters, which estimate the weighted average of the intensities of all of the pixels that belong to the processing window, which then use the information that is extracted from the local paths [12].

A review of the existing literature led to the conclusion that the current techniques for the analysis of power mode images are still at an initial stage. As was shown in [4], mostly gray-scale images are primarily being binarised using a threshold that is chosen by the examiner and classified into three categories that express the degree of the inflammatory changes that are observed [13,14] as depicted in Fig. 1.

Only a few reports have concentrated on the overall number of pixels that depict significantly increased values of the power signal [4,15]. This situation leads to the large variability and subjectivity of the classifications and makes the examination results highly dependent on the experience of the examiner and the technical conditions of the examination.

Setting the global binarisation threshold is crucial for the analysis of power Doppler images. If the threshold is too high, only a few small regions will be detected, whereas a threshold level that is too low will cause large image areas to be extracted and the analysis will be influenced by the non-uniformity of the image background. This difficulty was addressed in [16], in which it was proven that even simple, heuristic methods increased the efficiency of the examinations, thus enabling the objective tracking of the patients treatment records over a long period of time, especially when the composite Disease Activity Score (DAS) is used [17].

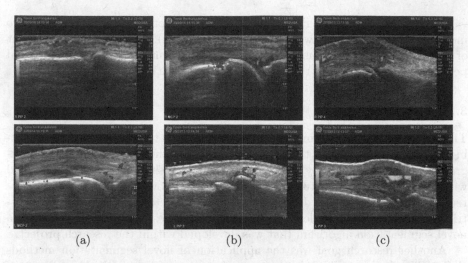

(a)   (b)   (c)

Fig. 1. Semi-quantitative grading of the power Doppler intensity with three levels: (a) mild flow, single Doppler signals, (b) moderate flow, many Doppler regions detected, (c) intense flow, many Doppler regions covering the synovium (Color figure online)

3   Research Goals

The main research goal was to develop a computer aided diagnostic system that would permit the automated assessment of synovitis activity. To achieve this goal, a collaboration of doctors, specialists in rheumatology and IT specialists in the field of machine learning and image processing was established under the Polish-Norwegian research project called MEDUSA. From one side, new image processing techniques have been created, including new methods that are able to detect multiple types of local features. The MEDUSA project provides a reference frame that helps to normalise measurements that are performed on ultrasound images with respect to changes in the probe position and the joint articulation. Machine-learning methods were used to train the feature detectors and the assessment function on the ultrasound images of synovitis cases, which were annotated and scored by medical experts. An additional research result is the novel visualisation methods for power Doppler images, which are intended to aid examiners. The result of the MEDUSA project is a prototype of a software system, which will be useful for medical personnel and will help in diagnosing rheumatic diseases.

3.1   Data Collection and Verification of Results

Ultrasound images were collected from patients with chronic arthritis during routine visits at the rheumatology department of Helse Frde. Preliminary analysis was conducted by trained, experienced personnel. Approval was obtained from the Ethics Committee and the patients were informed and signed consent forms before data collection. The medical usefulness of the automatic method was validated in clinical trials. Patients with diverse chronic types of arthritis were included (primarily those with rheumatoid arthritis). The systems assessment of synovitis activity was compared with the assessment prepared by the medical personnel and the systems results were correlated with clinical parameters.

3.2   Image Processing

The subjectivity of the examination can be alleviated to a large extent by adopting a segmentation scheme that is able to eliminate the influence of image background non-homogeneity by incorporating local thresholding schemes combined with the morphological features of the high-intensity of the power Doppler image regions. These features can describe the shape parameters and granulometric distribution of the detected blobs and they can be successfully incorporated into a novel segmentation algorithm that was developed within the research project.

Another research goal was the application of novel segmentation methods that utilise the generalised distance transforms, combined with the elements of the level-set theory developed by S. Osher and J. Sethian [18]. Part of the project covered the application of the Local Binary Patterns technique to effectively perform the segmentation by exploiting the textural image features [19].

The family of new segmentation techniques permit the fast and accurate extraction of the high intensity regions of the power mode images, which increases the objectiveness of the examination results and enables the progress of the treatment to be tracked over a long time period.

3.3   Visualisation

The power mode Doppler technique is significantly more sensitive than the standard colour mode and therefore is preferred when weak signals need to be detected and visualised [13]. However, the colour Doppler displays tend to be complex and confusing due to the strong colour variations that are caused by changes in the Doppler shift and flow direction.

Another important aspect of the proposed research project focused on the visualisation of the information that was acquired from the power mode images. Typically, the image intensity is transformed into colour using the appropriate pseudocolour look-up tables. The coloured regions are superimposed on a standard B-mode image, which completely occludes the information that is carried by the reference image as is shown in Fig. 1.

A novel method was suggested that extends the approach developed within the previous research project, which was focused on image colourisation [20] and applied them in such a way that the information from the power-mode is automatically embedded into the B-mode images. In this way, the examiner has access to the fused information that is acquired from the two image modalities.

3.4   A Machine-Learning Approach to the Automated Assessment of Joint Synovitis Activity

The proposed approach can be used to construct an adaptable function that is trained to assess the degree of synovitis activity from an ultrasound image of a joint using machine-learning methods. We call this function a synovitis estimator. The estimator was trained using ultrasound images of joints along with their synovitis assessment activity scores, which were provided by expert examiners, with the aim of the estimator being able to approximate the average expert score. The estimator is built with the help of features that were extracted from an image using processing methods such as those described above.

While many of the common techniques for image-based recognition use global features that preserve only a small part of image information, one goal of the proposed approach is to use as much of the relevant information as possible from images in a form that is invariant to the transformations that result from different probe placements, joint articulations and inter-patient differences. In order to achieve this goal, we used an articulated model that contained the image features that are related to parts of the skeleton, such as the bone edges and joints. A method of registration that was appropriate for the articulated models, such as the Constellation method, the optimisation of graph-cuts [21], Laplacian eigenfunctions [22] or stochastic graphs [23] were the basis of the learning algorithm that was used for the inference of the class model for each joint. This inference process is a supervised learning approach that relies on annotations that identify the desired features, which are added to the test images by trained examiners. Registration of an image of a joint with a class model brings the image into a common frame of reference. The measurements that are performed on the image relative to the class model reduce variability with respect to different probe placements and joint articulations and intensity normalisation with respect to inter-patient differences. The synovitis estimator was constructed using the methods for learning multivariate functions, such as multivariate regression. To ensure a greater reliability, multiple partial estimators were used, each based on a different type of image measurement. The final, integrated synovitis estimator combines the outputs of the partial estimators into the final assessment score using a method such as boosting [24].

3.5   Software Development

According to the requirements that were given, the proposed solution was based on the client-server model. For performance reasons and to ensure flexibility, the C++ programming language was selected. Multi-Core and GPU programming was used for accelerated computation. This software runs in multiplatform environments, in particular Linux and Windows platforms with possible extensions to mobile platforms (data browsing on tablets and smartphones and inter-user communication). The application has a modular structure, which permits its capabilities to be extended easily using a dedicated plug-in system. Exchanging/publishing plug-ins between users is also supported. This speeds up the diagnosis and elimination of problems and thus allows the development of the dynamic platform to continue.

4   Results and Discussion

Before any algorithm for the automatic description of synovitis activity can be evaluated, a database of ultrasound images, which has been annotated by expert physicians, is necessary. It is crucial that the gathered data consists of images that present ultrasound data, but that also contains the power mode Doppler information. Additionally, each USG image should be supported by coefficients for locating the skin border, bone areas and joint placement and a description of the finger joint synovitis region. It is also indispensable that the inflammation activity be classified and the images annotated with a number from 0 to 3, where 3 means the highest possible level of inflammation and 0 stands for no inflammation.

In order to simplify the work of expert physicians when collecting the data, a system supporting such data acquisition was developed. Its details are given in [25], while the database itself can be accessed at the MEDUSA project web page. The available database consists of almost 1000 images with a 737×562 pixel resolution, which are divided into three groups: images where two bones are visible, images with a single bone and images of low-quality data.

After the data was gathered, several methods for object segmentation were introduced within the project. Their results may also find many interesting applications for visualisation purposes. The automatic image colourisation described

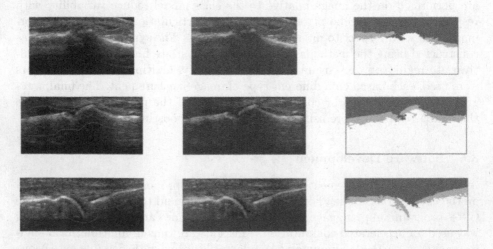

Fig. 2. Exemplary results of the segmentation of joint synovitis USG images: (right) original grayscale USG image with the colour scribbles indicated, (middle) colourised image and (right) the results of the colourisation-based segmentation (left) [26]. (Color figure online)

in [26] is a semi-automatic method for the detection of the bone regions. It uses scribbles (lines) that are drawn in different colours by the user to distinguish between various tissues (see Fig. 3 left). These are then used as the starting point for a colourisation method that colours the image using the forces that were calculated from the original image content, thereby resulting in images such as those presented in middle of Fig. 3. This method is very good for extracting the bone region as is presented in Fig. 3 right. An alternative technique for automatically locating the bone area was described in [27], where the concept of a confidence map was exploited to extract the region that was of interest. This approach is characterised by a very high correspondence between the annotated bone regions and those that are found using this method, unlike the methodology that was suggested in [28], in which the image content is analysed statistically in order to segment the image into blobs. Then, using the placement information of the blobs and exploiting some known topology of the USG image, the location of the skin border, description of the bone area, calculation of the joint coordinates and the estimation of the finger joint synovitis region are performed. Figure 3. presents the implemented system along with its performance compared to manual annotations. According to the experiments that were performed, this solution performs very well and can work to support the manual annotation of USG images.

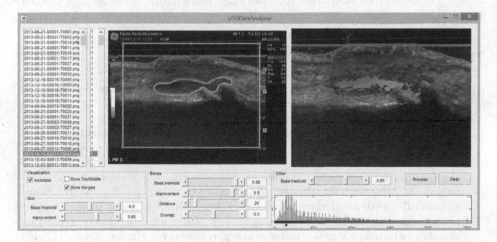

Fig. 3. USGDataAnlyser software. The image in the middle presents the original data with the annotations that were prepared by an expert physician. The image on the right presents the automatically generated output of the statistical transformation. In both cases, the red line represents the skin border, the pink line depicts the bones and the green represents the joint synovitis inflammation [28]. (Color figure online)

5   Conclusions

This work presents a brief overview of a computer aided diagnostic system that allows the automated assessment of the severity of synovitis, which was accomplished during the realisation of the MEDUSA project. All of the aspects concerning ultrasound image processing that were addressed in order to create software for the automatic annotation and classification of USG data are discussed. The presented software enables the ultrasound images presenting finger joint synovitis inflammation with the manual annotation prepared by an expert physician to be collected. Details concerning the database are also given. Finally, several algorithms that support the semi-automatic or automatic detection of the bone region were described as well as a system that uses the statistical data processing approach in order to automatically localise the regions of interest are presented.

The presented topics are only part of the research that is being conducted within this project; however, they indicate how advanced the research is in comparison to the goals of the project. Future research will concentrate on the automatic classification of the inflammatory state of finger joint synovitis.

Acknowledgement. This research obtained funding from the Norwegian Financial Mechanism 2009–2014 under Project Contract No. Pol-Nor/204256/16/2013. The ultrasound images for the MEDUSA project were created at the Section for Rheumatology; Department for Neurology, Rheumatology and Physical Medicine, 238 Central Hospital, Forde, Norway.

References

1. Ostergaard, M., Szkudlarek, M.: Ultrasonography: a valid method for assessing rheumatoid arthritis? Arthritis Rheum. **52**(3), 681–686 (2005)
2. Kamishima, T., Tanimura, K., Henmi, M., Narita, A., Sakamoto, F., Terae, S., Shirato, H.: Power Doppler ultrasound of rheumatoid synovitis: quantification of vascular signal and analysis of interobserver variability. Skeletal Radiol. **38**(5), 467–472 (2009)
3. Kamishima, T., Sagawa, A., Tanimura, K., Shimizu, M., Matsuhashi, M., Shinohara, M., Hagiwara, H., Henmi, M., Narita, A., Terae, S., Shirato, H.: Semi-quantitative analysis of rheumatoid finger joint synovitis using power Doppler ultrasonography: when to perform follow-up study after treatment consisting mainly of antitumor necrosis factor alpha agent. Skeletal Radiol. **39**(5), 457–465 (2010)
4. Albrecht, K., Muller-Ladner, U., Strunk, J.: Quantification of the synovial perfusion in rheumatoid arthritis using Doppler ultrasonography. Clin. Exp. Rheumatol. **25**(4), 630 (2007)
5. Hoskins, P., Martin, K., Thrush, A.: Diagnostic Ultrasound: Physics and Equipment. Cambridge University Press, Cambridge (2010)
6. Buades, A., Coll, B., Morel, J.M.: A review of image denoising algorithms, with a new one. Multiscale Model Simul. **4**(2), 490–530 (2006)
7. Strong, D., Chan, T.: Edge-preserving and scale-dependent properties of total variation regularization. Inverse Prob. **19**, 165–187 (2003)

8. Perona, P., Malik, J.: Scale-space and edge detection using anisotropic diffusion. IEEE Trans. PAMI **12**(7), 629–639 (1990)
9. Smolka, B., Plataniotis, K.N., Venetsanopoulos, A.N.: Nonlinear techniques for color image processing, in nonlinear signal and image processing. In: Barner, K.E., Arce, G.R. (eds.) Theory, Methods and Applications, pp. 445–505. CRC Press, Boca Raton (2004)
10. Smolka, B., Venetsanopoulos, A.N.: Noise reduction and edge detection in color images. In: Lukac, R., Plataniotis, K.N. (eds.) Color Image Processing: Methods and Applications, pp. 75–100. CRC Press, Boca Raton (2007)
11. Barner, K.E., Arce, G.R.: Nonlinear Signal and Image Processing. CRC Press, Boca Raton (2004)
12. Malik, K., Smolka, B.: Modified bilateral filter for the restoration of noisy color images. In: Blanc-Talon, J., Philips, W., Popescu, D., Scheunders, P., Zemčík, P. (eds.) ACIVS 2012. LNCS, vol. 7517, pp. 72–83. Springer, Heidelberg (2012)
13. McDicken, W.N., Anderson, T.: The difference between colour Doppler velocity imaging and power Doppler imaging. Eur. J. Echocardiogr. **3**, 240–244 (2002)
14. Naredo, E., Collado, P., Cruz, A., et al.: Longitudinal power Doppler ultrasonographic assessment of joint inflammatory activity in early rheumatoid arthritis: predictive value in disease activity and radiologic progression. Arthritis Rheum. **57**, 116–124 (2007)
15. Qvistgaard, E., Rogind, H., Torp-Pedersen, S., et al.: Quantitative ultrasonography in rheumatoid arthritis: evaluation of inflammation by Doppler technique. Ann. Rheum. Dis. **60**, 690–693 (2001)
16. Cloutier, G., Qin, Z., Garcia, D., et al.: Assessment of arterial stenosis in a flow model with power Doppler angiography: accuracy and observations on blood echogenicity. Ultrasound Med. Biol. **26**(9), 1489–1501 (2000)
17. van Riel, P.L., Schumacher, H.R.: How does one assess early rheumatoid arthritis in daily clinical practice? Best Pract. Res. Clin. Rheumatol. **15**, 67–76 (2001)
18. Osher, S., Sethian, J.A.: Level Set Methods: Evolving Interfaces in Geometry, Fluid Mechanics, Computer Vision and Materials Sciences. Cambridge University Press, Cambridge (1996)
19. Ojala, T., Pietikinen, M., Harwood, D.: A comparative study of texture measures with classification based on feature distributions. Pattern Recogn. **29**, 51–59 (1996)
20. Kawulok, M., Kawulok, J., Smolka, B.: Discriminative textural features for image and video colorization. In: IEICE Transactions on Information and Systems, vol. E95-D, no. 7 (2012). Kawulok, M., Smolka, B.: Texture-adaptive image colorization framework. EURASIP J. Adv. Sig. Process. **99** (2011)
21. Chang, W., Zwicker, M.: Automatic registration for articulated shapes. In: Computer Graphics Forum (Proceedings of SGP 2008), vol. 27, p. 5 (2008)
22. Mateus, D., Horaud, R.P., Knossow, D., Cuzzolin, F., Boyer, E.: Articulated shape matching using Laplacian eigenfunctions and unsupervised point registration. In: Proceedings of IEEE CVPR, June 2008
23. Segen, J.: Inference of stochastic graph models for 2D and 3D shape. In: Proceedings of NATO Advanced Research Workshop on Shape in Picture, Driebergen, Netherlands (1992)
24. Schapire, R.E.: The boosting approach to machine learning: an overview. In: Workshop on Nonlinear Estimation and Classification, Mathematical Sciences Research Institute (MSRI), Berkeley, CA (2003)

25. Kulbacki, M., Segen, J., Habela, P., Janiak, M., Knieć, W., Fojcik, M., Mielnik, P., Wojciechowski, K.: Collaborative tool for annotation of synovitis and assessment in ultrasound images. In: Chmielewski, L.J., Kozera, R., Shin, B.-S., Wojciechowski, K. (eds.) ICCVG 2014. LNCS, vol. 8671, pp. 364–373. Springer, Heidelberg (2014)
26. Popowicz, A., Smolka, B.: Bilateral filtering based biomedical image colorization. In: Tavares, J.M.R.S., Natal, A.M.J. (eds.) Computational Vision and Medical Image Processing, pp. 163–169. CRC Press, Boca Raton (2016)
27. Radlak, K., Radlak, N., Smolka, B.: Automatic detection of bones based on the confidence map for Rheumatoid Arthritis analysis. In: Tavares, J.M.R.S., Natal, A.M.J. (eds.) Computatiion Vision and Medical Imge Processing, pp. 215–220. CRC Press, Boca Raton (2016)
28. Nurzynska, K., Smolka, B.: Automatic finger joint synovitis1 localization in ultrasound images. In: Proceedings of SPIE, vol. 9897, pp. 98970N–98970N-11, April 2016. doi:10.1117/12.2227638

Intersection Method, Union Method, Product Rule and Weighted Average Method in a Dispersed Decision-Making System - a Comparative Study on Medical Data

Małgorzata Przybyła-Kasperek$^{(\boxtimes)}$ and Agnieszka Nowak-Brzezińska

Institute of Computer Science, University of Silesia,
Będzińska 39, 41-200 Sosnowiec, Poland
{malgorzata.przybyla-kasperek,agnieszka.nowak}@us.edu.pl

Abstract. In this article a dispersed decision-making system, that was proposed in the previous paper of one of the authors, is used. In the system four selected fusion methods were used. The aim of the paper is to compare the efficiency of inference of these methods for knowledge base from the medical field. Two medical data sets from the UC Irvine Machine Learning Repository were used. Note that these databases were used in a dispersed form, which means that one knowledge base was transformed into a set of knowledge bases. This application was intended to reflect a situation in which many medical centers independently collect knowledge from one field and then we want to use all of this knowledge at the same time in the process of inference.

Keywords: Dispersed decision-making system · Fusion method · Intersection method · Union method · Product rule · Weighted average method

1 Introduction

The use of dispersed knowledge is a very important issue. Sometimes knowledge is gathered independently in different data centers. In order to use knowledge stored in separate knowledge bases, special methods designed for this purpose are needed. Issues concerning dispersed knowledge are also useful, when we have access to a huge data set and it is possible to divide this set into smaller knowledge bases in order to perform inference in a reasonable time. Such situations often arise in the medical field, and therefore the use of dispersed knowledge is so important in this area.

The concept of inference based on dispersed knowledge is being considered by one of the authors for several years [10–14,16]. An approach that is used in this paper was proposed in the article [12]. The aim of the paper is to examine the use of four fusion methods - intersection method, union method, product rule and weighted average method in the system. Comparison of the effectiveness of inference of these methods is made on the basis of two medical data sets.

© Springer International Publishing Switzerland 2016
N.T. Nguyen et al. (Eds.): ICCCI 2016, Part II, LNAI 9876, pp. 451–461, 2016.
DOI: 10.1007/978-3-319-45246-3_43

The issue of combining classifiers is a very important aspect in the literature [2,4–6]. The aim of the issue is to improve the quality of the classification by combining the results of the predictions of the base classifiers. One of the basic questions is what combination rule to use. In the literature, many different methods have been proposed [1–7]. In this article four selected fusion methods are considered. Also different methods of exploring knowledge are analyzed in the papers [8,9,15].

The paper is organized as follows. The second section briefly describes the dispersed decision-making system. The third section describes the fusion method that are used. The fourth section shows a description and the results of experiments carried out using two medical data sets from the UC Irvine Machine Learning Repository: Lymphography and Primary Tumor data set are used. Finally, a short summary is presented in Conclusion Section.'

2   Basic Concepts of Dispersed Decision-Making System

A dispersed decision-making system, which is used in the paper, was proposed by one of the authors in the article [12]. The main assumptions of the system are very briefly described below. We do not give a detailed definition because it is not the subject of this paper. A detailed description of the system can be found in the paper [12].

The knowledge is stored in a dispersed form, which means in a set of decision tables. Dispersion of the knowledge is not performed by the system, we assume that the set of tables is accumulated in this way and pre-specified. However, one condition must be satisfied by local decision tables - they must store the knowledge from one domain, which is reflected by common decision attributes. Each local decision table $D_{ag} = (U_{ag}, A_{ag}, d_{ag})$ is managed by one agent, this agent is a resource agent ag. Our goal is to determine groups of resource agents that are homogeneous. These groups are formed from the agents who classify the test object in a similar way. Two steps are carried out in order to create groups. In the first stage initial clusters are defined. In the second stage the negotiation process is realized. For more details, please refer to the paper [12]. The final form of clusters are obtained when the second step is completed. A superordinate agent is defined, if cluster contains at least two resource agents. This agent is called a synthesis agent and is designated by as_j, where j is the number of cluster. The synthesis agent, as_j, has access to a aggregated decision table which is determined based on the knowledge of the resource agents that belong to its subordinate group. A formal definition of a dispersed decision-making system is as follows.

Definition 1. *A dispersed decision-making system with dynamically generated clusters is* $WSD_{Ag}^{dyn} = \langle Ag, \{D_{ag} : ag \in Ag\}, \{As_x : x \text{ is a classified object}\}, \{\delta_x : x \text{ is a classified object}\}\rangle$ *where Ag is a finite set of resource agents;* $\{D_{ag} : ag \in Ag\}$ *is a set of decision tables of resource agents;* As_x *is a finite set of synthesis agents defined for clusters dynamically generated for the test object x,*

$\delta_x : As_x \rightarrow 2^{Ag}$ is a injective function that each synthesis agent assigns a cluster generated due to classification of the object x.

Local decisions are generated based on the aggregate knowledge of one cluster. When this is accomplished, inconsistencies in the knowledge stored in different knowledge bases can occur. In order to resolve this problem the approximated method of the aggregation of decision tables was proposed in previous papers [10–12]. This method is also used in the paper. The aim of this method is to aggregate the decision tables from one cluster in one coherent decision table. In order to construct an aggregated decision table only the relevant objects from the decision tables from one cluster are used. Then the fusion method is used and global decisions are taken.

3   Fusion Methods

In the literature [6,17], fusion methods are classified into three types - Type 1 the abstract level, Type 2 the rank level and Type 3 the measurement level. In Type 1 each base classifier generates a class to which the observation belongs. In Type 2 each base classifier generates a set of classes ordered by the plausibility that they are the correct labels. In Type 3 each base classifier generates a vector that represents the probability of an observation belonging to different decision classes. In this article, we explore two methods from the rank level (the intersection method and the union method), and two methods from the measurement level (the product rule and the weighted average method). These methods are discussed in the papers [3,5,6].

The results of prediction of synthesis agents have different forms - vectors of ranks or vectors of probabilities, depending on the type of fusion method that is considered. The results of prediction are generated as follows.

In both cases, for Type 2 and Type 3, in the first step a vector of values $[\mu_{j,1}(x), \dots, \mu_{j,c}(x)]$ is defined for each j-th cluster, where c is the cardinality of a set of values of the decision attribute. The agents from the cluster j classify the test object x to the decision class v_i with a certain confidence factor, which is expressed by $\mu_{j,i}(x)$. This value is defined based on the relevant objects that are selected from the decision class v_i of the aggregated decision table of cluster j. m_2 relevant objects are chosen, this parameter value is adjusted experimentally. Then the average value of the similarity between the test object and the relevant objects is calculated. The results of prediction in the measurement level are in the form of these vectors.

In the case of the rank level a vector of the rank is specified. Ranks are determined based on the vector, which is defined above. The values of the decision attribute that have the maximum value μ receive Rank 1. The next most certain decisions receive Rank 2, etc. According to this description, for each j-th cluster, the vector of rank $[r_{j,1}(x), \dots, r_{j,c}(x)]$ will be specified based on the vector $[\mu_{j,1}(x), \dots, \mu_{j,c}(x)]$.

Below the fusion methods will be described.

The intersection method. This method belongs to the group of reduction methods. The aim of this method is to generate the smallest set of decisions that have the greatest probability that the correct class belongs to this set. This method requires a training process which is as follows. In the first stage each resource agent classifies objects from the training set. For this purpose the leave-one-out method is performed on the training set and the vector of rank is defined for each resource agent. This vector is generated in an analogous manner as was described above. At the beginning the vector of values μ is determined, and then based on it ranks are generated. Then the ranks that were given to any true class by the agent are analysed and the lowest rank are selected. In the next step the minimum of the ranks assigned to the resource agents from one cluster is chosen. This value is a threshold value for the cluster. The global decisions determined for the test object by the dispersed decision-making system is equal to the intersection of large neighborhoods taken from each cluster. The thresholds that were calculated define the sizes of the neighborhoods. The neighborhoods are designated by the worst-case behavior of the cluster and because of this they may be very large. This causes that sometimes a very large set of decisions is generated by the system. To overcome this drawback, the modification was applied. Only when the worst rank is assigned to a certain small percentage of cases (this percentage is designated by the parameter u) it can be the threshold value.

The union method. Like in the intersection method, the aim of the union method is to reduce the number of classes in the output without losing the true class. The union method also requires training, which consists in determining the threshold value for each classifier. This method relies on calculating the union of small neighborhoods that contain decisions whose ranks are above a threshold value. The thresholds on the ranks are selected using a max-min procedure. Note that the max-min procedure requires that the same sets of objects must be included in the decision tables of the resource agents. The max-min procedure is performed as follows. Firstly, each of the resource agents classifies the training objects by the leave-one-out method in the same way as described in the intersection method. Ranks that were assigned to the correct classes by the resource agents are stored in a matrix. The columns of the matrix correspond to the resource agents and the rows correspond to the training objects. Another matrix is defined based on this matrix. Firstly, the minimum rank of each row is selected. Only the minimum value is left in the rows of the new matrix, while the remaining cells are set to 0. The maximum value of the j-th column of the new matrix is the threshold value for the j-th resource agents. Based on the thresholds of the resource agents the threshold on the ranks for each cluster is determined by calculating the minimum thresholds on the ranks that were assigned to the resource agents belonging to a cluster. The decisions defined by the dispersed system for the test object is equal to the union of the neighborhoods taken from each cluster and determined by the threshold value.

The effectiveness of the union method is sensitive to outlying worst cases. During the experiments, it turned out that in many cases generated set was too large. Therefore, it was assumed that only when the worst rank is assigned to a certain small percentage of cases (this percentage is designated by the parameter u) it can be the threshold value. Another modification which aim is to reduce the generated set is that only the synthesis agents that made unambiguous decisions were considered. That is, all of the ranks above the threshold value were assigned to only one decision value. In extreme cases in which all of the synthesis agents made ambiguous decisions, we took into account only those agents who gave a rank above the threshold value to the minimum number of decisions.

The product rule. The product rule belongs to the measurement level method. In the product rule the product of the probability values μ is determined for each decision class. The set of decisions that have the maximum of the products of the probability values is the result that is generated by the dispersed system. As it is known the most pessimistic prediction has large impact on the output of the product rule. If one agent defines a probability equal to 0 for the decision, then the value of the product for this decision is also equal to 0, regardless of the values assigned by other agents. This is a veto mechanism because one agent is decisive. In order to prevent such situations modification was adopted. All probabilities equal to 0, were replaced by 10^{-3}.

The weighted average method. In the weighted average method the value is determined for each decision

$$\sum_{j=1}^{L} \omega_j \mu_{j,i}(x), \tag{1}$$

where ω_j is the weight for j-th synthesis agent. It was proven [7] that if the base classifiers are independent, it is optimal to adopt weights that are proportional to the expressions $\omega_j \propto \frac{1-e_{as_j}}{e_{as_j}}$, where e_{as_j} is the individual error rate of the j-th synthesis agent. The error rate of the synthesis agent is estimated based on the error rates of the resource agents that belong to its subordinate cluster. At first, the error rate e_{ag_i} for each resource agent $ag_i \in Ag$ was determined based on the training set. When the set of synthesis agents As_x was designated for a test object x, the error rate of the synthesis agent $as_j \in As_x$ was determined as follows: $e_{as_j} = \frac{1}{card\{\delta_x(as_j)\}} \sum_{ag_i \in \delta_x(as_j)} e_{ag_i}$. The set that is generated by the dispersed system is a set of decisions that have the highest value calculated by Formula 1.

4   Experiments

In the experimental part, a comparison of four fusion methods based on the dispersed medical data is made. This comparison is carried out using a dispersed decision-making system with a dynamic structure. The aim of the experiment is to compare the methods individually as well as two groups of methods - from the rank level and from the measurement level.

4.1 Datasets

The experiments were carried out using data sets from the UC Irvine Machine Learning Repository (archive.ics.uci.edu/ml/): Lymphography data set, Primary Tumor data set. These data are from the medical field and were obtained in the University Medical Centre, Institute of Oncology, Yugoslavia (M. Zwitter and M. Soklic provided this data). In lymphography, a contrast dye is injected into lymph vessels. These vessels can be visualized on X-ray picture. Lymphography is especially useful for cancers of the lymphatic system diagnosis. In the Primary Tumor data set the decision attribute specifies where (of 22 organs) the cancer cells are located. These data sets from the medical field were chosen because they have a lot of conditional attributes and a lot of decision classes. A large number of attributes is essential in the dispersion process. The dispersed decision-making system, which is used, sometimes generates a set of decisions, which is justified when a large number of decision classes occurs.

Each data set was divided into two disjoint subsets: a training set and a test set. The data sets have the following numeric properties: Lymphography: # The training set - 104; # The test set - 44; # Conditional - 18; # Decision - 4; Primary Tumor: # The training set - 237; # The test set - 102; # Conditional - 17; # Decision - 22.

The next step of data preparation consists in dispersion of datasets. The training set was divided into a set of decision tables. Different divisions were considered - with 3, 5, 7, 9 and 11 resource agents (decision tables). The following designations are used for these systems: WSD_{Ag1}^{dyn} - 3 resource agents; WSD_{Ag2}^{dyn} - 5 resource agents; WSD_{Ag3}^{dyn} - 7 resource agents; WSD_{Ag4}^{dyn} - 9 resource agents; WSD_{Ag5}^{dyn} - 11 resource agents. The dispersion of the data set was performed as follows. The cardinality of set of conditional attributes in each decision table of resource agent was determined, and the number of common conditional attributes of decision tables was defined. Then attributes are distributed randomly but in such a way that the above conditions were met. The decision attribute in decision tables is the same as the decision attribute in the data set. Each universe of decision tables includes all objects from the data set. We assume that in the decision tables the identifiers of objects are not stored. Thus, after dispersion, identifying of objects from different decision tables is not possible. This approach reflects a situation in which many medical centers collects knowledge from the same field. Then the knowledge accumulated in several decision tables is used at the same time in the process of inference. The authors are aware that the big bias based on random dispersion of attributes may occur in the experiments. The influence of random dispersion of attributes on the results will be investigated in a future work.

Some of the considered fusion methods have the property that the final decision may have ties. In order to analyze these properties the appropriate classification measures were applied. These measures are as follows: *estimator of classification error e* - if the correct decision of an object belongs to the set of global decisions generated by the system, then the object is properly classified; *estimator of classification ambiguity error e_{ONE}* - if only one, correct decision was generated to an object, then the object is properly classified; *the average size of the global decisions sets* $\overline{d}_{WSD^{dyn}_{Ag}}$ *generated for a test set.*

The following designations were used in the description of the experimental results: m_1 - parameter which are used in the process of cluster generation, determines the number of relevant objects that are selected from each decision class of the decision table; p - parameter which occurs in the definition of relations between agents; $A(m)$ - the approximated method of the aggregation of decision tables; $C(m_2)$ - the method of conflict analysis (the intersection method, the union method, the product rule and the weighted average method), with parameter which determines the number of relevant objects that are used to generate decision of one cluster.

4.2 Results

The parameters optimization was performed at the beginning of experiments. Tests for different parameter values were carried out: $m_1 \in \{1, \ldots, 13\}$, $m, m_2 \in \{1, \ldots, 10\}$ and $p \in \{0.05, 0.1, 0.15, 0.2\}$. From all of the obtained results, one was selected that guaranteed a minimum value of estimator of classification error (e), while maintaining the smallest possible value of the average size of the global decisions sets ($\overline{d}_{WSD^{dyn}_{Ag}}$). In tables presented below the best results, obtained for optimal values of the parameters, are given. In the tables the following information is given: the name of dispersed decision-making system (System); the selected, optimal parameter values (Parameters); the algorithm's symbol (Algorithm); the three measures discussed earlier $c, e_{ONE}, \overline{d}_{WSD^{dyn}_{Ag}}$; the time t needed to analyse a test set expressed in minutes. The tables show the results for four different fusion methods, the best results in terms of the measures e and $\overline{d}_{WSD^{dyn}_{Ag}}$ are bolded.

The results of the experiments with the Lymphography data set are presented in Table 1. Based on the results for the Lymphography data set it can be concluded that the methods from the measurement level produce unambiguous results (the average size of the global decision sets is very close to or equal to 1), while the methods from the rank level produce ambiguous results. Another conclusion is that, the methods from the rank level (the intersection method and the union method) produce better results than the methods from the measurement level (the product rule and the weighted average method). However, the improvement is usually achieved with greater ambiguity. Moreover, the weighted average method is better than the product rule. The union method is better for smaller number of resource agents, while the intersection method is better for larger number of resource agents. To summarize, in the case of the Lymphography data set, the best methods are the intersection method and the union method. Among the optimal values of the parameters we have $p = 0.05$ but very different values of the parameters m, m_1 and m_2 (as was shown in Table 1).

The results of the experiments with the Primary Tumor data set are presented in Table 2. The Primary Tumor data set has 22 decision classes and because of that even results with the average number of global decisions sets less than 4

Table 1. Summary of experiments results with the Lymphography data set

System	Parameters	Algorithm	e	e_{ONE}	$\overline{d}_{WSD_{Ag}}$	t
Intersection method						
WSD_{Ag1}	$m_1 = 5,\ p = 0.05,\ u = 7\%$	$A(1)C(1)$	0.114	0.205	1.091	0.01
WSD_{Ag2}	$m_1 = 10,\ p = 0.05,\ u = 5\%$	$A(1)C(1)$	0.114	0.409	1.295	0.01
WSD_{Ag3}	$m_1 = 10,\ p = 0.05,\ u = 3\%$	$A(1)C(1)$	0.136	0.295	1.159	0.01
WSD_{Ag4}	$m_1 = 13,\ p = 0.05,\ u = 1\%$	$A(3)C(7)$	**0.091**	0.500	**1.432**	0.03
WSD_{Ag5}	$m_1 = 13,\ p = 0.05,\ u = 3\%$	$A(1)C(1)$	**0.136**	0.500	**1.364**	0.26
Union method						
WSD_{Ag1}	$m_1 = 13,\ p = 0.05,\ u = 4\%$	$A(10)C(10)$	**0.045**	0.409	**1.386**	0.01
WSD_{Ag2}	$m_1 = 7,\ p = 0.05,\ u = 1\%$	$A(3)C(3)$	**0.091**	0.455	**1.364**	0.01
WSD_{Ag3}	$m_1 = 10,\ p = 0.05,\ u = 1\%$	$A(1)C(1)$	0.136	0.364	1.227	0.01
WSD_{Ag4}	$m_1 = 1,\ p = 0.05,\ u = 1\%$	$A(1)C(1)$	0.136	0.409	1.318	0.01
WSD_{Ag5}	$m_1 = 13,\ p = 0.05,\ u = 1\%$	$A(1)C(1)$	0.159	0.523	1.386	0.25
Product rule						
WSD_{Ag1}	$m_1 = 4,\ p = 0.05$	$A(1)C(2)$	0.136	0.136	1	0.01
WSD_{Ag2}	$m_1 = 1,\ p = 0.05$	$A(2)C(3)$	0.136	0.136	1	0.01
WSD_{Ag3}	$m_1 = 4,\ p = 0.05$	$A(5)C(9)$	0.136	0.136	1	0.01
WSD_{Ag4}	$m_1 = 1,\ p = 0.05$	$A(6)C(10)$	0.159	0.159	1	0.11
WSD_{Ag5}	$m_1 = 13,\ p = 0.05$	$A(1)C(3)$	0.205	0.455	1.250	0.26
Weighted average method						
WSD_{Ag1}	$m_1 = 2,\ p = 0.05$	$A(3)C(8)$	0.136	0.136	1	0.01
WSD_{Ag2}	$m_1 = 4,\ p = 0.05$	$A(3)C(7)$	0.114	0.114	1	0.01
WSD_{Ag3}	$m_1 = 12,\ p = 0.05$	$A(5)C(9)$	**0.114**	0.114	**1**	0.02
WSD_{Ag4}	$m_1 = 12,\ p = 0.05$	$A(6)C(8)$	0.114	0.114	1	0.02
WSD_{Ag5}	$m_1 = 1,\ p = 0.05$	$A(9)C(9)$	0.205	0.205	1	0.27

are interesting. As can be seen, for the Primary Tumor data set all considered methods (from the rank level and from the measurement level) generate ambiguous decisions, which means that the average size of the global decisions sets is significantly greater than 1. The averages are more or less equal for all methods. It was noted that the number of ties depends on the number of classifiers used - the more resource agents are present, the more ties occur. Based on the results it can be concluded that the methods from the rank level achieved the best results. Moreover, the intersection method produces much better results than the product rule and the weighted average method. To summarize, in the case of the Primary Tumor data set, the best methods are the intersection method and the union method. Among the optimal values of the parameters most frequently we have the values $1, 2$ and 3 for the parameter m and the values 1 and 4 for

Table 2. Summary of experiments results with the Primary Tumor data set

System	Parameters	Algorithm	e	e_{ONE}	$\overline{d}_{WSD_{Ag}}$	t
Intersection method						
WSD_{Ag1}	$m_1 = 7, p = 0.15, u = 10\%$	$A(2)C(6)$	0.294	0.971	3.686	0.02
WSD_{Ag2}	$m_1 = 1, p = 0.05, u = 8\%$	$A(3)C(5)$	**0.225**	0.961	**3.451**	0.06
WSD_{Ag3}	$m_1 = 1, p = 0.15, u = 8\%$	$A(8)C(3)$	**0.314**	0.990	**3.892**	0.58
WSD_{Ag4}	$m_1 = 1, p = 0.15, u = 8\%$	$A(2)C(4)$	**0.304**	0.971	**3.873**	0.22
WSD_{Ag5}	$m_1 = 1, p = 0.15, u = 8\%$	$A(3)C(4)$	0.324	0.980	3.863	1.49
Union method						
WSD_{Ag1}	$m_1 = 1, p = 0.05, u = 3\%$	$A(3)C(8)$	**0.265**	1	**3.490**	0.03
WSD_{Ag2}	$m_1 = 1, p = 0.2, u = 1\%$	$A(2)C(5)$	0.265	1	3.725	0.04
WSD_{Ag3}	$m_1 = 1, p = 0.1, u = 1\%$	$A(2)C(4)$	0.373	1	3.588	0.11
WSD_{Ag4}	$m_1 = 1, p = 0.2, u = 1\%$	$A(2)C(2)$	0.373	0.873	3.333	0.19
WSD_{Ag5}	$m_1 = 1, p = 0.2, u = 1\%$	$A(3)C(2)$	**0.314**	0.912	**3.990**	1.45
Product rule						
WSD_{Ag1}	$m_1 = 1, p = 0.15$	$A(3)C(1)$	0.382	0.794	2.627	0.01
WSD_{Ag2}	$m_1 = 1, p = 0.1$	$A(2)C(1)$	0.333	0.824	3.294	0.02
WSD_{Ag3}	$m_1 = 10, p = 0.05$	$A(1)C(2)$	0.353	0.892	3.882	0.03
WSD_{Ag4}	$m_1 = 1, p = 0.05$	$A(1)C(3)$	0.324	0.882	3.951	0.04
WSD_{Ag5}	$m_1 = 1, p = 0.15$	$A(3)C(2)$	**0.314**	0.912	**3.990**	1.45
Weighted average method						
WSD_{Ag1}	$m_1 = 7, p = 0.15$	$A(3)C(1)$	0.373	0.775	2.618	0.02
WSD_{Ag2}	$m_1 = 4, p = 0.2$	$A(7)C(1)$	0.324	0.794	3.216	0.02
WSD_{Ag3}	$m_1 = 4, p = 0.1$	$A(1)C(6)$	0.353	0.873	3.843	0.05
WSD_{Ag4}	$m_1 = 4, p = 0.15$	$A(1)C(5)$	0.324	0.892	3.902	0.07
WSD_{Ag5}	$m_1 = 1, p = 0.2$	$A(2)C(2)$	0.324	0.902	3.961	1.02

the parameter m_1. The parameters p and m_2 have very different optimal values depending on the system, which is considered (as was shown in Table 2).

Based on the results of the experiments for both data it can be concluded that the considered methods from the rank level achieve better results than the considered methods from the measurement level. In most cases, the intersection method produces the best results. In the group of methods from the measurement level the weighted average method achieves better results than the product rule. However, the weighted average method is more computationally complex.

5   Conclusions

In this article, two fusion methods from the rank level and two fusion methods from the measurement level were used in the dispersed decision-making system. Dispersed medical data were used in the experiments: Lymphography data set and Primary Tumor data set. The conclusions, that were reached based on the results of experiments are as follows. The considered methods from the rank level achieve better results than the considered methods from the measurement level. From the measurement level the best method is the weighted average method. In the rank level slightly better method is the intersection method.

References

1. Fumera, G., Roli, F.: Performance analysis and comparison of linear combiners for classifier fusion. In: Proceeding of 16th International Conference on Pattern Recognition, Canada (2002)
2. Gatnar, E.: Multiple-model approach to classification and regression. PWN, Warsaw (2008). (in Polish)
3. Ho, T.K., Hull, J.J., Srihari, S.N.: Decision combination in multiple classifier systems. IEEE Trans. Pattern Anal. Mach. Intell. **16**(1), 66–75 (1994)
4. Kittler, J., Hatef, M., Duin, R.P.W., Matas, J.: On combining classifiers. IEEE Trans. Pattern Anal. Mach. Intell. **20**(3), 226–239 (1998)
5. Kuncheva, L., Bezdek, J.C., Duin, R.P.W.: Decision templates for multiple classifier fusion: an experimental comparison. Pattern Recogn. **34**(2), 299–314 (2001)
6. Kuncheva, L.: Combining Pattern Classifiers Methods and Algorithms. Wiley, Chichester (2004)
7. Littlestone, N., Warmuth, M.: The weighted majority algorithm. Inf. Comput. **108**(2), 212–261 (1994)
8. Nowak-Brzezińska, A., Rybotycki, T.: Visualization of medical rule-based knowledge bases. J. Med. Inform. Technol. **24**, 91–98 (2015)
9. Nowak-Brzezińska, A., Wakulicz-Deja, A.: Exploration of Knowledge bases inspired by rough set theory. In: Proceedings of the 24th International Workshop on Concurrency, Specification and Programming, pp. 64–75 (2015)
10. Przybyła-Kasperek, M., Wakulicz-Deja, A.: Application of reduction of the set of conditional attributes in the process of global decision-making. Fund. Inform. **122**(4), 327–355 (2013)
11. Przybyła-Kasperek, M., Wakulicz-Deja, A.: Global decision-making system with dynamically generated clusters. Inf. Sci. **270**, 172–191 (2014)
12. Przybyła-Kasperek, M., Wakulicz-Deja, A.: A dispersed decision-making system - The use of negotiations during the dynamic generation of a systems structure. Inf. Sci. **288**, 194–219 (2014)
13. Przybyła-Kasperek, M., Wakulicz-Deja, A.: Global decision-making in multi-agent decision-making system with dynamically generated disjoint clusters. Appl. Soft Comput. **40**, 603–615 (2016)
14. Przybyła-Kasperek, M., Wakulicz-Deja, A.: The strength of coalition in a dispersed decision support system with negotiations. Eur. J. Oper. Res. **252**(3), 947–968 (2016). http://dx.doi.org/10.1016/j.ejor.2016.02.008

15. Simiński, R., Nowak-Brzezińska, A.: Goal-driven inference for web knowledge based system. In: Wilimowska, Z., Borzemski, L., Grzech, A., Swiatek, J. (eds.) ISAT 2015. AISC, vol. 432, pp. 99–109. Springer, Switzerland (2015)

16. Wakulicz-Deja, A., Przybyła-Kasperek, M.: Application of the method of editing and condensing in the process of global decision-making. Fund. Inform. **106**(1), 93–117 (2011)

17. Xu, L., Krzyzak, A., Suen, C.Y.: Methods of combining multiple classifiers and their application to handwriting recognition. IEEE Trans. Syst. Man Cybern. **22**, 418–435 (1992)

The Application of the Region Growing Method to the Determination of Arterial Changes

Ewelina Sobotnicka[1], Aleksander Sobotnicki[1(✉)], Krzysztof Horoba[1], and Piotr Porwik[2]

[1] Institute of Medical Technology and Equipment,
118 Roosevelt Street, 41-800 Zabrze, Poland
{esobotnicka, aleksander.sobotnicki,
kris}@itam.zabrze.pl
[2] Institute of Computer Science, University of Silesia,
39 Bedzinska Street, 41-200 Sosnowiec, Poland
piotr.porwik@us.edu.pl

Abstract. The paper discusses segmentation of medical images depicting aortic aneurysms and atherosclerotic changes in coronary vessels. The region growing method was deployed for segmentation. Before segmentation with the afore-mentioned method, the images were subjected to edging in order to acquire significant information, such as the size of the analyzed structure and pixel distribution. Edging paired with the region growing method ensures proper isolation of pixels with the same intensity, without the unwanted pixel overflow. In order to verify the method, results obtained by various authors were referred to, and a statistical analysis was performed to calculate the Dice coefficient.

Keywords: Segmentation · Region growing · Aortic aneurysms · Coronary vessels

1 Introduction

Acute and chronic diseases of the aorta and coronary vessels are a significant diagnostic issue in contemporary cardiology. The incidence of these diseases is growing due to the ageing of the population on the one hand, and owing to the fact that their detection and diagnostics are becoming more and more efficient on the other [2, 13]. Computed tomography (CT) and magnetic resonance (MR) play a significant role in non-invasive imaging diagnostics. Owing to the increasing accessibility of these methods, they are also used in cardiology to detect aneurysms and atherosclerotic changes in coronary vessels. Regardless of the applied imaging method each segment of the aorta must be measured, and the irregular segments must be identified [1, 20]. The irregularities within the segments may be evaluated with image processing based on commonly used segmentation methods.

Proper segmentation of medical images faces several challenges [10, 18]. The first issue that needs to be addressed is the fact that several anatomical structures are not

© Springer International Publishing Switzerland 2016
N.T. Nguyen et al. (Eds.): ICCCI 2016, Part II, LNAI 9876, pp. 462–471, 2016.
DOI: 10.1007/978-3-319-45246-3_44

homogeneous when it comes to the spacial repeatability of the pixels. It is also difficult to assign pixels to specific regions. Poor quality and low contrast images are another frequently encountered issue. What is more, 3D images contain varying shapes and textures of the analyzed images as a consequence of acquisition of images which depict the same dynamic structure for a period of time [12]. Hence, a specific segmentation method may work for a single (specific) structure/sequence of images and may not deliver the expected results for other structures/sequences [3–5]. Various segmentation methods have been discussed and developed over the years in scientific papers in an attempt to cope with these issues [7, 11]. Differing approaches were presented, consisting in the detection of edges or regions. They resulted in images segmented with varying precision and speed [15–17].

The paper demonstrates the results of segmentation of the aorta and coronary vessels with the region growing method. An image database containing the discussed structures was created for the purpose of the analysis.

The paper is mainly aimed at the preliminary evaluation of the usefulness of the region growing algorithm for the automatic detection of atherosclerotic changes in coronary vessels and aortic aneurysms. The article is divided into parts describing the methodology, experiment and its results.

2 Methodology

Segmentation of images depicting aortic aneurysms with the region growing method was build on images from 10 patients diagnosed with various imaging methods: Computed Tomography Angiography (CTA), High-Resolution Computed Tomography (HRCT), Multiplanar Reconstruction (MPR) and Volume Rendering Technique (VRT). Table 1 presents the imaging methods along with the analyzed images from two patients.

Images demonstrating atherosclerotic changes in coronary vessels were also segmented. The images come from 10 patients diagnosed with various imaging methods: digital subtraction angiography (DSA) and coronary angiography (CorCTA). Table 2 shows the imaging methods along with the analyzed images coming from two patients.

Based on scientific publishing [6, 8, 9], this paper divides the images showing the aorta and coronary vessels into segments in order to isolate specific features from them before the segmentation process itself. The edging technique was used to divide the images into segments. As a result, outlines of objects can be obtained, within which an analysis is carried out in the process of determining the pixels meeting the region growing criterion. This approach minimizes pixel overflow during the operation of the region growing algorithm. At the initial processing stage the images were subjected to edging, and next to segmentation with the region growing method. Image processing was carried out in the MATLAB environment with the use of algorithms developed by the authors.

Table 1. Examples of imaging methods for two patients, presenting the examinations of aorta used for the segmentation process with the region growing method.

Patient Imaging method	Patient I	Patient X
CTA (Computed Tomography Angiography)		
HRCT (High-Resolution Computed Tomography)		
MPR (Multiplanar Reconstruction)		
VRT (Volume Rendering Technique)		

2.1   Edging and Identification of Image Features

During identification of the features of the objects in the image, symmetry of the analyzed structures must be determined. Shape coefficients [9] are applied to reflect a selected object feature. The tested case under scrutiny uses the Blair-Bliss coefficient which defines the level of separation of the region and is independent of the size of the analyzed object:

$$R_{BB} = \frac{A}{\sqrt{2\pi \sum_i r_i^2}} \tag{1}$$

where:

A is the cross-section area of the object

r_i is the distance of the pixel of the object from the middle of the analyzed structure.

Deblurring with the use of the Prewitt's filter was applied to isolate the edges in the analyzed structures. If the symmetry of the object is determined and its edges are deblurred, matching pixels with the same intensity level can be isolated from the background and thus the seed point necessary to initiate segmentation with the region growing method can be selected. Figure 1 demonstrates examples of the edging process results and identification of the features of the objects. Various colours in the presented images stand for sets of pixels with various intensity levels and make it easier to depict their distribution in examples of images.

Table 2. Examples of imaging methods for two patients, presenting the examinations of coronary vessels used for the segmentation process with the region growing method.

Imaging method ╲ Patient	Patient XI	Patient XX
DSA (digital subtraction angiography)		
CorCTA (coronary angiography)		

2.2 Region Growing Method

The image is a certain distribution of intensity representing several objects or classes of objects (multi-shaded images), and in the case of binary images an image is a set of several separate homogeneous elements and the background [10, 19]. Segmentation is used for the simplification of image description by reducing the excessive information contained in the image.

Segmentation is also used for grouping the pixels with similar intensities or for merging pixels which describe individual elements. The purpose of segmentation is to process the data available in the image so as to obtain a division which will contribute to the recognition and interpretation of the objects in the image. As a result of segmentation, the pixels are divided into several separate classes. If the region growing

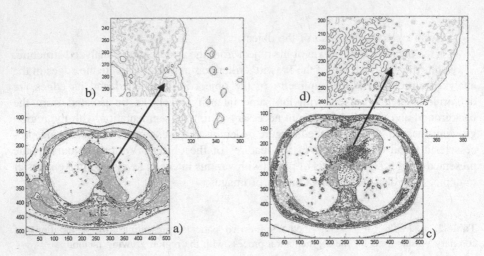

Fig. 1. Two examples of edging. (a) and (c) edging results. (b) and (d) enlarged image fragments.

method is deployed for segmentation, the classes are regions included or not included in the analyzed structure during the intensity check for individual pixels at the region growing stage.

First of all, the region growing method requires the middle of the segment in the considered images of the aorta or coronary vessels to be provided. The middle is understood as the segmentation seed point. Next, pixels are matched iteratively, within the boundaries outlined earlier in the edging process. Neighbouring pixels (4-neighborhood) are checked in each iteration. They are isolated from the analyzed structure in such a way that pixels are matched with pixels with the same intensity level. The valid similarity criterion requires a difference in intensity between the neighbouring pixels and the average intensity value in the analyzed (current) region. If the intensity of the neighbouring pixels is within the 5 % range of the average intensity of the region, pixels are added to this region.

3   Experiments and the Results

The implemented region growing algorithm was tested for images from 20 examinations. The analysis was carried out for images in the DICOM format. Images obtained during a CT examination were used for the segmentation of the aorta structure, whereas images obtained during digital subtraction angiography and coronary angiography were used for the segmentation of coronary vessels. Altogether, 200 images were subjected to segmentation with the implementation algorithm of the region growing method.

In order to verify the operation of the developed algorithm, the image analyzed in paper [14], Fig. 2(a) was used as a reference. The obtained results of segmentation with the region growing method for this specific image were compared with the results obtained by the authors in paper [14]. Figure 2(a) depicts the original image used for

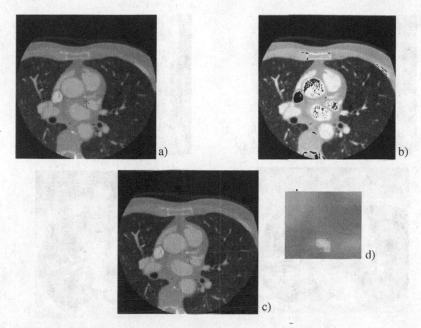

Fig. 2. Comparison of the segmentation results. (a) The original image used for segmentation in paper [14], (b) the final image obtained in [14], (c) the image obtained after segmentation, with the use of the developed algorithm of the region growing method, (d) enlargement of the segmented image.

segmentation in [14], in which the region to be segmented was marked. Figure 2(b) demonstrates the segmentation result obtained by the authors in paper [14]. On the other hand, Fig. 2(c) shows the result obtained with the use of the developed implementation algorithm of the region growing method. The region was enlarged to improve visualization of the outlined region Fig. 2(d). The presented images indicate similar segmentation results. If the region growing algorithm proposed by the authors is used, an image similar to the original image is obtained, without blurred structures, without pixel overflow and with precise indication of the pixels with the same intensity level.

Moreover, the operation of the algorithm was verified with two images, belonging to the authors, where clinical experts identified the position of pathological places, Figs. 3(a)(c). The places segmented by means of the implementated algorithm of the region growing method are marked in white and indicated with arrows.

An analysis of the obtained results proves that the applied segmentation algorithm works correctly. The segmentation process delivered regions indicated beforehand by clinical experts. Regions consisting of pixels with the same intensity level were additionally included. After the correct operation of the algorithm was verified, 200 images showing aortic aneurysms and atherosclerotic changes in coronary vessels were subjected to segmentation. Owing to a large number of images, the paper depicts only two typical segmentation results, Fig. 4. A basic statistical analysis was carried out; the Dice similarity coefficient was calculated. The Dice coefficient makes it possible to

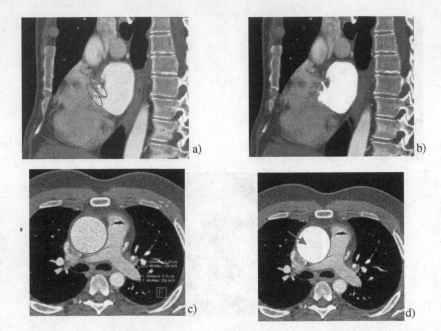

Fig. 3. Results of segmentation with the region growing method. (a) and (c) Original images with marked pathological places. (b) and (d) Images segmented with the developed algorithm of the region growing method.

determine the precision with which the analyzed structures were segmented. The Dice coefficient was calculated by means of the formula [18]:

$$Dice = \frac{2|im2 \cap im1|}{|im2| + |im1|} \tag{2}$$

where: im1 – the binary mask of the original structure;
im2 – the binary mask of the segmented structure.

The edging process delivered precise outlines of the structures in the images in Fig. 4, which were subsequently subjected to segmentation with the region growing method. It can be seen that the region growing method applied for the segmentation of both coronary vessels and the aorta results in a precise isolation of the structure.

Similar results were obtained for the remaining images being tested. This is confirmed by the average value of the Dice coefficient calculated for all images. The average value of the Dice coefficient for the segmented images of the aorta amounts to 0.8866, and for images showing coronary vessels it amounts to 0.8728. The calculated average value of the coefficient is high, which provides evidence for the efficiency of the developed implementation algorithm of the region growing method applied to the segmentation of the structures in question. Tables 3 and 4 demonstrate the values of the coefficient calculated for images obtained during 4 examinations.

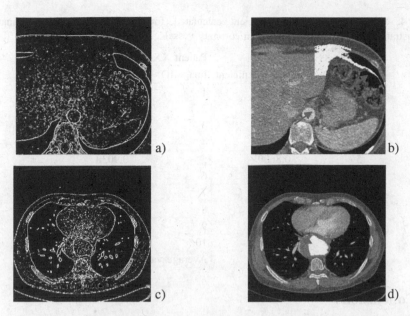

Fig. 4. Results of segmentation with the region growing method. (a) Preliminary image processing aimed at creating the edges: image for the segmentation of coronary vessels, (c) image for the segmentation of aorta. (b) Results of segmentation: of the coronary vessels, (d) of the aorta.

Table 3. Value of the Dice coefficient calculated for two examples of examinations demonstrating aortic aneurysms.

Patient I		Patient X	
Image ID	Dice coefficient	Image ID	Dice coefficient
1	0.9870	1	0.8471
2	0.8440	2	0.9032
3	0.9041	3	0.9032
4	0.9870	4	0.9065
5	0.8609	5	0.9038
6	0.8440	6	0.8471
7	0.8609	7	0.8452
8	0.9041	8	0.9054
9	0.8989	9	0.9054
10	0.9038	10	0.8471
Average value	0.89947	Average value	0.8814

Table 4. Value of the Dice coefficient calculated for two examples of examinations demonstrating atherosclerotic changes in coronary vessels.

Patient XI		Patient XX	
Image ID	Dice coefficient	Image ID	Dice coefficient
1	0.9870	1	0.9032
2	1.0000	2	0.9049
3	0.9038	3	0.9065
4	0.9870	4	0.9032
5	0.8609	5	0.9038
6	0.8440	6	0.8471
7	0.8609	7	0.8452
8	0.8609	8	0.8471
9	0.8989	9	0.9054
10	1.0000	10	0.6073
Average value	0.9203	Average value	0.8573

4 Conclusions

The developed implementation algorithm of the region growing method may be applied to the segmentation of hardly visible pathological changes in the images of the aorta and coronary vessels. Owing to the iterative pixel check, the probability of omitting significant information has been reduced to a minimum. Depending on the quality of the analyzed image, as well as the region subjected to the analysis, the obtained results are characterized by varying levels of precision. In spite of differences in the precision of the segmentation process, the obtained results are satisfactory, which is confirmed by the average value of the Dice coefficient of 0.8866 for images showing the aorta, and of 0.8728 for images of coronary vessels. Segmentation results make it possible to reliably classify the outlined structure as a pathological or healthy region.

There are plans to create a larger set of images for segmentation at the next stages of the works, as well as to expand the developed algorithm in order to further improve the efficiency of segmentation and the application usability of the method.

References

1. Al-Agamy, A., Osman, N., Fahmy, A.: Segmentation of ascending and descending aorta from magnetic resonance flow images. In: Biomedical Engineering Conference (CIBEC), pp. 41–44 (2010)
2. Avila-Montes, O., Kurkure, U., Nakazato, R., Berman, D., Dey, D.: Segmentation of the thoracic aorta in noncontrast cardiac CT images. IEEE J. Biomed. Health Inform. **17**(5), 936–949 (2013)
3. Ben Ayed, I., Wang, M., Miles, B., Garvin, G.J.: TRIC: trust region for invariant compactness and its application to abdominal aorta segmentation. In: Golland, P., Hata, N., Barillot, C., Hornegger, J., Howe, R. (eds.) MICCAI 2014, Part I. LNCS, vol. 8673, pp. 381–388. Springer, Heidelberg (2014)

4. Babin, D., Devos, D., Pizurica, A.: Robust segmentation methods with an application to aortic pulsewave velocity calculation. Comput. Med. Imaging Graph. **38**, 179–189 (2014)
5. Benmansour, F., Cohen, L.: A new interactive method for coronary arteries segmentation based on tubular anisotropy. In: International Symposium on Biomedical Imaging: From Nano to Macro. IEEE (2009)
6. Bruijne, M., Ginneken, B., Niessen, W., Loog, M., Viergever, M.: Model-based segmentation of abdominal aortic aneurysms in CTA images. In: SPIE, vol. 5032 (2003)
7. Chen, S., Wang, T., Lee, W.: Coronary arteries segmentation based on the 3D discrete wavelet transform and 3D neutrosophic transform. Biomed. Res. Int. **2015**, 1–9 (2015). doi:10.1155/2015/798303. Article ID 798303. PMID 25648181
8. Egger, J., Freisleben, B., Setser, R., Renapuraar, R.: Aorta segmentation for stent simulation. In: CI2BM09 – MICCAI (2011)
9. Goyal, P., Goyal, K., Gupta, V.: Calcification detection in coronary arteries using image processing. Int. J. Adv. Res. Comput. Sci. Softw. Eng. **3**(8), 279–284 (2013). ISSN: 2277 128X
10. Goyal, P., Gupta, V., Goyal, K.: Segmentation of coronary arteries of heart. Int. J. Adv. Electr. Electron. Eng. **2**(1), 93–98 (2013). ISSN: 2319-1112
11. Kovacs, T.,Cattin, P., Alkadhi, H.: Automatic segmentation of the vessel lumen from 3D CTA images of aortic dissection, Bildverarbeitung für die Medizin, pp. 161–165. Springer (2006)
12. Lara, D., Faria, A., Araújo, A., Menotti, D.: A novel hybrid method for the segmentation of thr coronary artery tree in 2D angiograms. Int. J. Comput. Sci. Inf. Technol. (IJCSIT) **5**(3), 45–64 (2013). doi:10.5121/ijcsit.2013.5304
13. Macía, I., Legarreta, J.H., Paloc, C., Graña, M., Maiora, J., García, G., de Blas, M.: Segmentation of abdominal aortic aneurysms in CT images using a radial model approach. In: Corchado, E., Yin, H. (eds.) IDEAL 2009. LNCS, vol. 5788, pp. 664–671. Springer, Heidelberg (2009)
14. Okusz, I., Ünay, D., Kadpasaoglu, K.: A hybrid method for coronary artery stenoses detection and quantification in CTA images. In: MICCAI Workshop 3d Cardiovascular Imaging: A MICCAI Segmentation (2012)
15. Ozkan, H.: Segmentation of ascending and descending Aorta in CTA images. Int. J. Med. Health, Biomed. Bioeng. Pharm. Eng. **6**(5), 451–453 (2012)
16. Piekar, E., Momot, A.: Gradient and polynomial approximation methods for medical image segmentation. J. Med. Imaging Health Inf. **5**, 1337–1349 (2015)
17. Piekar, E., Szwarc, P., Sobotnicki, A., Momot, M.: Application of region growing method to brain tumor segmentation-preliminary results. J. Med. Inf. Technol. **22** (2013). ISSN 1642-6037
18. Roussona, M., Baib, Y., Xua, C., Sauera, F.: Probabilistic minimal path for automated esophagus segmentation. In: SPIE, vol. 6144 (2006)
19. Tek, H., Zheng, Y., Gulsun, M., Funka-Lea, G.: An automatic system for segmenting coronary arteries from CTA. In: MICCAI Workshop on Computing and Visualization for Intravascular Imaging (2011)
20. Zhao, F., Zhang, H., Wahle, A.: Automated 4D Segmentation of Aortic Magnetic Resonance Images. In: BMVC, pp. 247–256 (2006)

Mining Medical Knowledge Bases

Agnieszka Nowak-Brzezińska[✉], Tomasz Rybotycki, Roman Simiński,
and Małgorzata Przybyła-Kasperek

University of Silesia, ul. Bankowa 12, 40-007 Katowice, Poland
agnieszka.nowak@us.edu.pl
http://zsi.ii.us.edu.pl/~nowak/

Abstract. In this work, the topic of applying clustering as a knowledge extraction method from real-world data is discussed. The authors propose hierarchical clustering and treemap visualization techniques for knowledge base representation in the context of medical knowledge bases, for which data mining techniques are successfully employed and may resolve different problems. The authors analyze the impact of different clustering parameters on the result of searching through such a structure. Particular attention was also given to clusters description. The authors examined how selected inter-cluster and inter-object similarity measures influence clusters representatives.

Keywords: Cluster analysis · Clusters visualization · Rule-based knowledge bases · Data mining · Cluster representatives

1 Introduction

The number of medical expert systems is growing and thanks to progress in the key areas, such as knowledge acquisition, model-based reasoning, and systems integration for clinical environments, their efficiency is getting better everyday. It is essential for physicians to understand the current state of such research as well as remaining theoretical and logistic barriers, before full potential of these systems can be used and new patterns can be discovered. Among many other methods, doctors can use the visualization and analysis of medical data for the purpose of extracting new and potentially valuable knowledge, both common and unusual. Issues that affect the difficulty of research are both an excessive amount of available information and their complicated structure (both in terms of high dimensionality and used data types). In this paper, a specific type of knowledge representation - rules (denoted as Horn clauses) - is considered. Unfortunately, if we use different tools for automatic acquisition and/or extraction of rules [1], their number grows rapidly. For modern problems, knowledge bases (KB) can count up to hundreds or thousands of rules, which results in an enormous number of possible inference paths. In such cases, a knowledge engineer cannot be absolutely sure that all possible rule interactions are legal and lead to the expected results. The big size of the KB causes problems with inference efficiency and interpretation of its results. Even for a domain expert it is difficult

© Springer International Publishing Switzerland 2016
N.T. Nguyen et al. (Eds.): ICCCI 2016, Part II, LNAI 9876, pp. 472–481, 2016.
DOI: 10.1007/978-3-319-45246-3_45

to analyze the presented knowledge if the number of elements to examine is too high. In such cases, clustering rules and a visualizing resultant structure can be helpful. That is why the authors propose a method to change a KB from a set of unrelated rules to groups of similar rules (using cluster analysis techniques). In this approach, besides the information about the rules in each cluster, the visualization of clusters is generated. Such representation of the KB, especially in specific areas (like medicine), can be very helpful for an expert in exploring the given domain. The paper consists of 5 sections. In Sect. 2, medical knowledge-based systems are described. Section 3 contains the description of the software created by the authors in order to achieve grouping and graphical representation of data, along with the descriptions of the cluster analysis idea for rules and visualization methods for a hierarchical data structure. The experiments with the analysis of their results are considered in Sect. 4. Section 5 contains the summary of the paper.

2 Medical Knowledge Bases

Data mining in medicine is an important part of biomedical informatics (intensive applications of computer science to these fields, whether at the clinic, the laboratory or the research center). The healthcare industry produces constantly growing amount of data stored in different forms (data matrices, complex relational databases, pictorial materials, time series etc.) and can benefit in various ways from deep analysis of data stored in their databases, which results in numerous applications of various data mining tools and techniques in this field. Efficient analysis requires not only knowledge of data analysis techniques but also involvement of medical knowledge and close cooperation between data analysis experts and physicians. Mined knowledge can be used in various areas of healthcare (research, diagnosis, and treatment). One of the many possible applications of the proposed idea (clustering a big set of rules and vizualizing it) is to help knowledge engineers in managing the KBs. When rules are partitioned into a smaller number of groups, and such groups are labelled with their representatives, it is possible to explore very crucial knowledge about the given domain. When these groups are graphically represented, it is easier to discover some unusual or redundant cases as well as some patterns in rules. It is an element of knowledge bases validation and verification tasks. When monitoring a patient, computer-assisted support can be of significant help because some dangerous events (for the patient) can be very rare and therefore difficult to identify in the continuous stream of data. With the increasing number of clinical databases it is likely that machine-learning applications will be necessary to detect rare conditions and unexpected outcomes. Today, the most popular applications utilize techniques based on different data mining algorithms from which clustering (dividing a set of objects into homogeneous clusters) is one of the most interesting. Very often doctors are confronted with unique situations. Therefore, systems that are intended to support doctors have to take such cases into account. Medical practitioners who work with time series on a daily basis rarely take advantage of the wealth of tools that the data mining community has made available. Possible reasons for this discrepancy are the bewildering number of parameters, the specialized hardware and/or software, and the long time of learning required by the time series data mining tools [2].

Verification of KBs is a critical step to ensure the quality of a knowledge-based system. The success of these systems depends heavily on how qualitative the knowledge is. However, manual verification is cumbersome and error-prone, especially for large KBs. Verification of completeness and consistency in knowledge-based systems is a topic that was early recognized and elaborated in many papers, e.g., [3,6]. In a knowledge-based system-engineering context, verification of the KB is the process of ensuring the quality of the KB, while validation is the process of checking that the knowledge of the human experts is accurately represented in the KB (if there are any semantic or syntactic inconsistencies, they affect the completeness and consistency of the KB). The KB needs to be complete, consistent (free of conflicting, identical, and subsumed rules) and to avoid rare rules as well as redundant ones [6]. The use of dispersed knowledge in medicine was considered in the paper [10]. If the KB contains a lot of rules, it is difficult to analyze in this context. A data mining technique, for example clustering methods with visualization techniques, also enables grouping similar rules and visualizing the created structure. Thanks to this, knowledge engineers can explore the domain knowledge with experts' support and improve the system's performance.

3   CluVis

CluVis (Cluster Visualizer) [11] is an application designed and implemented by the authors, which is meant to cluster rules and visualize the resultant structure of the grouping. It was created because an overview of the literature revealed that there is no application that would be able to group and visualize a large set of rules (such as some KBs). There were, however, applications able to visualize data (in the form of XML files or hierarchical ontologies) and some works like [9], where authors used the DBSCAN density-based algorithm to discover trends and relations between objects in databases and visualized discovered patterns using treemaps. Unfortunately, no application was able to work on raw KBs generated by RSES [1]. The main advantage of CluVis is that it not only works on such raw KBs but also combines functionalities of data mining and visualization applications.

CluVis implements the AHC algorithm described in Sect. 3.1 with every inter-cluster and inter-object similarity measures described in it. It also features two clustering validation measures based on popular validation indexes - Dunn's [8,11] and Davies-Bouldin's. It can visualize the resultant structure of grouping using two different treemap algorithms Circular Treemap and Rectangular Treemap both mentioned in Sect. 3.2. CluVis's additional features were described thoroughly in [8,11]. It is available in both English and Polish. Its code and any other information are currently available at: https://github.com/Tomev/CluVis.

3.1   Grouping Algorithm

Whenever any set of objects is examined, some kind of organization has to be proposed. One of the oldest and most popular forms of organization is grouping/clustering. It is an unsupervised process, which means that the raw nature of objects is the most important factor during clustering.

There are many different methods of grouping. In this work, the hierarchical approach was used [16]. It seeks to create a tree-like structure that would represent the whole grouping process. In this paper, the most popular, agglomerative approach was proposed, as it seemed more natural in the context of examined objects (rules). The authors selected a clustering algorithm known as Classical Agglomerative Hierarchical Clustering (AHC) Algorithm [8,11], which is based on similarity analysis and merges two most similar groups into one during each iteration step. It is well known in the literature and described in e.g. [8,11,16].

Main advantage of hierarchical clustering is that it doesn't impose on any special method of describing clusters similarity and thus can be applied to every kind of objects. Many distance and similarity measures are already known in the literature [7]. Considering complexity of examined objects (rules), in this paper three most popular similarity measures used to handle data consisting of mixed types (such as rules) were used: Gower's similarity coefficient [11], SMC and WSMC/Jaccard Coefficient [7,11]. Different inter objects similarity measures will produce different results (hierarchies) in most of the cases. Therefore, it is important to use more than just one algorithm parameters setting in the process of extracting knowledge from data. Additionally, four popular inter-cluster similarity measures - Single Link (SL), Complete Link (CoL), Average Link (AL) and Centroid Link (CL) - were used. There is, however, one crucial difference. In this work, centroid is considered to be cluster's representative (described in Sect. 3.1) as it aims to be the most average/centered rule. The reason for that is non-trivial means to define the geometrical center of the cluster of rules.

Cluster's Representation. It is very important for data to be presented in the most friendly way. Sole visualization of clustering (described further in Sect. 3.2) is not enough, as it would only reduce the whole pattern discovery to examining an accumulation of shapes. There are many methods of creating representatives. Its creation algorithm has been presented as pseudocode 1.

Pseudocode 1. Representative creation algorithm.
Input: cluster c, threshold t [%]
Output: cluster representative r

1. In the cluster attributes set A, find only these attributes that can be found in t objects, and put them in set A'
2. **FOR EACH** attribute a in A'
3. **IF** a is symbolic count modal
4. **IF** a is numeric count average value
5. **RETURN** attribute-value pairs from A' **AS** r

In this work, each cluster is described by its representative. A representative is simply an average object. As one can see, the representative created this way consists only of these attributes that are considered the most important for the whole cluster (they are common enough). This way the examined group is well described by the minimal number of variables.

3.2 Visualization Algorithms

Considering that the resultant structure of the used clustering algorithm is tree-like it is important to display not only groups, but also their hierarchies. The most common structure that is capable of such a task is dendrogram. Unfortunately, the authors were unable to use it due to its shortcomings [8]. There are, however, many other methods of visualizing treelike structures, one of them being treemaps. In this work, Rectangular Treemap (with the slice-and-dice deployment method) [12] and Circular Treemap [15] were used.

Many different visualization features can be used to represent information. In this work, two factors were used: the size and the colour of a shape. The bigger the cluster is (the more rules it contains), the bigger will be the shape representing it. Moreover, sizes were divided into 6 groups: $>= 75\%$, $75 - 50\%$, $50 - 25\%$, $25 - 10\%$, $< 10\%$ of the whole set and separate group for single object. Each of these groups has its own color, based on the colors of the rainbow: purple, blue, green, yellow, orange, and crimson (red is used to highlight a hovered shape) respectively. To distinguish different sizes of the same group (for example clusters with sizes 95% and 76%) different shades were used - the bigger the cluster, the lighter the shade. Sample visualization (using circle algorithms) can be seen in Fig. 1.

It shows the partition of a Pima dataset with 457 rules in 10 clusters using the following parameters: similarity measure - Gower, clustering method - complete linkage, representative strategy - 25%). The biggest cluster ($J448$) contains 128 rules whereas the smallest ($J441$) only 7. Interestingly enough, there is no singular (outlier) cluster (every rule was clustered).

Despite all the advantages that treemaps have (such as being able to visualize large data structures, having multiple variations of shapes, and laying algorithms) they also have some flaws. Just as the majority of other methods, treemaps tend to become less readable as the size of the visualized dataset grows. It is a common problem, hence few methods to deal with it have been invented. One of them, used in this work, is responsive visualization, which means that a user is able to interact with shapes displayed on the screen. CluVis allows a user to visualize objects (or groups) that that selected cluster consists of by clicking on it. Thanks to that operation, a smaller number of shapes is drawn, and less objects must be visualized, which results in improved readability of the visualization. Moreover, each treemap algorithm has some unique problems, which was covered in [8]. Therefore it is important not to rely on one visualization technique.

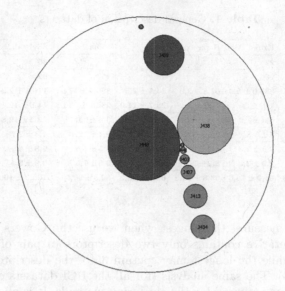

Fig. 1. Sample visualization generated by CluVis.

4   Experiments

The main goal of this article is to demonstrate that clustering big datasets and their visualization can improve interpretations of knowledge hidden in the data, particularly medical data. It is worth mentioning that real medical KBs usually contain a large set of rules and it is necessary to cluster rules before any deep exploration. Such rules clusters are further visualized in a specific structure (rectangular or circular one). Besides drawing of the groups, their descriptions are even more important. Although the authors are familiar mostly with medium-sized KBs, they had a chance to work with KBs consisting of over 4000 rules [13]. Created clusters may vary depending on the measure of similarity or clustering method that is used. To make the groups well separated and their representatives properly described, it is very important to use the exact measures to testify the quality of the created structure of objects (measured using the MDI and MDBI [8]). An interesting thing was to check whether some measures or methods of inner/between clustering are typical for achieving more or less unclustered rules (small, singular groups) or to get groups with better/worse quality. During the experiments all these issues have been examined. It was very important to check if different similarity measures (or different methods of clustering) change the quality of clustering and the length of the representatives of the created groups of rules. Representatives of groups are important because a too short/general or too detailed representative may be difficult to interpret. The authors are interested in verifying if certain measures (or clustering methods) are typical for long representatives and some for short ones. The smallest number of singular clusters is achieved when using the CoL method, while the largest when using the SL method. The (average) shortest representative is achieved

Table 1. General description of datasets

KB	Spect	Kruk	Hepa	Post	Pima	Soybean	Echo
a	23	23	20	9	9	36	13
b	67	200	35	46	457	62	63
c	$125,8 \pm 1,8$	$385,0 \pm 5,0$	$62,8 \pm 3,0$	$84,5 \pm 2,5$	$886,5 \pm 17,6$	$116,1 \pm 2,2$	118 ± 2
d	$42,2 \pm 18,5$	116 ± 55	$17 \pm 8,4$	$28,7 \pm 11,2$	$280,6 \pm 136,1$	$44,6 \pm 12,1$	$36,7 \pm 16,4$
e	$14,8 \pm 8,7$	$1,9 \pm 1,9$	$8,4 \pm 4,5$	$6,4 \pm 2,6$	$3,4 \pm 2.0$	$23,3 \pm 12,8$	$5 \pm 1,5$
f	$4,8 \pm 2,6$	$4 \pm 5,8$	$2,5 \pm 2,9$	$3,6 \pm 2,4$	$9,4 \pm 12,1$	$4,3 \pm 2,7$	$3,1 \pm 2,5$
g	$1,4 \pm 5,8$	$2,9 \pm 2,3$	$8,2 \pm 2,4$	$5,5 \pm 1,7$	$3,5 \pm 1,5$	$9,8 \pm 5$	$5,4 \pm 1,3$
h	$13,7 \pm 5,8$	$2,9 \pm 2,3$	$8,2 \pm 2,4$	$5,5 \pm 1,7$	$3,5 \pm 1,5$	$9,9 \pm 5,4$	$5,4 \pm 1,3$
i	$108,3 \pm 113,3$	$66,0 \pm 76,6$	$63,3 \pm 64,3$	$66,6 \pm 68,3$	$89,8 \pm 90,5$	$105,3 \pm 100,5$	$52,5 \pm 53,2$

using the SMC measure, the longest when we use the Gowers measure. The shortest representative contains only two descriptors (a pair of the attribute and its value) while the longest may contain 35 of the descriptors (maximum in examined KB). The same analysis (for all the UCI datasets calculating the average value) was prepared for the clustering methods. It is obvious now that when the single linkage or centroid linkage method is used, the length of the representative is the shortest and when the complete linkage method is used, the representative is the most detailed.

At first, for 7 different datasets (Kruk, Spect, Soybean, Echocardio, Hepa, Post oper, Pima), by multiple clusterings with different parameters, the authors analyzed the influence of the data size (attributes number and rules) on different parameters, like the number of the created groups of rules, the quality of such groups, and the length of their description (representatives). The results (mean $\pm$ SD (standard deviation) and the interval of both the minimum and maximum values of the analyzed parameters) are presented in Table 1 a - Attributes, b - Objects, c - Groups, d - the size of the biggest cluster, e - representative's length, f - ungrouped objects, $g - i$ - smallest/average/biggest representatives sizes.

One of the main goals of this research is to measure the influence of the size of a cluster representative on many important parameters, like the biggest cluster size and the representative's length, the number of ungrouped objects. Also the biggest, smallest and average representative sizes were calculated. The authors proposed to change the % of features common for every rule in a given cluster to be its representative using the algorithm presented in Sect. 3.1. It means that the 50 % method (in Table 2) is responsible for the case in which every rules cluster has got a representative created in the way that at least half of rules had a given part of the representative. It is obvious that the more % is necessary to cover, the shorter the representative is and there are not so many such clusters (and they are usually not so big).

We may see that the differences between the representative methods (proposed in this research) are not statistically significant. It is worth mentioning that the higher the % necessary to cover by features to be included in the cluster's representatives, the less is the number of single (loose, outlier) clusters. It was also interesting for the authors to check the differences between similarity

Table 2. Methods for clusters representatives

	25 %	50 %	75 %	100 %	p
BiggestClusterSize	81,5 ± 104,1	81,6 ± 105,1	80,3 ± 101,9	79,9 ± 104,0	0,998474
BiggestClusterRepLength	13,0 ± 10,2	11,7 ± 10,1	9,4 ± 8,5	1,9 ± 1,5	0,000000
Ungrouped objects	5,0 ± 6,2	4,7 ± 6,1	4,2 ± 5,6	4,2 ± 5,7	0,506154
BiggestRepSize	7,4 ± 3,8	8,1 ± 5,3	6,1 ± 4,2	5,0 ± 3,6	0,000000
SmallestRepSize	8,8 ± 5,4	8,1 ± 5,3	6,1 ± 4,2	5,0 ± 3,6	0,000000
AverageRepSize	78,2 ± 81,4	79,3 ± 84,4	79,9 ± 88,6	77,9 ± 88,0	0,996248

measures for rules (Gower, SMC, WSMC) and rules clusters (SL, CL, CoL, AL) in the context of the biggest cluster size and its representative's length as well as the biggest and the smallest representative lengths (see Fig. 2).

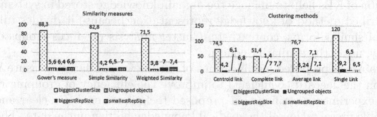

Fig. 2. Similarity and clustering methods - analysis

The analysis confirms statistical significance of the differences between similarity measures. The lowest number of small (single) clusters is created when we use the WSMC measure and the highest number when we use the Gower's measure.

When rules are clustered into groups, many possible partitions may be created. Some of them are optimal, the other ones are not. One of the most interesting features to analyze was the number of unclustered (single) rules, which may be an unusual medical case worth further analysis. The results of comparison of average number of unclustered rules according to different similarity measures (G, SMC or WSMC) or methods of clustering (SL, CoL, AL, CL) are presented in Fig. 2. It was noticed that for CoL, the number of ungrouped rules is few times smaller than for all other methods. It seems that it (CoL) generates many large clusters and minimalizes the number of outliers. There is a correlation between the number of small clusters created during the AHC algorithm in accordance with the used method of calculating inter-cluster and inter-object similarity. No matter which similarity measure we use, the number of small clusters is quite similar. The only difference is that the CoL method provides a minimum number of small aggregates (probably due to the fact that it allows at the same time for creating a greater number of large clusters). Very preliminary consultation with experts allows us to hope that the rules discovered in the visualization of medical data will contribute to the effective induction of knowledge in the examined area. Detection of abnormal cases (visualized as small circles in Fig. 2 indicates that there are rare cases and with different description from most of the accumulated knowledge. This usually signals the need for a deeper exploration of the examined areas especially in the part recognized as unusual.

5   Summary

The aim of this paper was to discuss the topic of applying data mining and visualization techniques to medical KBs. The AHC algorithm and treemap visualization techniques were introduced. In the author's opinion, clustering a large set of objects (rules in this case) is not enough when exploring such an enormous amount of data in order to find some hidden knowledge in it. The extraction of valuable knowledge from large data sets can be difficult or even impossible. Modularization of KBs help to manage the domain knowledge stored in systems using the described method of knowledge representation because it divides rules into groups of similar forms, context, etc. Cluster analysis produces groups of rules naturally, using the similarity concept.

The authors proposed modifications of the known similarity measure WSMC (based on the Jaccard coefficient) to improve the efficiency of grouping rules.

The experiments verified that proposed techniques enable a clear and comprehensible presentation of the medical knowledge hidden in the data. The parameters like inter-object similarity measures or inter-cluster similarity methods (SL, CoL, etc.) influence the cluster size or its representatives length. They also confirmed (it is known in the literature) that the SL clustering can produce straggling clusters, called chaining, where clusters may be forced together due to single elements being close to each other, even though many of the elements in each cluster may be very distant from each other. CoL tends to find compact clusters, however it suffers from a different problem. It pays too much attention to outliers, points that do not fit well into the global structure of the cluster. The authors propose to use clusters of rules and visualize them using treemap algorithms and hope that this two-phase way of rules representation allows the domain experts to explore the knowledge hidden in these rules quicker and more efficiently than before. Using this solution in such a specific domain like medicine brings hope that it will be easier to find some characteristics in presented diseases or to discover unusual symptoms, which will lead to predicting some serious diseases and preventing the development of distressing symptoms, often saving human lives. In the future, the authors plan to extend the software's functionality, especially in the context of parameters used in clustering and visualizing procedures, as well as importing other types of data sources. It would be easier then to support human experts in their everyday work by using the created software (CluVis) in work with many expert systems.

Acknowledgement. This work is a part of the project "Exploration of rule knowledge bases" founded by the Polish National Science Centre (NCN: 2011/03/D/ST6/03027).

References

1. Bazan, J., Szczuka, M.S., Wróblewski, J.: A new version of rough set exploration system. In: Alpigini, J.J., Peters, J.F., Skowron, A., Zhong, N. (eds.) RSCTC 2002. LNCS (LNAI), vol. 2475, pp. 397–404. Springer, Heidelberg (2002)
2. Berka, P., Rauch, J., Zighed, D.A.: Medical Information Science Refererence. Hershey, New York (2009)
3. Buchanan, B.G., Shortliffe, E.H.: Rule-Based Expert Systems the MYCIN Experiments of the Stanford Heuristic Programming Project. Addison-Wesley Publishing Company, Reading (1984)
4. Doreswamy M. G., Hemanth. K.S.: A study on similarity measure functions on engineering materials selection. Soft Comput. Appl. 1(3) (2011)
5. Lichman, M.: UCI machine learning repository. University of California (2013). http://archive.ics.uci.edu/ml
6. Nguyen, T., Perkins, W., Laffey, T., Pecora, D.: Knowledge base verification. AI Magaz. 8(2), 69–75 (1987)
7. Nowak-Brzezińska, A., Jach, T.: Wnioskowanie w systemach z wiedza niepewna. Studia Informatica. Wydawnictwo Politechniki Slaskiej, Gliwice (2011)
8. Nowak-Brzezińska, A., Rybotycki, T.: Visualization of medical rule-based knowledge bases. J. Med. Inf. Technol. 24, 91–98 (2015)
9. Nowak-Brzezińska, A., Xięski, T.: Exploratory clustering and visualization. Procedia Comput. Sci. 35C, 1082–1091 (2014). Elsevier
10. Przybyła-Kasperek, M., Wakulicz-Deja, A.: Global decisions taking on the basis of dispersed medical data. In: Ciucci, D., Inuiguchi, M., Yao, Y., Ślęzak, D., Wang, G. (eds.) RSFDGrC 2013. LNCS, vol. 8170, pp. 355–365. Springer, Heidelberg (2013)
11. Rybotycki, T.: Wizualizacja struktur hierarchicznych dla regulowych baz wiedzy, Engineer Thesis. Sosnowiec (2015)
12. Shneiderman, B.: Tree visualization with tree-maps: 2-d space-filling approach. Trans. Graphics (TOG) 11(1), 92–99 (1992). Association for Computing Machinery, New York
13. Siminski, R., Xięski, T.: Physical knowledge base representation for web expert system shell. In: Kozielski, S., Mrozek, D., Kasprowski, P., Malysiak-Mrozek, B., Kostrzewa, D., Mangai, J.A. (eds.) BDAS 2016. CCIS, vol. 613, pp. 558–570. Springer, Heidelberg (2016). doi:10.1007/978-3-319-34099-9_43
14. Turban, E., Aronson, J.E.: Decision Support Systems and Intelligent Systems, 6th edn. Prentice International Hall, Hong Kong (2001)
15. Wetzel, K.: Pebbles - using circular treemaps to visualize disk usage (2004)
16. Wierzchoń, S., Kłopotek, M.: Algorithms of Cluster Analysis Wyd. IPI PAN, Warszawa (2015)

The Matching Method for Rectified Stereo Images Based on Minimal Element Distance and RGB Component Analysis

Paweł Popielski[1]([✉]), Robert Koprowski[1], Zygmunt Wróbel[1],
Sławomir Wilczyński[2], Rafał Doroz[1], Krzysztof Wróbel[1], and Piotr Porwik[1]

[1] Institute of Computer Science, University of Silesia,
Będzińska 39, 41-200 Sosnowiec, Poland
pawel.popielski@us.edu.pl
[2] Department of Basic Biomedical Science,
School of Pharmacy with the Division of Laboratory Medicine,
Medical University of Silesia, Katowice, Poland

Abstract. A common problem occurring in medical practice is the localization of veins and arteries. To determine the location of these elements, it is not necessary to have a complete 3D model. A much better solution is preliminary segmentation yielding the contour of veins, and further search for stereo correspondence already in binary images. The computational complexity of this approach is much smaller, which guarantees its fast operation. The disparity matrix is created according to the principle that the most likely correct distance between the same elements in the left and right images is the minimum value. Then, the adjacent RGB components surrounding the elements aspiring to be homologous are analysed. The operation of the method is illustrated on the basis of the authors' own images as well as standardized images. In addition, its operation was compared with three recognized and widely used algorithms for image matching. The effectiveness of the new method reaches less than 94 % of correctly matched pixels with a standard deviation of 1.5 pixels and operation time of 90 ms.

Keywords: Disparity · Stereo correspondence · Stereovision

1 Introduction

In medical practice, there is often a need for locating veins or arteries [1]. Subcutaneous placement of vessels often prevents their visual location. Angiography [2] or a vein illuminator [3] allow only for the acquisition of a flat image of veins. Stereo visual imaging can enable to create from such images the point cloud representing the three-dimensional contour of veins or arteries. The three-dimensional image obtained in this way enables to specify the depth and location of vessels, which can be particularly useful in intraoperative navigation systems.

The problem of locating vessels can be solved by using appropriate segmentation [4] and binary image matching. The contour of vessels is obtained during

© Springer International Publishing Switzerland 2016
N.T. Nguyen et al. (Eds.): ICCCI 2016, Part II, LNAI 9876, pp. 482–493, 2016.
DOI: 10.1007/978-3-319-45246-3_46

segmentation and stored as a binary image. The proposed method seeks correspondence in the thus prepared binary images. Additional versatility of the method will be tested on standardized images.

The methods based on stereovision [5] guarantee non-invasiveness and lack of contact with the patient because they only use reflected light. Another advantage of stereovision is fast measurement at proper illumination of the scene below 100 ms. Stereovision can also be applied in the examination of faulty posture [6,7], metabolic diseases [8], and plastic surgery [9–11]. Stereovision builds a three-dimensional object model based on the known geometry of the system of two imaging devices. The key step is the search for correspondence between the left and right images of a stereo pair in order to calculate the coordinates of the point cloud on the basis of the coordinate differences.

Binarization of images, despite the loss of information about colour, greatly simplifies the image [4], which accelerates the analysis. Relatively quick search for stereo correspondence enables to implement the new method even on mobile devices.

Given the above, it was decided to develop a quick method for binary image matching. The next section presents the state of the art covering selected issues of image matching significant for the new method.

2   State of the Art

Currently in stereovision, correspondence seeking is realized for rectified images, which greatly speeds up the whole process of preparing the cloud of points. The best optimized algorithm for image rectification was developed by Bouguet [12]. Owing to the properties of the fundamental matrix [13–15], the search for correspondence for an element in row m of the left image only involves searching the row m of the right image, and not the whole right image as in the case of rectified images.

The matrix $D(m,n)$ coded as a Disparity Map or Parallax Matrix [16] contains the result of seeking stereo correspondence. Each matrix cell contains the difference in the position between homologous points.

Knowing the disparity matrix, the coordinates of the point cloud are calculated from the simple triangulation [16].

The early methods for correspondence seeking involved searching the right image with a mask representing a part of the surroundings of an element in the left image. The response of the similarity function determined whether the search was successful. The best known early method is normalized cross-correlation (NCC), which will be compared to the proposed new method [17].

A very interesting method for seeking stereo correspondence is the calculus of variations (VAR), which is very effective in matching various types of images [18]. It treats a stereo pair as a vector field. The method is based on minimization of the energy functional by solving the corresponding Euler - Lagrange equation, which is represented by a system of partial differential equations. The variational algorithm optimized by Ralli [19] was used in the comparison.

The quasi global method based on Information Theory (SGBM) exhibits very high efficiency when it comes to image matching. The cost function was chosen in such a way so as to take into account the Mutual Information carried by a stereo pair [20].

These three methods for image matching, namely NCC, VAR and SGBM, provide a platform for the comparison of the proposed new method.

3   Material

A set of 21 images borrowed from Middlebury College stereovision laboratory acquired with the method described in [21] then published inter alia in [22, 23] were selected as standardised images. They provide a comparative basis for all the new methods for seeking stereo correspondence. These images have disparity arrays prepared by the authors, which are ground truth. The standardized images used in the tests have a unified height of 370 pixels and various widths based on theirs aspect ratio. The authors' own images were also included in the tests (Fig. 1). Veins is a segmented image of veins of the forearm acquired using the AccuVein AV400 vein illuminator cropped to resolution of 113×84 pixels.

Models of solid figures were prepared from cardboard, i.e. a regular cube and piramid. Then images were registered with the stereovision rig consisting of two web cams. The object distance was adjusted to 50 cm. Both images were cropped and resized to $m \times n$ resolution where m means number of rows and n means number of columns. For a pyramid the most adequate resolution was 114×200 and for a cube 163×200. The clipping criterion was to preserve as many tiny elements in the images as it was possible and, at the same time, minimize the time spent on creating the disparity map D_T.

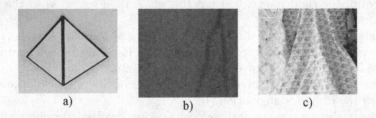

a)   b)   c)

Fig. 1. Examples of analysed images (a) Pyramid, (b) Veins, (c) Cloth4.

The stereovision head used to acquire authors' own images consists of two Logitech C920 HD Pro Webcams. They are able to acquire images with a maximal resolution of 1080×1920 pixels. Very good image quality is provided by the Tessar lens system from Carl Zeiss factory. The horizontal arm with shiftable carriages enables to move the cameras away from each other from 10 to 90 cm.

4   Proposed Method

The proposed method (MEDRGB) is a modification and development of the previously proposed method [24] and includes the analysis of rgb components. The rectified left $O_z^{(L)}(m,n)$ and right $O_z^{(P)}(m,n)$ images are binarized according to [25] using the Sauvola adaptive method [26] modified by Shafait [27] (Fig. 2). The resulting binary images $O_b^{(L)}(m,n)$ and $O_b^{(P)}(m,n)$ are the basis for calculating the difference between horizontally adjacent pixels, calculated for each row (1, 2).

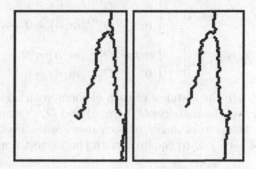

Fig. 2. Left and right binarized veins images.

$$O_r^{(L)}(m,n) = O_b^{(L)}(m,n+1) - O_b^{(L)}(m,n). \qquad (1)$$

$$O_r^{(P)}(m,n) = O_b^{(P)}(m,n+1) - O_b^{(P)}(m,n). \qquad (2)$$

for $n \in (1, N-1)$,
where
m – number of the next row,
n – number of the next column,
N – number of image columns.

The resulting images $O_{r(m,n)}^{(L)}$ and $O_r^{(P)}(m,n)$ show the object edges represented by the values $\{-1, 1\}$. The values $+1$ are located on the edges resulting from changes in the brightness from dark to light (as viewed from the left). The values -1 are located on the edges resulting from changes in the brightness from light to dark. Next, the image is modified in such a way that the values $+1$ only remain, i.e.:

$$O_k^{(L)}(m,n) = \begin{cases} 1 \; for & O_r^{(L)}(m,n) = 1 \\ 0 & \text{the others} \end{cases} \qquad (3)$$

$$O_k^{(P)}(m,n) = \begin{cases} 1 \; for \; & O_r^{(P)}(m,n) = 1 \\ 0 & \text{the others} \end{cases} \tag{4}$$

for $n \in (1, N-1)$.

In the thus obtained images $O_k^{(L)}(m,n)$ and $O_k^{(P)}(m,n)$, the values $+1$ coincide with the inner edges of objects with greater intensity. On this basis, the matrices $O_o^{(L)}(m,n)$ and $O_o^{(P)}(m,n)$ were created, containing information about the position of $+1$ in each row.

$$O_o^{(L)}(m,n) = \begin{cases} n \; for \; & O_k^{(L)}(m,n) = 1 \\ 0 & O_k^{(L)}(m,n) = 0 \end{cases} \tag{5}$$

$$O_o^{(P)}(m,n) = \begin{cases} n \; for \; & O_k^{(P)}(m,n) = 1 \\ 0 & O_k^{(P)}(m,n) = 0 \end{cases} \tag{6}$$

In order to calculate the distance of each element with all other elements in the corresponding rows, the matrices $O_o^{(L)}(m,n)$ and $O_o^{(P)}(m,n)$ are modified in such a way so that the zero elements in each row are removed. Then zeros are added at the end of each row to equate it with the longest non-zero row.

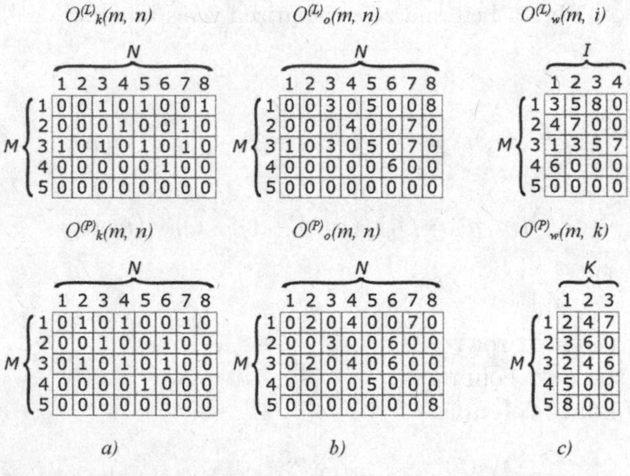

Fig. 3. The successive stages of creating matrices $O_w^{(L)}(m,i)$ and $O_w^{(P)}(m,k)$, (a) zero–one matrices, (b) early distance matrices, (c) final distance matrices

In Fig. 3 I, K are the numbers of the columns of the matrices $O_w^{(L)}(m,i)$ and $O_w^{(P)}(m,k)$ respectively and

$$I = \max_m \left(\sum_{n=1}^{N} O_k^{(L)}(m,n) \right). \tag{7}$$

$$K = \max_m \left(\sum_{n=1}^{N} O_k^{(P)}(m,n) \right). \tag{8}$$

Then the lengths on the right image are substracted from the lengths on the left image for each epipolar line. The result is the distance between homologous points in the left image and all the points in the right image.

$$O_{odl}(m,i,k) = O_w^{(L)'}(m,i,k) - O_w^{(P)'}(m,i,k). \tag{9}$$

where

$$O_w^{(L)'}(m,i,k) = [1,1,\ldots,1]_{(1 \times K)}^T \cdot O_w^{(L)}(m,i). \tag{10}$$

$$O_w^{(P)'}(m,i,k) = O_w^{(P)}(m,k)^T \cdot [1,1,\ldots,1]_{(1 \times I)}. \tag{11}$$

It is assumed that disparity D_m corresponds to the minimum distance calculated between the element in the left image and all the elements in the right image for a given row.

$$D_m(m,i) = \min_k \left(O_{odl}(m,i,k) \right). \tag{12}$$

A disparity matrix of a size compatible with a stereo pair is needed for further calculations.

$$D(m,n) = \begin{cases} D_m(m,i) \text{ for } O_k^{(L)}(m,n) \neq 0 \\ 0 \qquad\qquad \text{ the others} \end{cases} \tag{13}$$

for $m \in (1,M)$, $n = O_w^{(L)}(m,i)$, $i \in (1,I)$.

Then the vector components rgb are subtracted from each other for the pixel preceding each element of the disparity matrix $D(m,n)$ in the left image $O_z^{(L)}(m,n-1)$ and the right image $O_z^{(P)}(m,n-1)$. Similarly, the same procedure applies to the pixels following each element of the disparity matrix.

$$\Delta_{rgb}(w) = |O_z^{(L)}(m,n-1) - O_z^{(P)}(m,n-1)|. \tag{14}$$

where

$$\begin{aligned} O_z^{(L)}(m,n-1) &= (r,g,b), \\ O_z^{(P)}(m,n-1) &= (r,g,b), \\ w &\in (1,3), \\ r,g,b &\in\, <0,255> \end{aligned}$$

Then the vector $R(w)$ is created in such a way that if the absolute difference after a given rgb component is greater than the adopted threshold $proxDiff$, the value 1 is assumed for the component. Otherwise, 0 is assumed. In the tests, it was assumed that $proxDiff = 8$.

$$R(w) = \begin{cases} 1 \text{ for } \Delta_{rgb}(w) > proxDiff \\ 0 \qquad \Delta_{rgb}(w) \leq proxDiff \end{cases} \tag{15}$$

If the sum of the elements of the vector $R(w)$ is greater than the threshold $winP$

$$\sum_{w=1}^{3} R(w) > winP. \tag{16}$$

the disparity is considered erroneous and it is replaced in the matrix $O_{odl}(m, i, k)$ with the value symbolising infinity, and then the minimum distance is recalculated according to the formula (12).

In another case, the value of disparity is considered to be adequate and is recorded in the disparity matrix $D(m, n)$. In the algorithm it was assumed that $winP = 0$ and $Infinity = 10000$.

Similarly, the same procedure applies to the pixels following each element of the disparity matrix $D(m, n)$ in the left image $O_z^{(L)}(m, n + 1)$ and the right image $O_z^{(P)}(m, n+1)$. The final decision is the logical sum of the analysis results of the preceding and following elements of the disparity matrix.

Once all the elements of the disparity matrix $D(m, n)$ are agreed, it is now possible to establish the coordinates of the point cloud by similar triangles according to [16].

5   Results

Three criteria were adopted to evaluate the results. The effectiveness of matching δ_r according to [28] was formulated as:

$$\delta_r = \sqrt{\frac{1}{(M \cdot N)} \sum_{m=1}^{M} \sum_{n=1}^{N} |D(m, n) - D_T(m, n)|^2}. \tag{17}$$

where $D(m, n)$ – calculated disparity map, $D_T(m, n)$ – ground truth disparity map.

The next criterion, called the percentage of mismatches δ_D, was formulated as:

$$\delta_D = \frac{1}{M \cdot N} \sum_{m=1}^{M} \sum_{n=1}^{N} L_D^{(L)}(m, n). \tag{18}$$

where

$$L_D^{(L)} = \begin{cases} 1 \text{ for } |D(m, n) - D_T(m, n)| > p_D \\ 0 \qquad\qquad \text{the others} \end{cases}$$

p_D – allowable threshold of mismatches.

The allowable threshold of the matching error p_D was adopted arbitrarily as 2 pixels. The threshold value was determined based od the maximal real size of the pixel, which fluctuates around $80 \, \mu m$.

The time t was taken as the third criterion characterising the algorithm.

The algorithm was tested on a desktop computer with Intel Core i5 3.1 GHz processor. Implementation was made in MATLAB® Version: 8.6.0.267246 (R2015b) Image Processing Toolbox Version 9.3.

The results of the above three criteria for a total of 21 standardized images as well as Pyramid, Cube and Veins are available at https://goo.gl/ZQQXnk. Some selected representative values are presented in tables (Tables 1, 2, 3 and 4). Although the implementation of the algorithm MEDRGB is made in the interpretable code, due to low computational complexity, calculations were made very quickly achieving times below 1 s for the authors' own images (for the computer specifications given above). The minimum time of 50 ms was observed for Pyramid. For the standardized images, the times are mostly below 2 s, reaching a minimum value of 1.84 s for Bowling1 and a maximum value of 2.44 s for Cloth4.

Table 1. Some selected representative results of matching images using the proposed algorithm.

	Pyramid	Cube	Veins	Bowling1	Cloth4
δ_r [pix]	4.2	7.1	1.5	17.8	29.1
δ_D [%]	15.9	30.9	6.2	55.3	83.7
t [s]	0.05	0.12	0.09	1.84	2.44

The authors' own images are focused on matching images after segmentation with the extracted specific elements. An example is the image Veins, where the location of the veins was of interest. Such images are characterized by only a few edges which have to be matched. The image Cloth4 is characterized by an extreme number of small items, in the form of a patterned fabric texture that poses the greatest difficulty for the proposed method MEDRGB, which is reflected in the longest time of operation.

The standard deviation reaches the lowest value of 1.5 pixels for Veins, where only the outline of the veins remains after segmentation. The highest value of 29.1 pixels is again characteristic for the image Cloth4 with the patterned fabric texture. The image Veins has the smallest area of mismatches, only 6.2 %. The highest value of 83.7 % is characteristic for Cloth4. The patterned fabric texture with many small elements arranged next to each other effectively impedes matching for the proposed method. Another factor that hinders matching is the unevenly lit scene.

It should be borne in mind here that it is very difficult to obtain 100 % matching efficiency. There are always areas in the left image that are not visible in the right image, for instance, the area around the nose or along the shades and edges.

6   Discussion

In order to compare the proposed method MEDRGB with three commonly used methods for seeking stereo correspondence NCC, VAR and SGBM, presented in Sect. 2, the evaluation criteria used in Sect. 5 were adopted. VAR and SGBM methods owe their speed to a compiled code, which interferes with the comparison of the test results. However, due to the significance of these methods, it was decided to include them in the comparison. For NCC, a 3–pixel mask was used. In terms of the operation time (Table 2), the best method is SGBM reaching times of 100 ms and less. However, it should be noted that the uncompiled code of the proposed method MEDRGB is characterized by slightly longer operation times and takes second place. NCC, which for standardized images reaches times of 200 seconds for a 3–pixel mask, is the slowest. In terms of the standard deviations (Table 3), the proposed algorithm MEDRGB takes first place, reaching 1.5 pixels for Veins. However, for the standardized images the values of standard deviation are higher, but still comparable with other methods. The highest values for standardized images are reached by the proposed method and NCC.

The comparison of the percentage of the mismatched area (Table 4) indicates that the proposed method MEDRGB works best for the authors' own images and SGBM for the standardized images. The worst is NCC, where in the case of the authors' images, 100 % of the surface has a difference in disparity greater than the adopted threshold of 2 pixels. For the standardized images, the highest percentage of mismatches is reached by NCC and MEDRGB.

For the authors' images, NCC, VAR and SGBM methods reach almost a hundred percentage of mismatches. The segmented images consist of black and white areas. As a result, these images as very homogeneous and it makes match-

Table 2. Comparison of operation times t with other algorithms in seconds.

	Pyramid	Cube	Veins	Bowling1	Cloth4
MEDRGB	0.05	0.12	0.09	1.84	2.44
NCC	18.81	26.83	2.44	201.91	210.93
VAR	0.40	0.55	0.51	4.04	4.31
SGBM	0.01	0.02	0.12	0.07	0.07

Table 3. Comparison of standard deviations δ_r in pixels.

	Pyramid	Cube	Veins	Bowling1	Cloth4
MEDRGB	4.2	7.1	1.5	17.8	29.1
NCC	16.6	22.1	10.0	23.9	15.7
VAR	14.9	15.2	10.0	13.1	7.4
SGBM	15.5	22.0	10.0	7.9	5.9

Table 4. Comparison of the percentage of mismatches δ_D.

	Pyramid	Cube	Veins	Bowling1	Cloth4
MEDRGB	15.9	30.9	6.2	55.3	83.7
NCC	98.2	97.7	100.0	74.1	61.7
VAR	100.0	100.0	100.0	63.3	20.9
SGBM	98.6	98.1	98.3	19.2	4.2

ing impossible for methods based on the diversity of images. This explains such poor results of these methods.

Large values of mismatches of the MEDRGB for the standardized images arise from the nature of the method itself, which matches contours, outlines of objects created in the process of segmentation. The elements significant from the point of view of a particular method application are here subject to extraction. An example here may be segmentation performed in order to determine the position of the veins under the skin, coronary veins in angiography or the human figure for the purpose of rehabilitation.

7   Conclusion

The proposed new method MEDRGB can be used in matching medical images, where it is necessary to segment the elements significant from the point of view of the performed medical procedure, such as the localization of invisible subcutaneous veins or angiography, obtaining a spatial image of the vessel contour. Other methods match the whole flat images, which prevents the separation of the spatial image of vessels only from the surrounding background. A special feature of the method is its low time complexity so the presented algorithm is fast.

Further work will be aimed at improving the efficiency of matching for binary and standardized images, as well as accelerating the matching algorithm. It should be noted, however, that in its current form, the proposed method can achieve results comparable to the ones obtained using other recognised methods.

References

1. Shahzad, A., Saad, M., Walter, N., Malik, A., Meriaudeau, F.: A review on subcutaneous veins localization using imaging techniques. In: Current Medical Imaging Reviews, pp. 125–133. Bentham Science Publishers (2014)
2. Mohareb, M.M., Feng, Q., Cantor, W.J., Kingsbury, K.J., Ko, D.T., Wijeysundera, H.C.: Validation of the appropriate use criteria for coronary angiography: a cohort study. Ann. Intern. Med. **162**, 549–556 (2015)
3. Sanchez-Morago, G.-V., Sanchez Coello, M.D., Villafranca Casanoves, A., Cantero Almena, J.M., Migallon Buitrago, M.E., Carrero Caballero, M.C.: Viewing veins with AccuVein AV300. Rev. Enferm. **33**, 33–38 (2010)

4. Wójcicka, A., Jędrusik, P., Stolarz, M., Kubina, R., Wróbel, Z.: Using analysis algorithms and image processing for quantitative description of colon cancer cells. In: Piętka, E., Kawa, J., Wieclawek, W. (eds.) Information Technologies in Biomedicine, Volume 3. AISC, vol. 283, pp. 383–392. Springer, Heidelberg (2014)

5. Patias, P.: Medical imaging challenges photogrammetry. Virtual Prototyp. Bio Manuf. Med. Appl. **56**, 45–66 (2008)

6. Golec, J., Ziemka, A., Szczygiel, E., Czechowska, D., Milert, A., Kreska-Korus, A., Golec, E.: Photogrametrical analysis body position in hips osteoarthrosis. Ostry Dyur. **5**, 1–7 (2012)

7. Golec, J., Tomaszewski, K., Maślon, A., Szczygiel, E., Hladki, W., Golec, E.: The assessment of gait symmetry disorders and chosen body posture parameters among patients with polyarticular osteoarthritis. Ostry Dyur. **6**, 91–95 (2013)

8. Mitchell, H.L.: Applications of digital photogrammetry to medical investigations. ISPRS J. Photogramm. Remote Sens. **50**, 27–36 (1995)

9. D'Apuzzo, N.: Measurement and modeling of human faces from multi images. Int. Arch. Photogramm. Remote Sens. **34**(5), 241–246 (2002)

10. D'Apuzzo, N.: Automated photogrammetric measurement of human faces. Int. Arch. Photogramm. Remote Sens. **32**(B5), 402–407 (1998)

11. Walczak, M.: 3D measurement of geometrical distortion of synchrotron-based perforated polymer with Matlab algorithm. In: Pietka, E., Badura, P., Kawa, J., Wieclawek, W. (eds.) ITIB 2016. AISC, vol. 471, pp. 245–252. Springer, Switzerland (2016)

12. Bouguet, J.-Y.: Complete Camera Calibration Toolbox for Matlab (1999). http://www.vision.caltech.edu/bouguetj/

13. Hartley, R., Zisserman, A.: Multiple View Geometry in Computer Vision. Cambridge University Press, Cambridge (2003)

14. Forsyth, D.A., Ponce, J.: Computer Vision: A Modern Approach. Prentice Hall Professional Technical Reference (2002). ISBN: 0130851981

15. Stockman, G., Shapiro, L.G.: Computer Vision, 1st edn. Prentice Hall PTR, Upper Saddle River (2001). ISBN: 0130307963

16. Kraus, K.: Photogrammetry: Geometry from Images and Laser Scans. Walter de Gruyter (2007)

17. Lewis, J.P.: Fast normalized cross-correlation. Vis. Interface **10**(1), 120–123 (1995)

18. Kosov, S., Thormählen, T., Seidel, H.-P.: Accurate real-time disparity estimation with variational methods. In: Bebis, G., Boyle, R., Parvin, B., Koracin, D., Kuno, Y., Wang, J., Silva, C.T., Coming, D., Wang, J.-X., Wang, J., Pajarola, R., Lindstrom, P., Hinkenjann, A., Encarnação, M.L. (eds.) ISVC 2009, Part I. LNCS, vol. 5875, pp. 796–807. Springer, Heidelberg (2009)

19. Ralli, J., Diaz, J., Ros, E.: Spatial and temporal constraints in variational correspondence methods. Mach. Vis. Appl. **24**, 275–287 (2011)

20. Hirschmuller, H.: Stereo processing by semiglobal matching and mutual information. IEEE Trans. Pattern Anal. Mach. Intell. **30**, 328–341 (2008)

21. Scharstein, D., Szeliski, R.: High-accuracy stereo depth maps using structured light. In: 2003 IEEE Conference Computer Vision Pattern Recognition, Proceedings, vol.1, pp. 195–202 (2003)

22. Scharstein, D., Pal, C.: Learning conditional random fields for stereo. In: Proceeding of IEEE Computer Society Conference on Computer Vision and Pattern Recognition (2007)

23. Hirschmuller, H., Scharstein, D.: Evaluation of cost functions for stereo matching. In: Proceeding CVPR, pp. 1–8 (2007)

24. Popielski, P., Wrobel, Z., Koprowski, R.: The fast matching algorithm for rectified stereo images. In: Pietka, E., Badura, P., Kawa, J., Wieclawek, W. (eds.) Information Technologies in Medicine. AISC, vol. 471, pp. 107–118. Springer, Switzerland (2016)
25. Jedzierowska, M., Wrobel, Z., Koprowski, R.: Imaging of the anterior eye segment in the evaluation of corneal dynamics. In: Pietka, E., Badura, P., Kawa, J., Wieclawek, W. (eds.) ITIB 2016. AISC, vol. 471, pp. 63–73. Springer, Switzerland (2016)
26. Sauvola, J., Pietikainen, M.: Adaptive document image binarization. Pattern Recogn. **33**, 225–236 (2000)
27. Shafait, F., Keysers, D., Breuel, T.M.: Efficient implementation of local adaptive thresholding techniques using integral images. In: Yanikoglu, B.A., Berkner, K. (eds.) Document Recognition and Retrieval, pp. 1–6. International Society for Optics and Photonics (2008)
28. Scharstein, D., Szeliski, R.: A taxonomy and evaluation of dense two-frame stereo correspondence algorithms. Int. J. Comput. Vis. **47**, 7–42 (2002)

KBExplorator and KBExpertLib as the Tools for Building Medical Decision Support Systems

Roman Simiński[✉] and Agnieszka Nowak-Brzezińska

Institute of Computer Science, University of Silesia, Sosnowiec, Poland
roman.siminski@us.edu.pl
http://ii.us.edu.pl

Abstract. Medicine had been considered a good domain in which the concepts of rule-based decision support system could be applied. The early medical decision support systems were designed over forty years ago. Since that time, many different methods were proposed in the decision support area. Regardless of the development of different non knowledge-based methods, the rule representation and inference on the rules bases are still popular. In this paper the KBExplorator system and KBExpertLib software library are introduced in the context of medical decision support system implementation. The KBExplorator system may be considered as tool for building medical knowledge bases. This system is designed for knowledge engineers and domain experts, which are responsible for creating the knowledge base for particular problem. The software library KBExpertLib is a tool for programmers to develop software which utilize the knowledge bases designed with use of KBExplorator. The main properties and methods of practical usage of KBExplorator and KBExpertLib are described as well as experiments focused on the software effectiveness evaluation.

Keywords: Decision support system · Rule knowledge base · Inference

1 Introduction

Many of medical support systems arose out of earlier expert systems research, where the aim was to build a computer program that could simulate human thinking. Medicine had been considered a good domain in which these concepts could be applied. The rules can then be used to perform inference in order to reach appropriate conclusion or to confirm selected goals. The first medical expert systems were designed over forty years ago. MYCIN which used knowledge base and inference to identify bacteria causing severe infections, was developed over five or six years in the early 1970s at Stanford University [1]. ONCOCIN was an advanced expert system for clinical oncology that has been under development at Stanford University School of Medicine since 1979 [2]. Since that time, advances in both computer technology and software design have permitted the development of more sophisticated and specialized medical expert systems [3], most of them are designed as decision support systems. In

© Springer International Publishing Switzerland 2016
N.T. Nguyen et al. (Eds.): ICCCI 2016, Part II, LNAI 9876, pp. 494–503, 2016.
DOI: 10.1007/978-3-319-45246-3_47

the course of forty years, many additional artificial intelligence methods have been employed [4]. Complete list of research fields applied in medical systems is of course very long [5]. Unlike knowledge-based medical decision support systems, some of the nonknowledge-based decision support systems use a form of artificial intelligence – artificial neural networks and genetic algorithms are two well-known types of nonknowledge-based systems.

Regardless of the development of different nonknowledge-based methods, the rule representation and inference are still popular. Recent years have brought a renaissance of rules representation for knowledge bases. Currently, the rules are considered as standard result form of data mining methods, rules are again an important and useful material for constructing knowledge bases for different types of decision support systems. This work presents practical results of research focused on the development of the new method and tools for building knowledge-based decision support systems. Current works are connected with previous research, also in medical applications [6–8]. The KBExplorator system and KBExpertLib software library are introduced in the context of medical decision support system implementation. The main properties and methods of practical usage are described as well as experiments focused on the software effectiveness evaluation.

2   Related Works

When early medical decision support systems were constructed, their implementations typically began with classical, multi-purpose programming languages (C, C++) or artificial intelligence dedicated languages (LISP, Prolog). Several tools and languages are available for developing medical decision support systems [9]. The detailed discussion and comparison of modern tools goes beyond the scope of this study. Some aspect of such review can be found in [10,11], in this work only basic information is presented.

The Acquire system [12] provides an ability to develop web-based user interfaces through a clientserver development kit that supports Java and ActiveX controls. The ExSys system provides the Corvid Servlet Runtime and implements the Exsys Corvid Inference Engine as a Java Servlet. Developed at NASA, the C Language Integrated Production System (CLIPS) is a rule-based programming language useful for creating expert systems [13]. Jess is a popular rule engine for the Java platform. JESS or Java Expert System Shell are the skeleton of expert systems developed by Sandia National Laboratories. Jess is written in Java, it is possible to run code in this language using Jess. It uses a syntax similar to Lisp [14]. It is compatible with both the Windows and Unix systems. Rules written using Jess are saved in the form of an XML file which must contain a *rule-execution-set* element [15].

Another commercial expert system building tool is XpertRule, which offers a Knowledge Builder Rules Authoring Studio. The eXpertise2Go's Rule-Based Expert System provides free building and delivery tools that implement expert systems as Java applets, Java applications and Android apps [16]. Next system is Drools, a Business Rules Management System solution. It provides a core Business Rules Engine, a web authoring and rules management application (Drools Workbench) and an Eclipse IDE plugin for core development [17]. Still the PC-Shell expert system shell can be used in medical applications [18].

This work introduces another decision support system building tool. The KBExplorator system, it is the web application and it allows the user to create, edit and share rule knowledge bases. The KBExpertLib is a software library, it allows the programmers to use different kinds of inference within any software projects implemented in Java programming language. This library is able to run inference on rule knowledge bases stored in the KBExplorator database or saved locally in the XML files. The KBExplorator system and the KBExpertLib have been designed at the University of Silesia, Institute of Computer Science, in 2015–2016. This is rising software, at the current stage of software development it is too early to perform comparative tests with other packages (eg. JESS). For this reason, this work is focused on the current software properties, comparative tests are planned at the end of the experiments focused on functional properties and effectiveness of KBExplorator and KBExpertLib.

3   Methods

This section presents three main issues – the background information, main properties of the KBExplorator system and KBExpertLib library and selected practical issues related to their medical applications.

3.1   Background Information

The methodological assumption and theoretical description of research realized within the *KBExplorator* and KBExpertLib project can be found in [6,10,19], this section presents only summary of the necessary information. In the software considered in this work, the knowledge base is a pair $\mathcal{KB} = (\mathcal{R}, \mathcal{F})$ where $\mathcal{R}$ is a non-empty finite set of rules and $\mathcal{F}$ is a finite set of facts. $\mathcal{R} = \{r_1, r_2, \ldots, r_n\}$. Each rule $r \in \mathcal{R}$ will have a form of Horn's clause: $r : p_1 \wedge p_2 \wedge \cdots \wedge p_m \rightarrow c$, where m—the number of literals in the conditional part of rule r, and $m \geq 0$, p_i—i-th literal in the conditional part of rule r, $i = 1 \ldots m$, c—literal of the decisional part of rule r. We will also consider the *facts* as clauses without any conditional literals. The set of all such clauses f will be called *set of facts* and will be denoted by $\mathcal{F}$: $\mathcal{F} = \{f : \forall_{f \in \mathcal{F}} \ cond(f) = \emptyset \wedge f = concl(f)\}$. The rule's literals will be denoted as pairs of attributes and their values. Let A be a non-empty finite set of conditional and decision attributes. For every symbolic attribute $a \in A$ the set V_a will be denoted as the set of values of attribute a. The literals of the rules from $\mathcal{R}$ are considered an attribute-value pair (a, v), where $a \in A$ and $v \in V_a$. We also consider numeric attributes and literals representation dedicated for them: attribute-relation-value.

In the previous works, we introduced approach which assumes that the rule knowledge base is decomposed into the groups of rules, called *rules partitions* [10, 19]. Rules partitions terminologically correspond to the mathematical definition of the partition as a division of a given set into the non-overlapping and non-empty subset. The groups of rules which create partition are pairwise disjoint and utilize all rules from $\mathcal{R}$. The main results of the conception of rules partitions mentioned above are modified forward and backward algorithm [10,19] as well as the proposals for new algorithms [11,20]. The proposed modifications of the classical inference algorithms are based on information extracted from the groups of rules generated by the two selected partitioning strategy [10].

3.2 KBExplorator and KBExpertLib—General Description

The theoretical background presented in the previous section is the base for software implementation. At the current stage of development process, two main software components are implemented:

- The KBExplorator system is the web application. It allows the user to create, edit and share rule knowledge bases. Each user can register his own account, knowledge bases created by the users are stored in the KBExplorator data bases and they are accessible from any standard web browser software. Stored knowledge bases may be shared between registered system's users, it is also possible to download any stored knowledge base as the XML file. Currently available system functions allow users to create and edit attributes and their properties as well as rules and rules' properties (Fig. 1 presents an idea of KBExplorator role).
- The KBExpertLib is the software library, which allows the programmers to use different kinds of inference within any software projects implemented in Java programming language. This library is able to run different kinds of inference (classical and modified forward and backward algorithms) on rule knowledge bases stored in the KBExplorator database or saved locally in the XML files (Fig. 1 presents an idea of KBExpertLib utilisation).

The KBExplorator performs their function irrespective of the operating system and browsers running on the client side. Proposed system allows working without the need to be installed in user's computer. The KBExplorator is convenient for use, the knowledge engineers and domain experts can access it in any part of the world at any given time. Unlike the desktop applications, proposed system do not have to be installed as it runs on a dedicated web server. The time and trouble required for installing a software are also done away with. The KBExplorator works on multiple platforms, it only requires web browser and it is compatible with most of the computer operating systems. Of course, KBExplorator requires access to the Internet, but this requirement does not seem bothersome. The package KBExpertLib may be used on the server side, but the main scope of its application are client side desktops application, implemented in Java, also on mobile devices. The KBExpertLib is object oriented library, library's classes are divided into the packages: kbcore—the main, essential classes, kbinfer—classes providing classical and modified inference algorithms, kbpartition—classes allowing decomposition of rule bases, and kbtools—additional tool classes.

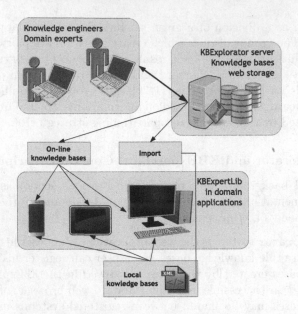

Fig. 1. The idea KBExplorator and KBExpertLib applications

We also began the work on the desktop, ready to off-line work version of the KBExplorator called KBExploratorDesktop. This system uses KBExpertLib library, it is implemented as JavaFX program. During the preparation of this work, KBExploratorDesktop was in the early prototype phase and for this reason its description is omitted. The KBExplorator system and the KBExpertLib library were implemented and currently intensive tests and experiments are performed. Public access to the system and software library is expected in the summer of 2016.

3.3   Practical Issues

The KBExplorator offers functionalities typical for web applications (registration, login). Knowledge management functions are also implemented typically, all operations performed on the rule knowledge base are intuitive. The limited size of this publication does not allow the authors to present more detailed, practical information about KBExplorator. Due its intuitive organisation, only more detailed description of programming with the KBExpertLib issues will be presented—two typical scenario of forward and backward inference. An example of forward inference activation in the console IO Java program is as follows:

```
import kbcore.*;
import kbinfer.*;
...
// Create knowledge base object
KBKnowledgeBase base = new KBKnowledgeBase();
// Create knowledge base loader
KBDataBaseLoader kbLoader = new KBDataBaseLoader();
```

```
// Load knowledge base from KBExplorator server
kbLoader.loadKnowledgeBase( "virus_infection", base );
// Add starting facts
base.addFactFromText( "cough", "=", "wet" );
base.addFactFromText( "temperature", ">=", "38.0" );
// Create object for inference
KBForwardInferer infer = new KBForwardInferer( base );
// Run inference with rules selection strategy
infer.classicInference( KBInferer.RuleSelStrategy.LAST_RULE );
// Chect out for new facts
if( infer.newFactInfered )
    System.out.println( "New facts available" );
else
    System.out.println( "No new facts" );
```

When we want to use local XML file rather than on-line knowledge base we simply write:

```
kbLoader.loadKnowledgeBaseFromXML( "infekcje_wirusowe.xml", base );
```

Analogous example for the backward inference is as follows (some unimportant lines and comments have been omitted):

```
KBKnowledgeBase base = new KBKnowledgeBase();
KBDataBaseLoader kbLoader = new KBDataBaseLoader();
if( kbLoader.loadKnowledgeBaseFromXML("infekcje_wirusowe.xml", base ) )
{
    KBBackwardConsoleInferer infer = new KBBackwardConsoleInferer( base );
    // Create inference goal
    KBLiteral goal = base.makeLiteralFromText( "disease", "=", "flu" );
    // Run backward inference
    if( infer.classicInferenceWithGoal( goal ) )
    {
        System.out.println( "Goal confirmed" );
        if( infer.newFactInfered )
            System.out.println( "New facts available" );
    }
    else
        System.out.println( "Goal unconfirmed" );
}
```

The KBExpertLib also provides modified forward and backward algorithms [10,19], as well as the proposals for new algorithms [11,20]. Some of them are intensively studied and practical result will be presented in the future publications.

4   Experiments and Discussion

The KBExplorator system and KBExpertLib are actually extensively evaluated on the real-world knowledge bases, consisted of 4438 rules. This rules base was duplicated (with a random rules modification) to obtain a larger base, counting 22190. It was not possible to obtain any larger rule base for experiments. This work presents only selected part of the experiments on KBExplorator system and KBExpertLib library.

4.1   KBExplorator—Selected Experimental Results

The main goal of the experiments concerning the KBExplorator system was the evaluation of rules retrieving effectiveness for the forward and the backward inference performed by the server side. The KBExplorator system's rule bases are physically stored in the relational database, each rule base's data are divided into the several entities. It should be noted that the size of the entities will grow when the users will add new knowledge bases and will create the new rules for particular knowledge base. Each rules searching request requires the execution of a SQL queries, which retrieve data from proper entities. In order to evaluate queries' execution time, a server side script was made, which sends the SQL query to the database engine and measures the response time using the proper time measurement functions. To rate the effectiveness of information retrieval, some descriptive statistics as the minimum, maximum or average as well as the median and standard deviation values (of the time duration of obtaining every rule from the knowledge base) were calculated.

The summarized results from the first experiment are presented in the Table 1. We can observe that there is almost no noticeable time difference between the two analysed rules sets, regardless of the rules set size or the order of rules being retrieved. Surprisingly, the average and median time for larger base proved to be lower than for small base. The relatively short rule retrieval time is caused by the usage of several column indexes, the rule order has a noticeable impact on maximal retrieval time.

Table 1. Rules retrieval time (in seconds) based on their identifiers

Case	Rules count	Minimum	Maximum	Mean	Median	σ
1	4438	0,00016	0,00091	0,00028	0,00024	0,00010
2	22190	0,00016	0,00335	0,00025	0,00020	0,00010

The experiment's results presented in the Table 2 summarize the second experiment, the main goal was to analyse the rules retrieval time based on their conclusions, a list of rules' conclusions was obtained from the database. Therefore, the results presented in Table 2 concern the time from gathering the relevant (to a given conclusion) rule identifiers to obtaining all information for those rules.

Table 2. Rules retrieval time (in seconds) based on their conclusions

Case	Rules count	Minimum	Maximum	Mean	Median	σ
1	4438	0,00034	0,03927	0,00098	0,00061	0,00158
2	22190	0,00136	0,20918	0,00489	0,00270	0,00890

It is worth to notice that results for rules sets counting maximum 22190 rules were acceptable from the users point of view, especially when backward inference was considered—rules retrieval time was negligible. Practical verification of the proposed relational model confirmed findings from previous works which were focused on inference control strategies. The selection of rules should be performed with time efficiency adequate to the needs and requirements of the algorithms implemented in such expert systems.

4.2 KBExpertLib—Selected Experimental Results

The main goal of the experiments concerning the KBExpertLib library was the evaluation of effectiveness of the forward and the backward inference performed by the client side, running within the desktop applications. The experimental research is focused on the time efficiency of the basic KBExpertLib operations. The estimation of the memory occupation of the library data structures is also presented. The experiment was performed on the knowledge bases described in the previous section, and two additional real-world knowledge bases were considered. Both bases are dedicated for supporting evaluation of sales representatives, first base consisted of 416 rules, second consisted of 1199 rules. In contrast to the previously described experiments, local copies of knowledge bases was used, saved as text in the XML files.

The main goal of first experiment was the time efficiency evaluation of loading the rules from XML files into the internal, object oriented data structures defined in the package kbcore as well as the estimation of memory occupation for such data structures. Second experiment was focused on the time efficiency of forward inference, performed on the rules loaded in to the kbcore objects allocated in the memory. Two forward inference algorithms were examined. All experiments were run on a typical PC desktop computer: Intel i5 2.5 GHz processor, 16 GB of RAM, classical mechanical hard disk, 64-bit Windows 10 operating system. Each experiment was repeated at least ten times, results were averaged. The Table 3 presents final results of the experiments.

The results of forward inference are satisfactory. Even for the largest knowledge base, the average inference time was less than one second. It may be considered as acceptable for most typical applications of knowledge based systems.

Table 3. The summarized KBExpertLib experiment results

Case	Rules count	Base loading time [s]	Memory occupation [B]	Depth-first forward inference time [ms]	Breadth-first forward inference time [ms]
1	416	0,312	95964	81,23	41,53
2	1119	1,487	285316	138,67	78,21
3	4438	27,39	1197148	186,69	372,36
4	22190	670,8	4646348	934,35	301,46

The memory usage for data structures which hold the rules seems to be reasonable. For 22190 rules the data structures occupy less than 5 MB of memory. Unfortunately, the XML parsing time efficiency is disappointing. Only for small rules sets parsing time is acceptable, for 22190 rules parsing time exceeded 11 min—other format for local rule bases representation should be considered.

5   Conclusions

In the context of medical applications of KBExplorator system and software library KBExpertLib, it is possible to point out two main scopes of their utilization. The KBExplorator system may be considered as tool for building medical knowledge bases. For this reason, this system is designed for knowledge engineers and domain experts, who are are responsible for creating the knowledge base for particular problem. The software library KBExpertLib may be considered as a tool for programmers to develop software which utilize the knowledge bases designed with the use of KBExplorator. The main scope of KBExpertLib usage is client side desktop application, implemented in Java. Programming with KBExpertLib is easy, library provides simple classes which encapsulate all knowledge oriented actions.

The package KBExpertLib may be used on the server side, but the main scope of its application are client side desktop application. The experiments on the forward inference allow to accept the efficiency of the current algorithms' implementations. The implementation of RETE version of forward inference algorithm and the comparison with JESS implementation is considered in the near future. However, even for the largest knowledge base, the average inference time was less than one second. Also, the rules loaded into memory consume a little amount of the memory.

This is promising regarding the use of the KBExpertLib on mobile devices. Mobile devices are especially helpful in regard to chronic health diseases because it frees physicians from routine office visits while still providing data on patient conditions. This helps doctors focus office care on those requiring more detailed medical assistance. Embedded decision support subsystem remote monitoring devices can be considered as an interesting tool for early warning and "first aid"' until the patients need more detailed care. The implementation of the decision support systems connected with the chronic diseases remote monitoring seems to be a very important and interesting domain of applications.

Acknowledgments. This work is a part of the project "Exploration of rule knowledge bases" founded by the Polish National Science Centre (NCN: 2011/03/D/ST6/03027).

References

1. Shortliffe, E.: Computer-Based Medical Consultations: MYCIN, vol. 2. Elsevier, Amsterdam (2012)
2. Shortliffe, E.H.: Medical expert systems-knowledge tools for physicians. West. J. Med. **145**(6), 830 (1986)
3. Miller, R.A.: Medical diagnostic decision support systems-past, present, and future: a threaded bibliography and brief commentary. J. Am. Med. Inform. Assoc. **1**(1), 8 (1994)
4. Liao, S.H.: Expert system methodologies and applications a decade review from 1995 to 2004. Expert Syst. Appl. **28**(1), 93–103 (2005)
5. Hudson, D.L.: Medical expert systems. In: Wiley Encyclopedia of Biomedical Engineering (2006)
6. Nowak-Brzezińska, A., Simiński, R.: Knowledge mining approach for optimization of inference processes in medical rule knowledge bases. J. Med. Informatics Technol. **20**, 19–27 (2012)
7. Nowak-Brzezińska, A., Rybotycki, T.: Visualisation of medical rule-based knowledge bases. J. Med. Informatics Technol. **24**, 91–98 (2015)
8. Plinta, R., Sobiecka, J., Drosdzol-Cop, A., Nowak-Brzezińska, A., Skrzypulec-Plinta, V.: Sexuality of disabled athletes depending on the form of locomotion. J. Hum. Kinet. **48**(1), 79–86 (2015)
9. Berner, E.S.: Clinical Decision Support Systems. Springer, New York (2007)
10. Simiński, R.: Multivariate approach to modularization of the rule knowledge bases. In: Gruca, A., Brachman, A., Kozielski, S., Czachórski, T. (eds.) ICMMI 2015, pp. 473–483. Springer International Publishing, Cham (2016)
11. Simiński, R., Nowak-Brzezińska, A.: Goal-driven inference for web knowledge based system. In: Wilimowska, Z., Borzemski, L., Grzech, V., Świątek, J. (eds.) Information Systems Architecture and Technology: Proceedings of 36th International Conference on Information Systems Architecture and Technology-ISAT 2015-Part IV, vol. 432, pp. 99–109. Springer, Switzerland (2016)
12. Acquired Intelligence: Acquired Intelligence Home Page. http://aiinc.ca. Accessed Mar 2016
13. CLIPS: CLIPS Home Page. http://www.clipsrules.net. Accessed Mar 2016
14. JESS: JESS Information. http://herzberg.ca.sandia.gov. Accessed Mar 2016
15. Canadas, J., Palma, J., Túnez, S.: A tool for mdd of rule-based web applications based on owl and swrl. In: Knowledge Engineering and Software Engineering (KESE6), p. 1 (2010)
16. eXpertise2Go: eXpertise2Go Home Page. http://expertise2go.com. Accessed March 2016
17. DROOLS: DROOLS Home Page. www.drools.org. Accessed Mar 2016
18. Simiński, R., Michalik, K.: The hybrid architecture of the ai software package sphinx. In: Proceedings of International Conference: Colloquia in Artificial Intelligence, CAI 1998 (1998)
19. Nowak-Brzezinska, A., Siminski, R.: New inference algorithms based on rules partition. In: Proceedings of the 23th International Workshop on Concurrency, Specification and Programming, Chemnitz, Germany, 29 September–1 October 2014, pp. 164–175 (2014)
20. Siminski, R., Wakulicz-Deja, A.: Rough sets inspired extension of forward inference algorithm. In: Proceedings of the 24th International Workshop on Concurrency, Specification and Programming, Rzeszow, Poland, 28–30 September 2015, vol. 2, pp. 161–172 (2015)

Building Medical Guideline for Intensive Insulin Therapy of Children with T1D at Onset

Rafał Deja[✉]

Department of Computer Science, The University of Dabrowa Gornicza,
Cieplaka 1c, Dabrowa Gornicza, Poland
rdeja@wsb.edu.pl

Abstract. Establishing the therapy of the juvenile diabetic patient at onset is a challenge. We propose a new way for building medical guideline to support physicians. The course of the treatment is mainly described by the sequence of insulin dosage. However the daily insulin dose is prescribed based on the glycemia levels from the previous day/days and on the other hand is verified by the glycemia levels in the following day. To generate medical guidelines we discover sequential patterns from the treatment sequences and supplement them with patterns of medical examinations and interventions. Before mining sequential patterns we group the sequences with respect to patient's medical data influencing the course of the disease by applying k-means clustering.

1 Introduction

A medical guideline is usually a document with the aim of guiding decisions and criteria regarding diagnosis, management, and treatment in specific areas of healthcare [5,11]. There are many researches concerning the construction of medical guideline in many fields of medicine. For example, medical guideline are delivered by the World Health Organization (WHO). The other examples of medical guideline for the disease of diabetes mellitus are given in [1,3].

In this paper we address the problem of generating medical guideline for intensive insulin therapy of children with Type 1 Diabetes. We propose to build medical guideline in the form of sequential patterns using the repository of historical medical records.

The idea of mining diabetes data using sequential patterns approach was proposed in [10]. The discovered patterns were applied to construct a decision tree, expressing also the possible flow of medical events. The final interpretation of the tree and the possible therapy was left to physicians. The limitation of that approach is the lack of reliable evaluation of patterns with respect to individual patients [10]. There are also other methods proposed with similar purpose, the review of them is available in [7,12]. However, none of the available guideline is able to accurately represent all features of diabetes mellitus disease.

In our previous works we have examined the application of the template-based patterns for the decision support and prediction in the therapeutic procedure [6]. The discovered template-based patterns were used to recommend the

© Springer International Publishing Switzerland 2016
N.T. Nguyen et al. (Eds.): ICCCI 2016, Part II, LNAI 9876, pp. 504–514, 2016.
DOI: 10.1007/978-3-319-45246-3_48

insulin dosage on the basis of the known value of blood glucose level. Afterward we extended the patterns into the whole initial period of therapy, we have introduced the notion of differential sequences [4]. This approach allowed to discover the pattern of therapy and recommend most suitable one for a new case.

In this study we advance our previous methodology. The proposed approach consists of the following steps:

- clustering of patient's data to obtain groups of patients with the similar medical characteristics,
- discovery of sequential patterns within the sequences of the treatment course separately for every of the previously obtained clusters,

The outcome of our approach is a medical guideline, specific for every group of diabetic patients. In this way, the physician receives information about the possible courses of treatment and the way it is changing.

1.1   Medical Background

At the first day of treatment the physician determines with a patient daily energy requirements (meals sizes and number) and considering other factors prescribes the daily insulin dose. The main factors influencing insulin dosage despite the amount of meals are: the patient weight, age, sex, state at admission (presence of ketoacidosis), additional infection, psycho-emotional condition etc. [1,3]. Later on, the proper concentration of glucose level is adaptively obtained mainly with modifying the dosage of insulin. The overall daily insulin dosage is usually divided into two parts; first part is a long-lasting insulin so called basal and the second part so called premeal insulin which is short acting one. It is said that the proper ratio is 30 % to 70 % respectively [1]. The basal insulin is usually served in the evening (one injection a day) and ensures the correct level of glucose in the night and before meals.

Table 1. The attributes used in the study (static data)

Attribute	Medical meaning
Age	The patient age at onset
Sex	Male (1) or female (0)
Weight	The weight at onset
C-peptyde	Insulin secretion
CRP	Certificate of infection, 1 or 0
PH	ACID based balance

Each day of hospitalization the treatment is verified and is changed up to the patient stable state and finally the patient is released. The strict medical procedure cannot be defined mainly because of numerous factors that can influence the glycemia level [1]. Thus to the large degree it is based on physician's experience [3]. The factors (attributes) considered in the study are presented in Table 1. The other data considered in the study concern the results of glycemia measurements (blood glucose levels) and insulin doses - here are called treatment data.

2   Building Medical Guideline

2.1   Data Preparation

In the first stage of our approach, we cluster static patients' data presented in Table 1 to obtain patient cohorts with similar medical characteristics. Numeric values of those features are normalized to the $[0, 1]$ interval. Min-max normalization [8] is used assuming that the minimum and maximum values of every feature are retrieved from data. Afterward, the data gathered from all patients are clustered using k-means algorithm.

On the other hand the data concerning glycemia and insulin dose were generalized using the physician's indication. The level of fasting glucose level was interpreted in accordance with accepted medical standards [1, 3]. The interpretation of the blood glucose level and the corresponding discrete values is presented in Table 2.

Table 2. Blood glucose level and its interpretation

BGL [mg/dl]	Interpretation	Discrete value
<70	hypoglycemia	1
[70, 90]	normoglycemia	2
(90, 200]	mild-hyperglycemia	3
>200	hyperglycemia	4

The insulin dose was standardized by applying the patient weight: the given insulin ratio is referred to every 10 kg of the patient's weight and rounded. For example the patient 566 was admitted 10 units of insulin. Since the patient weight was 30 kg then insulin dosage is rounded to 3 unit (referred to 10 kg of weight).

2.2   Clustering

As previously explained, the approach presented in this paper consists of two steps: the clustering based of static data and mining sequential patterns. The simple k-means clustering algorithm proposed by McQueen [9] has been applied. The data used for clustering are described in Table 1. The number of clusters (the parameter of the clustering algorithm) has been deduced experimentally (see Sect. 3). Choosing the cluster number we considered the following objective: from one side having big number of small clusters can cover each patient specificity, but from the other perspective the mined patterns from within such clusters are of small generality.

2.3   Discovering Sequential Patterns

For discovering medical guideline we apply sequential patterns introduced by Agraval in [2]. The Apriori algorithm is also the the bases for our approach. For the purpose of our research we introduce the proper sequences notion.

The sequence for each patient is constructed from the set E of items that are of two types: the glycemia levels denoted with g and the basal insulin doses d. The set of items for a given day dt of therapy consists of five items $E_{dt} = \{g1, g2, g3, g4, d\}$ e.i. four glycemia measurements and one basal insulin dose. Thus the daily sequence a_{dt} is the sequence of items from E_{dt} ordered according to the event time. (Nocturnal glycemia considered in the study is usually measured at 0 A.M., 3 A.M., 5 A.M. and 7 A.M. The basal insulin dose is delivered at 10 P.M. within the day.) Since we consider the whole therapy the patient j sequence consists of m daily sequences $s^j = \langle a_1^j, a_2^j, ..., a_m^j \rangle$, where m is the number of days of hospitalization.

The sequence $a = \langle a_1, a_2, ..., a_n \rangle$ is properly contained in the other sequence $b = \langle b_1, b_2, ..., b_m \rangle$, i.e. $a \subseteq b$ if there exist integer numbers $i_1, i_2, ..., i_n$ such that $a_1 = b_{i_1}, a_2 = b_{i_2}, ..., a_n = b_{i_n}$. The sequence a is called a subsequence of b, and the sequence b is called a supersequence of a.

The proper daily sequence is any subsequence that contains the insulin dosage $ap_{dt} = \langle e_1, .., e_n \in E_{dt} : \exists e_n = d \rangle$. Similarly the proper patient sequence is any subsequence consisting of proper daily sequences e.i. $sp^j = \langle ap_1^j, ..., ap_m^j \rangle$, where ap_k^j is the proper daily sequence of patient j in day k.

Suppose the collection of sequences $S = \{s^1, s^2, ..., s^m\}$ is available, where s^j denotes patient j sequence (consisting of s_{dt}^j daily subsequences). The support (1) of sequence a is a fraction of sequences in S that contain a.

$$sup(a) = \frac{card(s^j \in S | a \subseteq s^j)}{card(S)} \tag{1}$$

The objective is to determine such proper sequences $a \subseteq s^j$ that are sequentially contained in any sequences from S, i.e., with the support $sup(a) \geq sup_{min}$, where sup_{min} is a threshold given by experts.

Let $C = \{c_1..c_n\}$ denotes a set of clusters obtained using k-means algorithm. According to the previously introduced notation, $S_{cluster} = \{s^1, s^2, ..., s^m\}$ denotes the set of patients sequences that belong to cluster $cluster$.

The goal is to mind sequential patterns $a \subseteq s^j$ in any sequences from $S_{cluster}$ with the support $sup(a) \geq sup_{min}$, where sup_{min} is a threshold given by experts. The discovered sequential patterns constituting the medical guideline for the

---

Algorithm 1. Building medical guidelines

---

    input : $P = \{p_1, p_2, ..., p_n\}$ the set of vectors describing n patients by static
            features,
            $S = \{s^1, s^2, ..., s^n\}$ the set of patients' sequences,
            c - number of clusters.
    output: $A = \{a_1, a_2, ..., a_m\}$ the set of patterns
    function GenerateGuideline(P, S, c) {
    $A \leftarrow \emptyset$; /* set of sequential patterns */
    $C(P) = \text{Kmeans}(P, c)$; /* clustering static data */
    foreach $cluster \in C(P)$ **do**
        | /* discovery of sequential patterns */
        | $A_{cluster} = \text{SequentialPatterns}(S_{cluster})$;
        | /* accumulation of sequential patterns */
        | $A \leftarrow A + A_{cluster}$;
    end
    return A;
    }

---

physician. The algorithm for building the medical guideline is described with Algorithm 1.

The set of vectors describing patients (static data), the set of sequences representing the treatment path and the number of clusters are the algorithm inputs. The static data are used for patients clustering by applying k-means algorithm. For each cluster we mine sequential patterns from the set of corresponding sequences $S_{cluster}$. The final outcome of Algorithm 1 is the set A of all discovered sequential patterns returned by the *SequentialPatterns* function. The details of this function are given as Algorithm 2.

We start with retrieving the building blocks of sequences i.e. retrieving the proper subsequences for each day of the patient therapy. When the support of that building block is greater than the required level $psup_{min}$ the sequence is stored in the collection ap_k^i. Then we create sequences by joining the building blocks from different days enhancing their width w (the number of building blocks used) up to the number of therapy days dt. We are verifying the support of the sequence and only these with accepted support survive to the next iteration, and they are stored in the patterns collection A.

The algorithm time complexity depends on the number of unique elements (proper subsequences) N_1, the number of therapy days dt and the minimal support. This threshold restricts the number of sequences that survive to the next iteration. Let denote with N_i the number of elements in iteration i. Since the complexity for generating set of size m is $O(N^m)$ therefore, the total time complexity would be calculated as $O(N_1^2 + N_2^3 + .. + N_{dt-1}^{dt})$.

Algorithm 2. Discovery of sequential patterns

input : $S = \{s^1, s^2, ..., s^n\}$ the set of patients' sequences,
$\qquad$ $psup_{min}$ - the acceptable support level
output: $A = \{a_1, a_2, ..., a_m\}$ the set of patterns
$A \leftarrow \emptyset$;
foreach $s^i \in S$ **do**
$\quad$ Let $ap_k^i[]$ denotes the collection of proper subsequences for day k
$\quad$ Let $numDt$ denotes number of days of the therapy of patient i
$\quad$ **for** $k{=}1$ to $numDt$ **do**
$\qquad$ $ap_k^i[] \leftarrow$ find all proper daily subsequences for day k
$\qquad$ **foreach** *subsubsequence in the collection* $ap_k^i[]$ **do**
$\qquad\quad$ calculate the support level $ap_k^i[]$
$\qquad\quad$ **if** $Support(ap_k^i[j]) > psup_{min}$ **then**
$\qquad\qquad$ | store its support level
$\qquad\quad$ **end**
$\qquad\quad$ **else**
$\qquad\qquad$ | remove the subsequence from collection
$\qquad\quad$ **end**
$\qquad$ **end**
$\quad$ **end**
$\quad$ $a^i(1) \leftarrow ap_k^i$;
$\quad$ **for** $w{=}2$ to $numDt$ **do**
$\qquad$ $a^i(w) \leftarrow$ build sequneces using w subsequences from $a^i(w-1)$
$\qquad$ **foreach** *sequence in the collection* **do**
$\qquad\quad$ calculate the support level $a^i(w)[]$;
$\qquad\quad$ **if** $Support(a^i(w)[j]) > psup_{min}$ **then**
$\qquad\qquad$ | $A \leftarrow A + a^i(w)[j]$;
$\qquad\quad$ **end**
$\qquad\quad$ **else**
$\qquad\qquad$ | remove the subsequence from collection
$\qquad\quad$ **end**
$\qquad$ **end**
$\qquad$ **if** *collection* $a^i(w)$ *is empty* **then**
$\qquad\quad$ | break ;
$\qquad$ **end**
$\quad$ **end**
end

2.4 Application of the Guideline

The sequential patterns are applied in medical practice in the following way:

1. The patient newly admitted to the hospital is assigned to one of the discovered clusters.
2. The sequential patterns for a given cluster are presented to physicians as the medical guideline dedicated to the considered patient (that is the possible course of the therapy).

3. When the patient's therapy is established the possible course can be limited. Only the patterns that are supported by the already established initial part of the therapy serve as a guideline.

3   Experiments

For the validation and evaluation of the proposed approach we collected data of 102 children with onset of diabetes type 1 treated in Deptartment of Diabetology at the Silesian Medical University in Katowice, Poland. For each child the hospital treatment course (several days) from the medical history has been gathered and structured to form a database table.

3.1   Clustering

As previously explained, the number of clusters has been selected experimentally considering two factors: the average support of the mined sequential patterns through all the clusters, and the number of objects in the clusters.

The onset (static) data gathered from all patients has been clustered using k-means clustering algorithm with different number of clusters (from 2 to 8). Before clustering the data were normalized as described in Sect. 2. Then we generated the sequential patterns as descried in Experiment 1 and calculated the best support level for each cluster. Finally the partition is chosen with the highest average of the support level and the minimum number of objects greater than 5. The results of the experiments are presented in Table 3. According to results the best partitioning is obtained when dividing into 6 clusters - we used this partitioning in all further experiments. When dividing into 6 clusters the distribution of cases with respect to the clusters varies from 5.8 % for the cluster 5 to 45.1 % for the cluster 3.

Table 3. Experiments with different clustering

No of clusters	2	3	4	5	6	7	8
Average support	0.27	0.25	0.28	0.30	0.35	0.34	0.35
Min. no of objects	23	23	6	6	6	6	5

3.2   Experiment 1

In the first experiment the sequences are created only from basal insulin doses taken from the whole period of the therapy. The patterns not shorter then 3 items with the highest support for each cluster are presented in Table 4.

As expected the patients' treatment differs between the clusters. In cluster 1 the patients are usually admitting the insulin dose around 3 units per 10 kg of patient weight and in cluster 2 around 4 units. In clusters 3 and 4 the doses are changing from 3 units to 2 during the treatment. Similarly in cluster 6 the doses

Table 4. Treatment sequential patterns

Cluster	Patterns	Supp [%]	Cluster	Patterns	Supp [%]
1	333	0.62	4	222	0.29
	332	0.38		332	0.24
	333333	0.25		211	0.24
2	4444	0.3	5	111	0.33
3	333	0.26		344	0.33
	221	0.24	6	433	0.33
	222	0.24		444	0.3

are changing from 4 to 3 units per 10 kg. In the cluster 5 we have the patients that do not fit to the other clusters: either the patients doses stay at the level of 1 unit per 10 kg of weight or they increasing during therapy from 3 to 4. It is interesting to notice that the insulin doses are usually decreasing during the therapy.

3.3   Experiment 2

In the experiment described above we considered the course of treatment described with the insulin doses. In the current experiment we are trying to find out how the glycemia level is influencing the therapy path. Thus the nocturnal glycemia measurements are generalized, converted into the sequence and mined as described in Sect. 2. The mined sequential pattern for each cluster is presented in Table 5

The set of longest patterns with support greater or equal to 0.06 are presented in Table 5. It is obvious that the support of such patterns is lower than in the previous experiment. Each pattern consists of the sequence of glycemia levels and insulin doses marked with d. The glycemia levels are the discrete values of blood glucose generalized as described in Sect. 2.

Table 5. Therapy patterns

Cluster	Patterns	Supp [%]	Cluster	Patterns	Supp [%]
1	3 d3 2333 d3	0.09	4	2 d2 222 d2	0.09
	2 d3 2 d3	0.09		2 d2 2222 d2	0.09
2	223 d4 2222 d4	0.08	5	3 d3 33 d4 2222 d4	0.09
3	2 d2 2 d2	0.08	6	2 d4 2 d4	0.09
	222 d2 2 d2	0.06			
	22 d2 222 d2	0.06			

Unfortunately the mined patterns with the given minimal support level are short. However we can observed the treatment path with the example of the pattern from cluster 5. The patient nocturnal glycemia probably was vering but initially was recognized in one of measurements as above normal (3 = mild-hyperglycemia). In the next day the mild-hyperglycemia remained in two of measurements, thus physician decided to increase the basal insulin dose from 3 into 4 units per 10 kg of patient weight. In the next day we could observe the normal glycemia level in all measurements thus the dose remains the same for the next day.

The mined patterns presented above afirm the patterns from Experiment 1. The treatment path can be analyzed dipper and the process of taking therapeutic decision is discovered.

3.4 Validation

The results of experiments have been validated. We divided randomly the set of patients into the training set and the test set respectively (20 % of objects). The training set was subjected of k-means clustering grouping objects into 6 clusters. The objects from the test sets were assigned to the clusters using lazy classifier (object is assigned to the cluster with the shortest Euclidean distance to the center of the cluster).

Using the training set the sequential patterns were mined as described in Experiment 1 with the given minimal support level. For each patient from the test set we build the sequence describing the patient therapy. We are then calculating whether the patient sequence is supported by the mined patterns (i.e. whether the guideline fit to the patient) for each cluster. The evaluation coefficient Ec can be calculated as the number of supported patients (with medical guideline) divided by the number of patients from the test set for a given cluster. The validation results for each cluster are presented in Table 6.

Table 6. Sequential patterns mined from training set

Cluster	Pattern	Support	Cluster	Pattern	Support
1	333	0.71	4	344	0.33
	3333	0.43		222	0.33
	333333	0.29	5	33	0.5
2	332	0.27		43	0.5
	211	0.27	6	3333	0.29
3	221	0.32			
	222	0.25			
	333	0.25			

In Table 6 we presented the patterns with the highest support for each cluster when considering the sequences of the therapy with basal insulin. The average

Table 7. Test set evaluation

Cluster no.:	1	2	3	4	5	6
Number of test cases	1	5	11	0	3	0
Average Ec	1	0.25	0.82	-	1	-

evaluation coefficients Ec for each cluster are collected in Table 7. Thus 77 % of new cases are supported with the patterns mined from the training set.

4   Conclusions

In this paper we presented the way the medical guideline has been build. The idea was to support the physician in determining the therapy for children with type 1 diabetes. First the new patient is classified into one of the cluster based on his/her clinical data like weight, age, the level of insulin secretion etc. Based on historical data we have mined the sequential patterns from the treatment path of the similar patients i.e. belonging to the same cluster. The patterns consist not only the insulin doses but also the nocturnal glycemia measurements. Assuming similar cases are treated in a similar way the mined sequential patterns represent the most probable treatment path for a new patient. After obtaining some initial results from the new patient treatment (eg. after one day) the sequential patterns are restricted to these supported by the patient sequence. We have found the sequential patterns concept very useful in supporting the physician with the probable treatment path for a new patient.

References

1. ADA: American diabetes association. Standards of Medical Care in Diabetes-2012 Diabetes Care **35**, pp. 11–63 (2012). doi:10.2337/dc12-s011
2. Agrawal, R., Srikant, R.: Mining sequential patterns. In: Yu, P.S., Chen, A.L.P. (eds.) ICDE, pp. 3–14. IEEE Computer Society (1995)
3. Couper, J., Donaghue, K.: ISPAD clinical practice consensus guidelines. phases of diabetes in children and adolescents. Pediatr. Diabetes **12**, 13–16 (2009)
4. Deja, R., Froelich, W., Deja, G.: Differential sequential patterns supporting insulin therapy of new-onset type 1 diabetes. Biomed. Eng. Online **14**(1), 13 (2015)
5. Field, M., Lohr, K.: Guidelines for Clinical Practice: From Development to Use. Institute of Medicine, National Academy Press, Washington, D.C (1992)
6. Froelich, W., Deja, R., Deja, G.: Mining therapeutic patterns from clinical data for juvenile diabetes. Fundamenta Informaticae **127**(1), 513–528 (2013)
7. Gani, A., Gribok, A.V., Lu, Y., Ward, W.K., Vigersky, R.A., Reifman, J.: Universal glucose models for predicting subcutaneous glucose concentration in humans. IEEE Trans. Inf. Technol. Biomed. **14**(1), 157–165 (2010)
8. García, S., Luengo, J., Herrera, F.: Data Preprocessing in Data Mining. Intelligent Systems Reference Library, vol. 72. Springer, Switzerland (2015)

9. MacQueen, J., et al.: Some methods for classification and analysis of multivariate observations. In: Proceedings of the fifth Berkeley Symposium on Mathematical Statistics and Probability, vol. 1, pp. 281–297, Oakland (1967)
10. Rahaman, S., Shashi, M.: Sequential mining equips e-health with knowledge for managing diabetes. Int. J. Inf. Process. Manage. **2**(3) (2011)
11. Rosique, R.: Care Pathways: The basics. Asian Hospital and Healthcare Management (2008)
12. Stahl, F., Johansson, R.: Diabetes mellitus modeling and short-term prediction based on blood glucose measurements. Math. Biosci. **217**(2), 101–117 (2009)

A New Personal Verification Technique Using Finger-Knuckle Imaging

Rafal Doroz[1]($\boxtimes$), Krzysztof Wrobel[1], Piotr Porwik[1], Hossein Safaverdi[1],
Michal Senejko[1], Janusz Jezewski[2], Pawel Popielski[1], Slawomir Wilczynski[3],
Robert Koprowski[1], and Zygmunt Wrobel[1]

[1] Institute of Computer Science, University of Silesia, Sosnowiec, Poland
{rafal.doroz,krzysztof.wrobel,piotr.porwik,hossein.safaverdi,
pawel.popielski,robert.koprowski,zygmunt.wrobel}@us.edu.pl,
michal759@gmail.com
[2] Institute of Medical Technology and Equipment (ITAM), Zabrze, Poland
janusz.jezewski@itam.zabrze.pl
[3] Department of Basic Biomedical Science,
Medical University of Silesia, Sosnowiec, Poland
http://zsk.tech.us.edu.pl, http://biometrics.us.edu.pl

Abstract. This paper focuses on automatic pattern-based extracting of biometric features where finger-knuckle images are analyzed. Knuckle images are captured by digital camera, and then by the image processing techniques the most relevant features (patterns) are discovered and extracted. Knuckle-based images were filtered by the Hessian filters. It enabled to enhance image regions with image ridges. In the next stage similarity of images were computed by the Normalized Cross-Correlation algorithm. Ultimately, similarities were classified by the k-NN classifier. The discovered features belong to so-called human physical features, which involves innate human characteristics. Physical biometric features can often be gathered with specialized hardware, needing only software for analysis. That capacity makes such biometrics simpler.

We conducted a variety of experiments and showed advantages and disadvantages of the approaches with promising results.

Keywords: Biometrics · Finger-Knuckle imaging · Cross-correlation · k-NN

1 Introduction

It is easily to show that utilization of biometric systems help in limitation of access to different resources [1,7,8]. These systems can work in either verification or identification mode. Security of systems is still problematic; therefore single biometric modalities are insufficient in professional applications. For this reason new modalities are constantly sought, or some well-known biometrics are improved and modified. It leads to new biometric multi-modalities, which have demonstrated in many researches [4–6,17,18]. The overall performance of

© Springer International Publishing Switzerland 2016
N.T. Nguyen et al. (Eds.): ICCCI 2016, Part II, LNAI 9876, pp. 515–524, 2016.
DOI: 10.1007/978-3-319-45246-3_49

these recognition systems significantly improves accuracy of biometric application compared to conventional single-based biometric modalities.

Some biometric techniques like fingerprint, iris, face or signature recognition [4,8,18] are well known and their advantages as well as disadvantages are discovered and announced in the literature. In this paper human finger-knuckle based analysis will be described. This biometric feature is relatively new and poorly understood, so it can be used in future as a primary technique as well as multi-modal biometric technique.

The finger-knuckle analysis has been proposed in many papers. In [1] knuckle based features were analyzed by both Hidden Markov Models (HMM) and Support Vector Machine (SVM) classifiers. In [9] knuckle images were represented by own Author's coding system, where Radon transform, Principal Component Analysis (PCA), Independent Component Analysis (ICA) and Linear Discriminant Analysis (LDA) have been employed. In researches, devoted finger-knuckle analysis, also other well known techniques have been previously proposed - the Gabor filtering [23], surface curvature analysis [22], as well as texture analysis [7] and SIFT method [12].

Knuckle-based analysis, in combination with advanced image processing technique, allows to recognize furrows and ridges on the knuckle skin surface [21]. These features are unique for each person - therefore it can be incorporated into the new set of biometric features.

Image of the finger-knuckles can be captured by photo camera. Details of such image have been presented in Fig. 1

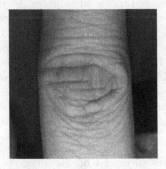

Fig. 1. The finger knuckle image.

It should be noticed that it is contact less acquisition method. Because finger knuckles belong to the physical features [8,21], temporary emotional states of analyzed person do not affect the measurement - which is an important beneficial phenomenon.

2   Proposed Method

In this paper we propose a new, based on knuckles, biometric features extraction method. Selected features will be used in the person verification process. Proposed approach is realized in few stages:

- Acquisition of the knuckle images by means of the specialized device. Images are stored in the database.
- Finger knuckle patterns determination. This task will be conducted on the basis of the Hessian filtering.
- Determination of the similarities between knuckle-based images. It will be done by means of the Normalized Cross Correlation technique [10,19].
- Final verification decision. Decision produces by the k-NN classifier.

The proposed finger-knuckle based verification system is shown in Fig. 2.

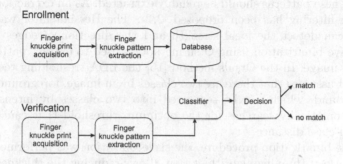

Fig. 2. Block diagram of the finger-knuckle based verification system.

The successive steps of the proposed method are described in detail in the following subsections.

3   Finger-Knuckle Image Acquisition

The acquisition was carried out by means of the specialized device. This device consists of box with limited space where camera and LED-type constant lights

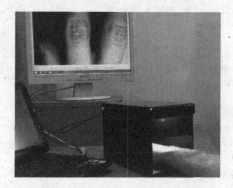

Fig. 3. Finger knuckle measuring station.

are installed. Additionally, the actual image can be observed in the on-line mode on the computer screen. It was presented in Fig. 3.

Only index finger knuckle was fixed and then captured. Exemplary image of the finger knuckle is presented in Fig. 4(a).

4   Finger Knuckle Pattern Extraction

The presence of joints in the finger region forms the flexion on the outer surface of the skin, which creates the dermal patterns consisting of lines, wrinkles, contours, etc. These patterns should be reliably extracted. As an extraction method the Hessian filtering has been proposed [2,3]. The Hessian filter was applied because it can detect the local strength and the direction of edges and lines [14]. Adaptive binarization using Otsu method is used to extract patterns from the Hessian image. In the Otsu's method [15] the LDA tresholding technique is employed. It assumes that there are two classes in an image, foreground (object) and background, which can be separated into two classes by intensity. Otsu method automatically searches for the optimum threshold that can maximize the between-class distance.

After the binarization procedure, the skeletonization was performed. It consists in the use of thinning algorithms that allows reducing the thickness of lines in the image. It realizes by the Pavlidis's thinning algorithm [16]. After thinning the line thickness is equal to one pixel. All these stages are presented in the Fig. 4. Finally, the extracted features are available (Fig. 4(c)).

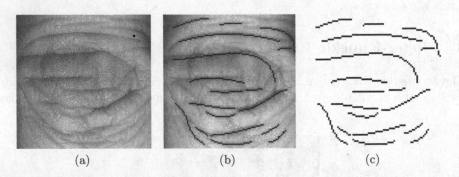

(a)   (b)   (c)

Fig. 4. (a) Captured finger knuckle image, (b) knuckle patterns imposed on the skin, (c) knuckle physical features which will be analyzed.

The result of the acquisition stage is to collect the reference knuckle image patterns from each person. A registered image of a given person has assigned as a unique identifier $ID = A, B, C, ...$ and the set $\Phi^{ID} = \{Im_{ID}^1, Im_{ID}^2, ..., Im_{ID}^n\}$ of reference knuckle images is stored in the biometric database.

5   Verification Process

The purpose of knuckle verification is to classify the input (test) knuckle as a legitimated or not legitimated. The verification process involves two major phases: training and prediction. An unknown person claims identity (let it be $ID = Q$) and provides a knuckle image denoted as Im^* to verify. After that, the training process of the classifier is performed for person Q. The training set consists of two sets: G and F. The set G contains values of the similarity coefficients calculated between the pairs of all reference knuckle images from the user Q.

$$G = \{sim(Im_Q^i, Im_Q^j)\}, \quad i, j = 1, ..., n, \quad Im_Q^i, Im_Q^j \in \Phi^Q, \tag{1}$$

where, Im_Q^i, Im_Q^j represent images of the person Q, and the $sim(Im_Q^i, Im_Q^j)$ represents the similarity between these images.

The set F is constructed on the basis of both legitimated and illegitimated knuckle images. The illegitimated samples are randomly selected from the legitimated users other than Q.

$$F = \{sim(Im_Q^i, Im_E^j)\}, \quad i, j = 1, ..., n, \quad Im_Q^i \in \Phi^Q, Im_E^j \notin \Phi^Q. \tag{2}$$

It should be noted that number of elements in the both G and F sets is equal to $n!/(n-2)!$.

In the next step all elements of the set G are assigned to the class c_1, whereas all elements of the set F belong to the class c_2.

Similarities between images belonging to the G and F sets were computed on the basis of the Normalized Cross-Correlation (NCC) technique [13]. The NCC has been commonly used as a metric to evaluate the degree of similarity (or dissimilarity) between two compared images [13,20].

Similarity $sim(Im_X, Im_Y)$ between images was performed on the basis of non-overlapping square images. Idea of this method is presented in the Fig. 5. The image Im_X is divided into sub-images. Length of the square's side is a parameter and can be changed. These changes impact on the accuracy of the method and will be presented in this paper later. Each sub-image is treated as a template T_k. Next, attempt to find the template T_k inside the image Im_X was taken.

Similarity between a given template T_k and image Im_Y was calculated as follows:

$$ncc(T_k, Im_Y) = \max \left(\frac{\sum_{(i,j) \in T_k} Im_Y(x+i, y+j) \cdot T_k(i,j)}{\sqrt{\sum_{(i,j) \in T_k} Im_Y^2(x+i, y+j)} \cdot \sqrt{\sum_{(i,j) \in T_k} T_k^2(i,j)}} \right),$$

$$\tag{3}$$

$$x = 1, \ldots, w, y = 1, \ldots, h.$$

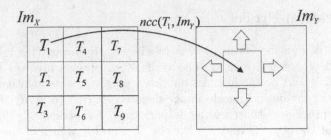

Fig. 5. The template T_1 searching inside of the image Im_Y.

where, Im_Y is the image under examination, of size $w \times h$, the T_k is the template of size $m \times m$. The template size $m \times m$ is smaller than the scene image size $w \times h$. It should be noted that similarity coefficient is not symmetrical, so $sim(Im_X, Im_Y) \neq sim(Im_Y, Im_X)$.

The final similarity between the images Im_X and Im_Y is determined from the formula:

$$sim(Im_X, Im_Y) = mean\{ncc(T_1, Im_Y), ..., ncc(T_k, Im_Y)\}, \qquad (4)$$

where, $T_{i=1,...,k} \subset Im_X$.

The verified knuckle image Im^* is compared with one randomly selected reference knuckle $Im_Q^i \in \Phi^Q$ belongs to the person who the verified person claims to be (on our example Q):

$$d^Q = sim(Im^*, Im_Q^i), \qquad (5)$$

where, Im^* is the knuckle image to be verified and Im_Q^i is a randomly selected original knuckle image of the person Q.

Next, the verified knuckle image Im^* is classified as genuine or forgery by means of the k-NN classifier. Classification of the image Im^* is done by measuring the Euclidean distance between classified objects d^Q and training objects from the sets G and F, respectively. The image Im^* is assigned to the majority

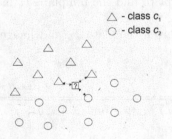

Fig. 6. Classification rule provided by k-NN approach with $k = 3$ case. In this case, the test point "?" would be labeled the class c_1.

class, what determines the k-NN classifier. The number k of the k-NN classifier was established on the basis of the formula [11]:

$$k = \sqrt{card(G) + card(F)}. \tag{6}$$

The k-NN classification principle was presented in Fig. 6.

If the object belongs to the class c_1 it means that verified knuckle image Im^* comes from legitimated person, otherwise this object is illegitimated.

6   Researches

Effectiveness of the proposed method was experimentally verified. During investigations 150 images of finger knuckles were gathered. For each person 5 knuckle images were captured - it means that the database comprises images coming from 30 persons (5 different images per person).

Training and testing was performed using a leave one out methodology. That is, we leave one image as an example, train the classifier using the remaining, and then verify the left-out sample. For each user in the database all experiments were repeated 10 times to provide better statistical accuracy and then the average values of evaluation metrics for all 10 trials are calculated. All tabulated data show mean values and standard deviations for each experiment.

System accuracy was validated using four common factors: FAR (False Acceptance Rate), FRR (False Rejection Rate), Overall Accuracy (ACC) and AER (Average Error Rate):

$$FAR = \frac{\text{number of forgeries accepted}}{\text{number of forgeries tested}} \cdot 100\,\%, \tag{7}$$

$$FRR = \frac{\text{number of genuine rejected}}{\text{number of genuine tested}} \cdot 100\,\%, \tag{8}$$

$$ACC = \frac{\text{number of persons correctly recognized}}{\text{number of persons tested}} \cdot 100\,\%, \tag{9}$$

$$AER = \frac{FAR + FRR}{2} \cdot 100\,\%. \tag{10}$$

The aim of the first experiment was to determine the effectiveness of the proposed verification method. For every user, the learning sets G and F have been formed. These collections are built on the basis of the four image-patterns come from a given person. Remaining images were used in the testing procedure. Efficiency of the proposed approach was checked for different values of the parameter T. This parameter was introduced in the previous section. The results obtained are shown in Table 1.

In the next experiment it has been checked how the training set size, containing knuckle images, affects the accuracy of the classifier. Taking into account the size of the databases, the amount of knuckles was changing between two to three. The results obtained are shown in Table 2.

Table 1. Knuckle verification results for different pattern size, ± standard deviation.

Length of the square side in the template T [px]	FAR [%]	FRR [%]	AER [%]	ACC [%]
10	17.75 ± 1.63	27.61 ± 2.11	22.68 ± 2.90	82.05 ± 5.79
20	10.26 ± 0.43	15.00 ± 1.15	12.63 ± 1.47	89.64 ± 6.85
30	6.04 ± 0.37	10.95 ± 0.76	8.50 ± 0.88	93.85 ± 6.37
40	4.30 ± 0.23	8.19 ± 1.05	6.24 ± 1.09	95.96 ± 7.53
50	4.82 ± .29	12.38 ± 1.31	8.60 ± 0.75	95.02 ± 6.16
60	4.73 ± 0.33	15.47 ± 1.27	10.10 ± 0.89	95.05 ± 8.75
70	6.10 ± .57	18.80 ± 0.98	12.45 ± 1.11	93.65 ± 7.48
80	7.78 ± 0.53	22.61 ± 1.89	15.20 ± 1.85	91.92 ± 7.94
90	12.35 ± 1.07	30.23 ± 2.64	21.29 ± 1.72	87.29 ± 6.43
100	17.07 ± 1.27	36.90 ± 2.79	26.98 ± 2.20	82.53 ± 6.27

Table 2. The size of training datasets influence on the classifier accuracy.

Number of images in both F and G learning datasets	FAR [%]	FRR [%]	AER [%]	ACC [%]
2 and 2	7.56 ± 0.53	29.43 ± 2.85	18.49 ± 1.34	94.38 ± 6.86
6 and 6	4.00 ± 0.09	20.23 ± 1.86	12.11 ± 1.59	95.66 ± 6.12
12 and 12	4.30 ± 0.23	8.19 ± 1.05	6.24 ± 1.09	95.96 ± 7.53

The results of our experiments have been also expressed with detection error trade-off (DET) curves (Fig. 7). In this curves the mutual correlation between FAR and FRR errors is depicted.

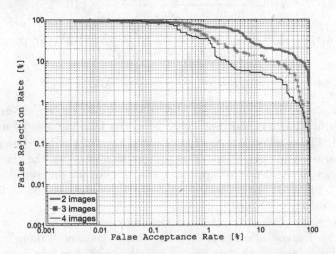

Fig. 7. DET curves showing the verification results.

Table 3. The average time of one object preparation in learning and testing mode.

Classifier mode	Average time [s]
Training	0.182
Testing	0.037

In addition to performance of the proposed method, time complexity factor was computed. The average knuckle image verification time was measured when conducting the experiments. This factor was calculated separately for the learning and testing mode (Table 3).

In experiment, the measurement of time was estimated by a PC class computer equipped with Intel Core i7-3770 processor, 3.40 GHz, 16 GB RAM, and Windows 7 × 64 operating system.

7   Conclusions

In this paper we proposed a new method of verification on the basis of the personal finger-knuckle images. From the experiments carried out follows that finger-knuckle based technique is a promising method for biometric solutions. In contrast to other solutions, knuckle features extraction is based on Hessian filtration and Normalized Cross Correlation. In future the larger database of knuckles will be prepared and we will propose other features extraction methods as well as other classifiers will be incorporated into investigations. It allows to construct the efficient biometric classification method.

References

1. Ferrer, M.A., Travieso, C.M., Alonso, J.B.: Using hand Knuckle texture for biometric identifications. IEEE Aerosp. Electron. Syst. Mag. **21**(6), 23–27 (2006)
2. Iwahori, Y., Hattori, A., Adachi, Y., Bhuyan, M.K., Woodham, R.J., Kasugai, K.: Automatic detection of polyp using Hessian Filter and HOG features. Procedia Comput. Sci. **60**(1), 730–739 (2015)
3. Jin, J., Yang, L., Zhang, X., Ding, M.: Vascular tree segmentation in medical images using Hessian-based multiscale filtering and level set method. Comput. Math. Methods Med. **2013**, 502013 (2013)
4. Kasprowski, P.: The impact of temporal proximity between samples on eye movement biometric identification. In: Saeed, K., Chaki, R., Cortesi, A., Wierzchoń, S. (eds.) CISIM 2013. LNCS, vol. 8104, pp. 77–87. Springer, Heidelberg (2013)
5. Koprowski, R., Teper, S.J., Weglarz, B., Wylegała, E., Krejca, M., Wróbel, Z.: Fully automatic algorithm for the analysis of vessels in the angiographic image of the eye fundus. Biomed. Eng. Online **11** (2012)
6. Koprowski, R., Wilczynski, S., Wrobel, Z., Kasperczyk, S., Blonska-Fajfrowska, B.: Automatic method for the dermatological diagnosis of selected hand skin features in hyperspectral imaging. Biomed. Eng. Online **13** (2014)

7. Kumar, A., Ravikanth, C.: Personal authentication using finger Knuckle surface. IEEE Trans. Inf. Forensics Secur. **4**(1), 98–110 (2009)

8. Kumar, A., Wang, B.: Recovering and matching minutiae patterns from finger Knuckle images. Pattern Recogn. Lett. **68**, 361–367 (2015)

9. Kumar, A., Zhou, Y.: Human identification using Knuckle codes. In: Proceedings BTAS 2009, pp. 98–109 (2009)

10. Lewis, J.P.: Fast normalized cross-correlation. Vis. Interface **10**(1), 120–123 (1995)

11. Li, B., Wang, K., Zhang, D.: On-line signature verification based on PCA (Principal Component Analysis) and MCA (Minor Component Analysis). In: Zhang, D., Jain, A.K. (eds.) ICBA 2004. LNCS(LNAI, LNBI), vol. 3072, pp. 540–546. Springer, Heidelberg (2004)

12. Morales, A., Travieso, C.M., Ferrer, M.A., Alonso, J.B.: Improved finger-Knuckle-print authentication based on orientation enhancement. Electron. Lett. **47**(6), 380–382 (2011)

13. Nakhmani, A., Tannenbaum, A.: A new distance measure based on generalized Image Normalized Cross-Correlation for robust video tracking and image recognition. Pattern Recogn. Lett. **34**(3), 315–321 (2013)

14. Nitsch, J., Klein, J., Miller, D., Sure, U., Hahn, K.H.: Automatic segmentation of the Cerebral Falx and adjacent Gyri in 2D ultrasound images. Bildverarbeitung für die Medizin 2015: Algorithmen - Systeme - Anwendungen, pp. 287–292. Springer, Heidelberg (2015)

15. Otsu, N.: A threshold selection method from gray-level histograms. IEEE Trans. Syst. Man Cybern. **9**(1), 62–66 (1979). http://dx.doi.org/10.1109/tsmc.1979.4310076

16. Pavlidis, T.: A thinning algorithm for discrete binary images. Comput. Graph. Image Process. **13**(2), 142–157 (1980)

17. Porwik, P., Doroz, R.: Self-adaptive biometric classifier working on the reduced dataset. In: Polycarpou, M., de Carvalho, A.C.P.L.F., Pan, J.-S., Woźniak, M., Quintian, H., Corchado, E. (eds.) HAIS 2014. LNCS, vol. 8480, pp. 377–388. Springer, Heidelberg (2014)

18. Porwik, P., Doroz, R., Wrobel, K.: A new signature similarity measure. In: 2009 World Congress on Nature and Biologically Inspired Computing, NABIC 2009 - Proceedings, pp. 1022–1027 (2009)

19. Wei, S.D., Lai, S.H.: Fast template matching based on normalized cross correlation with adaptive multilevel winner update. IEEE Trans. Image Process. **17**(11), 2227–2235 (2008)

20. Di Stefano, L., Mattoccia, S., Tombari, F.: An algorithm for efficient and exhaustive template matching. In: Campilho, A.C., Kamel, M.S. (eds.) ICIAR 2004. LNCS, vol. 3211, pp. 408–415. Springer, Heidelberg (2004)

21. Usha, K., Ezhilarasan, M.: Finger Knuckle biometrics - a review. Comput. Electr. Eng. **45**, 249–259 (2015)

22. Woodard, D.L., Flynn, P.J.: Finger surface as a biometric identifier. Comput. Vis. Image Underst. **100**(3), 357–384 (2005)

23. Xiong, M., Yang, W., Sun, C.: Finger-Knuckle-print recognition using LGBP. In: Liu, D., Zhang, H., Polycarpou, M., Alippi, C., He, H. (eds.) ISNN 2011, Part II. LNCS, vol. 6676, pp. 270–277. Springer, Heidelberg (2011)

Automatic Detection of Fetal Abnormality Using Head and Abdominal Circumference

Vidhi Rawat[1(✉)], Alok Jain[2], Vibhakar Shrimali[3],
and Abhishek Rawat[4]

[1] Department of Biomedical Engineering,
Samrat Ashok Technological Institute, Vidisha, India
Vidhi_pearl@rediffmail.com
[2] Department of Electronics and Instrumentation Engineering,
Samrat Ashok Technological Institute, Vidisha, India
[3] Department of Electronics and Communication Engineering,
G. B. Pant Government Engineering College, Delhi, India
[4] Department of Electrical Engineering, IITRAM, Ahmedabad, India

Abstract. In present scenario women's are suffering from thyroid, diabetes and high blood pressure and therefore early detection and diagnosis of fetal abnormality can save lives and reduce cost of treatment. In this paper we propose an artificial neural network (ANN) based method for the detection of fetal abnormality in 2-D ultrasound images of 14–40 weeks. The accurate values of fetal anatomical structures are found by segmentation techniques and these values transferred to neural model for detection of possible abnormalities from 2D fetal ultrasound images. The ANN model is able to find Intrauterine Growth Retardation (IUGR) and abnormal fetus using head and abdominal circumference.

Keywords: Segmentation · Ultrasound imaging · Artificial neural network · Fetus abnormality · IUGR

1 Introduction

Ultrasound (US) imaging is considered to be one of the safest, non- invasive nature compared to other imaging modalities, such as Computed Tomography (CT) and Magnetic Resonance Imaging (MRI) for an obstetrician and gynecologist [1]. The US image is formed by the echoes received by the transducer. The image obtained might not have the expected anatomical significance and can be distorted or incomplete. It can be affected by signal dropouts, attenuation, missing boundaries, shadows and speckle making it, one of the most challenging modalities in medical science. A great deal of expertise is needed to get early detection of abnormality using this imaging technique. In particular, delay in diagnosis and lacks of clarity of images, are major issues.

As per the records of Consortium on National Consensus for Medical Abortion in India, every year an average of about 11 million abortions take place and around 20,000 women die due to abortion related complications [2]. Accurate fetal biometric parameters measurements are one of the most important factors for high quality obstetrics health care. The fetal biometric parameter is Biparietal Diameter (BPD), Head Circumference (HC),

N.T. Nguyen et al. (Eds.): ICCCI 2016, Part II, LNAI 9876, pp. 525–534, 2016.
DOI: 10.1007/978-3-319-45246-3_50

Abdominal Circumference (AC), Femur Length (FL). Fetal 2D US biometrics have been extensively used to estimate its size and weight, gestational age and identify fetal abnormalities [3]. Among these parameters, AC and HC are best correlated with the fetal growth and has been found to be the first biometric measure to indicate Intrauterine Growth Retardation (IUGR) as reported in [4–6].

Some semi-automatic segmentation methods previously developed in the fetal imaging field for the estimation of biometric parameters [7–11]. These methods were based on the morphological operator, active contour, Hough transform, deformable models, or machine learning approaches. Some author also segmented fetal head [8–14] and fetal abdomen [15–18]. Carneiro et al. [19] used a constrained boosting tree classifier to segment structure of interest and to reproduce the standard biometric measurement. A commercial system has developed and patented that integrated into Siemens software, called Auto OB [20]. In clinical practice that system is used by the radiologist for fetal biometry measurements. But presently, the radiologist indicates the head and abdomen contour for measuring the circumference which is subjective, tedious and time consuming task. In addition, manual contour as shown in Fig. 1, extraction is influenced by the variability of the human observer, which limits its reliability and reproducibility [19, 20]. In this work, an ANN model for early and accurate detection of head and abdomen circumference has been proposed to identify abnormal fetus. The abnormality of the fetus is detected by artificial neural network (ANN) model. The block diagram of proposed approach is shown in Fig. 2.

The paper is organized as follows; Sect. 2 describes the methodology for detection of fetal abnormality. Section 3 describes structure of ANN and Sect. 4 cover important result and discussions. Finally, Sect. 5 concludes the work that is presented within this paper and suggests further work.

2   Methodology for Detection of Fetal Anatomy

The US image is formed by the echoes received by the transducer, the image obtained might not have the expected anatomical significance and can be distorted or incomplete. So first we have to filter and segment these images before use. These images are in DICOM format as received from US. In this problem we are applying median filter, Segmentation of ROI and Feature extraction for the primary treatment of US images which are briefly described below.

2.1   Median Filtering Despeckling Filter

The reduction of speckle noise is an important process to increase the quality of US images. Image variances or speckle is a granular noise that inherently exists in and degrades the quality of the active images. Speckle filtering consists of moving a kernel over each pixel in the image, doing a mathematical computation on the pixel values under the kernel and replacing the central pixel with the calculated value. By applying the filter a smoothing effect is achieved and the speckle becomes less obtrusive [22].

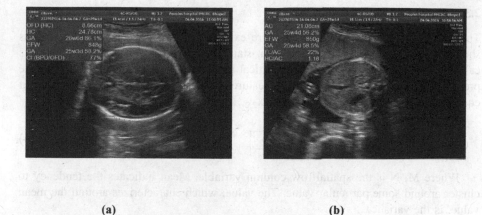

(a)   **(b)**

Fig. 1. (a) Fetal Head Biometric Measurements: Manual Contour (yellow dotted line) for Head Circumference (HC), (b) Fetal abdomen Biometric Measurement: Manual Contour (yellow dotted line) for abdomen Circumference (AC). (Color figure online)

$$F(u, v) = \sum_{x=0}^{M-1} \sum_{y=0}^{N-1} f(x,y) e^{-2\pi\left(\frac{ux}{M} + \frac{vy}{N}\right)} \tag{1}$$

$$g(x,y) = \sum_{x=0}^{M-1} \sum_{y=0}^{N-1} F(u, v) e^{-2\pi\left(\frac{ux}{M} + \frac{vy}{N}\right)} \tag{2}$$

Where $F(u, v)$ is the image function having u and v is the spatial variable and $g(x,y)$ is the filtered image.

2.2 Segmentation of Region of Interest

Speckle free image is passed through the gradient vector flow (GVF) algorithm for segmentation of fetal abdomen contour [23] as shown in Fig. 3. The Gradient vector flow field is the vector field $r(X, Y) = (p(X, Y) + q(X, Y))$ that minimizes the energy functional.

$$E = \iint \mu(p_x^2 + p_y^2 + q_x^2 + q_y^2) + |\nabla f|^2 + |r - \nabla f| dxdy. \tag{3}$$

Where $f(x,y)$ is the edge map of the image. When $|\nabla f|$ is small, the energy is dominated by sum of the squares of the partial derivatives of the vector field, yielding a slowly varying field. On the other hand, when $|\nabla f|$ is large, the second term dominates the integrand, and is minimized by setting $r = \nabla f$. The parameter μ is a regularization parameter, set according to the amount of noise present in the image.

2.3   Features Extraction

The AC and HC features of the fetus are extracted from 11 weeks to 40 weeks as shown in Fig. 3. For normalize the data, statistical mean and standard deviations is calculated. The variations of these statistical parameters give some intimation to the presence of a fetal abnormality. These features can be found by the following set of equations: The mean of the values $x_{1,1}$, $x_{2,2}$,.... $x_{m,n}$ is:

$$\bar{x} = \frac{1}{MN} \sum_{i=1}^{n} \sum_{j=1}^{m} x_{i,j} \tag{4}$$

Where M, N is the spatial row column variable. Mean indicates the tendency to cluster around some particular value. The value, which characterizes around the mean value, is the variance.

$$Var = \frac{1}{(N-1)(M-1)} \sum (x_{i,j} - \bar{x}) \tag{5}$$

3   Artificial Neural Network (ANN)

The artificial neuron is inspired from real biological neuron model, which is formed by dendrites (inputs), body, and axon (output), input neuron consists of inputs x = [x1 x2... xn]^T (i.e., the features vector), the weight vector w = [w1.1,w1.2, w1.n], the bias b, the summation which performs a linear combination of inputs and the transfer function f which produces the scalar output y [26]. The neuron output is calculated as

$$y = f(wx^T + b) \tag{6}$$

In the multilayered neural networks (MLPNN), input layer is the first layer in which the number of its neurons is equal to the number of selected specific features. The output layer is the last layer, which determines the desired output classes as shown in Fig. 4. The intermediate hidden layer may increase the MLPNN's capability and is most useful for nonlinear systems. An MLPNN trained in back-propagation mode can be used detect the abnormality of the fetus. The generalized delta rule [27] involves minimizing an error term defined as:

$$E = 0.5 * \sum_{k} (t_k - y_k)^2 \tag{7}$$

In Eq. (7), t_k and y_k are the targets and actual output vectors corresponding to the input vector k, respectively. The generalized delta rule implements a gradient descent in E to minimize that error as follows:

$$\Delta w_{jk} = -\alpha \frac{\partial E_k}{\partial w_{jk}} \tag{8}$$

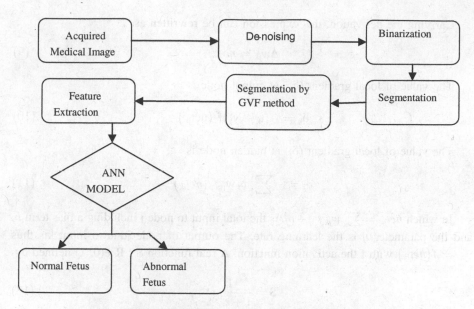

Fig. 2. Block diagram of proposed approach

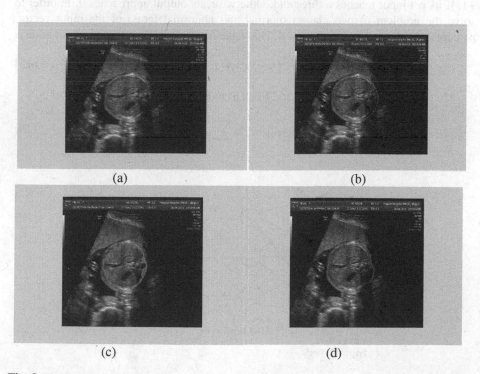

Fig. 3. (a) Original 24 week fetus abdomen image; (b) Initial contour; (c) Contour at iteration 20; (d) Final result.

Carrying the derivation, this expression can be rewritten as

$$\Delta w_{jk} = \eta \delta_k z_j \qquad (9)$$

The value of local gradient (δ_k) at output node is

$$\delta_k = -(t_k - y_k)f'(net_{jk}) \qquad (10)$$

The value of local gradient (δ_j) at hidden node is

$$\delta_j = -\sum_k \delta_k w_{jk} f'(net_{jk}) \qquad (11)$$

In which $net_{jk} = \sum_k w_{jk} y_{jk} + b_j$, is the total input to node j including a bias term b_j and the parameter η is the learning rate. The output of node jdue to input kis thus $y_{jk} = f(net_{jk})$ with f the activation function. A real function S : R $\in$(0, 1) defined by:

$$S = \frac{1}{1 + e^{-cx}} \qquad (12)$$

The constant c can be selected arbitrarily. This function forces a neuron to output +1, if its net input reaches a threshold. Otherwise, its output approaches 0. In order to solve the problem of two classes (normal and abnormal fetus) of the input vector, perceptron with two neurons are used. The value of local gradient (δ_k) at output node is

$$\delta_k = (t_k - y_k)y_k(1 - y_k) \qquad (13)$$

The value of local gradient (δ_j) at hidden node is

$$\delta_j = y_k(1 - y_k)\sum_k \delta_k w_{jk} \qquad (14)$$

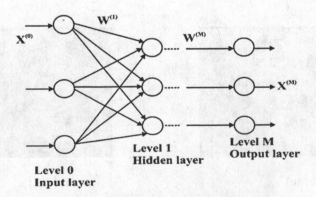

Fig. 4. MLPNN Architecture.

Finally, an additional momentum term can be added to the learning equation resulting is:

$$\Delta w_{ji}[n] = \eta \delta_{jk} y_{jk} + \mu \Delta w_{ji}[n-1] \tag{15}$$

Where μ is the momentum rate and η is the learning factor. At each iteration, the weights are thus modified as follows:

$$\Delta w_{ji}[n+1] = w_{ji}[n] + \Delta w_{ji}[n] \tag{16}$$

4 Problem Formulation

Initially we collect US image from ultrasound in DICOM format. For the preprocessing of these raw images we apply median filter, Segmentation of ROI and Feature extraction. These process help us to find accurate noise free data. The data transferred for the training and validation our ANN model. After proper training this model can automatically predict accurate of fetus status. The real medical fetus ultrasound image features are trained and tested by back- propagation algorithm. The data is collected after applying the GVF segmentation algorithm [23–25]. Hence GVF snake converge to the real AC and HC boundary precisely. The maximum diameter of that contour is measured in pixels, then pixel count is divide by a factor, which depends on the resolution of the monitor, to obtain the diameter in millimeter. The AC and HC parameters of the fetus from 14 weeks to 40 weeks is given to the ANN model. After the training of neural network, data is tested for AC and HC parameters of 36 to 40 weeks and compared the results of actual and target output. The output value of the ANN's detects the normal and abnormal (IUGR) fetus.

5 Result and Discussion

Total 120 US images used for the evaluation of the training and testing of ANN model. The data set is obtained by proper preprocessing of US images and then measuring the features from images. The target and the actual output of ANN model are compared and the error is calculated as given in Tables 1 and 2. The error minimization curve at each epoch of the neural network during learning is shown in Fig. 5. As the number of neurons exceed then the mean square error is reduced. Table 3 presents certain final parameters of the successfully trained neural network. The experimental data sets were divided into corresponding two fetus state (abnormal and normal) by one expert radiologist physician with full accuracy. Experimental results proved good precision and effectiveness of our recognition algorithm in clinical studies. After the proper training ANN model distinguish normal and abnormal fetus very rapidly.

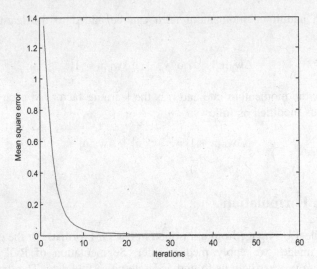

Fig. 5. Neural network learning curve.

Table 1. Neural network final training parameters.

Input nodes	2
Hidden nodes	100
Output nodes	1
Learning rate	0.4
MSE	0.0001
Iterations	6000
Training time (seconds)	2015
Run time (seconds)	0.01

Table 2. Examples of the values corresponding to the Head Circumference.

S. No	Gestational age (weeks)	Head Circumference (mm)	Target output (1- normal 0-abnormal)	Actual output	Error
1	14	80	0	0.047	−0.047
2	15	108	1	0.789	0.211
3	16	128	1	0.656	0.343
4	18	120	0	0.021	−0.021
5	20	170	1	0.612	0.387
6	22	188	1	0.797	0.202
7	24	220	1	0.923	0.076
8	25	231	1	0.894	0.105
9	26	200	0	−0.129	0.129
10	30	210	0	−0.125	0.125
11	32	288	1	0.752	0.247
12	34	305	1	0.912	0.087

Table 3. Examples of the values corresponding to the Abdomen Circumference

S. No	Gestational age (weeks)	Abdominal Circumference (mm)	Target output (1- normal 0-abnormal)	Actual output	Error
1	12	63	0	0.027	−0.027
2	14	84	1	0.6784	0.3216
3	15	96	1	0.7568	0.2432
4	16	106	1	0.7881	0.212
5	18	131	1	0.7456	0.2544
6	20	151	1	0.6879	0.3121
7	24	201	1	0.8765	0.1235
8	25	212	1	0.8190	0.181
9	26	223	0	−0.2340	0.2340
10	30	262	0	−0.2451	0.2451
11	32	283	1	0.8764	0.1236
12	34	305	1	0.8224	0.1776

References

1. Duncan, J., Aayche, N.: Medical image analysis: progress over two decades and challenges ahead. IEEE Trans. Pattern Anal. Mach. Intell. **22**, 85–106 (2000)
2. Current status of abortion in India, Consortium on National Consensus for Medical Abortion in India, 11 December 2009. http://www.aiims.ac.in/aiims/events/Gynaewebsite/ma_finalsite/report.htm. Accessed 11 Dec 2009
3. Hearn-Stebbins, B.: Normal fetal growth assessment: a review of the literature and current practice. J. Diagn. Med. Sonography **11**(4), 176–187 (1995)
4. Gurgen, F., Onal, E., Varol, F.G.: IUGR detection by ultrasonography examinations using neural networks. IEEE Eng. Med. Biol. Mag. **16**, 55–58 (1997)
5. Chauhan, S.P., Cole, J., Sanderson, M., Magann, E.F., Scardo, J.A.: Suspicion of intrauterine growth restriction: use of abdominal circumference alone or estimated fetal weight below 10 %. J. Matern. Fetal Neonatal Med. **19**(9), 557–562 (2006)
6. Campbell, S., Wilkin, D.: Ultrasonic measurement of the fetal abdomen circumference in the estimation of fetal weight. Br. J. Obstet. Gynaecol. Weight **82**(9), 689–697 (1975)
7. Rueda, S., Fathima, S., Knight, C.L., Yaqub, M., Papageorghiou, A.T., Foi, A., Maggioni, M., Pepe, A., Tohka, J., Stebbingand, R.V., McManigle, J.E., Ciurte, A., Bresson, X., Sun, C., Ponomarev, V., Gelfand, M., Kazanov, M., Wang, C., Chen, H., Peng, C., Hung, C., Noble, J.A.: Evaluation and comparison of current fetal ultrasound image segmentation methods for biometric measurments: a grand challenge. IEEE Trans. Med. Imaging **33**(4), 797–813 (2013)
8. Lu, W., Tan, J., Floyd, R.: Automated fetal head detection and measurement in ultrasound images by randomized Hough transform. Ultrasound Med. Biol. **31**(7), 929–936 (2005)
9. Yu, J., Wang, Y., Chen, P.: Fetal ultrasound image segmentation system and its use in fetal weight estimation. Ultrasound Med. Biol. **46**, 1227–1237 (2008)
10. Chalana, V., Winter, T.C., Cyr, D.R., Haynor, D.R., Kim, Y.: Automatic fetal head measurements from sonographic images. Acad. Radiol. **3**(8), 628–635 (1996)

11. Pathak, S.D., Chalana, V., Kim, Y.: Interactive automatic fetal head measurements from ultrasound images using multimedia technology. Ultrasound Med. Biol. 23(5), 665–673 (1997)

12. Foi, M.M., Pepe, A., Tohka, J.: Head contour extraction from the fetal ultrasound images by difference of gaussians revolved along elliptical paths. In: Proceedings of Challenge US: Biometric Measurements from Fetal Ultrasound Images, ISBI 2012, pp. 1–3 (2012)

13. Stebbing, R.V., McManigle, J.E.: A boundary fragment model for head segmentation in fetal ultrasound. In: Proceedings of Challenge US: Biometric Measurements from Fetal Ultrasound Images, ISBI 2012, pp. 9–11 (2012)

14. Sun, C.: Automatic fetal head measurements from ultrasound images using circular shortest paths. In: Proceedings of Challenge US: Biometric Measurements from Fetal US Images, ISBI 2012, pp. 13–15 (2012)

15. Nithya, J., Madheswaran, M.: Detection of intrauterine growth retardation using fetal abdominal circumference. In: International Conference on Computer Technology and Development (2009)

16. Yu, J., Wang, Y., Chen, P., Shen, Y.: Fetal abdominal contour extraction and measurement in ultrasound images. Ultrasound Med. Biol. 34(2), 169–182 (2008)

17. Ciurte, A., Rueda, S., Bresson, X., Nedevschi, S., Papageorghiou, A.T., Alison Noble, J., Cuadra, M.B.: Ultrasound image segmentation of the fetal abdomen: a semi-supervised patch-based approach. In: Proceedings of Challenge US: Biometric Measurements from Fetal Ultrasound Images, ISBI 2012, pp. 13–15 (2012)

18. Holcomb, W.L., Mostello, D.J., Gray, D.L.: Abdominal gestational age birth weight in diabetic pregnancy. J. Clin. Imaging 24, 1–7 (2000)

19. Carneiro, G., Georgescu, B., Good, S., Comaniciu, D.: Detection and measurement of fetal anatomies from ultrasound images using a constrained probabilistic boosting tree. IEEE Trans. Med. Imaging 27(9), 1342–1355 (2008)

20. Carneiro, G., Georgescu, B., Good, S.: Knowledge-based automated fetal biometrics using syngo Auto OB measurements. Siemens Medical Solutions (2008)

21. Sarris, C.I., Chamberlain, P., Ohuma, E., Roseman, F., Hoch, L., Altman, D.G., Papageorghiou, A.T., International Fetal and Newborn Consortium for the 21st Century International Fetal and Newborn Consortium for the 21st Century (INTERGROWTH-21st): Intra- and inter observer variability in fetal ultrasound measurements. Ultrasound Obstet. Gynecol. 39(3), 266–273 (2012)

22. Rawat, V., Jain, A., Shrimali, V.: Investigation and assessment of disorder of ultrasound B-mode images. Int. J. Comput. Sci. Inf. Secur. 7(2), 252–256 (2010)

23. Rawat, V., Jain, A., Shrimali, V.: Automatic assessment of fetal biometric parameter using GVF snakes. J. Biomed. Eng. Technol. Indersci. 12(4), 221–233 (2013)

24. Kass, M., Witkin, A., Terzopoulos, D.: Snakes: active contour models. Int. J. Comput. Vis. 1, 321–331 (1988)

25. Chenyang, X., Prince, J.L.: Snakes, shapes and gradient vector flow. IEEE Trans. Image Process. 7, 359–369 (1998)

26. Demuth, H., Beale, M., Hagan, M.: Neural Network Toolbox 6 User's Guide. The MathWorks Inc, Natick (2008)

27. Rumelhart, D.E., McClelland, J.L.: The PDP Research Group Parallel Distributed Processing; Explorations in the Microstructure of Cognition, vol. I. MIT Press, Cambridge (1986)

Low Resource Language Processing

On One Approach of Solving Sentiment Analysis Task for Kazakh and Russian Languages Using Deep Learning

Narynov Sergazy Sakenovich
and Arman Serikuly Zharmagambetov[(⊠)]

"Alem Research" LLP, Dostyk Avenue 132, Office 13,
050051 Almaty, Kazakhstan
sergazy@gmail.com, armanform@gmail.com

Abstract. The given research paper describes modern approaches of solving the task of sentiment analysis of the news articles in Kazakh and Russian languages by using deep recurrent neural networks. Particularly, we used Long-Short Term Memory (LSTM) in order to consider long term dependencies of the whole text. Thereby, research shows that good results can be achieved even without knowing linguistic features of particular language. Here we are going to use word embedding (word2vec, GloVes) as the main feature in our machine learning algorithms. The main idea of word embedding is the representations of words with the help of vectors in such manner that semantic relationships between words preserved as basic linear algebra operations.

Keywords: NLP · Sentiment analysis · Deep learning · Machine learning · Text classification

1 Introduction

In recent years, there is an active trend towards using various machine learning techniques for solving problems related to Natural Language Processing (NLP). One of these problems is the automatic detection of emotional coloring (positive, negative, neutral) of the text data, i.e. sentiment analysis. The goal of this task is to determine whether a given document is positive, negative or neutral according to its general emotional coloring. We don't perform sentiment analysis related to particular object, i.e. it is not an aspect based sentiment analysis. Therefore, we deleted from our dataset document with mixed sentiment. Nevertheless, analyzing general sentiment of a document is difficult task by itself. The difficulty of sentiment analysis is determined by the emotional language enriched by slang, polysemy, ambiguity, sarcasm; all this factors are misleading for both humans and computers.

The high interest of business and researchers to the development of sentiment analysis are caused by the quality and performance issues. Apparently the sentiment analysis is one of the most in-demand NLP tasks. For instance, there are several international competitions and contests [1], which try to identify the best method for sentiment classification. Sentiment analysis had been applied on various levels, starting

© Springer International Publishing Switzerland 2016
N.T. Nguyen et al. (Eds.): ICCCI 2016, Part II, LNAI 9876, pp. 537–545, 2016.
DOI: 10.1007/978-3-319-45246-3_51

from the whole text level, then going towards the sentence and\or phrase level. In general, importance of solving this problem is considered in [16].

It is obvious that similar sentimental messages (text, sentence…) can have various thesauruses, styles, and structure of narration. Thus, points corresponding to the similar messages can be located far away from each other that make the sentiment classification task much harder [2]. The scientific novelty of the paper is in applying Word2Vec algorithm [3] in the sentiment classification task for Kazakh and Russian languages and use this word vectors as input to deep recurrent neural networks to deal with long term dependency of the textual document.

2 Related Works

The study of sentiment analysis has relatively small history. Reference [4] is generally considered the principal work on using machine learning methods of text classification for sentiment analysis. The previous works related to this field includes approaches based on maximum relative entropy and binary linear classification [5] and unsupervised learning [6].

Most of these methods use well known features as bag-of-words, n-grams, tf-idf, which considered as the simplest one [7]. But as show the experiment results the simple models often works better than complicated ones. Reference [7] use distant learning to acquire sentiment data. Additionally, since they mostly work with movie comments and tweets, they used additional features as ending in positive emoticons like ":)" ":-)" as positive and negative emoticons like ":(" ":-(" as negative. They build models using Naive Bayes, Maximum Entropy and Support Vector Machines (SVM), and they report SVM outperforms other classifiers. In terms of feature space, they try a Unigram, Bigram model in conjunction with parts-of-speech (POS) features. They note that the unigram model outperforms all other models.

The other feature is syntactic meta information. It's obvious that the recursively enumerable grammar describes the most complete of any natural language. The computational performance of the best syntactic parser of context free grammar is linear, so syntactic information is expensive for the sentiment analysis task. However, those experiments that involved dependency relations showed that syntax contributes significantly to both Recall and Precision of most algorithms. For the task of text classification in general see [8–10] deal with a task sentiment classification based on syntactic relations. In reference [11] it was shown that POS-tagging and other linguistic features contributes to the classifier accuracy. The experiments were conducted on feedback data from Global Support Services survey.

Sentiment analysis using recurrent and recursive neural networks described in [17–19]. Important fact is that authors could not find previous works related to automatic sentiment classification for Kazakh language.

3 Dataset

The labeled by human data set consists of ~30,000 news articles in Russian language, specially selected for sentiment analysis, which consist of 11,286 neutral, 10,958 positive and 7756 negative articles. The sentiment of each document can be one of the following: positive, negative, neutral. 18,000 reviews from this dataset were chosen as training data, 6,000 as cross validation dataset and 6,000 as test dataset. Furthermore, ~10000 (3021 positive, 2548 negative, 4431 neutral) news articles in Kazakh language were labeled in order to train sentiment classifier. Each entry on this dataset consists of the following field:

- Id - Unique ID of each review.
- Sentiment - Sentiment of the review: 1 for positive reviews, 0 for negative reviews, 2 to the neutral reviews.
- Text - Text of the document (on Kazakh or Russian language).

The goal is to increase the accuracy (precision and recall) in sentiment classification of test dataset.

Furthermore, we used ~70 GB of plain text data in Russian language and ~10 GB plain text data in Kazakh language in order to train word embedding by unsupervised method. These texts we obtained from open electronic libraries, news articles, crawled web sites, etc.

4 Learning Model

The first step of learning model is unsupervised training of word embedding. Word2vec, published by Google in 2013, is a neural network implementation that learns distributed representations for words. Distributed word vectors, i.e. word embeddings are powerful and can be used for many applications, particularly word prediction and translation. It accepts large un-annotated corpus and learns by unsupervised algorithms. There are two different architectures of Word2Vec algorithm. At Fig. 1 continuous bag of words (CBOW) architecture presented, the purpose of such network topology assumes mapping from context to particular term and vice versa at Fig. 1 skip-gram architecture that maps particular term to its context.

We used "gensim" [13] python library with build-in Word2Vec model. It accepts large textual dataset for training. As was mentioned above, 70 GB and 10 GB raw data in Russian and Kazakh languages, respectively. The following options were used while training for Word2Vec: 300 dimensional space, 40 minimum words and 3 words in context. The vector representation of word has a lot of advantageous. It raises the notion of space, and we can find distance between words and finding semantic similar words. The simple result that can be obtained for such vector presented in Table 1.

Finally, map with a word as key and N dimensional vectors as value is obtained from abovementioned word2vec algorithm. Next, these vectors will be used in classification task. But before, each article should be preprocessed.

The preprocessing includes the following: (1) The all HTML tags, punctuations, were removed by "Beautiful Soup" python library. There are HTML tags such as

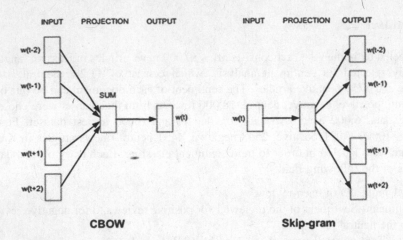

Fig. 1. The architectures of Word2Vec model. The CBOW architecture predicts the current word based on the context (**left**). The skip-gram predicts surrounding words given the current word (**right**). Taken from [3].

Table 1. Semantic similar words to Russian word - 'man'

Words	cos dist
Woman	0,6056
Guy	0,4935
Boy	0,4893
Men	0,4632
Person	0,4574
Lady	0,4487
Himself	0,4288
Girl	0,4166
His	0,3853
He	0,3829

"< br/>", abbreviations, punctuation - all common issues when processing text from online. (2) Moreover, numbers and links were replaced by tags NUM and LINK, respectively. (3) Removing stop words. Conveniently, there is Python package - Natural Language Toolkit (NLTK) [12] that removes stop words with built in lists. (4) Lemmatization of each word. For Russian language was used lemmatization tool "Mystem" [23], which was developed by Yandex. For Kazakh language there is no lemmatization. In future works we plan to implement morphological lemmatization tool for Kazakh language.

Next, each word is mapped to vector and for one document we get a sequence of N dimensional vectors which will be given as input to LSTM recurrent neural network.

Long Short-Term Memory (LSTM) is a special type of recurrent neural networks which was invented to consider sequential dependencies as a set of words in some text. Furthermore, LSTM overcomes common problems of RNN as exploding gradient or

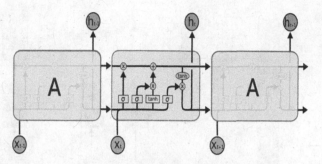

Fig. 2. The architecture of LSTM. Instead of one output it uses three gates: input, forget, output. Taken from [24].

vanishing gradient. To overcome this drawback LSTM uses additional internal transformation, which operate with memory cell more cautious represented in Fig. 2. For detailed information how LSTM works you can refer to [20].

For sentiment classification task we compared several model architectures with LSTM, which are presented in Figs. 3, 4, 5 and 6. General idea for all schemes is the same – input word vectors are processed via LSTM units, and then outputs from these units go further via vanilla neural networks or logistic regression unit. Below, in results section we give comparison of results of these various neural networks architectures.

Training algorithm was implemented using Theano [21] and Lasagne [22] packages for python language. C-extension for python (cython [15]) and GPU were used for acceleration and efficient calculations. In our experiments, using GPU gives up to 6–7 times faster calculation compared with CPU usage with multithread. Abovementioned packages give opportunity to easily implement various deep learning algorithms

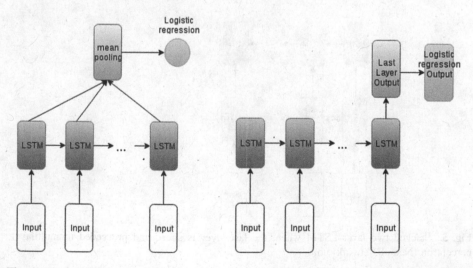

Fig. 3. Mean pooling from each output of LSTM followed by logistic regression (**left**). Stacked LSTM where the last layer is sliced and proceed to logistic regression unit (**right**).

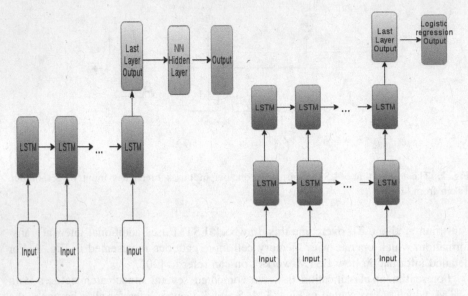

Fig. 4. Stacked LSTM where the last layer is sliced and proceeded to multilayer perceptron (Neural Network) unit **(left)**. Stacked two layer LSTM where the last layer is sliced and proceeded to logistic regression unit **(right)**.

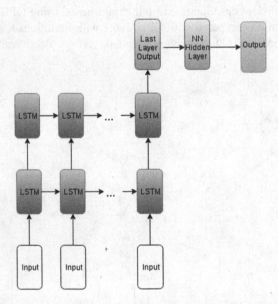

Fig. 5. Stacked two layer LSTM where the last layer is sliced and proceeded to multilayer perceptron (Neural Network) unit.

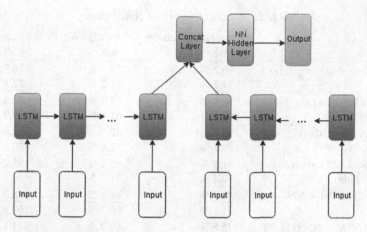

Fig. 6. Stacked bidirectional LSTM where the last layers of each direction are concatenated and proceeded to multilayer perceptron unit.

including LSTM, multilayer perceptron, etc. Also they have extension to use GPU to optimize computations.

Sigmoid and tanh functions were used as LSTM internal activation functions. Similarly, softmax was used as activation function for logistic regression and Neural Networks units. Advantage of softmax functions is that it gives correct generalization of the logistic sigmoid to the multinomial case:

$$h_i = \frac{e^{a_i}}{\sum_{j=0}^{N} e^{a_j}} \tag{1}$$

We used categorical cross entropy to define the loss function which should be optimized while training:

$$J = -\sum_{i=0}^{N} \sum_{j=0}^{m} y_j^{(i)} * \log(h_j^{(i)}) \tag{2}$$

5 Results and Discussions

Table 2 summarizes the results on sentiment classification. As mentioned above there are 30,000 news articles in Russian language and 10,000 news articles in Kazakh language were chosen for train and test data.

It can be seen that the best model for Russian sentiment analysis is - "Stacked two layer LSTM with one hidden layer Neural Networks" (Fig. 5) and for Kazakh language is - "LSTM with mean pooling and logistic regression unit" (Fig. 4). Another important fact is that sentiment analysis for Kazakh language shows worse result. Probably, it can be explained due to relatively small amount of training data set and lack of lemmatization.

Table 2. Results of sentiment classification

Method	Average precision (ru/kz)	Average recall (ru/kz)	Accuracy (ru/kz)
Mean pooling + log. reg. (Fig. 3 - left)	80.2 % *73.2 %*	73.7 % 72.2 %	76.3 % *72.8 %*
Stacked LSTM + log.reg (Fig. 3 - right)	81.3 % 69.1 %	77.2 % 61.3 %	80.8 % 67.3 %
Stacked LSTM + NN (Fig. 4 - left)	78.2 % 70.1 %	82 % *72.6 %*	82.8 % 70.5 %
Stacked two LSTM + log.reg (Fig. 4 - right)	81.6 % 70.4 %	85.7 % 71.3 %	85.2 % 70.9 %
Stacked two LSTM + NN (Fig. 5)	*84.5 %* 71.1 %	*86.4 %* 66.7 %	*86.3 %* 69.8 %
Biderect. LSTM + NN (Fig. 6)	76.8 % 62.9 %	69.3 % 61.2 %	71.1 % 62.3 %

The given work shows that deep recurrent neural networks can be efficiently applied to the task of sentiment classification. Particularly, LSTM shows stable results even for long sequential data as words or sentences in a news article. Additionally, word embedding helps extract semantic relations between words which have effect to training process. Future works will be dedicated to improvement of sentiment classification by studying deeply long term dependencies in the text document and by extracting syntax relations. Neural Turing Machines, adversarial neural networks will be considered instead of or jointly with recurrent relation. Moreover, aspect based sentiment classification task will be studied.

References

1. Chetviorkin, I., Braslavskiy, P., Loukachevich, N.: Sentiment analysis track at ROMIP 2011. In: International Conference "Dialog 2012": Computational Linguistics and Intellectual Technologies, Bekasovo, pp. 1–14 (2012)
2. Pak, A.A., Narynov, S.S., Zharmagambetov, A.S., Sagyndykova, S.N., Kenzhebayeva, Z.E., Turemuratovich, I.: The method of synonyms extraction from unannotated corpus. In: DINWC 2015, Moscow, pp. 1–5 (2015)
3. Mikolov, T., Chen, K., Corrado, G., Dean, J.: Efficient estimation of word representations in vector space. In: Workshop at ICLR, Scottsdale, AZ, USA (2013)
4. Bo, P., Lee, L.: A sentimental education: sentiment analysis using subjectivity summarization based on minimum cuts. In: ACL (2004)
5. Joachims, T.: Text categorization with support vector machines: learning with many relevant features. In: Nédellec, C., Rouveirol, C. (eds.) ECML 1998. LNCS, vol. 1398, pp. 137–142. Springer, Heidelberg (1998)
6. Turney, P.D.: Thumbs up or thumbs down? semantic orientation applied to unsupervised classification of reviews. In: 40th Annual Meeting of the Association for Computational Linguistics (ACL 2002), Philadelphia, Pennsylvania, pp. 417–424 (2002)

7. Go, A., Bhayani, R., Huang, L.: Twitter sentiment classification using distant supervision. Technical report, Stanford (2009)
8. Furnkranz, J., Mitchell, T., Riloff, E.: A case study in using linguistic phrases for text categorization on the WWW. In: AAAI/ICML Workshop on Learning for Text Categorization, pp. 5–12 (1998)
9. Caropreso, M.F., Matwin, S., Sebastiani, F.: A learner-independent evaluation of the usefulness of statistical phrases for automated text categorization. In: Chin, A.G. (ed.) Text Databases and Document Management: Theory and Practice, pp. 78–102. Idea Group Publishing, USA (2001)
10. Nastase, B., Shirabad, J.S., Caropreso, M.F.: Using dependency relations for text classification. In: 19th Canadian Conference on Artificial Intelligence, Quebec City, pp. 12–25 (2006)
11. Gamon, M.: Sentiment classification on customer feedback data: noisy data, large feature vectors, and the role of linguistic analysis. In: COLING 2004, Geneva, pp. 841–847 (2004)
12. Natural Language Toolkit. http://www.nltk.org/
13. Gensim: Topic modeling for humans. https://radimrehurek.com/gensim/
14. Sci-kit: Machine learning in python. http://scikit-learn.org/stable/
15. Cython: C-Extensions for Python. http://cython.org/
16. Liu, B.: Sentiment analysis and opinion mining. Synth. Lect. Hum. Lang. Technol. 5(1), 1–167 (2012)
17. Maas, A.L., Daly, R.E., Pham, P.T., Huang, D., Ng, A.Y., Potts, C.: Learning word vectors for sentiment analysis. In: 49th Annual Meeting of the Association for Computational Linguistics: Human Language Technologies, pp. 142–150. Association for Computational Linguistics (2011)
18. Tarasov, D.S.: Deep recurrent neural networks for multiple language aspect based sentiment analysis of user reviews. In: Dialog 2015, Moskow (2015)
19. Socher, R., Perelygin, A., Jean, Y.W., Chuang, J., Manning, C.D, Ng, A.Y., Potts, C.: Recursive deep models for semantic compositionality over a sentiment treebank. In: Conference on Empirical Methods in Natural Language Processing (EMNLP), pp. 1642–1656. Citeseer, Seattle (2013)
20. Hochreiter, S., Schmidhuber, J.: Long short-term memory. J. Neural Comput. 9(8), 1735–1780 (1997)
21. Theano: Framework for python. http://deeplearning.net/software/theano/
22. Lasagne: Framework for python. https://github.com/Lasagne/Lasagne
23. Mystem: Morphology analysis tool. https://tech.yandex.ru/mystem/
24. Understanding LSTM Networks. Colah's personal blog. http://colah.github.io/posts/2015-08-Understanding-LSTMs/

Morphological Transducer for Mongolian

Zoljargal Munkhjargal, Altangerel Chagnaa[✉], and Purev Jaimai

Department of Information and Computer Science,
National University of Mongolia, 14200 Ulaanbaatar, Mongolia
{zoljargal,altangerel,purev}@num.edu.mn

Abstract. This paper describes the development of finite state morphological transducer for Mongolian and presents some issues in Mongolian morphology, linguistic issues encountered and how they were dealt with. The work done here includes all the morphophonological rules needed for all Mongolian nominal and verb. Nominal morphotactic is implemented completely and verbal morphotactic covers one level continuation lexica. An evaluation is done via analysis on two separate corpora, which shows high-level and medium-level coverage respectively. It is more elaborate and accurate than previous implementations of its kinds.

Keywords: Mongolian morphology · Transducer · Two-level rule

1 Introduction

The morphological transducer is an important part in modern applications. It is particularly important for morphologically complex languages [2]. Morphological transducer is researched broadly and implemented in most languages. However, in less-studied and low-resource languages such as Mongolian, there is not enough research and implementations. There were some works on the morphological analyzer with smaller scale in both lexicon and tests. This paper describes the development of morphological transducer using free/open-source platform HFST.

The paper is composed of five main parts, which similarly follows the structure of [9]. First, background section gives some information about Mongolian language, and morphological transducers. Subsequent sections talks about issues encountered with the morphotactics and morphophonology. Finally, evaluation results are given.

2 Background

2.1 The Mongolian Language and Scripts

Mongolian is an agglutinative language that a word is inflected for rich suffix chains in the verbal and nominal domains. As to its origins, the Mongolian language belongs to the Altaic language family, and is most closely related with

© Springer International Publishing Switzerland 2016
N.T. Nguyen et al. (Eds.): ICCCI 2016, Part II, LNAI 9876, pp. 546–554, 2016.
DOI: 10.1007/978-3-319-45246-3_52

Turkish, Manchurian-Tungus, Korean, and Japanese. In terms of structure and an agglutinative paradigm, Turkic languages are most closely related. Mongolian nominal is inflected in several cases, as well as in number and possessive. Due to the language's agglutinative nature, each of these cases is expressed with a distinct bound morpheme suffix, forming in some cases chains of 5 or more suffixes. On the other hand, nouns are never prefixed. Mongolian verbal is inflected with aspect, voice, personal, temporal endings, subordinating conjunctives, as well as noun inflections.

There are a number of closely related varieties of Mongolian: Khalkha or Halha, the official language of Mongolia, and Oirat, Chahar and Ordos, spoken mainly in the Inner Mongolia, a part of China, and Buryat and Kalmyk, spoken in the Russian Federation. Today, Mongolian has approximately 8 million speakers around the world including Mongolia (2.7 mln), Inner Mongolia in China (3.38 mln) and Russia (0.5 mln).

According to Tsevelyn Shagdarsuren [8], the scripts created and used by the nomadic Mongols date back to ancient times. There are at least nine or ten such scripts. Following Fig. 1 shows examples written in those scripts.

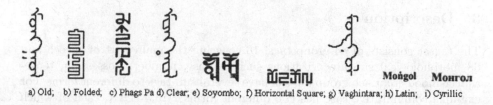

a) Old;   b) Folded;   c) Phags Pa d) Clear;   e) Soyombo;   f) Horizontal Square;   g) Vaghintara; h) Latin;   i) Cyrillic

Fig. 1. Word "Mongol" written in Mongolian scripts.

Nowadays, Mongolian is written on two official scripts: the (new) Cyrillic Mongolian Script and the (old) Mongolian Script. But, today, the first one is predominately used in everyday life and also in digital environment. Thus, it is used also for most of research on local language processing and resource development including works described in this paper. Here after we refer new Cyrillic Mongolian as Mongolian.

The (old) Mongolian alphabet was adopted from the Uighur alphabet in the 12th Century and was used in Mongolia until 1941 when a new Cyrillic-based alphabet was adapted. In 1990, the Mongolian Government decided to bring back the (old) Mongolian alphabet to official use. Throughout this period, the (old) Mongolian alphabet is still used in Inner Mongolia. Currently, Mongolian grammar follows the orthographic dictionary of Damdinsuren Tsend which was compiled in 1983. It has categorized nominal into 29 sub lexicon, and verbs into 18 sub lexicons, according to their inflectional behavior.

2.2 Morphological Transducers

Here we give general overview on transducers and previous implementation attempts in Mongolian. Since the first introduction of morphological transducers with two level rule formalism, it has been successfully implemented to many languages several different languages such as Finnish, English, Turkish, Japanese, etc. [2–4].

The very first attempt to use two level rule formalism in Mongolian was [7], where authors applied Mongolian morphophonologic rules with non-Unicode tool PC-KIMMO. A C++ program is developed to map the 8 bit 1 or 2-byte Cyrillic Mongolian codes into corresponding English alphabet characters. With 36 morphophonological rules and 6,199 nouns, 18,551 verbs, 4,516 adjectives, it had about 63 % of coverage on small corpus.

Afterwards, this work was extended by adding 84 rules containing 25 subsets [1]. It was tested only on 2,000 dictionary entries, but not on real world data. It lacked to handle ambiguous word forms (see more detail on Sect. 4.2).

These works were relatively small according to their lexicons, rule sets and also size of the test corpus.

3 Description

The tagset consists of 84 group tags, 16 covering the main part of speech and 68 morphological subcategorizations such as case, number, possession, tense-aspect-mood etc. Each group tag can have several sub-tags to different forms. For example commutative case has two different forms тЙй,лУгАА, each of which has its own tag. Number of sub-tags sums up 195. Tag boundary is represented by plus '+' symbol.

4 Morphotactics

We have focused on only inflectional suffixes in this work. Morphotactics for derivational suffixes are left untouched, since derived words can be put into the lexicon directly. Another reason is adding morphotactics for derivational process is complex and needs further development. Note that the fundamental morphophonological rules needed for derivational morphotactics are already defined.

Morphotactics for Mongolian nominal and verb are shown in Figs. 2 and 3 respectively [5].

The nominal includes a large word classes, namely nouns, proper nouns, adjectives, adverbs, numerals and pronouns. With this morphotactic, one nominal can have about 300 different inflections, which is complete for a nominal.

Mongolian verb is inflected with aspect, voice, conjunctive, personal, temporal endings, as well as determining suffixes, followed by noun inflections. Without derivational suffixes, this morphotactic is complex enough, which can produce

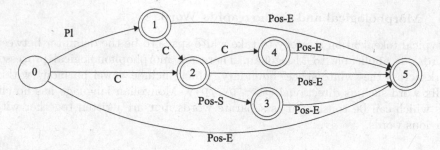

Fig. 2. A Simplified FSA for the nominal morphotactics. Here is *Pl*: plural, *C*: case, *Pos-E*: possessive ending and *Pos-S*: possessive suffix

about 24 thousand possible inflections for a single verb. If we include derivational suffixes the complexity becomes very huge. This paper implemented only one level inflection on a verb, which means continuation lexicon is not complete. We think this is pretty enough for initial steps on real world data. Because, statistics on corpus shows the low percentage of verbs inflected with more than one suffix among all verbs (65,299 verbs were inflected with more than one suffix among 516,246 verbs in NUM corpus [6]).

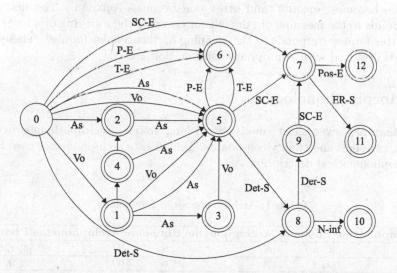

Fig. 3. A Simplified FSA for the verbal morphotactics. Here is *As*: aspect, *Vo*: voice, *SC-E*: subordinating conjunctive ending, *P-E*: personal ending, *T-E*: temporal ending, *Pos-E*: Possessive ending, *Det-S*: determining suffix, *ER-S*: emotion representing suffix, *Der-S*: derivational suffix and *N-inf*: noun inflection

4.1   Morphological and Orthographic Words

A typical tokenization strategy to take white space to be the delimiter between 'words', is applicable to Mongolian. There are morphophonological processes work across the 'white space' boundary, which include vowel harmony within suffix word such as directional case: "руу/рүү" Mongolian language has no clitics which can be considered as separate words, but are written together with previous words.

4.2   Ambiguities of Orthographic Word Forms

One of the morphotactic challenges met in defining a finite-state transducer for Mongolian is that it is even impossible to rely only on orthographic word forms. For example, theoretically, two words "дарга" and "чарга" should have same behavior on the same inflectional suffixes, but they don't. For example, with dative case (+Dat/Д), the first word has the form "даргад", whereas the latter word has "чарганд" with extra "н". This is because of their original old script forms differ, where "дарга" is written as "daruG-a" and "чарга" is written as "ciruGa".

Also, semantically different words have same word form, such as "өрх". This word has two different forms when it is affixed with genitive case (+Gen). The first one becomes "өрхний" and other one becomes "өрхийн". The first form corresponds to the meaning of "the felt for covering the roof-ring of a tent/ger". The latter form corresponds to the meaning of "household, family". These kind of words arranged with appropriate continuation lexica.

5   Morphophonology

Mongolian has seven short vowels, seven long vowels, five diphthongs, five 'y'-vowels, two signs and twenty consonants. Table 1 lists archiphonemes used in the morphophonological description.

Table 1. Archiphonemes in Mongolian

Archiphoneme	Phoneme	Archiphoneme	Phoneme	Archiphoneme	Phoneme
Б	б, в	А	а, э	Й	а, э, о
Д	д, т	Э	а, о	Ө	а, и, о
Ж	ж, ч	И	а, и	Ү	а, э, о, ө
Ы	и, й, ы	У	у, ү	Я	я, ё, е
Ъ	ъ, ь				

5.1 Vowel Harmony

In Mongolian, there are several vowel archiphonemes {Y}, {Й}, {У}, {Ы}, {А} and {Я}. According to rule of vowel harmony, final vowel in a root of a word harmonizes vowels in a suffix (Tables 2 and 3). For example, underlying form оруул-н{Y} realizes into surface form оруулна.

Table 2. Vowel harmony for archiphoneme {Y}

Presenting vowel	Result	Presenting vowel	Result
а, аа, ай я, яа	а	и, ий	э
у,уу, уй, юу	а	ө, өө, ео	ө
э, ээ, эй	э	о, оо, ой, ё, ёо	о
Y, YY, Yй, юу	э		

Table 3. Vowel harmony for archiphoneme {Я}

Presenting vowel	Result	Presenting vowel	Result
а, аа, ай я, яа	я	и, ий	е
у,уу, уй, юу	я	ө, өө, ео	е
э, ээ, эй	е	о, оо, ой, ё, ёо	ё
Y, YY, Yй, юу	е		

5.2 Voicing Assimilation

In Mongolian archiphonemes {Д}, {Б} and {Ж} are defined for voicing assimilation. These are not realized by two-level rules, since separate continuation lexica are defined.

5.3 Vowel Dropping

There are two cases to drop a vowel. First case is for an epenthetic vowel. But, it is not possible to identify epenthetic vowels with two-level rule. This was dealt by defining it in the lexicon and then it is then decided to drop or leave by appropriate two-level rules. For example, word 'саат{а}л' with the underlying form 'саат{а}л+{Ы}{Ы}г' realizes into surface form 'саатлыг'.

Second case is to drop final short vowel of root when suffixed by suffix which starts with a long vowel. For example, underlying form 'дарга+{Y}{Y}р' realizes into surface form 'даргаар'. This process is implemented with two-level rules.

5.4   Soft Sign Changing

In Mongolian soft sign 'ь' changes to 'и' in several cases. When a suffix starting with a vowel is added to a root which ends with a soft sign, it changes to 'и'. For example, underlying form 'хонь+{Y}р' realizes into surface form 'хониор'.

5.5   Sign Symbol Insertion

When 'Y'-vowel starting suffix is added to consonant ending root, one of two sign symbols is inserted as a buffer, considering vowel harmony rules. For example, underlying form 'хам+{Я}' realizes into surface form 'хамъя'.

5.6   Irregular Plurals

Because of historical base, some words have irregular plural forms. хөгшин+Pl -> хөгшид, өвгөн+Pl -> өвгөд. They are also arranged by directly adding to the lexicon. өвгөн+N+Pl -> өвгөд.

6   Statistics

Morphotactic contains a total of 221 continuation lexica, modeling the morphotactics. The phonological rules file contains 55 rules. Table 4. gives approximate number of stems in each main word class.

Table 4. Number of stems in lexicon by major part-of-speech.

Part of Speech	Number of stems
Nominal (noun, adjective etc.)	26,077
Verb	13,800
Others (closed class words)	1,907

7   Evaluation

It is practically efficient to measure the quality of morphological transducer over the real language data. Hence, we have evaluated the analyzer by calculating the naïve coverage and mean ambiguity on two corpora of Mongolian. The first one is NUM corpus, which is manually spell checked, POS tagged corpus, available for research purpose. NUM corpus includes texts on various styles, such as online news, literature, law, newspaper etc. Second one is a database dump of the Mongolian Wikipedia, dated 2016-03-05. Two test corpora are not touched, with the exception of non-Cyrillic word removal from Wikipedia corpus.

To calculate verb coverage in test corpora, POS tagger [10] is used.

7.1 Coverage and Ambiguity

To calculate the naïve coverage of the analyzer, above mentioned two corpora was used. The Wikipedia corpus was processed with Wikipedia_Extractor[1] to extract sentences and removed all the non-Cyrillic entries.

The transducer couldn't analyze 261,015 words out of 2,638,6221 words and found 1.60 mean ambiguity in NUM corpus. 2.37 morphemes are found per word on average. In Wikipedia corpus mean ambiguity was 1.42 and morphemes per word was 3.25 (Table 5).

Mean ambiguities on both corpora have not much differences.

The reasons of lower naive coverage 81.2 % in Wikipedia than naive coverage 90.1 % in NUM corpus might be misspelled variations of words, out of vocabulary words such as proper names, recently entered words into usage etc., since it is a community created corpus.

The reasons for average morphemes per word is higher in Wikipedia are: (a) that its verbal coverage was lower than NUM's, (b) Wikipedia has more inflected words.

Table 5. Naive coverage and mean ambiguity of the analyzer over two test corpora

Corpus	Tokens	Known	Naïve coverage(%)	Mean ambiguity	Verb coverage(%)
Mongolian Wikipedia	4,258,510	3,458,739	81.2	1.42	14.5
NUM corpus	2,638,621	2,377,606	90.1	1.60	19.0

8 Conclusion and Future Work

In this paper we presented morphological transducer for Mongolian. The coverage on the two real-world test corpora is adequate. Except for a few known cases such as out of vocabulary and misspelled words, transducer is analyzing pretty well. Future works include enriching vocabulary with mainly proper nouns, and completing verb morphotactics.

Acknowledgments. This work was partially funded by Young researcher grant of National University of Mongolia (P2016-1118) and HERP research grant (#20).

References

1. Altangerel, C., Adiyatseren, B.: Two level rules for mongolian language. In: Proceedings of the the 7th Multimedia and Information Technology Application (MITA2011), pp. 130–133 (2011)
2. Çöltekin, C.: A freely available morphological analyzer for Turkish. In: Proceedings of the 7th International Conference on Language Resources andEvaluation (LREC 2010), pp. 820–827 (2010). http://www.lrec-conf.org/proceedings/lrec2010/summaries/109.html

---

[1] http://medialab.di.unipi.it/wiki/Wikipedia_Extractor.

3. Karttunen, L., Beesley, K.R.: Twenty-five years of finite-state morphology. Inquiries Into Words, a Festschrift for Kimmo Koskenniemi on his 60th Birthday, pp. 71–83 (2005)
4. Koskenniemi, K.: Two-level morphology: a general computational model for word-form production and generation. Publications of the Department of General Linguistics, University of Helsinki. Helsinki: University of Helsinki (1983)
5. Munkh-Uchral, E., Coimaa, S.: Horvuuleh programd zoriulsan mongol helnii sudalgaa. In: Ph.D. thesis. Ulaanbaatar, Mongolia (2010)
6. Purev, J., Odbayar, C.: Part of speech tagging for mongolian corpus. In: The 7th Workshop on Asian Language Resources, Singapore (2009)
7. Purev, J., Tsolmon, Z., Altangerel, C., Cheolyoung, O.: Pc-kimmo-based description of mongolian morphology. vol. 1, pp. 41–48 (2005). http://www.jips-k.org/dlibrary/JIPS_v01_no1_paper8.pdf
8. Shagdarsuren, T.: Mongolchuudiin useg bichigyn tovchoon, usegzuin sudalgaa. In: Bibliotheca Mongolica, Monograph: Tomus I, Centre for Mongol Studies, National University of Mongolia, pp. 300–320 (2001)
9. Washington, J., Ipasov, M., Tyers, F.: A finite-state morphological transducerfor kyrgyz. In: Chair, N.C.C., Choukri, K., Declerck, T., Doğan, M.U.,Maegaard, B., Mariani, J., Moreno, A., Odijk, J., Piperidis, S. (eds.)Proceedings of the Eight International Conference on Language Resources andEvaluation (LREC 2012). European Language Resources Association (ELRA), Istanbul, Turkey, May 2012
10. Zoljargal, M., Purev, J.: Mongolian trigram part of speech tagger. In: Proceedings of the the 7th Multimedia and Information Technology Application (MITA 2011), pp. 161–163 (2011)

An Approach of Automatic Extraction of Domain Keywords from the Kazakh Text

Yermek Alimzhanov[✉] and Madina Mansurova

Al-Farabi Kazakh National University, Al-Farabi av. 71, 050040 Almaty, Kazakhstan
aermek81@gmail.com

Abstract. In this paper we consider the approach of automatic extraction of domain keywords from the Kazakh Text based on statistical methods of natural language processing. The proposed approach can be used to build domain dictionaries and thesauri without manual work of domain experts. Results of experiments on a corpus of texts from a Kazakh book and online websites demonstrate that applying latent semantic analysis to keywords extraction significantly decreases information noise and strengthens the words relations.

Keywords: Natural language processing · Latent semantic analysis · Domain knowledge

1 Introduction

Automatic extraction problem of keywords and collocations from the text arise in many areas like information retrieval, information extraction, sentiment analysis, word segmentation, ontology engineering, etc. Volumes and dynamics of information in these areas was making actual this problem. Extracted keywords and collocations can be used to build and improve terminological resources and efficiently process documents in information retrieval systems (indexing, summarization and classification).

Keywords extraction problem have been investigated by many authors during the last years and several method have been suggested [1–5], well known two approaches are linguistic, statistical, spectral and hybrid methods. Also in the last few years several term extraction tools have been developed, but most of them are language-dependent [6]. And we still don't have such tools for texts in Kazakh language. But many investigations have been made related to natural language processing tasks for Kazakh texts (e.g., [7–9]).

The basic idea of statistical methods lies on the assumption that words differently distributed among relevant and irrelevant documents of the certain domain. For example, in [10], the method based on a measure of mutual information was used to identify collocations (to compute the co-occurrence of words). In [11], χ^2 statistical criterion of co-occurrence was used for the same purpose. In [12], the likelihood ratio criterion was used for extraction of collocations. Statistical methods have well proven themselves in automatic extraction of terms and collocations.

© Springer International Publishing Switzerland 2016
N.T. Nguyen et al. (Eds.): ICCCI 2016, Part II, LNAI 9876, pp. 555–562, 2016.
DOI: 10.1007/978-3-319-45246-3_53

The objective of this work is to study possibilities of statistical methods for automated extraction of keywords from domain texts. To achieve this goal we need to perform the following tasks:

1. Preparing a training collection that contains two classes of documents - relevant and irrelevant to domain;
2. Creating the dictionary on the basis of training collection;
3. Identifying keywords from the dictionary of the domain;
4. Building the semantic relations between extracted keywords;
5. Checking the validity of the created dictionary (to assess its applicability for the automatic classification of domain documents).

2   Keywords Extraction and Identification of Their Relations

2.1   Preparing a Training Collection of Documents

Preparing collection of documents is a selection of documents from a variety of documents available to the researcher. The important point of preparation of this collection is to split documents to the relevant and irrelevant ones. Since from quality of this collection depends the result of whole work, formation of collection may be an iterative process. During the experiments our collections can be replenished and adjusted, which corresponds to the increasing nature of the training, it is the accumulation of knowledge through trial and error.

2.2   Creating the Dictionary of Words

By dictionary from the collection of documents we mean the set of all words of the collection reduced to canonical form without affixes. The construction of the dictionary consists of two stages: tokenization, i.e. the breaking of the texts of documents on the minimum lexical components of words (tokens), and lemmatisation, that is bring the highlighted words to normal forms (lemmas). Tokenization is the initial step in automatic text processing and it is known, that errors made at this stage significantly affect the further processing.

2.3   Identifying Domain-Specific Keywords

For pruning uninformative terms we use the χ^2 criterion, the value of which allows to know how independent a certain term and domain. The choice of this criterion due to its simplicity and versatility allow to use it in large quantities of various tasks associated with verification of agreement of model and experimental data.

$$\chi^2 = \frac{|C_d| \cdot (P_1 \cdot N_0 - P_0 \cdot N_1)^2}{\prod_{i=0}^{1}(P_i + N_i)\prod_{X=P}^{N}(X_1 + X_0)}, \tag{1}$$

where $|C_d|$ is number of documents in collection;

P_i are number of relevant documents, for $i = 1$ there documents contain certain term and for $i = 0$ conversely;

N_i are number of irrelevant documents, for $i = 1$ – documents contain certain term and for $i = 0$ conversely.

This criterion we will calculate for each incoming words into the dictionary. We will include the word in the dictionary, if the criterion value exceeds a certain positive number, which we will call threshold. The value of the threshold can be varied, the higher the value, the better the dictionary.

2.4   Building the Semantic Relations

We will establish associative links between the extracted keywords based on latent semantic analysis [13]. Latent semantic analysis is a method of identifying relations between text collections and terms found in these documents. The method is based on the principles of factor analysis: identification of latent relationships between the studied objects and reducing their number by combining groups of interdependent objects. This method is purely statistical-algebraic one, because it does not use any ontologies to analyze links.

To perform latent semantic analysis on a set of document collection, we build the matrix "terms-by-documents", where rows correspond to keywords in the dictionary and columns to the documents. At the intersection of rows (word) and column (document) we assign the frequency of occurrences of this word in this document. Each word in the matrix will be a row vector and each document - column vector. According to the theorem about singular decomposition this matrix can be represented as the product of three matrices:

$$A = U S V^T, \tag{2}$$

where A is original matrix $(m \times n, m > n)$, U is an $m \times n$ orthogonal matrix, S is a $n \times n$ diagonal matrix, which elements on main diagonal ordered in descending order.

Such decomposition has the property that if in matrix S in (2) we keep only k largest singular values, and in matrices U and V we preserve just correspond to these values of columns and rows, the product of the resulting matrices (matrix A') would be the best approximation of a matrix of rank k to original matrix A (Fig. 1).

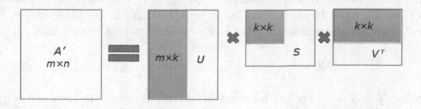

Fig. 1. Approximation of the original matrix by a matrix of a smaller rank

The resulting matrix A' (Fig. 1) more accurately reflects the structure of relations in the original collection [13]. On the basis of this matrix we can evaluate the proximity between words in the collection as distance between the corresponding vectors using cosine measure

$$\cos(\bar{x}, \bar{y}) = \frac{\bar{x} \cdot \bar{y}}{|x| \cdot |y|} \tag{3}$$

or Euclidian measure

$$|\bar{x} - \bar{y}| = \sqrt{\sum_i (x_i - y_i)^2}, \tag{4}$$

where x, y are rows and columns of matrix A'.

3   Validation of the Dictionary and Experiments

To check the validity of the obtained keywords we use it to automatically classifying documents into two classes: the class of domain documents and documents which not relevant to the domain. As the classification method we use Rocchio algorithm [14]. The essence of Rocchio algorithm lies in calculating proximity between the document vectors and centroid of the class of documents relevant to domain. As centroid we understood the average vector of the class

$$\bar{\mu}_D = \frac{1}{|TD|} \sum_{i;d \in TD} \bar{d}_i, \tag{5}$$

where TD is subset of documents collection which relevant to the domain, d_i is documents from the set TD.

The document will be considered as belonging to the domain, if the distance between the class centroid and the document vector will be less than a certain value, which we shall call the radius of the class. The distance between vectors we understand Euclidean measure calculated by the formula (4).

The main indicators of quality of classification are the error of the first and second kind. First kind error or false pass (false negative) is registered when document relating to a specific class incorrectly not detected. Second kind error or false detection (false positive) is registered when document is not relevant to domain mistakenly relies as related. Let the experimental sample contains S documents, including SD documents related to the domain and SN documents do not belong to it. It is obvious that $S = SD + SN$. Using these values, one can calculate the percentage of error levels of the first and second kind.

$$nFN = \frac{FN}{SD} \times 100\,\%, \quad nFP = \frac{FP}{SN} \times 100\,\%$$

where nFN and nFP are the percentages of error levels of the first and second kind respectively; FN and FP the number of false passes and false detection respectively.

For experiments we have selected the domain "Computer science". We prepared a training collection of 500 documents, which are chapters of textbooks and pages published on the website. Half of documents belong to subject "Computer science" and another half about biographies of peoples from other domain.

As a result of tokenization and lemmatization from the texts of the training collection was generated dictionary with 12643 reduced to canonical form of words. To this words was applied χ^2 criterion, which allowed us to identify 384 keywords of a subject domain "Computer science". The threshold value of χ^2 criterion was taken equal to 22. In the Fig. 2 you can view first 10 words with the highest value criterion and the last 10 words with the lowest criterion value.

Keyword	χ^2	Keyword	χ^2
алгоритм	547,3	жүйе	26,6
есеп	478,2	белгі	26,4
элемент	275,6	теңдік	25,7
сан	261,7	ақиқат	25,5
мән	245,2	екілік	25,4
ағаш	168,1	ұзындық	25,2
сөз	167,7	жазу	24,6
тізім	159,3	құрал	24,3
талдау	157,4	процесс	23,2
айнымалы	154,8	тәсіл	22,2

(in English)

Keyword	χ^2	Keyword	χ^2
algorithm	547,3	system	26,6
problem	478,2	sign,symbol	26,4
element	275,6	equality	25,7
number	261,7	true	25,5
value	245,2	binary	25,4
tree	168,1	length	25,2
word	167,7	record	24,6
list	159,3	tool	24,3
analysis	157,4	process	23,2
variable	154,8	method	22,2

Fig. 2. First and last 10 keywords of a subject domain

Then we fulfilled the search of associative links between words in the dictionary. There was formed the matrix "terms-by-documents" dimension 384 (the number of terms in the dictionary) to 250 (by number of documents collection related to domain). In the cells of the matrix were recorded in the frequency of occurrence of terms in documents. With the help of singular value decomposition the obtained matrix was approximated by matrix of rank 165. In Figs. 3 and 4 you can see a part of matrix "terms-by-documents" before and after singular value decomposition.

terms \ documents	1	2	3	4	5	6	7
1	8	7	0	0	0	5	0
2	1	0	0	0	0	0	0
3	7	0	0	0	0	0	0
4	1	1	3	4	2	0	16
5	4	1	0	1	2	0	0
6	3	1	0	1	3	0	0
7	2	0	0	0	0	0	0

Fig. 3. Part of matrix "terms-by-documents" before singular value decomposition

terms \ documents	1	2	3	4	5	6	7
1	8,0520	7,0977	−0,0011	−0,1328	−0,0810	4,9294	0,0244
2	1,3677	−0,0972	0,0968	0,4338	−0,1015	0,0249	0,0005
3	6,3477	−0,0353	0,0826	0,1404	−0,1713	−0,1026	−0,1166
4	1,0124	1,0247	3,0212	4,0735	1,9890	0,0292	15,9325
5	4,1900	1,1507	−0,0763	1,1001	2,0669	−0,1085	−0,0437
6	2,8336	0,9321	0,0714	0,9574	3,0026	0,0182	0,0163
7	1,5220	0,3381	0,0642	0,0258	0,1325	−0,0916	0,0259

Fig. 4. Part of matrix "terms-by-documents" after singular value decomposition

As it can be seen from the Figs. 4 and 5 we are able to get rid of noise and increase significant relationships between terms and documents by approximation of the original sparse matrix of "terms-by-documents" with a smaller rank matrix. After this using the cosine measure, we calculated pairwise similarity between the terms which represented by the vectors and rows in the new matrix. A fragment of the matrix of pairwise proximity are presented in Fig. 5. Then we selected all word pairs from the resulting matrix which proximity value more than 0.5. Also selected pairs can be joined into semantic clusters to distinguish key concepts of domain and construct thesaurus for subject area.

terms	1	2	3	4	5	6	7
1	1	0,222758	0,300158	0,202998	0,385061	0,39617	0,280362
2	0,222758	1	0,199121	0,155349	0,337179	0,289496	0,19304
3	0,300158	0,199121	1	0,095526	0,197616	0,191545	0,270208
4	0,202998	0,155349	0,095526	1	0,151962	0,214588	0,13238
5	0,385061	0,337179	0,197616	0,151962	1	0,768539	0,557483
6	0,39617	0,289496	0,191545	0,214588	0,768539	1	0,586044
7	0,280362	0,19304	0,270208	0,13238	0,557483	0,586044	1

Fig. 5. Part of matrix "terms-by-terms" of pairwise proximity

We used obtained dictionary to classify 150 messages posted on the thematic forums. 75 posts are belonged to the category "Software" and another 75 posts – to the category "Miscellaneous". As we noted before we used the method of Rocchio with a radius of the class equal to 0.002. Number of false pass was

7 documents and number of false detections was 8 documents. Thus, the per-centage of errors of the first kind was 10 %, the percentage of the second kind errors - 11 %.

4   Conclusion

Despite the fact that the training document collection was not too large in size the generated dictionary is fairly well correlated with the selected subject area, which is confirmed by experiment.

In the further works we are planning to classify semantic relations identified between the terms of a thesaurus through the use of lexical and grammatical patterns.

Acknowledgements. In this work for tokenization and lemmatization of texts we used the morphological analyzer, courtesy of our colleague Kairat Koibagarov from Institute of Informational and Computational Technologies of Science Committee Ministry of Education and Science of Republic Kazakhstan, for which we express him our gratitude. Also this research work is doing in frame of project 5033/GF4 financed by MES of RK.

References

1. Bourigault, D., Jacquemin, C.: Term extraction+term clustering: an integrated platform for computer-aided terminology. In: Proceedings of the EACL (1999)
2. Xu, F., et al.: A domain adaptive approach to automatic acquisition of domain relevant terms and their relations with bootstrapping. In: LREC (2002)
3. Collier, N., Nobata, C., Tsujii, J.: Automatic acquisition and classification of terminology using a tagged corpus in the molecular biology domain. Terminology **7**(2), 239–257 (2002)
4. Kozakov, L., Park, Y., Fin, T., Drissi, Y., Doganata, Y., Cofino, T.: Glossary extraction and utilization in the information search and delivery system for IBM technical support. IBM Syst. J. **43**(3), 546–563 (2004)
5. Wermter, J., Hahn U.: Finding new terminology in very large corpora. In: Proceedings of the K-CAP 2005, Banff, Alberta, Canada, October 2-5 2005
6. Oliver, A., Vazquez, M.: TBXTools: a free, fast and flexible tool for automatic terminology extraction. In: Proceedings of Recent Advances in Natural Language Processing (RANLP 2015), pp. 473–479 (2015)
7. Yessenbayev, Z., Karabalayeva, M., Sharipbayev, A.: Formant analysis and mathematical model of Kazakh vowels. In: International Conference on Computer Modeling and Simulation (UKSIM), pp. 427–431 (2012)
8. Tukeyev, U.: Automaton models of the morphology analysis and the completeness of the endings of the kazakh language. In: Proceedings of the International Conference Turkic Languages Processing TURKLANG-2015, Kazan, Tatarstan, Russia, 17-19 September, pp. 91–100 (2015). (in Russian)
9. Sundetova, A., Tukeyev, U.: Automatic Detection of the Type of Chunks in Extracting Chunker Translation Rules from Parallel Corpora, Mevlana University, Konya, Turkey (2016)

10. Church, W.K., Hanks, P.: Word association norms, mutual information and lexicography. In: The 27th Meeting of the Association of Computational Linguistics, pp. 76–83 (1989)
11. Church, W.K., Gale, A.W.: Concordance for parallel text. In: The 7th Annual Conference of the UW Centre for New OED and Text Research, pp. 40–62. Oxford (1991)
12. Lin, D.: Extracting collocations from text corpora. In: Workshop on Computational Terminology, pp. 57–63. Montreal, Canada (1998)
13. Nugumanova, A., Bessmertny, I.: Applying the latent semantic analysis to the issue of automatic extraction of collocations from the domain texts. In: Klinov, P., Mouromtsev, D. (eds.) KESW 2013. CCIS, vol. 394, pp. 92–101. Springer, Heidelberg (2013)
14. Manning, C.D., Raghavan, P., Schutze, H.: An Introduction to Information Retrieval, p. 181. Cambridge University Press, Cambridge (2009)

Inferring of the Morphological Chunk Transfer Rules on the Base of Complete Set of Kazakh Endings

Ualsher Tukeyev(✉), Aida Sundetova, Balzhan Abduali,
Zhadyra Akhmadiyeva, and Nurbolat Zhanbussunov

Information Systems Department, Al-Farabi Kazakh National University,
Al-Farabi Avenue, 71, 050040 Almaty, Kazakhstan
ualsher.tukeyev@gmail.com, sun27aida@gmail.com,
{balzhan_5696,akhmadieva.22,nurbolat_03.93}@mail.ru
http://www.kaznu.kz

Abstract. In this paper we propose method of constructing a set of machine translation rules associated with the transfer of word's morphological structures of the rich morphology source language, such as the Kazakh language, into a syntactic structure of the inflectional target language, such as Russian or English. The proposed method is based on the definition of a complete set of endings types of the source language, on the base of which a complete set of structural transfer rules of source language word's morphological structures into target language phrase syntactic structure is constructed. The proposed method of the morphological chunk transfer rules inferring is shown in the examples of the Kazakh-Russian and Kazakh-English machine translation.

Keywords: Inferring · Machine translation · Chunk · Transfer rules · Endings

1 Introduction

The question of the sentences structural transfer in the rule-based machine translation (RBMT) can divided into two groups: the syntax structural transfer and the transfer of word's morphological structure into a phrase syntax structure. The second group of transfers usually occur when made machine translation of languages with complex morphology into languages with a simple morphology, for example, the Kazakh language into Russian or English. In this case, some source language word's morphological structures transformed into target language phrase's syntactic structure. We call this second group of structural transfers as "morphological chunk transfers".

This problem of morphological chunk transformation significantly affects to the quality of statistical machine translation, known as the problem of morphological segmentation of the SMT [1].

In this paper, we propose an approach of the inferring of the morphological chunk transfer rules based on a finding of complete set of endings for one of the languages in a pair language of machine translation. Then, the system of obtained morphological chunk transfer rules of machine translation will also be complete. The proposed

© Springer International Publishing Switzerland 2016
N.T. Nguyen et al. (Eds.): ICCCI 2016, Part II, LNAI 9876, pp. 563–574, 2016.
DOI: 10.1007/978-3-319-45246-3_54

approach to the inferring of the complete system of morphological chunk transfer rules considered for Kazakh-Russian and Kazakh-English languages pairs of machine translation.

2 Related Works

The question of automatic inferring of the structural rules of machine translation from one language to another are rather actual for machine translation systems based on grammatical rules (RBMT). This is due to the time-consuming process of drawing up the rules for RBMT.

Tasks of automatically inferring the structural rules of machine translation based on processing parallel corpora are considered in [2–5]

Probst K. in 2005 created method which release transfer rules from limited bilingual phrases [2]. Bilingual annotator collected segments and translated, and all segments will be analyzed that there are examples of important grammatical structures. This method needs analyzed hand-processed corpus of target language. In 2006 Caseli and others created method getting shallow transfer rules and bilingual dictionary from parallel corpora [3]. Sánchez-Martínez F. and Forcada M. L. in 2009 described automatically inferring of shallow-transfer machine translation rules from small parallel corpora [4]. The method of automatically derive shallow-transfer rules for sentences is realized by Victor M. Sánchez-Cartagena, Juan Antonio Pérez-Ortiz, Felipe Sánchez-Martínez [5].

The challenges of using an approach based on the alignment of parallel corpus, automatically generating the structural rules of machine translation, consist in the following:

- for obtaining sufficiently complete set of machine translation rules it is need to process very large parallel corpora;
- there are difficulties in the adaptation of existing methods for deep difference level languages such as Kazakh and English languages;

In this paper proposed the approach to a construction of machine translation rules, in particular, the structural transfer rules on the base of the endings complete set of one of language in a language pairs of machine translation [6, 7]. This approach allows generating a complete system of morphological chunk transfer rules for deep difference level languages pair of RBMT.

3 Complete Set of Kazakh Endings

Consider a system of Kazakh word endings: nominal endings (nouns, adjectives, numerals) and verbal endings (verbs, participles, gerunds, mood and voice). In this section, we consider the complete set of endings of Kazakh [7].

The nominal endings of the Kazakh language have four types base affixes: Plural affixes (denoted by K); Possessive affixes (denoted by T); Case affixes (denoted by C); Personal affixes (denoted by J).

Consider all types of base affixes placements (affixes sequence) variants: of the one type, of two the types, of the three types, and of the four types. Number of placements determined by the formula: $A_n^k = n!/(n-k)!$.

Then, the number of placements will be determined as follows:

$A_4^1 = 4!/(4-1)! = 4$, $A_4^2 = 4!/(4-2)! = 12$, $A_4^3 = 4!/(4-3)! = 24$, $A_4^4 = 4!/(4-4)! = 24$.

All possible placements number is 64.

Consider what placements are semantically valid. The endings placements of one type (K, T, C, J) are semantically valid.

The endings placements for two types are the following:

KT, TC, CJ, JK   **KC, TJ**, CT, JT   **KJ**, TK, CK, JC.

The analysis of the semantics of the two types of endings placements shows that bold placements are valid, and the remaining placements belongs to unacceptable.

Thus, the number of valid (correct) placements of two types of endings is 6.

The semantic permissible endings placements of the three types is 4 from possible 24 endings placements:

KTC, KTJ, TCJ, KCJ.

The semantic permissible ending's placements of the four types will be 1 (**KTCJ**) from possible 24 endings placements.

Total permissible ending's placements of one type of - 4, of two types - 6, of three types - 4, four types - one. So, the total number of valid types of ending's placements in the nominal words is 15.

The set of endings of Kazakh language to the verbal stems includes the following types: - system of verb endings; - system of participles endings; - system of verbal adverbs endings; - system of moods endings; - system of voices endings.

The system of endings of verbs include the following types: - tense (9 tenses); - person (4 types). Then, the number of possible types of verb endings is - 36.

The system of participle endings includes the following types: - participle's base affixes (denoted R); - plural affixes (denoted K); - possessive affixes (T); - case affixes (denoted C); - personal affixes (denoted J).

Then, having considered possible variants of affixes types sequences (participle's base affixes for all variants is the same) on the semantic permissibility:

– from one type affixes permissible sequences: **RK, RT, RC, RJ;**
– from two type affixes permissible sequences:

RKT, RTC, RCJ, RKC, RKJ;

– from three type affixes permissible sequences:

RKTC, RKCJ;

– from four type affixes permissible sequences: 0.

Thus, the quantity of permissible types of the endings of participles is 11.

Let's consider types of the endings of verbal adverbs. They are represented by the endings of transitive future time for which follows personal endings: PJ, where P - the base ending of a verbal adverb, J - personal endings. For the given class we shall allocate only the following base endings: - ghany,-geli, -qaly, -keli. Thus, we count, that quantity of types of the endings of a verbal adverb is 1.

Let's consider the endings of moods, namely, conditional, imperative, desirable. The endings of an indicative mood coincide with the endings of verbs in the present, the past and the future.

The type of the endings of declinations is similar adverbs, i.e. the base endings of moods which personal endings follow. Thus, we consider that there are three types of the endings of moods: conditional, imperative, desirable.

Types of the endings of voices, namely, reflexive, passive, joint and compulsory, also are determined under the previous scheme: the base endings of voices for which follow personal endings. Thus, types of the endings of voices are 4.

So, the total of types of the endings of words with verbal bases is 59.

The total of the endings of words with nominal bases plus total of types of the endings of words with verbal bases equal 74.

Statement 1. Derived system of endings types in Kazakh language is complete.

The proof of this statement is based on the derivation scheme by the consideration of all possible sequences of basic system of affixes types in Kazakh language and output the set of semantically valid endings types, which presented above.

4   Inferring Chunk Transfer Rules for Kazakh-English and Kazakh-Russian

The scheme of inference transfer rules based on the following:

- the source is a complete set of endings types of the Kazakh language;
- for each types of endings (template of morphological structure types of endings) of the Kazakh language is constructed a equivalent grammatical structure logical template (pattern) in the target language (for example, Russian and English languages);
- on the basis of grammatical structure logical template built template of program structure for transfer of the morphological structure of word's endings into the equivalent grammatical structure of the target language.

4.1   Construction of the Chunk Transfer Rule Logical Templates for Kazakh-English and Kazakh-Russian Words with Nominal Base

The following Table 1 presents the 15 types of endings of Kazakh language nominal bases words with examples and corresponding grammatical structure of these types in Russian and English with examples. In the logical template of the chunk transfer rules used tags of Apertium (http://www.wiki.apertium.org).

Table 1. Logical templates of the chunk transfer rules for Kazakh-English and Kazakh-Russian words with nominal base.

Kazakh types of endings	Examples in Kazakh	Appropriate grammatical structure template in Russian	Examples in Russian	Appropriate grammatical structure template in English	Examples in English
S-K	тэте-лер (tate-ler)	[N + PL]	тети (teti)	[N + PL]	aunt-s
S-T	тэте –м (tate-m)	[PosPRN] [N]	моя тетя (moya tetya)	[PosPRN] [N]	my aunt
S-J	тэте –мін (tate-ming)	[PRN] [N]	я тетя (ya tetya)	[PRN] [to be] [N]	I am aunt
S-C	тэте –ге (tate-ge)	[PR] [N]	к тете (k tete)	[PR] [N]	to aunt
S-K-T	тэте-лер-ім (tate-ler-im)	[PosPRN] [N + PL]	мои тети (moi teti)	[PosPRN] [N + PL]	my aunt-s
S-K-J	тэте-лер-міз (tate-ler-miz)	[PRN] [N + PL]	Мы тети (My teti)	[PRN] [to be] [N + PL]	we are aunt-s
S-K-C	Тэте-лер-ге (tate-ler-ge)	[PR] [N + PL]	к тетям (k tetyam)	[PR] [N + PL]	to aunt-s
S-T-J	Тэте-м-сіз (tate-m-siz)	[PRN] [PosPRN] [N]	Вы моя тетя (Vy moya tetya)	[PRN] [to be] [PosPRN] [N]	you are my aunt
S-T-C	Тэте-м-ге (tate-m-ge)	[PR] [PosPRN] [N]	К моей тете (K moei tete)	[PR] [PosPRN] [N]	to my aunt
S-J-K	Тэте-сін-дер (tate-sing-der)	[PRN] [N + PL]	Вы тети (Vy teti)	[PRN] [to be] [N + PL]	you are aunt-s
S-C-J	Тэте-дсн-сіц (tate-den-sing)	[PRN] [PR] [N]	Вы от тети (Vy ot teti)	[PRN] [to be] [PR] [N]	you are from aunt
S-K-T-J	Тэте-лер-ім-сіндер (tate-ler-im-singder)	[PRN] [PosPRN + PL] [N + PL]	Вы мои тети (Vy moi teti)	[PRN] [to be] [PosPRN + PL] [N + PL]	you are my aunt-s
S-K-T-C	Тэте-лер-ім-ге (tate-ler-im-ge)	[PR] [PosPRN + PL] [N + PL]	К моим тетям (K moim tetyam)	[PR] [PosPRN + PL] [N + PL]	to my aunt-s
S-K-C-J	Тэте-лер-ге-мін (tate-ler-ge-min)	[PRN] [PR] [N + PL]	Я к тетям (Ya k tetyam)	[PRN] [to be] [PR] [N + PL]	I am to aunt-s
S-T-C-J	Тэте-ң-нен-біз (tate-ng-nen-biz)	[PRN] [PR] [PosPRN] [N]	Мы от твоей тети (My ot tvoei teti)	[PRN] [to be] [PR] [PosPRN] [N]	we are from your aunt
S-K-T-C-J	Тэте-лер-ің-ге-міз (tate-ler-ing-ge-miz)	[PRN] [PR] [PosPRN + PL] [N + PL]	Мы к твоим тетям (My k tvoim tetyam)	[PRN] [to be] [PR] [PosPRN + PL] [N + PL]	we are to your aunt-s

4.2 Construction of the Chunk Transfer Rule Logical Templates for Participle in Kazakh-English and Kazakh-Russian

Below is an example of a forming of transfer rule logical template for a template of words morphological structure in Kazakh into Russian (Fig. 1). Translations of Russian words see in Table 2.

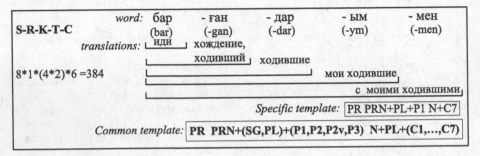

Fig. 1. Chunk transfer rules for participle, situation R-K-T-C

Table 2 presents the set of chunk transfer rule logical templates for participle in Kazakh-English and Kazakh Russian languages pair.

4.3 Construction of the Chunk Transfer Rule Logical Templates for Verbs in Kazakh-English and Kazakh-Russian Language Pairs

In modern Kazakh language the category of tense, according to its morphological features and the meaning, are divided into Past tense (Өткен шақ), Present tense (Осы шақ) and Future tense (Келер шақ). We can notice the same phenomenon in English language. English tense system includes four paradigmatic forms: Indefinite, Continuous, Perfect and Perfect Continuous. "Өткен шақ" (Past tense) is a grammatical form of a verb denoting the completed real events, actions in the past. The verb affixes created by the verb tenses. For example: V + A(PastCon) + (Sg,Pl) + (P1,P2,P2v,P3) Мен жаса-п отырдым (I was working). Table 3 presents the set of chunk transfer rule logical templates for verbs in Kazakh-English and Kazakh Russian languages pair.

Statement 2. The constructed system of chunk transfer rules of word's morphological structures in Kazakh language into equivalent phrase grammatical structures of target language is complete.

<u>Proof.</u> The proof of completeness of derived system of chunk transfer rules of word's morphological structures in Kazakh language into an equivalent phrase grammatical structures of target language based on the completeness of set of endings in Kazakh language (Statement 1) and the construction appropriate chunk transfer rule for each type of endings.

Table 2. Logical templates of the chunk transfer rules for participle Kazakh-English and Kazakh-Russian

Types of endings Kazakh	Examples in Kazakh	Grammatical structure in English	Examples in English	Grammatical structure in Russian	Examples in Russian
S-R-K	бар-ган-дар (bar-gan-dar)	[N + Pl]	Gones	[N + Pl]	Ходившие
S-R-T	бар-ган-ым (bar-gan-ym)	[PosPRN + (Sg,Pl) + (P1,P2,P3)] [N + (Sg,Pl)]	My going	[PosPRN + (Sg,Pl) + (P1,P2,P3)] [N + (Sg,Pl)]	Мое хождение
S-R-J	бар-ган-мын (bar-gan-myn)	[PRN + (Sg,Pl) + (P1,P2,P3)] [V(Past)]	I went	[PRN + (Sg,Pl) + (P1,P2,P3)] [V(Past)]	Я ходил
S-R-C	бар-ган-га (bar-gan-ga)	[Pr] [N + (C1,C2,C3,C4,C5,C6)]	to gone	[Pr] [N + (C1,C2,C3,C4,C5,C6)]	К ходившему
S-R-K-T	бар-ган-дар-ым (bar-gan-dar-ym)	[PosPRN + (Sg,Pl) + (P1,P2,P3)] [N + (Pl)]	My gones	[PosPRN + (Sg,Pl) + (P1,P2,P3)] [N + (Pl)]	Мои ходившие
S-R-K-J	бар-ган-дар-мыз (bar-gan-dar-myz)	[PRN + (Sg,Pl) + (P1,P2,P2v, P3)] [to be] [N + (Sg,Pl) + C1]	We are gones	[PRN + (Sg,Pl) + (P1,P2,P2v, P3)] [N + (Sg,Pl) + C1]	Мы ходившие
S-R-K-C	бар-ган-дар-га (bar-gan-dar-ga)	[PR] [N + (Pl) + (C1,C2,C3,C4, C5,C6)]	to gones	[PR] [N + (Pl) + (C1,C2,C3,C4, C5,C6)]	К ходившим
S-R-T-C	бар-ган-ым-нын (bar-gan-ym-nyng)	[PR] [PosPRN + (Sg) + (P1,P2, P3)][N + (C1,C2,C3,C4,C5, C6)]	of my gone	[PR] [PosPRN + (Sg) + (P1,P2, P3)] [N + (C1,C2,C3,C4,C5, C6)]	К моему хождению
S-R-C-J	бар-ган-да-мыз (bar-gan-da-myz)	[PR] [PosPRN + (Sg,Pl) + (P1, P2,P3)][N + (Sg) + (C1,C2, C3,C4,C5,C6)]	in our gone	[PRN + (Sg,Pl) + (P1,P2, P3)] [PR][N + (Sg) + (C1,C2, C3,C4,C5,C6)]	Мы у ходившего
S-R-K-T-C	бар-ган-дар-ым-мен (bar-gan-dar-ym-men)	[PR] [PosPRN + (Sg,Pl) + (P1, P2,P3)][N + (C2,C3,C4,C5,C6)]	with my gones	[PR] [PosPRN + (Sg,Pl) + (P1, P2,P3)][N + (C1,C2,C3, C4,C5,C6)]	С моими ходившими
S-R-K-C-J	бар-ган-дар-га-мын (bar-gan-dar-ga-myn)	[PRN + (Sg,Pl) + (P1,P2,P3)] [PR] [N + (Pl) + (C1,C2,C3, C4,C5,C6)]	I to gones	[PRN + (Sg,Pl) + (P1,P2,P3)] [PR] [N + (Sg,Pl) + (C1,C2, C3,C4,C5,C6)]	Я к ходившим

Table 3. Logical templates of the chunk transfer rules for verbs Kazakh-English and Kazakh-Russian.

	Languages	The tense of language and transliteration	Grammar structure for Kazakh, English, Russian	Examples for Kazakh, English, Russian and transliteration
1.	kaz	Жай нақ осы шақ (Zhai nak osyshak)	[PRN] [V + A(PresSm) + (Sg, Pl) + (P1,P2,P2v, P3)]	Мен істеп жүрмін (Men istep zhurmin)
	eng	Present Simple	[PRN] [V]	I work
	rus	Настоящее время	[PRN] [V + (Sg, Pl) + (P1,P2,P2v, P3)]	Я работаю
2.	kaz	Күрделі нақ осы шақ (Kurdeli nak osy shak)	[PRN] [V + A(PresComp) +(Sg,Pl) + (P1,P2, P2v,P3)]	Мен істеп жатырмын (Men istep zhatyrmyn)
	eng	Present Continuous	[PRN] [to be] [V + ing]	I am working
	rus	Настоящее длительное	[PRN] [V + (Sg, Pl) + (P1,P2,P2v, P3)]	Я работаю
3.	kaz	Ауыспалы осы шақ (Auyspaly osy shak)	[PRN] [V + A(PresNow) + (Sg, Pl) + (P1,P2,P2v, P3)]	Мен істедім (Men istedim)
	eng	Present Perfect	[PRN] [to be] [V + ed]	I have worked
	rus	Настоящее совершенное	[PRN] [V + (Sg, Pl) + (P1,P2,P2v, P3)]	Я работал
4.	kaz	Жедел өткен шақ (Zhedel otken shak)	[PRN] [V + A (PastOper) + (Sg, Pl) + (P1,P2,P2v, P3)]	Мен істедім (Men istedim)
	eng	Past Simple	[PRN] [V + ed]	I worked
	rus	Прошедшее время	[PRN] [V(Sg, Pl) + (P1,P2,P2v, P3)]	Я работал
5.	kaz	Бұрынғы өткен шақ (Buryngy otken shak)	[PRN] [V + A(PastOld) + (Sg, Pl) + (P1,P2,P2v, P3)]	Мен істегенмін (Men istegenmin)
	eng	Past Continuous	[PRN] [to be] [V + ing]	I was working
	rus	Прошедшее длительное	[PRN] [V + (Sg, Pl) + (P1,P2,P2v, P3)]	Я работал

(Continued)

Table 3. (*Continued*)

	Languages	The tense of language and transliteration	Grammar structure for Kazakh, English, Russian	Examples for Kazakh, English, Russian and transliteration
6.	kaz	Ауыспалы өткен шақ (Auyspaly otken shak)	[PRN] [V + A(PastMay) + (Sg, Pl) + (P1,P2,P2v, P3)]	Мен істеп отырдым (Men istep otyrdym)
	eng	Past Perfect	[PRN] [to be] [V +ed]	I had worked
	rus	Прош. совер-шенное время	[PRN] [V + (Sg, Pl) + (P1,P2,P2v, P3)]	Я работал
7.	kaz	Болжалды келер шақ (Bolzhaldy keler shak)	[PRN] [V + A(FutCast) + (Sg, Pl) + (P1,P2,P2v, P3)]	Мен істеймін (Men isteimin)
	eng	Future Simple	[PRN] [to be] [V]	I shall work
	rus	Будущее простое	[PRN] [to be] [V + (Sg,Pl) + (P1, P2,P2v,P3)]	Я буду работать
8.	kaz	Мақсатты келер шақ (Maksatty keler Shak)	[PRN] [V + A(FutObj) + (Sg, Pl) + (P1,P2,P2v, P3)]	Мен істеп отырмын (Men istep otyrmyn)
	eng	Future Continuous	[PRN] [to be] [be] [V + ing]	I shall be working
	rus	Будущее сложное	[PRN] [V + (Sg, Pl) + (P1,P2,P2v, P3)]	Я работаю
9.	kaz	Ауыспалы келер шақ (Auyspaly keler shak)	[PRN] [V + A(FutSub) + (Sg, Pl) + (P1,P2,P2v, P3)]	Мен істеймін (Men isteimin)
	eng	Future Perfect	[PRN] [to be] [have] [V + ed]	I shall have worked
	rus	Будущее совершенное	[PRN] [to be] [V + (Sg,Pl) + (P1, P2,P2v,P3)]	Я буду работать

5 Example of Program Realization of the Inferring of the Morphological Chunk Transfer Rules

To generate showed structures was created algorithm, which uses following data:

- name of structure, for instance, S-K, S-K-C, etc.;
- structure of bilingual phrase, where words on the left side is in source language and on the right side is in target language. Structure should show morphological information. For the experiments were used free/open-source Apertium platform to do morphological analysis (for structure: S-K → N+PL). Target and source language words are separated by pipeline and lemmas must be removed.

Algorithm for generating rules from the structures consists of next steps:

1. Create new file Template.txt. In this file, we write structure name: S-K, respectively template of bilingual phrase: <n> <pl> <nom>|<n> <m> <nn> <pl> <nom>. Templates, which belongs to the same type of structure, but have some differences between tags (for instance, different gender), are written as a list: each template per line.
2. The following steps are done automatically by application:
   (a) Reading file "template.txt";
   (b) Creating category from source language side of structure and attributes from target language side of structure;
   (c) Creating rule with pattern from category;
   (d) Assigning chunk name of rule from structure name;
   (e) Assigning output tags attribute's tags;
   (f) Creating file to write new rule: rule.xml

This algorithm was realized as application on Java, so it could be run on any platform that support Java, without any recompilation. The data is saved in "template. txt" and this file must be placed in the same folder as where application will be run.

Application creates one rule from structure of bilingual phrase, which is read from file, and names rule's chunk as a structure name.

```
<chunk name="__S-K__" case="caseFirstWord">
<tags>
<tag><lit-tag v="NP"/></tag>
</tags>
<lu><clip pos="1" side="tl" part="lem"/>
<clip pos="1" side="tl" part="_attr_n_"/></lu></chunk>
```

[Example of chunks from generated rule]

All tags of structure, described in the code above, will be transferred into next stage – "interchunk". Also, rule generates category and attribute, from which rule generates tags sequences of target language word and finds pattern of input source language word.

Table 4. Comparison of hand-written and generated rules

Rule	Input	Chunk	Output
For Kazakh-English			
Hand-written	тәтелер	^noun<SN><nom><pl>{^aunt<n><3>$}$	aunts
Generated from the structure	тәтелер	^__S-K__<NP>{^aunt<n><pl>$}$	aunts
For Kazakh-Russian			
Hand-written	тәтелер	^n<SN><nom><pl>{^тетя<n><f><aa><3><nom>$}$	тети
Generated from the structure	тәтелер	^__S-K__<NP>{^тетя<n><f><aa><pl><nom>$}	тети

As can be seen from the Table 4, rules, generated by our application, and hand-written rules show the equal translation. This application also could generate rules from the structures of English-Kazakh, Russian-Kazakh language pairs with the same results (Fig. 2).

Results of translation:

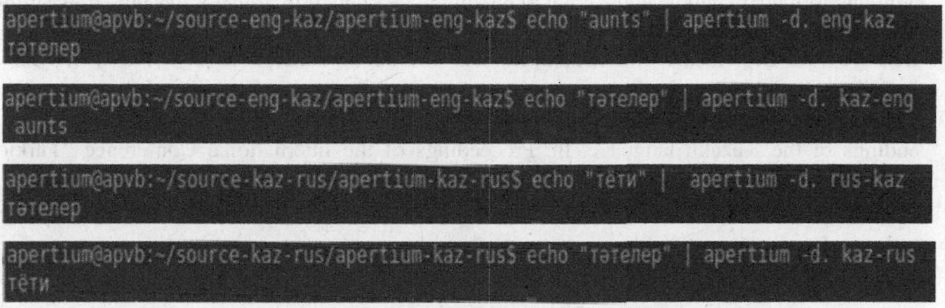

apertium@apvb:~/source-eng-kaz/apertium-eng-kaz$ echo "aunts" | apertium -d. eng-kaz
тәтелер

apertium@apvb:~/source-eng-kaz/apertium-eng-kaz$ echo "тәтелер" | apertium -d. kaz-eng
aunts

apertium@apvb:~/source-kaz-rus/apertium-kaz-rus$ echo "тёти" | apertium -d. rus-kaz
тәтелер

apertium@apvb:~/source-kaz-rus/apertium-kaz-rus$ echo "тәтелер" | apertium -d. kaz-rus
тёти

Fig. 2. Output after chunker rule

As a future work, it is planned that rules will be more general. To solve this problem, tags that are special for each part of speech, for instance, nouns in Russian have gender and animated or non-animated tags, will be get from word dynamically, without assigning it in rule by "<lit-tag>".

6   Conclusion and Future Work

In this paper we developed the method of construction the complete set of morphological chunk transfer rules for Kazakh-Russian and Kazakh-English language pairs of machine translation. Proposed method based on the complete set of endings types of the Kazakh language. The completeness of developed morphological chunk transfer rules based on the completeness of the endings types of the Kazakh language. Presented example of program realization of the inferring of morphological chunk rules in Apertium system.

The planned works in future relate with expansion of the automation constructing of morphological chunk transfer rules on the base of proposed method.

References

1. Makhambetov, O., Makazhanov, A., Yessenbayev, Zh., Sabyrgaliyev, I., Sharafudinov, A.: Towards a data-driven morphological analysis of Kazakh language. In: Proceedings of 2nd International Conference TURKLANG 2014, 6–7 November, pp. 91–96. Instanbul Technical University, Instanbul (2014)
2. Probst, K.: Automatically induced syntactic transfer rules for machine translation under a very limited data scenario. Ph.D. thesis, Carnegie Mellon University (2005)

3. Caseli, H.M., Nunes, M.G.V., Forcada, M.L.: Automatic induction of bilingual resources from aligned parallel corpora: application to shallow-transfer machine translation. Mach. Transl. **20**(4), 227–245 (2006). Published in 2008
4. Sánchez-Martínez, F., Forcada, M.L.: Inferring shallow-transfer machine translation rules from small parallel corpora. J. Artif. Intell. Res. **34**(1), 605–635 (2009)
5. Sánchez-Cartagena, V.M., Pérez-Ortiz, J.A., Sánchez-Martínez, F.: A generalised alignment template formalism and its application to the inference of shallow-transfer machine translation rules from scarce bilingual corpora. Comput. Speech Lang. **32**(1), 46–90 (2015)
6. Bektayev, K.: Big Kazakh-Russian and Russian-Kazakh dictionary, Almaty, "Altyn Kazyna". p. 704 (in Kazakh, Russian)
7. Tukeyev, U.: Automaton models of the morphology analysis and the completeness of the endings of the Kazakh language. In: Proceedings of the International Conference "Turkic Languages Processing" TURKLANG-2015 Kazan, Tatarstan, Russia, 17–19 September, pp. 91–100 (2015)

Author Index

Printed in the United States
By Bookmasters

Printed in the United States
By Bookmasters